Plant Virus, Vector
Epidemiology and Management

Plant Virus, Vector
Epidemiology and Management

S. Mukhopadhyay
Formerly Dean
Faculty of Agriculture
B.C. Agricultural University
West Bengal
India

CRC Press
Taylor & Francis Group
Boca Raton London New York

CRC Press is an imprint of the
Taylor & Francis Group, an **informa** business

Science Publishers
Enfield, New Hampshire

Taylor & Francis Group
6000 Broken Sound Parkway NW, Suite 300
Boca Raton, FL 33487-2742

First issued in paperback 2017

ISBN-13: 978-1-57808-674-0 (hbk)
ISBN-13: 978-1-138-11201-8 (pbk)

Cover illustration: Leaf hopper vector *Nephotettix* species (photograph by author). Background image of *Cucumovirus* (photograph courtesy of IACR, Rothamsted, UK).

Library of Congress Cataloging-in-Publication Data

Mukhopadhyay, S.

Plant virus, vector epidemiology and management / S. Mukhopadhyay. - - 1st ed.
    p. cm.

Includes index.

ISBN 978-1-57808-674-0 (hardcover)

1. Plant viruses. 2. Virus-vector relationships. 3. Virus diseases of plants. I. Title.

SB736.M85 2010

632'.8--dc22

2009051832

Visit the Taylor & Francis Web site at
http://www.taylorandfrancis.com

and the CRC Press Web site at
http://www.crcpress.com

Foreword

More than 25 years ago it was my pleasure to meet Prof Mukhopadhyay at Kalyani where he had established a very effective and productive research group investigating, particularly, the epidemiology of the disease known as Rice Tungro. This disease has an especially interesting aetiology, transmission and dispersal characteristics which his group helped clarify. Even then he was keen to provide a volume that would both demonstrate his enthusiasm and interest, and his view that the practical aspects of plant viruses and their vectors needed study.

Since then much progress has been made in the study of viruses but, arguably, more effort has been devoted to their molecular structures and genetics in the laboratory than to their study in the "real" world where complex interactions between viruses, vectors, plants and the environment occur. While it is clear that molecular methods have the potential to offer much in virus disease control it is equally certain that they will only make a contribution rather than being the solution to control.

Support and encouragement from the British Council-sponsored higher education Link Project with IACR-Rothamsted, now Rothamsted Research, which, as well as stimulating research, has delivered the valuable compendium of "Viruses of Crops and Weeds in Eastern India", and with the Natural Resources Institute, Chatham, UK, and now part of the University of Greenwich, has helped Prof Mukhopadhyay develop his ideas and discuss them with various interested parties and research groups.

This volume represents Prof Mukhopadhyay's view of the priorities in practical plant virus research and ways in which their control or management should be sought through an understanding of the practical and environmental aspects of the interactions of viruses with their vectors and their environment. Such a view will always be at the heart of virus management and control and I hope this volume will be instrumental in ensuring that others benefit from his knowledge and work.

R.T. Plumb
Rothamsted
UK

Preface

During the current decade two outstanding books on plant viruses have been published, "Matthews Plant Virology" in 2002 and "Principles of Plant Virology: Genomes, Pathogenicity and Virus Ecology" in 2007. Both these are unparallel resource books for plant virologists but these seem to fail to satisfy agricultural practitioners who need more practical information on the ways to combat the ever-increasing problems of virus diseases. Thrust areas of these books mostly appear to be on genes and genetic biotechnology with a view to produce transgenics against viruses in the future. A cursory survey on the research conducted during the last five years also shows that most of the papers published were on gene information technology. In the current perspective, therefore, requirements of agricultural practitioners are not properly addressed.

Virus diseases in fact are the result of interactions between the viruses, hosts, vectors and the environment particularly climate and weather. Vectors play a key role in the spread of virus diseases. Unfortunately very limited number of papers has been published during the last five years on vectors, their dispersal, movement and migration that largely depend upon climatic conditions. Epidemiology of different virus diseases also now needs to be revisited in view of global warming and climate change.

Moreover, transgenics alone may not be the ultimate solution as the virus genomes are continuously reassorting and mutating. Susceptibility of the vectors to the viruses, their transmission efficiencies in various hosts under different circumstances are also changing with time, cropping patterns and crop genetics. Incidence of new viruses is on the rise and no regional package is there for management of virus diseases with respect to climatic zones.

The primary focus of this book is the proper understanding of the vectors, their biology, dispersal, movement and migration, contemporary canvases of epidemiology, and the management of virus diseases keeping in view the globalization of agriculture as also the viruses and their quarantine requirements.

This book starts with a "Prelude", to briefly recapitulate the background knowledge on this unique form of life. Chapter 1 of this book also deals with the recapitulation of the current knowledge on "Nomenclature,

Taxonomy and Classification of Plant Viruses". In this chapter, early records, international consciousness, group system and modern systems of classification have been presented including an Appendix on the updated list of viruses. Chapters 2 & 3 of the book deal with the "Diversities of Physical and Chemical Structures of Viruses". One of the Appendices of Chapter 3 extensively reviews the "Genomic Organization and Expression". The next chapter (Chapter 4) deals with "Diagnostics of Viruses" depending on the differences of shape, sero-diagnosis, molecular diagnosis and diagnosis by biological methods. It also includes an Appendix on methods currently developed for routine diagnosis of some viruses.

Chapter 5 deserves special attention as it gives different unconventional approaches to "Vectors: Their Morphology, Biology and their Relationship with Viruses". Chapter 6 deals with "Dispersal, Movement and Migration of Vectors" including the methodologies to study these aspects particularly flight activity, atmospheric transport and long distance migration. Chapter 7 deals with "Epidemiolgy" in view of all the modern particularly ecological perspectives including the anticipated changes of the conventional approaches due to global warming. Chapter 8 deals with "Virus Diseases Management" including the management of viral sources, sources of resistance and the scope of minimizing the vector build up citing a few cases of field practices.

It is hoped that it may be a good resource book for practical agriculturists to view virus diseases in totality, diagnosis, vectors, and the environment.

May 2010   **S. Mukhopadhyay**

Acknowledgment

Unbound gratitude is accorded to Professor R.T. Plumb, the outstanding Plant virus epidemilogist, for enriching my knowledge on all aspects of viruses and improving the manuscript wherever required. His meticulous observations in general established his acute love for the subject, keen interest for the publication of this book and high spectra of his human values.

Technical support received from Dr. Sukumar Chakraborty, CSIRO Plant Industry, Queensland Bioscience Precinct, Australia, Professor F. Van den Bosch, Biomathematics and Bioinformatics Division, Rothamsted Research, Herpenden, UK and Professor R. Harrington of the same Institute, particularly for providing information on 'Climate Change and Virus and Vector Ecology' and 'Molecular and Ecological Epidemiology' is deeply acknowledged.

Thanks are extended to Dr. B.K. Das, Department of Entomology, Bidhan Chandra Krishi Viswavidyalaya, West Bengal, India, for supplying the photographs of insects used in this book. Services provided by Ms Carol Connet of the IACR-Rothamsted to process the Agreement and sending the images of 7 viruses to the Publisher and Mr. Tarun Sen of the Air Bridge Green, West Bengal for partly preparing the photographs in printable forms are also appreciated.

Heartfelt thanks are extended to my wife, Sujata, for sparing me from my duties at home almost for two years, Arnab, my eldest son for his hospitality and initiating the literature search for me through websites during my stay in the UK, Kaustav, my youngest son for playing a key role in the preparatory process by providing his help in computer applications, particularly scanning and setting the figures and looking after the preparation of the final manuscript.

Last but not the least in this line is Master Anish and Arush, my grandsons for carefully glueing the pages together of my first manuscript, perhaps to save it from misplacement or spoilage.

May 2010   S. Mukhopadhyay

Contents

Foreword   *v*
Preface   *vii*
Acknowledgment   *ix*
Prelude   *xiii*

1. Nomenclature and Classification   **1**
    A. Nomenclature   1
    B. Classification   3

2. Diversity of Physical Structure   **9**
    A. Simple Nucleic Acid Threads   10
    B. Particulate Structure   11

3. Diversity in Chemical Components and Genomic Structure   **25**
    A. Basic Components   25
    B. Diversities in Quantitative Presence of
       Important Components   25
    C. General Properties of Proteins and their Roles
       in Virus Structure   30
    D. Diversity in Genomic Structure   31

4. Plant Virus Diagnostics   **35**
    A. Identification and Detection   35

5. Vectors of Viruses   **67**
    A. Vectors: Morphology and Biology   68
    B. Vectors: Their Relation with Viruses   88

6. Dispersal, Movement and Migration of Vectors   **147**
    A. Dispersal and Flight Activity   148
    B. Atmospheric Transport and Migration of Vectors   161
    C. Dispersal of Vectors other than Insects   168

7. Plant Virus Epidemiology and Ecology   **175**
    A. Introduction   175
    B. Nature of Viruses and Their Epidemiological Relevance   176

C. Conventional Epidemiology 177

D. Ecological Epidemiology 188

E. Molecular Ecology and Epidemiology 193

F. Evolutionary Epidemiology 196

G. Ecological Genomics and Epidemiology 201

H. Global Warming and Epidemiology 202

I. Epidemiology of Some Internationally Important Diseases 208

(a) Barley Yellow Dwarf Virus (BYDV) 208

(b) Maize Streak Virus (MSV) 213

(c) Rice Tungro Disease (RTD) 216

(d) Citrus Tristeza Virus (CTV) 220

(e) Beet Curly Top Virus (BCTV) 224

(f) Tomato Yellow Leaf Curl Virus (TYLCV) 226

8. Management: Strategies and Tactics **251**

A. Integrated Pest Management (IPM) 252

B. Some Examples of Currently Operational IPM 297

C. Phytosanitation and quarantines 298

Appendix I 325

Appendix II 353

Appendix III 365

Appendix IV-VIII 469

Index 483

Color Plate Section 499

Prelude

Etymologically, virus means "Poison, venom, also a rammish smell as of the arm-pits, also a kind of a watery matter, whitish, yellowish and greenish at the same time, which issues out of ulcers and stinks very much; being induced with eating of poison of malignant qualities". Historically, this term has been used to denote the causal agents of a particular type of disease, which differs from those caused by other pathogens such as fungi, bacteria or nematodes. These causal agents, or viruses, were later found to be submicroscopic particulate structures mostly having cubical or helical symmetry being composed of only one type of nucleic acid and protected by a protein coat, some of which could be isolated in purified form and crystallized.

1. VIRUSES: MOLECULES OR ORGANISMS

In any discourse on viruses, an age-old question always arises; is a virus a *Molecule* or an *Organism*? This controversy began during the 1930s and lingers even today. The discovery that some viruses could be crystallized like a chemical molecule and yet retain their ability to cause disease contributed to this controversy and the question of whether viruses are living or non-living? That seemingly sterile fluids from diseased plants, animals or bacteria could be infective was not readily explained by knowledge at that time. At the beginning of the controversy, the *Organism* concept predominated. Pathologists were content to assume that viruses were essentially similar to bacteria as both of them multiply in host organisms and while doing so, occasionally change and produce progeny with new characters. Unusual inclusion bodies were also reported in cells of some virus infected plants and animals and similar inclusions were known in many animal diseases caused by organisms. The discovery of viruses, however, changed the contemporary concept of organisms as systems that possess genetic continuity and an evolutionary independence as viruses possess such continuity and are apparently independent of the host cells for their evolution. Thus, they are independent genetic systems having their own independent movement and survival.

After the successful isolation and purification of many plant viruses, the *"Molecular view"* of the nature of viruses started to gain ground. As distinct from normal organisms, viruses could be crystallized and reconstituted; they are chemically simple and possess regular internal structures. The chemical simplicity and regular internal structure are constant for all the viruses so far purified, even for those where crystallization was not possible. The *Molecular view* was further strengthened by the apparent homogeneity of the purified virus preparations.

As more and more information is obtained on physical, chemical and biological properties of viruses, the arguments and discussions about the nature of viruses move beyond molecules or organisms, living and non-living, and rather they appear to be at the threshold of life, linking living to the non-living and matter to organism. Viruses are not organisms since they do not possess independent metabolic activities. They are not molecules as they can change and mutate. They are living while within host cells and non-living outside those cells. Thus depending on the phase of their existence, they may either be an *organism* or a *molecule*. The essential organism and molecular features of viruses are summarized in Table P.1.

Table P.1 Molecular and Organism Properties of Viruses

Molecular properties	*Organism behavior*
Non-cellular	Replicates and maintains genetic continuity
Absence of energy releasing systems	Mutates
Uniform size and shape	Intercellular movement in host
Uniform chemical composition	Operates recognition system to select hosts and vectors
Can be crystallized	Show biological relation with vectors
Can be chemically dissociated and reconstituted	Possess own characteristic host range
Synthesis follows specific pathways	Survives beyond the death of parasitized cell

2. DEFINITION OF A VIRUS

The term virus describes a very dynamic entity. With advances in the knowledge of the nature of this entity, scientists started to define it according to their own perceptions and the knowledge at the time. In 1950, Bawden defined a virus as "obligate parasitic pathogen with a dimension of less than 200 mμ". Lwoff and Tournier in 1966 defined viruses as being composed of either deoxyribonucleic acid [DNA] or ribonucleic acid [RNA] and reproducing through the replication of the constituent nucleic acid. They further stated that viruses do not possess any energy producing or releasing system and do not multiply by binary fission. They further included in their definition that viruses are strictly intracellular and replicate within the host cells, using the host's ribosomes. Fraenkel-Conrat (1969) described viruses as infectious molecules of either DNA or RNA, normally en-coated by protein; after virus entry into the host cell, the nucleic acid takes control of the important organelles of the cell and replicates; in some cases, the viral genome becomes reversibly integrated in the host genome and thereby becomes cryptic or transforms the character of the host cell. Gibbs and Harrison (1976) defined viruses as "transmissible parasites whose nucleic acid genome is less than 3×10^8 daltons in weight and that need ribosome and other components of their host cells for multiplication" and Matthews (1992) defined viruses as "Sets of one or more genomic nucleic acid molecules, normally encased in a protective coat or coats of protein or lipoprotein which is able to mediate its own replication only within suitable host cells. Within such cells, virus replication is (1) dependent on the host's protein synthesizing machinery; (2) derived from pools of the required materials rather than from binary fission; and (3) located at the sites not separated from the host cell contents by a lipoprotein bi-layer membrane".

Viruses are very small obligate parasites that contain one to several hundred genes of their own which can mutate. As more knowledge on viruses is acquired, its definition continues to change.

3. VIRUSES AND OTHER INTRACELLULAR ORGANISMS

Viruses have several properties that are exclusive to them and several that they share with other intracellular obligate parasites. The exclusive properties are: (i) the constituent nucleic acids are either DNA or RNA; (ii) nucleic acids may be single or double stranded; (iii) a mature particle may contain other poly-nucleotides in addition to its own genomic ones; (iv) the genomic and/or other nucleotides may be distributed in one or more particles; (v) specific enzymes may be present in the particle for the replication of the genomic nucleic acid; (vi) many viruses usually multiply in the virus-induced area of the host cell; (vii) multiplication of some viruses requires the presence of other viruses.

The properties that viruses share, include (i) size; (ii) nature and quantity of genomic nucleic acid; (iii) presence of RNA/DNA; (iv) presence of envelope; (v) intracellular multiplication; (vi) absence of energy producing system; (vii) dependence on host cell's amino acid pool. There are several intracellular organisms of a similar size to viruses but the poxvirus is larger than the *Chlamydiae*. The nature, quantity and properties of the nucleic acids are more or less similar in both types of parasites, but intracellular organisms normally contain both types of nucleic acids. There are several viruses (complex viruses) that contain envelopes/membranes, as do intracellular organisms. Both viruses and other intracellular organisms multiply within cells. There are some intracellular microorganisms, such as *Chlamydiae*, which, like viruses, do not have any energy producing system and many bacteria also depend upon the amino acid pool of the host cells for their multiplication. Properties of viruses not found in any intracellular microorganism are that virus particles do not remain separated from the organelles of a cell at the time of their multiplication; viruses do not have any protein synthesizing system; viruses never multiply by binary fission.

4. INTRACELLULAR MICROORGANISMS ALLIED TO VIRUSES

The intracellular microorganisms that are often considered allied to viruses include *viroids, phytoplasmas, spiroplasmas, rickettsia* and *chlamydiae*. *Viroids* are normally called mini-viruses and found only in plants; Diener (1971), while searching for the pathogen of potato spindle tuber disease, discovered that free ribonucleic acid is the infectious agent for this disease and called it a *viroid*. Subsequently similar causal agents were found as the causal agents in several diseases that were previously thought to be virus disease. The RNA of the *viroids* so far known is single stranded and circular.

Doi *et al* (1967) while searching, by electron microscopy, for the causal agents of the so-called "yellows" and "little leaf" type of diseases, saw mycoplasma-like organisms in infected cells. *Mycoplasmas* are wall-free prokaryotic organisms, surrounded by an exterior membrane and contain both RNA and DNA and can be cultured in a cell-free medium. Mycoplasma-like organisms (MLO) now more commonly known as *Phytoplasmas*, cause a range of "yellows" type diseases in plants and are well known pathogens of animals and human beings. They are simple, pleiomorphic cells, normally 100-400 nm in diameter; some of the filamentous forms may be up to 1700 nm long. Their appearance ranges from simple or budding spheres, dumbbell-shaped, simple or branched filaments, occasionally connected in long chains and covered by a continuous membrane, and confined to the sieve tubes of infected plants. The cells of *mycoplasmas* are non-motile and grow in minute colonies with a central nipple in culture media; each cell is normally surrounded by single tri-laminar lipoprotein membrane, about 10 nm thick. A *mycoplasmal* or *phytoplasmal* cell contains ribosomes, RNA and double stranded circular DNA, the molecular weight of which is usually $4 \times 10^8 - 1 \times 10^9$ Da.

Mycoplasmas usually multiply by binary fission or by budding and may undergo structural changes during the course of the reproduction cycle. They can be distinguished from other pathogens by their sensitivity to tetracycline and resistance to the penicillin group of antibiotics. They are transmitted by insect vectors and can be transmitted by grafting.

Spiroplasmas, Rickettsiae and *Chlamydiae* are mostly found as pathogens of human beings and animals. Fudl-Allah *et al* (1972) and Saglio *et al* (1973) while working to diagnoze the pathogen of the citrus stubborn disease successfully isolated a wall-less prokaryote. Davis *et al* (1972) using dark-field and phase contrast microscopy, saw spiral bodies in the crude sap collected from *stunt* infected corn plants. The culture of the prokaryote collected from the *citrus stubborn* infected plants also showed spiral bodies (Cole *et al* 1973). These spiral bodies were designated as "*Spiroplasma*". Normally *spiroplasmas* are pleiomorphic, commonly helical and measure 150-200 nm in diameter and 3-15 nm long. Under unfavourable conditions, they may divide into small helices, asteroids or coccoid structures. The *spiroplasmas* are normally motile but, like viruses and *phytoplasmas*, are transmitted by vectors.

Rickettsia normally does not infect plants but are occasionally found in the vascular tissue of some plants. They are small rods (0.2-0.5 nm × 1-4 nm) bounded by a tri-laminar membrane and an additional cell wall. They can be cultured *in vivo* and transmitted by grafting and vectors. The number of proteins in these organisms also differs. The number of proteins found in Eubacteria (*Escherichia coli*), Mycoplasma, *Chlamydiae* (*Psittacosis*), a large virus infecting vertebrates (*Vaccinia virus*), a large virus infecting angiosperms (*wound tumour virus*), a small virus infecting angiosperms (*TMV*) and one of the smallest known plant viruses (*tobacco necrosis satellite virus*) are 4,100, 820, 660, 260, 12, 4 and 1 respectively (Matthews 1992).

Viruses have three-dimensional structure and their volume may be as large as 6×10^5 nm^3 and as small as 2×10^4 nm^3. The smallest plant virus may be as small as a messenger-RNA (mRNA) and the largest may be of the same size as that of the smallest cell.

5. HISTORICAL PERSPECTIVES OF THE NATURE OF VIRUSES

Although we know that virus diseases of humans, animals and plants have always been present, the science of virology and the understanding of the nature and function of viruses do not have a long history. The human disease called "*smallpox*" was described in China as early as the 10th century BC and "*yellow fever*" ravaged tropical Africa for many centuries. "*Jaundice of silkworms*, "*leaf-roll of potato*" and "*tulip flower breaks*" were known to people from at least the 16th century, but the history of virology, and plant virology in particular perhaps started in the 17th century through the accidental observation of some horticulturalists in Holland that they could transfer *tulip break* and *jasmine mosaic* by grafting (see McKay and Warner 1933).

Edward Jenner, in 1798, laid the foundation of virus research by successfully establishing the antigenic nature of a virus. It was possible to vaccinate humans against *smallpox* by inoculating them with extracts of *cowpox virus*. Pasteur (1884) made the next break-through in antigen research by culturing *Rabies virus* in tissues of laboratory animals. They were able to immunize human beings against that virus by inoculating the attenuated culture. Animal virologists also took the lead in the understanding the mode of transmission of viruses. The transmission of *Yellow fever* by mosquitoes was first observed as early as 1848 and Reed (1902) established the relationship between mosquitoes and *Yellow fever*. Experimental transmission of a plant virus (*Rice stunt virus*) was made by Hashimoto in 1884 (see Fukushi 1933) and confirmed by Fukushi (1933). While animal virologists remained busy with immunology, transmission and epidemiology, plant virologists concentrated their efforts on understanding the nature of viruses and were handicapped as there was no method for their artificial culture.

Mayer (1886) was the first to start research on the nature of plant viruses. He demonstrated the transfer of the mosaic symptom found in tobacco plants by transferring the juice from an infected to a healthy plant using a capillary tube and discovered the sap transmissibility of *Tobacco mosaic virus* (TMV). Sanarelli (1898) became the first person to successfully transmit *Rabbit Somatosis* virus by inoculating the extract from infected tissues to the healthy animals, but neither Mayer nor Sanarelli could find any pathogen in the sap or extract. Ivanowski (1892) was preoccupied with the supposed organismal nature of TMV and demonstrated that the infectious extract would pass through bacteriological filters. While disappointed by the results he unknowingly established the "Filterable nature" of a virus. Loeffler and Frosch (1898) made a similar observation with the *Foot and mouth disease* of animals but they went further and established the serial transmission of this virus and the multiplication of the virus in the host tissues; they also developed a quantitative assay method for the same virus. Beijerinck (1898), a Dutch scientist working with TMV, also made serial transmissions of TMV and established that TMV multiplied in tobacco plants but could not develop any method for its quantitative assay, as it could not be artificially cultured in tobacco tissues. He described the virus as a *"contagium vivum fluidum"* or an infectious living fluid. At the same time search for a method to quantitatively estimate a plant virus continued and Holmes (1929) discovered the "*local lesion*" reaction in certain hosts and used this method for the quantitative estimation of sap transmissible plant viruses.

As well as the fundamental research on understanding the nature of viruses, reports of their incidence continued and many viruses were reported from more than one thousand species of plants in many widely different families and genera. These viruses were identified by the symptoms they caused, their host ranges, characteristics of their transmission and several physical properties of the infectious sap and different virus-vector relationships without having any knowledge of the virus structure or organization.

Stanley (1935) made an important breakthrough in understanding of the nature of a virus when he was successful to isolate, purify and crystallize infectious protein from a sap extracted from the TMV infected plants. Bawden and Pirie (1936) discovered that TMV is actually an infectious nucleoprotein containing RNA coated by protein molecules; Schlesinger (1936) observed that Coli-phages contain DNA.

Thus virologists became confident that, whatever the host (animal, plant or bacteria), viruses consist only of infectious nucleoprotein, with either RNA or DNA but did not know how they looked like. Kausche *et al* (1939) used their newly constructed electron microscope to examine sap extracted from a tobacco plant infected by TMV. To their great surprise they found small rod like particles but did not prove that they were the infectious agent. As technology and instrumentation improved, virologists established that viruses are either rods or spheres. The shape of the particles suggested an organized structure. X-Ray crystallography revealed the arrangement of protein molecules in the rod shaped particles of TMV.

Delbruck and Bailey (1946), Hershey and Chase (1952) and Zinder and Lederberg (1952) discovered that viruses transfer genetic characters from one bacterium to another where they may be stabilized and carried through subsequent generations and called the process "*Transduction*". Though genetic studies, particularly on bacterial viruses, were making progress, the infectious principle of the virus remained unknown. In-depth studies on the infection of *Escherichia coli* by T2 phages, Hershey and Chase (1952) showed that it is the DNA of the phage that enters the bacterial cell and causes infection; the protein component of the phage stays on the bacterial cell wall. Fraenkel-Conrat and Williams (1955) also drew the same conclusion on the infectivity of the nucleic acid component through their historic "*Reconstitution of TMV*" by changing the pH and ionic strength of the isolation medium. Recombining the RNA of the one strain of TMV with the protein subunits of another, they obtained a particle that showed the properties of the strain from which the RNA was obtained. Gierer and Schramm (1956) took a direct approach to prove the infectivity of the RNA of TMV. They isolated the nucleic acid from the infected sap and inoculated it to healthy plants that became infected.

In plant and animal cells DNA is double stranded (and infective) whereas normal cellular RNA is single stranded (and non-infective). Bawden and his associates, while laying the foundation of nucleic acid research on TMV in 1936, found that virus RNA was infective and single stranded. The infective DNA of the coli-phages on the other hand, was double stranded (Hershey and Chase, 1952). Subsequent workers proved that there are double stranded infective RNAs in many plant viruses. Shepherd and Wakeman (1971) found double stranded DNA in *cauliflower mosaic virus*. Later, the presence of single stranded DNA in plant viruses was also demonstrated for *Mungbean yellow mosaic virus*. Kassanis (1962) demonstrated that not all viruses are independent in their multiplication; some depend on another virus which he called "satellite virus" (*Tobacco necrosis satellite virus*). It was also discovered that some viruses consisted of separate genomic pieces and these studies were further extended by Kaper and Waterworth (1977) who discovered a fifth piece of nucleic acid in multipartite *cucumber mosaic virus*, which they called "*Satellite RNA*". Recent research on the nature of viruses has mostly concerned with the molecular and genomic properties of the nucleic acid; replication strategies; detection at the molecular level; synthesis of the virus-specific proteins and their assembly into specifically organized structures; their antigenic properties and the genetics controlling their biological properties (Hull 2002). Other recent work with viruses is in their use as vehicles for the transfer of genes from one cellular system to another, the use of virus proteins as promoters for protein synthesis (Porta and Lomonosoff 1996, Scholthop *et al* 1996) and the use of the virus protein shells for the production of vaccines in plant cellular systems (Canizares *et al* 2005). So the wide-angle journey of the virology continues to diversify towards understanding the life at molecular level and its application for human and animal welfare.

References

Bawden, F.C. and Pirie, N.W. 1936. The isolation and some properties of liquid crystalline substances from solanaceous plants infected with three strains of tobacco mosaic virus. Proc. R. Soc. Lond. Ser. B. Biol. Sci. 123: 274-320

Bawden, F.C. 1950. Plant Viruses and Virus Diseases. Chronica Bot. Co. Waltham, Mass., USA

Beijerinck, M.W. 1898. Concerning a *Contagium Vivum Fluidum* as Cause of the Spot Disease of Tobacco Leaves. In: Selected Papers on Virology. N. Hahon (ed). Prentice-Hall Inc., Englewood Cliffs, N.J., USA. pp. 52-63

Canezares, M.C., Nicholson, L. and Lomonosoff, G.P. 2005. Use of viral vectors for vaccine production in plants. Immunol. Cell Biol. 33: 263-270

Cole, R.M., Tully, J.G., Popkin, T.J. and Bove, J.M. 1973. Morphology, ultrastructure and bacteriophage infection of the helical mycoplasma-like organism (*Spiroplasma citri* gen. nov, sp. nov) cultured from 'Stubborn' disease of citrus. J. Bacteriol. 115: 367-386

Davis, R.E., Worley, J.F., Whitcomb, R.F., Ishijima, T. and Steere, R.L. 1972. Helical filaments produced by mycoplasma-like organism associated with corn stunt disease. Science 176: 521-523

Delbrck, M. and Bailey, W.T. 1946 Induced Mutations in Bacterial Viruses. Cold Spring Harbour Symp. Quant. Biol. XI: 33-37

Diener, T.O. 1971 Potato spindle tuber Virus IV. A replicating low molecular weight RNA. Virology 45: 411-428

Doi, Y., Teranaka, M., Yora, K. and Asuyama, H. 1967. Mycoplasma or PLT group-like micro-organisms found in the phloem elements of plants infected with mulberry dwarf, potato witches' broom, aster yellows, or Paulownia witches' broom. Ann. Phytopathol. Soc. Japan 33: 259-266

Fraenkel-Conrat, H. 1969. The Chemistry and Biology of Viruses. Academic Press, New York, USA

Fraenkel-Conrat, H. and Williams, R.C. 1955 Reconstitution of Active Tobacco Mosaic Virus from its Infective Protein and Nucleic Acid Components. Proc. National Acad. Sci. Washington 41: 690-698

Fudl-Allah, A.E.A., Calavan, E.C. and Igwebe, E.C.K. 1972. Culture of a mycoplasm-like organism associated with stubborn disease of citrus. Phytopathology 62: 729-731

Fukusi, T. 1933. Transmission of the Virus through the Eggs of an Insect Vector. In: Selected Papers on Virology. N. Hahon (ed). Prentice-Hall Inc., Englewood Cliffs, N.J., USA. pp. 171-175

Gibbs, A.J. and Harrison, B.D. 1976 Plant Virology: The Principles. Edward, London, UK

Gierer, A. and Schramm, M. 1956. Infectivity of Ribonucleic Acid from Tobacco Mosaic Virus. Nature 177: 702-703

Hershey, A.D. and Chase Martha. 1952. Independent Functions of Viral Protein and Nucleic Acid in Growth of Bacteriophage J. Gen. Physiol. 36: 39-56

Holmes, F. O. 1929 Local Lesions in Tobacco Mosaic. Bot. Gaz. 87: 39-55

Hull, R. 2002. Matthews Plant Virology. Acad. Press, San Diego, USA

Ivanowski, D. 1892. Ueber die Mosaikkrankheit der Tabakspflanze. Bull. Acad. Imp. Sci. St. Petersberg (NW) 13: 65-70

Jenner, E. 1798. An Inquiry into the Causes and Effects of the *Variolae Vaccinae*, a Disease Discovered in Some of the Western Counties of England particularly Gloucestershire, and known by the Name of the Cow Pox. In: Selected Papers on Virology. N. Hahon (ed). Prentice-Hall Inc., Englewood Cliffs, N.J., USA. pp. 5-29

Kaper, J.M. and Waterworth, H.E. 1977. Cucumber mosaic virus associated RNA 5: Causal Agent for tomato necrosis. Science 196: 429-431

Kassanis, B. 1962. Properties and behaviour of a virus depending for its multiplication on another. J. Gen. Microbiol. 27: 477-488

Kausche, G.A., Pfankuch, E. and Rusca, H. 1939. Die Sichtbormachung von pflanzlichem Virus im Obermikrosk P. Naturwissenschaften 27: 292-299

Loeffler, F. and Frosch, F. 1898. Report of the Commission for Research on the Foot and Mouth Disease. Centrebl. F. Bakt. Prasit. Infekt. Part I, 23: 371-391

Lwoff, A. and Tournier, P. 1966. The classification of viruses. Annu. Rev. Microbiol. 20: 45-74

Matthews, R.E.F. 1992. Fundamentals of Plant Virology. Academic Press Inc., Harcourt Brace Jovanovich Publishers. San Diego, USA

Mayer, A. 1886. Concerning the Mosaic Disease of Tobacco. In: Selected Papers on Virology. N. Hahon (ed). Prentice-Hall Inc., Englewood Cliffs, N.J., USA. pp. 37-45

McKay, M.B. and Warner, M.F. 1933. Historical sketch of tulip mosaic or breaking: the oldest known plant virus disease. Nat. Hort. Mag. 3: 179-216

Pasteur, L.1884. A New Communication on Rabies. In: Selected Papers on Virology. N. Hahon (ed). Prentice-Hall Inc., Englewood Cliffs, N.J. pp. 30-36

Porta, C. and Lomonosoff, G.P. 1996. Use of viral replicons for the expression of gene in plants. Molec. Biotech. 5: 209-221

Reed, W. 1902 Recent Researches Concerning the Etiology, Propagation, and Prevention of Yellow Fever, by the United States Army Commission. J. Hyg. II(2): 101-119

Saglio, P., Hoptal, M., Lafleche, D., Dupont, G., Bove, J.M., Tull, Y.J.G. and Freundt, E.A. 1973. *Spiroplasma citri* gen and sp.n: a mycoplasma-like organism associated with "stubborn" disease of citrus. Int. J. Syst. Bacteriol. 23: 191-204

Sanarelli, G. 1898. Das Myxomatogene Virus. Beitrag zum studium der krankheitserreger ausserhalb der Si chtbaren. Zentr. Bakt. Parasit. I orig 23: 865-873

Schaffer, F.L. and Schwerdt, C.E. 1935. Crystallization of purified MEF-1 Poyomyelitis Virus. Proc. Nat. Acad. Sci. 41: 1020-1023

Schlesinger, M. 1936. The Feulgen Reaction of the Bacteriophage Substance. Nature 138: 1051

Scholthop, H.B., Scholthof, K.B.G. and Jackson, A.O. 1996. Plant virus gene vectors for transient expression of foreign proteins in plants. Annu. Rev. Phytopathol. 34: 299-323

Shephard, R.J. and Wakeman, R.J. 1971. Observation on the size and morphology of cauliflower mosaic virus. Phytopathology 61: 188-193

Sinsheimer, R.L. 1959. A Single-Stranded Deoxyribonucleic Acid from Bacteriophage, X 174. J. Mol. Biol. 1: 43-45

Stanley, W.M. 1935. Isolation of a crystalline protein possessing the properties of tobacco mosaic virus. Science 81: 644-645

Zinder, N.D. and Lederberg, J. 1952. Genetic Exchange in *Salmonella*. J. Bacteriol. 64: 679-699

1

Nomenclature and Classification

A. NOMENCLATURE

Before the discovery of virus particles, they were identified on the basis of their biological properties, particularly symptoms, host range, transmission and physical properties of the infectious sap. Viruses used to be denoted by the common name of their main host, or the host on which they were first identified, and the common term specifying the symptoms, as for example *Tobacco mosaic, Tobacco ring-spot, Tobacco leaf-curl, Aster yellows, Rice dwarf, Banana bunchy-top* etc. These approaches for identifying and naming a virus served a useful purpose but led to immense confusion and controversies. The same virus can infect many different hosts and produce different symptoms, and a single host may be infected by many viruses that produce different symptoms and is known by different names. *Tobacco mosaic virus* (TMV) produces mosaic symptoms in many plants and is recorded by a large number of different names, for example *Tomato mosaic* when it infects a tomato and *Brinjal streak mosaic* when it infects a brinjal. *Tobacco mosaic virus* and *Cucumber mosaic virus* produce typical mosaic symptoms in tobacco and cucumber respectively. When they simultaneously infect cucumber plants, an altogether different symptom appears in leaves which become horseshoe shaped. So naming a virus and virus nomenclature was, and remains, a serious concern.

1. Early Records

The system that was first proposed for virus nomenclature in 1927 consisted of three parts, the common name of the host on which the virus was first recorded, and the word virus followed by an arabic numeral to denote the chronological order of its detection. According to this system, *Tobacco mosaic virus* is called *Tobacco Virus 1*. It was later modified in 1937 replacing the common name of the host by its Latin name. So *Tobacco Mosaic Virus 1* became *Nicotiana Virus 1*. Subsequently, several attempts were made to introduce a binomial nomenclature using different generic and specific names in Latin, following the methods used for plants and animals. According to this system, *Nicotiana Virus 1* was renamed as *Marmor tabaci* where *Marmor* means mosaic and *tabaci* reflects the host. But these systems were not widely accepted by virologists and created more confusion.

To give a definitive name to a virus and in the hope of avoiding all future confusion, Gibbs (1968) proposed a 'Cryptogramic' system of nomenclature. According to this system, a cryptogram was to follow the common vernacular name of a virus. Four pairs of characters defined the cryptogram, each pair being the key characters of a virus (Appendix I.1). According to this system, *Tobacco mosaic virus* was referred as '*Tobacco mosaic virus R/1:2/5: E/E: s/x*'. It provides precise information on a few key characters of a virus but it failed to make any ground among the virologists, possibly because of its complexities, but it did stimulate futher discussion and approaches to the issues of virus nomenclature and taxonomy.

2. International Consciousness

The necessity of a scientific and acceptable nomenclature and classification of viruses was realized by the International Organization of Microbiologists as early as 1951 and a Subcommittee was constituted for this purpose at the International Microbiological Congress held in the same year. International efforts continued and the status of the Subcommittee was elevated to that of a full Committee in 1966 and was called the International Committee on Taxonomy of Viruses (ICTV) with different Subcommittees to deal separately with animal, plant, and bacterial viruses. Since then, a number of systems have been proposed but the animal, plant and bacterial virologists could not come to a uniform conclusion.

3. Group System of Nomenclature of Plant Viruses

The Group Concept of Nomenclature of plant viruses was first proposed by Harrison *et al.* (1971). To prepare the concept paper, the authors first collected all the documented data on the 630 viruses collated by Martyn (1968). These were sorted to obtain data useful for analysis. Out of the 630 viruses, 60 were suspected to be due to Mycoplasma-like organisms and the required information was not available for 457 viruses. Therefore only 113 viruses were available for analysis. They were separated into 16 groups by cluster analysis taking the type specimen into account. The characters they considered for analysis and the groups resulting from the analysis are given in Appendices I.2 and I.3. Sixty equally weighted independent qualitative characters were used to give satisfactory and statistically valid groupings. This concept provided a new dimension in the nomenclature of plant viruses, where a virus was denoted by the usual vernacular name followed by the group name. This system was approved by the ICTV in 1971. These groups are non-hierarchical and did not have any notion of species.

Subsequent workers gave much importance to the physical and chemical properties of the virus particles but the principle for naming a virus remains the same. Francki *et al* (1991) designated 37 groups and these were approved by the ICTV (Appendix I.4). In their groupings, *Tobacco necrosis virus* and *Tomato spotted wilt virus* of Harrison *et al* (1971) was distinctly grouped. Fauquet (1999) proposed to designate the viruses according to their morphology, physical properties of the virion, properties of the genome, including genome organization and replication, properties of proteins and lipids, and their antigenic and biological properties. But this did not significantly change the nomenclature.

Brunt *et al.*(1996) proposed a different nomenclature using the Virus Identification Data Exchange (VIDE) system, which used the Description Language of Taxonomy (DELTA). In their virus species description, the vernacular name comes first, followed by the name which has been used in ICTV publications or listed in Steadman's ICTV words (Calisher and Fauquet 1992). They changed the order of words in some names so that the host's name appears first.

B. CLASSIFICATION

Classification primarily provides proper understanding of organisms. The binomial system of nomenclature and classification in which relationships are determined by similarities and dissimilarities in selected properties or phylogeny has become the most widely used method and systems of classification that may be hierarchical or phylogenetic. The difference being in the use of the available properties of the organisms to separate them into species, genera, families, orders, classes and so on. All characters are considered in natural phylogenetic systems but the hierarchial system is based on only a few key characters. Systems of classification may be monothetic or polythetic depending upon the number of characters considered for systematic arrangement. In polythetic systems all possible characters are taken into consideration.

Attempts to classify plant viruses began as early as 1935 but Holmes (1939) proposed the first internationally accepted classification. This system was further improved in 1948 (Holmes 1948) and included all the viruses irrespective of their hosts using Latin names. All the viruses were placed under the Order 'Virales' divided into three Suborders, 'Phaginae', 'Phytopaginae' and 'Zoopaginae' depending upon the nature of the host (bacteria, plant or animal). Plant viruses were divided into six families on the basis of key symptoms. These were 'Chlorogenaceae' (producing yellows symptoms), 'Marmaraceae' (producing mosaic symptoms), 'Annulaceae' (producing ring-spot symptoms), 'Gallaceae' (Fiji disease of sugarcane), 'Acrogenaceae' (producing potato spindle tuber disease) and 'Rugaceae' (producing leaf curl diseases). Though, this classification has no relevance today, it is astonishing to see Holmes' vision of differentiating plant viruses simply on the basis of the symptoms.

Subsequently several systems of classification were proposed but none of them was internationally accepted. The next proposal that received approval of the ICTV was put forward by Lwoff *et al* (1962). They divided phylum 'Vira' into two subphyla, 'Deoxyvira' and 'Ribovira' depending on the type of the nucleic acid present. Each sub-phylum was further subdivided into 'Classes', 'Orders', 'Suborders', 'Families', 'Genera' and 'Species' depending on different physical, chemical and biological properties known at that time. This classification also did not differentiate animal, plant and bacterial viruses.

After the establishment of the Group Concept for naming plant viruses, several plant virologists started to propose a separate system of classification for plant viruses. Matthews (1981) proposed a monothetic hierarchical system to classify plant viruses using five key characters: (i) properties of the nucleic acid, (ii) number of nucleic acid strands present in the particle, (iii) presence or absence of envelope around the particles, (iv) number of genomic pieces in the nucleic acid, and (v) structure of the particles. This proposal received the approval of the ICTV and essentially classified all the groups of plant viruses known till that date.

The Plant Virus Subcommittee of the ICTV subsequently advocated the adoption of the 'Family-Genus-Species' concept for the classification of plant viruses. They preferred to rename the so-called 'Groups' and designate them as 'Families' and to put the related members as 'Genera'. Francki *et al* (1991) proposed such a system listing 37 groups described as 'Families/Groups'. Their classification was duly approved by the ICTV. In this classification, they considered the structure of the particles, the presence or absence of an envelope around the particles, the type of nucleic acid, number of nucleic acid strands and the number of genomic fragments as key characters. The approved 'Groups' were systematically arranged and named including two families (Reoviridae and Rhabdoviridae) and 33 groups. They reported 334 members or 'species', 320 probable members or 'deemed species' and the total number of members reported was 655 (Appendix I.4). The major difficulties were in separating one species from another. A virus normally consists of a set of genes that code for functional proteins. But in this perspective defining a 'species' becomes questionable. According to van Regenmortel (1990) a "Species represents a polythetic individual constituting a replicative lineage occupying a particular ecological niche." According to Matthews (1992), a virus 'Species' may be differentiated simply from the information on the coat protein of the particle supplemented by the information on the amino acids of the protein coat, their composition, sequence and immuno-responses. He further postulated possible separation of related strains, variants and pathovars of a 'Species' on the basis of nucleotide sequence of the genome, restriction endonuclease maps and the extent of cross-hybridization.

Fauquet (1999) defined a "Species" mostly on its genomic characters particularly genome rearrangement and sequence homology. He also included in the definition some biological properties, particularly serological relationship, vector transmission, host range, pathogenicity, tissue tropism and geographical distribution. He grouped the 'Species' into 'Genera' on the basis of virus replication strategy, genome organization, size, genome segments, sequence homology, and vector transmission. The 'Genera' were grouped into 'Families', 'Families' into 'Orders' on the basis of biochemical composition, replication strategy, particle-structure and genome organization. A strain may be a set of natural isolates that may have one or several characteristic properties. Strains normally have some stability over time whereas a pathotype is a collection of isolates of a single virus that have a similar behaviour with respect to host resistance.

Adopting these principles, van Regenmortel *et al* (2000) classified plant viruses where the viruses were first separated on the basis of nucleic acids (ds DNA-RT, ssDNA, dsRNA, ssRNA–, ssRNA +. They designated 4 orders (on the basis of the nature of the nucleic acids), 14 families and 71 genera. Fifty genera were assigned to families and the rest remained unassigned. They differentiated 657 species and 85 tentative ones. The total number of members reported was 742. The Families and the Groups/Genera are given in Appendix I.5 and Figure 1.1. Brunt *et al* (1996) and Hull (2002) have extensively described the properties of the Groups/Genera. Fauquet *et al* (2005) in the VIIIth Report of the ICTV designated 3 orders, 18 families, 9 sub-families, and 80 genera. They assigned 5450 viruses to 1950 species and classify them at the family and genus levels (Appendix I.6). The list of the currently accepted English names, families and abbreviations of the viruses is given in Appendix I.7.

It is interesting to note that Harrison *et al* (1971) based their groups mostly on biological properties, as information on the virus particles was very limited at that time. van Regenmortel *et al* (2000) and Fauquet *et al* (2005) based their groupings mainly on the properties of the virion (molecular mass, buoyant density, sedimentation co-efficient, stability under different conditions), genome (presence or absence of 5′ cap, presence or absence of 3′ terminal Poly (A) tract, nucleotide sequence, replication strategy, translation process, etc.), proteins (antigenic properties, epitope mapping), tissue tropism etc. However, these critical studies did not substantially change the groupings made by Harrison *et al* (1971) and only increased the number of the groups or genera.

Current studies however suggest the possibility of a phylogenetic classification of plant viruses. It may be presumed that viruses have originated following different evolutionary pathways and gradually take the current forms. The principal points of origin are: (i) pre-biotic RNAs, (ii) escape of plant host genes, (iii) transformation of transposons found in plants, animals and insects and (iv) degeneration of cellular components.

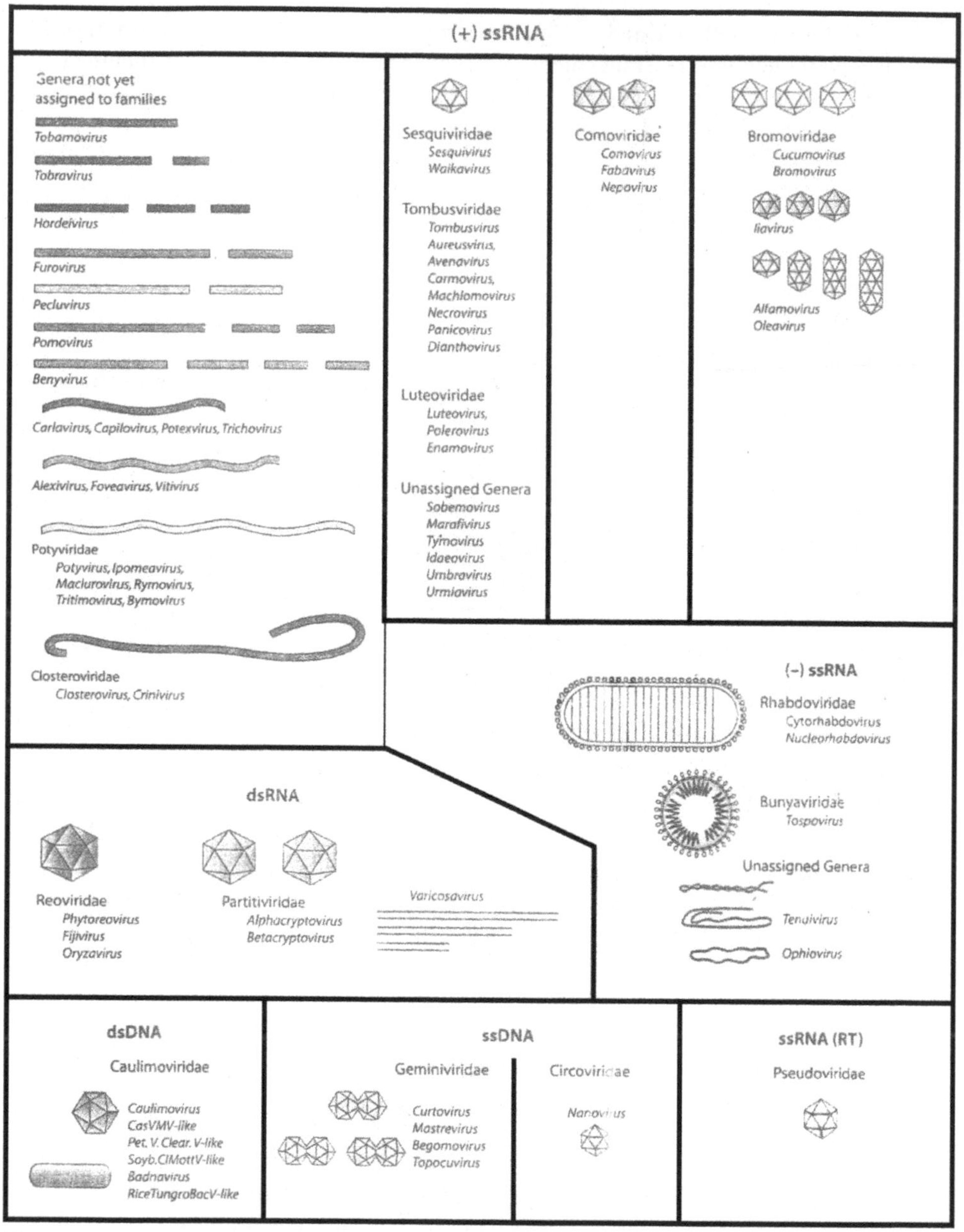

Figure 1.1 Morphological diversity of some genera belonging to different families opted (Agrios 2005 with permission from the Rights Department, Elsevier LTD)

Introns are often found in eukaryotic genes and splicing out the introns from RNA transcripts and legation of the extrons provide RNA the ability to produce new combinations of genes that later parasitize plant cells and undergo changes to produce new viruses. It is common that some viruses exchange materials with their hosts. Some changed materials may escape the host and transform into new viruses. DNA viruses perhaps originated in this way as the viral DNA can integrate with the host genome. A transposon called *Tnt 1*, 5334 nucleotides long, isolated from plants shows similar nucleotide sequences and open reading frames (ORF) that are found in *Drosophila* and are called *Popia*. Such transposons may be transformed into retroviruses with time that are capable of infecting plants, vertebrates and insects. There is evidence that large DNA viruses have originated due to degenerative processes in host cells. It appears that the scope of the evolution of DNA and dsRNA viruses is very limited. Accordingly limited numbers of DNA (excluding Begomovirus) and dsRNA viruses are found in nature. On the contrary, the scope of changes of ssRNA is enormous. Point-mutations occur very frequently in ssRNA, as do errors in the copying processes during genome replication. There may also be recombination, reassortment of pieces of segmented genes, loss of genetic material or acquisition of nucleotide sequences from unrelated viruses or host genomes. As a result, a large number of ssRNA viruses have evolved.

The classification presented by Fauquet *et al* (2005) reveals very interesting distribution of member (species) in different genera of viruses. Large numbers of members are found in *Begomovirus* (ssDNA) and *Potyvirus* (ssRNA). There are 112 *Begomoviruses* and 100 *Potyviruses*. Next are *Carlavirus* and *Nepovirus* that contain only 33 and 32 members respectively. There are 13 genera each of which contain only one member known so far. These are: 1. *Rice Tungrovirus* (RTV), 2. *Topocuvirus* (TPCTV), 3. *Babuvirus* (BBTV), 4. *Varicosavirus* (LBVaV), 5. *Oleavirus* (OLV-2), 6. *Machlomovirus* (MCMV), 7. *Panicovirus* (PMV), 8. *Enamovirus* (PEMV-1), 9. *Maculavirus* (GFkV), 10. *Mandarivirus* (ICRSV), 11. *Avenavirus* (ACSV), 12. *Petuvirus* (PVCV), and 13. *Cirevirus* (*Citrus leprosis*). Distribution of members (species) in other genera containing 12 to 27 members is given in Table 1.1 and the rest of the genera contain 2-9 members only. Thus, it is apparent that the genomes of monotypic genera are highly conserved. High potential variability of genomes may occur in the

Table 1.1 Genera Containing 12 To 27 Members (Species)

Potexvirus	27
Tobamovirus	22
Tymovirus	21
Alfamovirus	16
Ilarvirus	15
Comovirus	15
Tombusvirus	13
Sobemovirus	11
Pseudovirus	11
Badnavirus	10

genera where the number of species is high and the variation in genomes and evolution of new species/strains/pathotypes may be a continuously occurring natural process.

References

Agrios, G.N. 2005. Plant Pathology. Elsevier, Academic Press, San Diego, USA

Brunt, A.A., Crabtree, K., Dalwitz, M.J., Gibb, A.J. and Watson, L. 1996. Viruses of Plants. CAB International, Wellingford, UK

Calisher, C.H. and Fauquet, C.M. 1992. Steadman's ICTV Virus Words. Williams and Wilkins, Baltimore, USA

Fauquet, C.M. 1999. Taxonomy, classification and nomenclature of viruses. In: Encyclopedia of Virology. A. Granoff and R.G. Webster (eds.). Academic Press, New York, USA, pp. 130-156

Fauquet, C.M., Mayo, M.A., Maniloff, J., Desselberger, U. and Ball, L.A. (eds.). 2005. Virus Taxonomy. Classification and Nomenclature of Viruses. Eighth Report of the International Committee on the Taxonomy of Viruses. Elsevier/Academic Press, Amsterdam, The Netherlands

Francki, R.I.B., Fauquet, C.M., Knudson, D.L. and Brown, F. 1991. Fifth Report of the International Committee for Taxonomy of Viruses. Arch. Virol. Supplementum 2, Springer-Verlag, Wien, Austria

Gibbs, A.J. 1968. Crytptograms. In: Plant Virus Names. E.B. Martyn (ed.). Phytopathological Paper No. 9. Commonw. Mycol. Inst. Kew, Surrey, UK

Harrison, B.D., Finch, J.T., Gibbs, A.J., Hollings, M., Shepherd, R.J., Valenta, V. and Wetter, C. 1971. Sixteen Groups of Plant Viruses. Virology 45: 356-363.

Holmes, M.O. 1939. Handbook of Phytopathogenic Viruses. Burgess, Minneapolis, Minnesota, USA

Holmes, F.O. 1948. Bergey's Manual of Determinative Bacteriology. 6th edit. Suppl. No. 2 Williams and Wilkins, Baltimore, USA

Hull, R. 2002. Matthews Plant Virology. Academic Press, San Diego, USA

Lwoff, A., Horne, R. and Tournier, P. 1962. Cold Spring Harbour Symp. Quant. Biol. 27: 51-55

Martyn, E.B. (ed.) 1968. Plant Virus Names: An Annotated List of Names and Synonyms of Plant Viruses and Diseases. Phytopathological Papers No. 9, Supplement No. 1, Commonw. Mycol. Inst. Kew, Surrey, UK

Matthews, R.E.F. 1981. Plant Virology. Academic Press, New York, USA

Matthews, R.E.F. 1992. Fundamentals of Plant Virology. Academic Press, New York, pp. 351-374

Van Regenmortel, M.H.V. 1990. Virus species, a much overlooked but essential concept in virus classification. Intervirology 31: 241-254

van Regenmortel, M.H.V., Fauquet, C.M., Bishop, D.H.L., Carstens, E.B., Estes, M.K., Lemon, S.M., Maniloff, J., Mayo, M.A., McGeoch, D.J., Pringle, C.R., Wickner, R.B. (eds.). 2000. Virus Taxonomy: Classification and Nomenclature of Viruses. Seventh Report of the International Committee on Taxonomy of Viruses. Academic Press, San Diego, USA

2

Diversity of Physical Structure

Viruses show a remarkable diversity in structure from simple nucleic acid threads, particles containing either of the two nucleic acids with structures behaving like organic molecules to highly complex enveloped multi-layered particles. The simplest form is infectious nucleic acid or viroid. True viruses are particulate and structured. Structures are either helically or cubically symmetrical. Helically symmetrical viruses are anisometric particles. They are of different forms and sizes. Forms vary from short or medium length rods, to long or very long flexuous particles (Figures 2.1–2.4). These particles often form liquid crystals in which rods are regularly arrayed in two dimensions. X-ray crystallography is not possible with flexuous rods and complex particles. Cubical particles are isometric with icosahedral symmetry having 20 equal sides. Simple particles form true crystals. They normally do not widely differ in form and size. There is a unique case where particles consist of twinned or geminate icosahedra. Major differences among these viruses lie in the patterns of the assembly of protein subunits and their physical, biochemical, genomic and biological properties.

All types of particles may be mono-partite or multi-partite depending upon the type of the viruses. Some of the mono-partite viruses are: *Dianthovirus, Caulimovirus, Phytoreovirus, Carmovirus, Luteovirus, Potexvirus, Capillovirus, Carlavirus, Potyvirus* and *Closterovirus*. *Comovirus, Alphacryptovirus, Betacryptovirus, Tobravirus* and *Furovirus* are bipartite. Tri-partite viruses are: *Cucumovirus, Ilarvirus, Pomovirus, Begomovirus* and *Varicosavirus*. *Alfamovirus, Ourmiavirus* and *Benyvirus* are quadric-partite whereas *Tenuiviruses* are penta-partite.

Complex particles are basically isometric, helical or both but the pattern of the assembly of the protein subunits is complex and lipo-protein layers envelope these particles. The basic symmetry in Rhabdoviruses is helical whereas it is isometric in Tospoviruses.

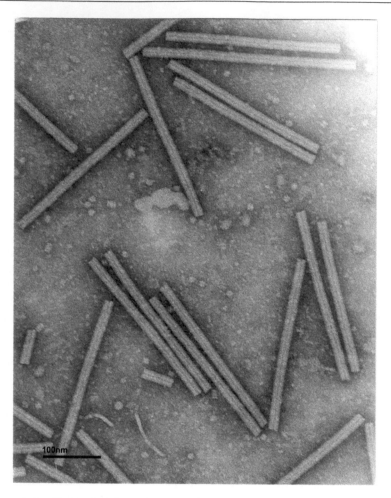

Figure 2.1 Short straight rod-shaped virus particle (*Tobamovirus*). Received from the Archives of the IACR, Rothamsted, UK

A. SIMPLE NUCLEIC ACID THREADS

Infectious nucleic acid threads are commonly known as "Viroids". The type of nucleic acid so far known in viroids is RNA and consists of a single molecular species that may occur in circular or linear forms autonomously replicating using the hosts' polymerase. There are a few groups of these viroids, mostly depending on the sequences of nucleic acid, nucleotide deletions and differences in their symptom expression. These are purely organic molecules but have lineage with viruses indicating the molecular origin of the latter.

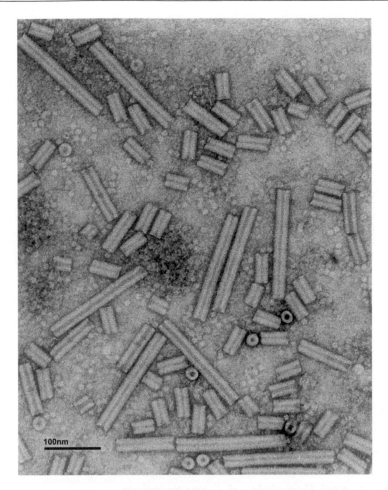

Figure 2.2 Medium straight rod-shaped virus particle (*Tobravirus*). Received from the Archives of the IACR, Rothamsted, UK

Viroid RNA has a series of regions comprising secondary structures separated by single chain loops that gives it *in vitro* the compact conformation of a rod having two loops at the ends.

B. PARTICULATE STRUCTURE

The unique feature of viruses is that their construction follows a general mathematical principle.

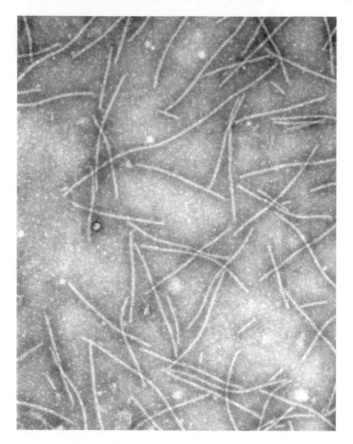

Figure 2.3   Long filamentous virus particle (*Potyvirus*)

1.   General Principle of Construction

In most of the viruses infectious nucleic acid is coated by protein that forms a protective structure. The assembly of the protein subunits around the nucleic acid is a process that widely differs in anisometric and isometric viruses. Basically the overall structure of a virus particle is a design of its protein subunits. Normally, a simple virus particle contains only one type of protein (sometimes two or three as in *Comovirus, Sequivirus, Phytoreovirus* respectively) or protein subunits also known as "structural units" or "capsids". These subunits are composed of identical protein molecules that are packed together in a regular manner in efficient protective designs. But construction of only a few efficient designs is possible, though many identical molecules are present. Construction of these designs depends upon the type of the assembly process. Self-assembly is a process akin to crystallization and is governed by the laws of statistical

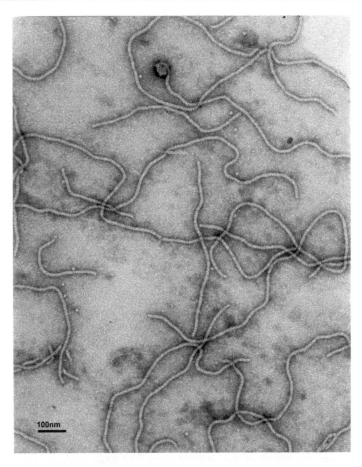

Figure 2.4 Very long filamentous flexuous virus particle (*Closterovirus*). Received from the Archives of the IACR, Rothamsted, UK

mechanics. A system of units with equivalent bonding properties will condense to form an ordered structure or set of ordered structures if the free energy of the ordered state is less than that of all other possible states. The most probable ordered structure is that in which the maximum number of most stable bonds are formed between the units. The necessary physical condition for the stability of any structure is that it is to be in a state of minimum free energy. When changing the environmental conditions of any ordered structure, dissociation can be induced without altering the integrity of the constituent components. Changing the environment back to the conditions that favour bond formation can restore the original organized structure.

An ordered structure built of identical units, always has some well-defined symmetry. Specific bonding between the units necessarily leads to a symmetrical structure. There will be only a limited number of ways in which any unit can be connected to its neighbours to form the maximum number of stable bonds. Most probable minimum energy designs for surface crystals constructed of a large number of units are tubes with helical or cylindrical symmetry and closed shells with polyhedral symmetry or polyhedral shells with cubic symmetry. Most virus particles, irrespective of their hosts (plant, animal or bacteria), are either helically constructed anisometric rods or cubically arranged isometric icosahedrons.

2.   Anisometric Particles

In helical particles, the nucleic acid thread is embedded on the protein subunits in an ordered manner (for figure, see Matthews 1992, Franklin 1955). The structure of *Tobacco mosaic virus* (TMV) shows the ordered arrangement of this process and a more or less similar process is found in other helical viruses.

(a)   Structure of TMV Particle

The TMV particle is a hollow cylinder 300 nm long with an inner diameter of 4 nm and an outer diameter of 18 nm. A particle contains 2130 copies of the viral coat protein (17495 daltons) protecting a single strand of RNA of 2×10^6 daltons (~6400 nucleotides). The protein subunits are arranged along the long axis of the particle as a right-handed one-start helix of 2.3 nm pitch and containing 16.34 subunits per turn (in the type strain). The polypeptide chain of the protein subunits is in four α helices. The RNA is embedded between successive turns of the helix at a distance of 4 nm from the axis. There are three contiguous nucleotides of the RNA chain for each successive protein subunit that determine the length of the virus particle. The basic structure repeats every 6.9 nm of its length or, in other words, in every three turns of the helix so there are 49 protein subunits in each repetition and the approximate number of turns in the TMV helix is 130 (for figure, see Franklin 1955).

The domains of the structure result from folding of its polypeptides which involve two types of arrangements: α-helices generated by the rotation of the polypeptide chain on itself, β-sheets in which the chain folds on itself turns and loops that connect the sheets and hence into globular mass. The secondary structures thus formed are not entirely reliable and they need to be confirmed by referring to proteins already known by means of crystallographic methods. Tertiary structures of a polypeptide chain may also be formed from the spatial organization of helices and sheets stabilized by non-covalent interactions between amino acids. A noteworthy feature in the aggregation of protein subunits is that both the N-and C-termini of the polypeptide chains are exposed at the surface of the particle and the protein subunits are tapered at the outside. Each subunit has two groves that together form a furrow and provide space for the embedment of the helical nucleic acid.

3. Isometric Particles

In isometric particles, there is no direct physical bonding between the protein subunits and the nucleic acids. The subunits organize themselves, following geometrical principles, around the nucleic acid to make the protective cover or capsid. The morphological units of this cover, or the capsomers, can be seen by electron microscopy or X-ray crystallography.

The basic structure of isometric viruses is an icosahedron that has 20 equilaterally placed triangular faces forming among them 12 vertices (Figure 2.5). This solid presents three series of axes of rotation generating symmetries: 5-fold axes pass through the vertices; 3-fold axes are located at the centre of the triangles and are perpendicular to the plane of the triangle; 2-fold axes pass through the midpoint of edges and are perpendicular to the edges. Each face

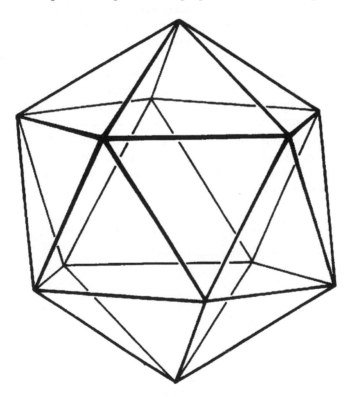

Figure 2.5 Diagrammatic representation of a cubical, isometric and icosahedral virus particle showing 20 equal sides. There are 12 vertices with fivefold rotational symmetry; the centre of each triangular face is on a threefold symmetry axis. Three structural units of any shape can be placed in identical positions on each face, giving 60 structural units (Matthews 1992, with permission from the Rights Department, Elsevier LTD)

is made up of three asymmetrical units. Geometrically, this type of icosahedral structure allows 60 identical subunits, each equilateral triangular face having 3 subunits constituting a closed surface, all having the same environment. These 60 subunits correspond through the 2-fold, 3-fold and 5-fold axes of rotation and in a virus particle they surround at the centre of the particle at fixed internal volume that shelter a polynucleotide.

Such geometrical assembly of 60 protein subunits is found in several viruses, viz. *Nepoviruses* and Satellite viruses. Geometrical clustering of more than 60 protein subunits is also possible. The surface of an icosahedron can be divided into a larger number of smaller equilateral triangles. The degree of such a subdivision is usually denoted by the term *Triangulation Number* (T), i.e. the number of triangles into which each icosahedral face has been divided. Mathematically T can have only certain values. For a simple icosadeltahedron the value of T would be 3. Each face of this structure has 9 and the whole structure would have 180 subunits. The shapes of the subunits are usually designated as banana or β-barrel. Under high-resolution electron microscopy these subunits do not appear identical because of their clustering in certain positions in the triangles. These clusters can be very clearly seen by electron microscopy. Depending on the nature of the viruses, the capsomers show different clustering patterns. Generally the clusters may be of two, three, five and six subunits designated as *dimers, trimers, pentamers* and *hexamers,* respectively.

Normally in icosahedrons, there are 12 vertices with 5-fold symmetry but there may also be 3-fold symmetry. Accordingly, their T values also differ (Figure 2.6). Different clusters of different numbers of subunits are found in the axes of these symmetries (Figure 2.7). As for example in *Tymoviruses* there are

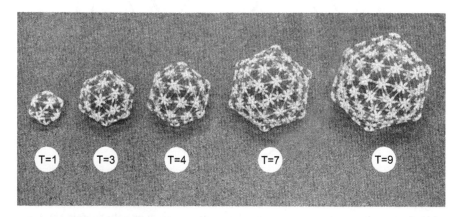

Figure 2.6 Models of icosadeltahedra for the first five triangulation numbers. These icosahedral surface lattices represent possible minimum energy designs for closed shells constructed from identical units (Caspar 1964); reprinted with permission of the University Press of Florida, USA

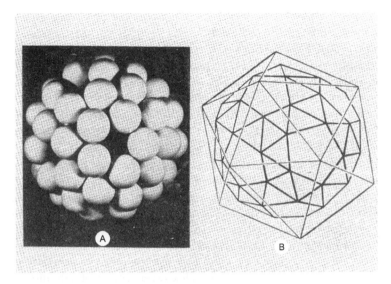

Figure 2.7 Clustering of protein subunits in a particle with icosahedral symmetry to form the protein shell in a spherical virus (Holton *et al* 1959). With permission from Holton C.S. Plant Pathology@1959 by the Board of Regents of the University of Wisconsin System, USA

12 groups of *pentamers* around the 5-fold symmetry and 20 groups of *hexamers* around the 3-fold symmetry axes. In more complex icosahedra, there may be twofold symmetry. In such cases the subunits adjust their bonding to adjust different symmetry-related positions in the shell as found in *Tombusvirus*. In viruses where more than one type of protein occurs as in the *Comoviruses* and *Reoviruses*, different polypeptides adjust themselves in a different symmetry environment with the shell. The geometry of these assembly processes has been studied and illustrated by several authors (Klug and Casper 1960, Casper 1964, Gibbs and Harrison 1976, Matthews 1981, Rossmann and Johnson 1989, Matthews 1992, Hull 2002, Astier *et al* 2007).

Protein subunits in capsomers and the capsomers in the capsids are stabilized by the formation of different types of bonds. Usually non-covalent bonds are prevalent in virus particles. There are two types of non-covalent bonds: polar (salt and hydrogen bonds) and non-polar (van der Waal's and hydrophobic bonds). The occurrence of these bonds varies in different viruses. Protein subunits in some viruses are electrovalently linked while in others they are hydrophobically bonded. Electrovalently bonded protein subunits can be easily dissociated and are not very stable. Nucleic acid-free shells are less stable than those containing the nucleic acid. Thus the type of nucleic acid-protein bonding may also contribute to the stability of the particles.

In isometric particles, the arrangement between the capsid and nucleic acid normally does not reflect the stability of the particles but the capsids in many cases are related to the replication of the nucleic acids. The nucleic acids may remain confined to radii inside the particle and there may be penetration of the protein subunits into the region where the nucleic acid is located. There are some spherical viruses the nucleic acids of which remain covered by double capsids made up of more than one type of protein subunits. In others the outer capsids have projecting spikes whereas the inner capsids have hidden spikes as in the *Phytoreoviruses* (for figure, see Hatta and Francki 1977).

The structure of an isometric virus may have simple morphological form as in *Cucumovirus* with round capsid containing 32 capsomers in T=3 (Figure 2.8). In some cases, the isometric particles may take a form of bacilliform particle as in the *Alfamoviruses* (Figure 2.9). The coat protein in this virus behaves as a water-soluble *dimer* stabilized by hydrophobic interactions between the molecules. This *dimer* is the morphological unit out of which the viral shells are constructed. In an extreme case an isometric virus may form twinned or geminate particle (Figure 2.10). There may also be complex arrangement of subunits with 12 pentamers and 60 hexamers in T= 7 as found in *Caulimovirus*. In a *Phytoreovirus* the isometric particle has undeveloped three-layered capsid shell with 260 trimers in T=13; inner shell with 60 dimers in T=1 arrangement.

4. Complex Viruses

The particles of some viruses are complex and cannot be easily differentiated into rods or spheres. The shape of these viruses is normally bacilliform or round. The basic feature of these particles is that they are covered by a lipoproteinaceous membrane; several types of proteins occur in the coat that are complexes with the membrane on the one hand and nucleic acids on the other; the nucleic acids may be helical or spherical and form the core of the particle.

Common complex plant viruses are the *Rhabdoviruses* and the *Tospoviruses*. *Rhabdoviruses* are bullet-shaped and rounded at both ends to give a bacilliform shape. The inner part constitutes the inner capsid formed by an RNA linked to a basic protein. This ribonucleoprotein is arranged in a helix. The outer coat is derived from cellular membranes; it contains two viral proteins, one of which is directed towards the exterior (Figure 2.11). *Tospovirus* particles are round and about 100 nm in diameter. The central core of the particle contains the RNA. A layer of dense material surrounded by a typical lipoprotein bi-layer membrane covers the RNA (Figure 2.12). Jackson *et al* (1999, 2005) reviewed the structure and function of different *Rhabdoviruses*.

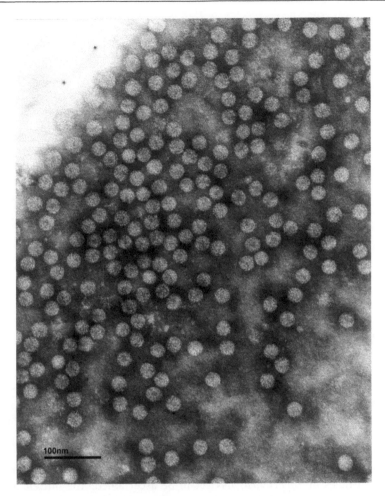

Figure 2.8 An isometric virus with a simple morphological form (round capsid, 32 capsomers, T=3): *Cucumovirus*. Received from the Archives of the IACR, Rothamsted, UK

5. Structural Diversity

Most of the viruses known so far (Fauquet *et al* 2005) are positive sense RNA (RNA+) and a few are negative sense RNA (RNA–), dsRNA, ssDNA and dsDNA. Structurally these viruses are of five types: isometric, geminate, rod, filamentous, and bacilliform. Particles occur as mono-, bi-, tri-, quadripartite and pentads. Approximately 50 viruses contain positive sense RNA, 20 of which are isometric, two are quasi-isometric (*Ilarvirus* and *Oleavirus*) and one is isometric with an envelope (*Tospovirus*). Approximately 16 isometric

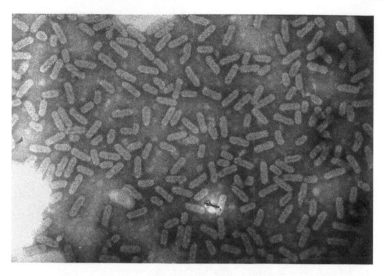

Figure 2.9   Isometric particle taking the form of a bacilliform virus (*Alfamovirus*)

particles are mono-partite, 5 are bi-partite (*Alfacryptovirus, Betacryptovirus, Fabavirus, Nepovirus, Idaeovirus*), 2 are tri-partite (*Bromovirus, Cucumovirus*), 1 is quadri-partite (*Oleavirus*) and 1 is a pentad virus (*Nanovirus*). There are 4 mono-partite virus particles that contain dsDNA with the reverse transcriptase enzyme (RT). These are *Caulimovirus, Soymovirus (Soybean chlorotic mottle virus), Cavemovirus (Cassava vein mosaic virus)* and *Petuvirus (Petunia vein clearing virus)*. Particles of four viruses are geminated or twinned (*Mastrevirus, Curtovirus, Begomovirus, Topocuvirus*); particles of *Mastrevirus* and *Curtovirus* are mono-partite whereas those of *Begomoviruses* are tri-partite; the nature of the particles of *Topocuvirus* has yet to be ascertained. The nucleic acid of all these viruses are ssDNA (circular). Another ssDNA (circular) containing isometric virus, the *Nanoviruses* are pentad viruses that have five particles. Isometric particles of three viruses are enveloped and double shelled with spikes and contain dsRNA (*Fijivirus, Oryzavirus, Phytoreovirus*).

There are approximately 8 rod-shaped viruses of which 1 is mono-partite (*Tobamovirus*), 3 are bi-partite (*Furovirus, Pecluvirus, Tobravirus*), 3 are tri-partite (*Varicosavirus, Pomovirus, Hordeivirus*) and 1 is quadri-partite (*Benyvirus*). Particles of all these viruses contain ssRNA.

Filamentous viruses are more common than rod-shaped ones. There are approximately 18 filamentous viruses of which 14 are mono-partite (*Closterovirus, Ipomovirus, Macluravirus, Potyvirus, Rymovirus, Tritimovirus, Allexivirus, Capillovirus, Carlavirus, Foveavirus, Mandarivirus, Potexvirus, Trichovirus, Vitivirus*), 2 are bi-partite (*Crinivirus, Bymovirus*) and 2 are quadri-partite (*Tenuivirus, Ophiovirus*).

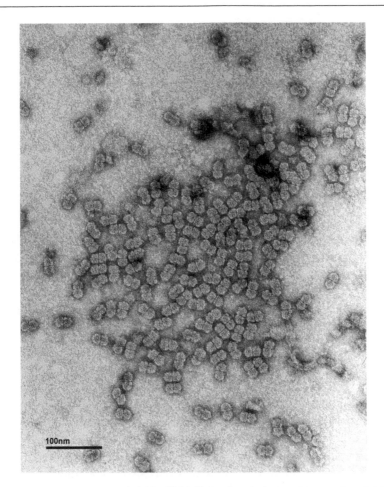

Figure 2.10 An isometric virus with twinned or geminate particle (*Geminivirus*). Received from the Archives of the IACR, Rothamsted, UK

There are 6 bacilliform viruses (*Badna virus, Tungro virus, Ourmia virus* and *Umbra virus*) and 2 are Rhabdo- or bullet/bacilliform viruses (*Cytorhabdovirus* and *Nucleorhabdovirus*. Particles of *Badna* and *Tungro* viruses are mono-partite and do not have an envelope. The particles of Rhabdoviruses are enveloped. The envelope is composed of two layers of lipo-protein membranes. The inner layer contains host-derived lipids penetrated into the G protein of surface layer (Jackson *et al* 1999, 2005).

The majority of the viruses have isometric particles. The particles of 38 viruses are isometric and 4 viruses have geminate particles.

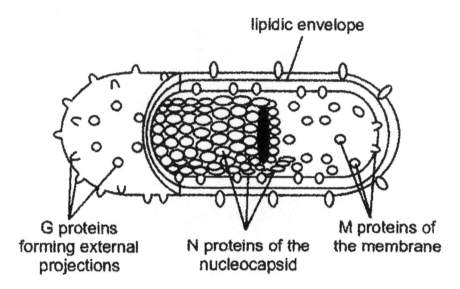

lipidic envelope

G proteins forming external projections

N proteins of the nucleocapsid

M proteins of the membrane

Figure 2.11 Bacilliform complex enveloped particle of *Rhabdovirus* with membrane protein, nucleocapsid proteins and proteins forming external projections (Astier *et al* 2001, source INRA 2001, France)

The structures of the coat protein of most of the viruses are not completely understood, but in general they contain only one type of protein except for *Marafivirus, Fijivirus, Oryzavirus* and *Phytorevirus. Marafivirus* coat protein contains two types of polypeptides whereas the coat protein of *Fiji, Oryza* and *Phytoreoviruses* contains 6-8 types of protein species.

The salient features of the structure of different viruses are given in Appendix II.1. The structures of viruses containing dsDNA-RT are either isometric or bacilliform and mono-partite. The structures of viruses containing ssDNA on the other hand are mostly geminate except one (*Nanovirus*) that is isometric and separated into five particles. The structures of most of the dsRNA viruses are isometric and double shelled; these are mono-partite and enveloped. A few are non-enveloped and bi-partite and one is rod-shaped and tri-partite. The structures of negative sense ssRNA containing viruses are very conservative and found only in rhabdo- or bullet/bacilliform viruses and two other viruses. For three ssRNA-RT viruses (*Metavirus, Pseudovirus* and *Seravirus*), the structural details have yet to be determined.

A lot of diversity is found in the structures of viruses containing positive sense ssRNA. The structures of most of these viruses are filamentous followed by isometric and rod-shaped particles. These may be the basic structures of viruses; others may arise by changes in the genomes or adaptation of viruses in plants that had altogether different primary hosts.

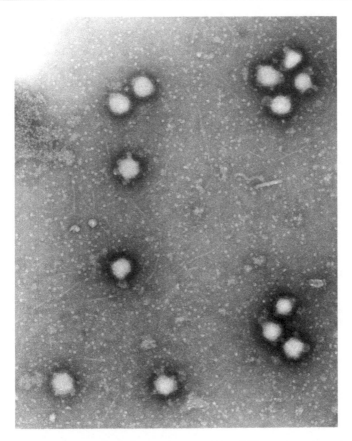

Figure 2.12 Round enveloped complex particle of *Tospovirus*. Received from the Archives of the IACR, Rothamsted, UK

References

Astier, S., Albouy, J., Maury, Y., Robaglia, C. and Lieoq, H. 2007. Principles of Plant Virology: Genome, Pathogenicity, Virus Ecology. Science Publishers Inc., Enfield, USA, pp. 1-34

Casper, D.L.D. 1964. Structure and function of regular virus particles. In: Plant Virology. M.K. Corbett and HD Sisler (eds.). University of Florida Press, Gainesville, Florida, USA, pp. 267-290

Fauquet, C.M., Mayo, M.A., Maniloff, J., Desselberger, U. and Ball, L.A. (eds.). 2005. Virus Taxonomy. Classification and Nomenclature of Viruses. Eighth Report of the International Committee on the Taxonomy of Viruses. Elsevier/Academic Press, Amsterdam, The Netherlands

Franklin, R.E.1955. Structure of tobacco mosaic virus. Nature 177: 379-381

Franklin, R.E., Caspar, D.L.D. and Klug, A. 1959. The structure of viruses as determined by X-ray diffraction. In: Plant Pathology—Problems and Progress. C.S. Holton (ed.). University of Wisconsin Press, Madison, Wisconsin, USA

Gibbs, A. and Harrison, B. 1976. Plant Virology: The Principles. Edward Arnold (Publishers) Ltd., London, WCIB 3 DQ, UK, pp. 47-67

Hatta, T. and Francki, R.I.B. 1977. Morphology of Fiji disease virus. Virology 76: 797-807

Hull, R. 2002. Matthew's Plant Virology. Academic Press, San Diego, USA

Jackson, A.O., Goodin, M., Moreno, I., Johnson, J. and Lawrence, D.M. 1999. *Rhabdoviruses* (Rhabdoviridae). In: Encyclopedia of Virology. 2nd edit. A. Granoff and R.G. Webster (eds.). Academic Press, San Diego, USA, pp. 1531-1541

Jackson, A.O., Dietzgen, R.G., Goodin, M.N., Bragg, J.N. and Deng, M. 2005. Biology of Plant Rhabdoviruses. Annu. Rev. Phytopathol. 43: 623-660

Klug, A. and Caspar, D.L.D. 1960. The structure of small viruses. Adv. Virus Res. 7: 225-325

Matthews, R.E.F. 1981. Fundamentals of Plant Virology, Academic Press, New York, USA

Matthews, R.E.F. 1992. Fundamentals of Plant Virology. Academic Press Inc. Harcourt Brace Jovanovich Publishers, San Diego, USA, pp. 67-90

Rossmann, M.G. and Johnson, J.E. 1989. Icosahedral RNA virus structure. Annu. Rev. Biochem. 58: 533-573

3

Diversity in Chemical Components and Genomic Structure

A. BASIC COMPONENTS

Virus particles are, in general, composed of two types of biological macromolecules, nucleic acid and protein. In complex viruses an additional component— lipid— also occurs. The amount of these substances present in the particles depends upon their architecture. In a general way it is considered that isometric particles normally contain 15 to 45% nucleic acid whereas the anisometric particles normally contain only 5% nucleic acid. The complex particles normally contain 20% lipid. In addition to these major components, particles of some viruses may contain traces of a few enzymes, polyamines and metallic ions having specific roles in replication and structural stabilization. However wide diversities exist in the nucleic acid and protein contents of the known viruses, but the reasons and effects of such diversity are yet to be explained.

B. DIVERSITIES IN QUANTITATIVE PRESENCE OF IMPORTANT COMPONENTS

The main differences in nucleic acid content are found in particles with different symmetries. In the rod-shaped and filamentous viruses, nucleic acid content normally ranges from 4% to 5%, with the exception of a few that contain a higher percentage of nucleic acid. As for example, *Tenui-* and *Potyviruses* contain 5.2 to 12% and 20-21% nucleic acid respectively. A much greater diversity is observed in the bacilliform viruses with the nucleic acid contents of *Alfamo-, Cytorhabdo-, Nucleorhabdo-* and *Ourmiaviruses* of 16%, 5%, 1-5% and 15-25% respectively. Not all the bacilliform viruses contain lipid, e.g. *Alfamo-* and *Ourmiaviruses*, whereas *Cytorhabdoviruses* and *Nucleorhabdoviruses* contain 25% and 15-37% lipid (Table 3.1).

Table 3.1 Diversities of Important Chemical Components found in Some Bullet, Bacilliform, Rod-shaped and Filamentous Viruses (Brunt *et al.* 1996, Hull 2002, Fauquet *et al* 2005, ICTVdB 2006)

Virus	Family	Shape	Nucleic acid content (%)	Protein content (%)	Lipid content (%)
1. *Alfamovirus* (ssRNA+)	Bromoviridae	Bacilliform/ Quasi-isometric	16	84	Nil
2. *Allexivirus* (ssRNA+)	Flexiviridae	Filamentous	5	95	Nil
3. *Ampelovirus*	Flexiviridae	Filamentous	–	–	–
4. *Badnavirus* (dsDNA)	Caulimoviridae	Bacilliform	–	–	–
5. *Benyvirus* (ssRNA+)	No family	Rod	–	–	–
6. *Bymovirus* (ssRNA+)	Potyviridae	Filamentous	5	95	Nil
7. *Capillovirus* (ss RNA+)	Flexiviridae	Filamentous	5	95	Nil
8. *Carlavirus* (ssRNA+)	Flexiviridae	Filamentous	5	95	Nil
9. *Closterovirus* (ssRNA+)	Closteroviridae	Filamentous	5-5.17	94.83-95	Nil
10. *Crinivirus* (ssRNA+)	Closteroviridae	Filamentous	5	95	Nil
11. *Cytorhabdovirus* (ssRNA–)	Rhabdoviridae	Bullet/Bacilliform	5	70	25
12. *Foveavirus* (ssRNA+)	Flexiviridae	Filamentous	–	–	–
13. *Furovirus* (ssRNA+)	No family	Rod	4-59	5-96	Nil
14. *Hordeivirus* (ssRNA+)	No family	Rod	3.8-5	95-96.2	Nil
15. *Ipomovirus* (ssRNA+)	Potyviridae	Filamentous	5	95	Nil
16. *Macluravirus* (ssRNA+)	No family	Filamentous	5	95	Nil
17. *Mandarivirus* (ssRNA+)	Flexiviridae	Filamentous	–	–	–
18. *Nucleorhabdovirus* (ssRNA–)	Rhabdoviridae	Bullet/Bacilliform	1-5	68-80	15-37
19. *Ophiovirus* (ssRNA–)	No family	Filamentous	–	–	–
20. *Ourmiavirus* (ssRNA+)	No family	Bacilliform	15-25	75-85	Nil
21. *Pecluvirus* (ssRNA+)	No family	Rod	4	96	Nil

22. *Pomovirus* (ssRNA+)	No family	Rod	–	–	–
23. *Potexvirus* (ssRNA+)	Flexiviridae	Filamentous	5-8	92-95	Nil
24. *Potyvirus* (ssRNA+)	Potyviridae	Filamentous	20-21	78-80	Nil
25. *Rymovirus* (ssRNA+)	Potyviridae	Filamentous	5	95	Nil
26. *Tenuivirus* (ssRNA+)	No family	Filamentous	5.2-12	88-94.8	Nil
27. *Tobamovirus* (ssRNA+)	No family	Rod	5	95	Nil
28. *Tobravirus* (ssRNA+)	No family	Filamentous	5	95	Nil
29. *Trichovirus* (ssRNA+)	Flexiviridae	Filamentous	5	95	Nil
30. *Tritimivirus* (ssRNA+)	Potyviridae	Filamentous	5	95	Nil
31. *Tungrovirus* (dsDNA-RT)	Caulimoviridae	Bacilliform	–	–	Nil
32. *Varicosavirus* (ssRNA–)	No family	Filamentous	–	–	–
33. *Vitivirus* (ssRNA+)	Flexiviridae	Filamentous	5	95	Nil

The nucleic acid content of isometric viruses on the other hand differs widely. Many have nucleic acid contents in the range 14-24%. The *Tospoviruses* contain only 5% nucleic acid. Higher contents of nucleic acid are found in a few viruses ranging from 28% to 46%. These viruses are *Tombusviruses* (28%), *Fabaviruses* (35%), *Comoviruses* (38%), *Tymoviruses* (39%), *Sequiviruses* (40%), *Oryzaviruses* (42%) and *Nepoviruses* (46%). Empty shells without any nucleic acid are also found in a few viruses such as *Faba-, Como-, Oryza-,* and *Nepoviruses*. Among the isometric and geminate viruses only the *Tospoviruses* contain lipid (20%) (Table 3.2).

Table 3.2 Important Chemical Components of Isometric and Geminate Viruses (Brunt *et al* 1996, Hull 2002, Fauquet *et al* 2005, ICTVdB Management 2006)

Virus	Family	Shape	Nucleic acid content (%)	Protein content (%)	Lipid content (%)
1. *Alphacryptovirus* (dsRNA)	Partitiviridae	Isometric	25	75	Nil
2. *Aureusvirus* (dsRNA)	Tombusviridae	Isometric	18	82	Nil
3. *Avenavirus* (ssRNA+)	Tombusviridae	Isometric	18	82	Nil

4. *Babuvirus* (ssDNA)	Nanoviridae	Isometric	–	–	Nil
5. *Begomovirus* (ssDNA)	Geminiviridae	Geminate or prolate	18-22	78-82	Nil
6. *Betacryptovirus* (dsRNA+)	Partitiviridae	Isometric	24	76	Nil
7. *Bromovirus* (ssRNA+)	Bromoviridae	Isometric	20-23.7	76-79	Nil
8. *Carmovirus* (ssRNA+)	Tombusviridae	Isometric	14-23	77-86	Nil
9. *Caulimovirus* (dsDNA-RT)	Caulimoviridae	Isometric	14.5-17	83-85.9	Nil
10. *Cavemovirus* (dsDNA-RT)	Caulimoviridae	Isometric	–	–	–
11. *Cheravirus* (ssRNA+)	No family	Isometric	–	–	–
12. *Comovirus* (ssRNA+)	Comoviridae	Isometric	0-38.6	62.82-100	Nil
13. *Cucumovirus* (ssRNA+)	Bromoviridae	Isometric	16-21.2	78.8-84	Nil
14. *Curtovirus* (ssDNA)	Geminiviridae	Geminate	–	–	Nil
15. *Dianthovirus* (ssRNA+)	Tombusviridae	Isometric	20-28	72-80	Nil
16. *Enamovirus* (ssRNA+)	Luteoviridae	Isometric	28	72	Nil
17. *Fabavirus* (ssRNA+)	Comoviridae	Isometric	0-35	67-100	Nil
18. *Fijivirus* (dsRNA)	Reoviridae	Isometric	–	– Nil	
19. *Idaeovirus* (ssRNA+)	No family	Isometric	24	76	Nil
20. *Ilarvirus* (ssRNA+)	Bromoviridae	Isometric	12-24	76-88	Nil
21. *Luteovirus* (ssRNA+)	Luteoviridae	Isometric	28-30	70-72	Nil
22. *Machlomovirus* (ssRNA+)	Tymoviridae	Isometric	18	82	Nil
23. *Macluravirus* (ssRNA+)	Tymoviridae	Isometric	3-4	96-97	Nil
24. *Marafivirus* (ssRNA+)	Tymoviridae	Isometric	25-30	67Nil	
25. *Mastrevirus* (ssDNA)	Geminiviridae	Geminate	19-20	80Nil	
26. *Metavirus* (ssRNA-RT)	Metaviridae	LRT-retrotransposon-like	–	–	–
27. *Nanovirus* (ssDNA)	Nanoviridae	Isometric	16-17	83-84	Nil

28. *Necrovirus* (ssRNA+)	Tombusviridae	Isometric	18-19	72-81	Nil
29. *Nepovirus* (ssRNA+)	Comoviridae	Isometric	0, 27 -40, 42-46	54-100	Nil
30. *Oryzavirus* (dsRNA)	Reoviridae	Isometric	0-42	58-100	Nil
31. *Oleavirus* (ssRNA+)	Bromoviridae	Isometric /Bacilliform	16	84	Nil
32. *Panicovirus* (ss RNA+)	Tombusviridae	Isometric	18	82	Nil
33. *Petuvirus* (dsDNA-RT)	Caulimoviridae	Isometric	–	–	Nil
34. *Phytoreovirus* (dsRNA)	Reoviridae	Isometric	20-22	78-80	Nil
35. *Polerovirus* (ssRNA+)	Luteoviridae	Isometric	30	70	Nil
36. *Pseudovirus* (ssRNA-RT)	Pseudoviridae	LRT -retrotransposon -like	–	–	–
37. *Sadwavirus* (ssRNA+)	No family	Isometric	0-37	63-100	Nil
38. *Sequivirus* (ssRNA+)	Sequiviridae	Isometric	40	60	Nil
39. *Sirevirus* (ssRNA-RT)	Pseudoviridae	Isometric	–	–	–
40. *Sobemovirus* (ssRNA+)	No family	Isometric	17-28	72-83	Nil
41. *Soymovirus* (dsDNA-RT)	Caulimoviridae	Isometric	16.1	83.9	Nil
43. *Tombusvirus* (ssRNA+)	Tombusviridae	Isometric	14-18	82-84	Nil
44. *Topocuvirus* (ssDNA)	Geminiviridae	Geminate	–	–	Nil
45. *Tospovirus* (ssRNA–)	Bunyaviridae	Spherical to pleomorphic, enveloped with surface projections	5	70	20
46. *Tymovirus* (ssRNA+)	Tymoviridae	Isometric	23-39	61-77	Nil
47. *Umbravirus* (ssRNA)	No family	Vesicle-like	–	–	–
48. *Waikavirus* (ss RNA+)	Sequiviridae	Isometric	0-42	58-100	Nil

There are other diversities as some of the viruses contain negative sense RNA (*Cytorhabdovirus, Nucleorhabdovirus, Varicosavirus. Ophiovirus*), and the nucleic acid of others contain reverse transcriptase enzyme (*Caulimovirus, Cavemovirus, Petuvirus, Soymovirus, Metavirus, Pseudovirus, Sirevirus*).

C. GENERAL PROPERTIES OF PROTEINS AND THEIR ROLES IN VIRUS STRUCTURE

Essentially proteins are amino acids linked together by polypeptide linkages (–CO–NH). These linkages are formed between the carboxyl group (–COOH) of one amino acid to the amino (–NH$_2$) group of another. Several peptides together make a polypeptide. A linear polypeptide has two terminals, N and C. N terminals usually occur at the left hand side whereas –C terminals occur at the right hand side. These terminals regulate the acidity and basicity of the amino acids. The properties of the polypeptides vary with the differences in the sequences of the acidic and basic amino acids of the polypeptides.

Natural, biologically active, protein molecules are not always simple linear polypeptides as they may be spirally or helically coiled to form secondary structures which may be further bended or folded to give rise to tertiary structures; interactions between them may give rise to quaternary structures. Formation of these complex structures from the native linear conformation contributes to their varied biological properties. Such complex protein molecules can be denatured and renatured by changing the environment in which they operate.

In principle viral coat proteins do not differ from those found in the cells of the organisms except in their amino acid sequences. The amino acid composition and amino acid sequences differ from virus to virus and are an important criterion for differentiating one virus from another. The size of plant viral proteins normally ranges from 150 to 600 or more amino acid residues. Normally, in simple virus particles, only one type of amino acid sequence is found, whereas in the particles of complex viruses, more than one polypeptide occurs and different types of amino acid sequences are observed depending upon the number of polypeptide species. Coat proteins of most of the viruses have acetylated N terminals. As for structures, mostly secondary and tertiary forms are found; amongst the secondary forms the α-helix arrangement is common in helical particles. The size of the protein, its structural conformation and amino acid sequence determine the properties of the coat protein and this contributes to the stability, antigenicity, transmissibility by vectors and several other properties of the virus particles.

D. DIVERSITY IN GENOMIC STRUCTURE

In contrast to plant or animal genomes, viral genomes may either be RNA or DNA. Usually the entire viral nucleic acid is genomic but there may be sub-genomic or non-genomic polynucleotides in some viruses. The genomes of several viruses may remain incomplete and depend on the genomes of some other viruses for their replication. In some RNA viruses, self-duplicating satellite-RNAs are found which are involved in pathogenesis or other functions. The size of the viral nucleic acids and their mode of occurrence differ widely. Most of the plant viruses contain one or multiple pieces of single-stranded RNA but few contain multiple pieces of double-stranded RNA. Plant viruses containing single- and double-stranded DNA are also found. Nucleic acids of viruses also widely differ in their molecular weights. These may be as low as 0.4×10^6 dalton (Satellite virus) and as high as 15.5×10^6 dalton (*Wound tumour virus*). There are many enigmatic properties of viral nucleic acids, the gradual understanding of which has opened new vistas in molecular genetics.

Single-stranded RNA may be positive sense (ssRNA+) or negative sense (ssRNA–). SsRNA+ genomes directly function as messenger RNAs (mRNA), but there are several ssRNA+ viruses that have small proteins covalently linked to the 5′ end of the RNA. Some ssRNA+ viruses require the presence of coat protein or coat protein mRNA to become functional as in *Alfamovirus*. ssRNA– genomes are non-infective by themselves and these viruses contain a viral coded enzyme that copies the genomic RNA to mRNA.

DNA in viruses may also be single (ssDNA) or double-stranded (dsDNA). These DNAs use host enzymes to produce mRNAs.

Structure

The genomes of all viruses have specific organizations for the efficient use of their nucleic acids. Their coding sequences are very closely packed with only a few non-coding nucleotides between the genes. There may also be overlapping of coding regions for two different genes.

Depending on the virus, genomes may be linear or circular, in one piece or several pieces. Each genome may contain functional or non-functional genes. The number of viral genes ranges from one (Satellite virus, Satellite tobacco necrosis virus) to 12 (Reoviruses). Genomes may also contain non-gene nucleotide sequence or *cis*-acting nucleotide sequences that operate recognition and control functions for virus replication. 5′ and 3′ termini of viral genomes may be of 2 different types. In some viruses these termini may contain specialized structures while others contain free sugar, hydroxyl or phosphate groups. The 5′ end of some virus genomes may have methylated bases or 'caps', *Tobamovirus*,

Tobravirus, Tymovirus, Bromovirus, Cucumovirus, and Alfamovirus contain 5′ capped RNA. The presence of a cap is normally related to infectivity, stability and efficiency of translation particularly in multi-partite genomes.

The 5′ end of the genome may also contain a small genome-linked virus protein or 'VPg' covalently bonded with the 5′ end. The genomes of *Potyvirus, Comovirus, Nepovirus, Sobemovirus,* and *Luteovirus* are normally attached to VPgs that usually play roles in the initiation of RNA synthesis. It is not necessary for translation of RNA. It is cleaved from a polyprotein in which it is part of the replication module (Fellers *et al* 1998).

The 3′ terminal ends of the genomes show different structures. In the case of those viruses, the genomes of which act as messengers (mRNA), the 3′ terminals have variable 'Poly-A' sequences. Such sequences are found in *Potyvirus, Potexvirus, Capillovirus, Comovirus, Nepovirus, Benyvirus* and *Trichovirus*. Poly-A sequences may have a role in the protein synthesizing systems. 3′ terminal sequences of the genomes of some viruses may fold into transfer RNA (tRNA)-like structures. tRNA at the 3′ end may act in accepting and donating an amino acid, facilitate translation and may act as a 'replicase' recognition site. Such structures are found in *Tobamovirus, Tymovirus, Hordeivirus, Bromovirus, Cucumovirus, Furovirus,* and *Tobravirus*. The genome size and other physico-chemical properties of different viruses vary widely (Appendices III.1-III.5). The molecular basis of these differences is yet to be properly understood. The organization of genomes and their expression in different genera are given in Appendix III.6.

The secondary structures in hairpins or stem-loop forms are constituted when a single chain presents complementary inverse sequences, which in pairing will form a stem in a double helix, separated by a few nucleotides forming a loop. These are sites of interaction with cellular or viral proteins, and with other nucleotide sequences. A pseudo-knot is formed when the loop of a hairpin pairs with a nearby or more distant nucleotide sequences, forming a tertiary structure. The tRNA-like structures carried by certain viral RNAs at the 3′ end are formed of hairpins and pseudo-knots, the latter being absent in cellular tRNA (Astier *et al* 2007).

References

Astier, S., Albouy, J., Maury, Y., Robaglia, C. and Leicoq, H. 2007. Principles of Plant Virology: Genome, Pathogenesis, Virus Ecology. Science Publishers Inc., Enfield, USA, pp. 11-34

Brunt, A.A., Crabtree, K., Dalwitz, M.J., Gibbs, A.J. and Watson, L. 1996. Viruses of Plants. CAB International, Wallingford, UK

Chung, K.-R. and Brlansky, R.H. 2006. Fact Sheet. Florida Cooperative Extension Service, Institute of Food and Agricultural Sciences, University of Florida, FL, USA

Fauquet, C.M., Mayo, M.A., Maniloff, J., Desselberger, U. and Ball, L.A. (eds.). 2005. Virus Taxonomy. Classification and Nomenclature of Viruses. Eighth Report of the International Committee on the Taxonomy of Viruses. Elsevier/Academic Press, Amsterdam, The Netherlands

Fellers, J., Wan, J., Hong, H., Collins, G.B. and Hunt, A.G. 1998. In vitro interaction between a potyvirus-encoded genome-linked protein and RNA-dependent RNA polymerase. J. Gen. Virol. 79: 2043-2049

Hull, R. 2002. Matthew's Plant Virology. Academic Press, San Diego. 5

ICTVdB-2006. The Universal Virus Databases, Version 4. C. Buchen-Osmond (ed.). Columbia University, New York, USA

ICTVdB Management 2006. In: ICTDdB—The Universal Database, Version 4. Buchen-Osmond (ed.). Columbia University, New York, USA

Locali-Fabris, E.C., Freitas-Astǔa, J., Souza, A.A., Tacit, M.A., Astǔa-Monge, G., Antonioli-Luizon, R., Rodrigues, V., Targon, M.L.P.N. and Machado, M.A. 2006. Complete nucleotide sequence, genomic organization and phylogenetic analysis of *Citrus leprosis* virus cytoplasmic type. J. Gen. Virol. 87: 2721-2729

4

Plant Virus Diagnostics

Diagnosis is the first requirement when confronted with a disease caused by an unknown pathogen, and this is especially crucial for diseases thought to be caused by viruses. There are about 80 recognized genera of plant viruses (see Appendix III.6) and the diagnostic features of most of these are known but at the specific level there is still more to be learnt. Combinations of physical, biological, biochemical and serological properties can normally ensure the proper detection and identification of a virus. The principal property is the physical one, the shape and size of the virus particle and particles may be enveloped or non-enveloped. The shape ranges bullet, bacilliform, short or long rods, flexuous filaments and isometric or even other variable shape.

Diagnosis involves two processes: *identification* and *detection.* There are various methods to identify a virus but detection can be done by sophisticated laboratory-based tests. Identification implies the presence of a virus and detection signifies the quantitative aspect of its presence as well. Recently Astier *et al* (2007) reviewed the diagnostic methods for plant viruses and provided very useful information in this context. In this chapter, an attempt has been made for precise identification and detection of viruses emphasizing more on practical points of view.

A. IDENTIFICATION AND DETECTION

(i) Shape Differences: Criteria for Primary Diagnosis

Among the currently recognized virus genera, the particles of five viruses are enveloped (*Cytorhabdovirus, Nucleorhabdovirus, Tospovirus, Umbravirus, Cilevirus*), the rest are non-enveloped. Enveloped particles usually appear as bullet-shaped. Particles of four genera (*Mastrevirus, Curtovirus, Begomovirus, Topocovirus*) are geminate; particles of six genera (*Badnavirus, Rice Tungro bacilliform virus, Alfamovirus, Oleavirus, Ourmiavirus, Soymovirus*) are bacilliform; particles of eight genera (*Varicosavirus, Tobamovirus, Tobravirus, Furovirus, Pecluvirus, Pomovirus, Benyvirus, Hordeivirus*) are short or long rods; particles of 19 genera (*Tenuivirus, Ophiovirus, Potyvirus, Ipomovirus,*

Macluravirus, Rymovirus, Tritimovirus, Bymovirus, Closterovirus, Crinivirus, Potexvirus, Carlavirus, Allexivirus, Capillovirus, Foveavirus, Trichovirus, Vitivirus, Mandarivirus, Ampelovirus) are filaments; and particles of the remaining genera are isometric. It is therefore apparent that the shapes of particles can narrow down the process of identification of possible viruses and provide an immediate cue for their sorting.

(ii) Detection of Shape Differences

1. Electron Microscopy

Transmission Electron Microscope (TEM) is used to detect the shape and some other properties of virus particles. Submicroscopic virus particles are highly magnified to make them visible. The extent of magnification depends upon the resolving power of the microscope. Proper handling of an electron microscope, however, depends much upon its operational principle. This principle of the operation of an electron microscope is given in Appendix IV.1, Figure 4.1.

Resolving and magnifying power

The electron microscope, that was first demonstrated in 1932, did not have much better resolving power more than that of a conventional light microscope but, with advances in technology the resolving power was gradually raised to 0.2 nm; the theoretical limit for the resolution, however, is 0.0016 nm). Normally the resolving power of an electron microscope is such that an object can be magnified up to 80,000×. With appropriate accessories the magnification may be raised from 250,000× to 400,000× depending upon the type of the microscope. However, for most work observing virus morphology and structure, 80,000× is adequate

Methodology

Materials for observation need to be put on support grids. A standard support grid is usually a copper grid of 200 mesh. Sometimes specimens are placed on supporting films (150-200 Å thick). The films may be made up of collodion (nitrocellulose) or formvar (polyvinyl formaldehyde), for special purposes, films of beryllium, aluminium oxide or even glass can be used.

Direct observation of the virus particles can be made under an electron microscope by placing a drop of the sap from an infected leaf on a film, adding a heavy metal stain (see below) and placing the film on the carbon-coated grid. It is the simplest method to observe the virus particles and is called "Leaf Dip Method".

In this method, the infected leaf (2-3 mm in diameter) is crushed in phosphate buffer (0.07 M, pH 6.5-7.0). The extract (10 μl) is placed on parafilm or a waxed slide. A carbon-coated grid (film-side downward) is then placed on the surface of the droplet, ensuring that the grid surface becomes wet (keeping for 2-10 minutes). The grid is then picked up and gently washed with a continuous flow of 10-15 drops of double distilled water to remove excess sap. The filmed surface of the grid is then placed briefly on a droplet, or a drop or two, allowed to run over the surface, of 2% uranyl acetate or phosphotungstic acid (pH 6.5-7.9). Excess stain is removed with a strip of filter paper. The grid is then dried and examined under the microscope. Depending upon the concentration of the virus particles in the sap, the shape of the particles can be determined.

There are several methods that can be used to increase the contrast of the object but are usually only applicable with isolated and purified viruses. These methods may be physical or chemical.

Physical method

A commonly used physical method is "Shadow Casting" but this is now rarely used for virus studies. In this technique the contrast of the specimen is enhanced by depositing heavy metal atoms from a directed beam produced by evaporation at an angle to the surface, thus producing a build up of metal atoms on one side of the specimen and a shadow on the other. Palladium, platinum, gold platinum alloy, are commonly used for this purpose.

Chemical method

Increased contrast of the specimen can be obtained by applying positive or negative electron stains. Positive stains combine chemically with the specimen and in negative staining the surrounding pores or interstices in the specimen are infiltrated with heavy unreactive compounds. Negative contrasts have certain advantages. Compounds that are commonly used as negative stains are phosphotungstic acid (PTA), ammonium molybdate, uranyl actate, uranyl formate, and methyl tungstate. The applicability of a stain is often specific to a virus; several stains may affect the virus structure. For example, sodium phosphotungstate breaks ribosomes and some viruses like the *Alfamoviruses*. In many isometric viruses, the application of stains can release the nucleic acid from the particles keeping the protein shell behind.

Virus particles can also be observed in infected tissues by staining ultra-thin sections of the tissues. To prepare such sections, cells are fixed with suitable chemicals like glutaraldehyde and osmium tetroxide. Fixed tissues are dehydrated, impregnated and mounted in hard plastics such as epoxyresin, araldite, or methacrylates. Thin sections (50-100 mm thick) are cut by a glass or diamond edge using an Ultra Microtome. These sections can then be stained and viewed under the microscope.

One of the most effective uses of the electron microscope in virus diagnosis is in Immunosorbent Electron Microscopy (ISEM). In this method virus-specific antibodies are used and the method has been described in Sero-Diagnosis.

2. Sero-diagnosis

Sero-diagnosis is normally based on the specific properties of the coat proteins.

Principles

Many substances particularly proteins and polysaccharides when introduced into an animal system, induce the production of homologous structures in the blood stream. Substances inducing such structures are *"Antigens"* and the products are *"Antibodies"*. The essential pre-requisites for such reactions are that the introduced substances should not be normal constituents of the animal to which they are introduced; molecular weights should be at least 10^4 and have definite molecular structures. Induced *antibody* usually neutralizes the introduced *antigen* and provides immunity to the animal. Serum of the animal containing the antibody is called *"Anti-serum"*. In many *antigen-antibody* reactions, associated cells often aggregate; such aggregation is called *"Agglutination"*. The concerned antigen and antibody are called *"Agglutinogen"* and *"Agglutinin"* respectively. This property has been widely used in serological diagnostic tests.

Production of antibodies

In *antigen-antibody* production system, antibodies are produced specific to the molecular conformity of the antigen. If the antigen is a mixture of molecular structures, the antibodies produced also have mixed structures.

   Antibodies are produced in animals by cells of their reticulo-endothelial system particularly by cells of lymphocytic and plasma cell series. *Antibodies* occur in all the circulating body fluids of immunized animals. They are generally found in the *"Globulin"* fraction of the blood serum proteins. This fraction is different from the globulins found in the normal serum. The most common type of *antibody* globulin is *"Immuno- or γ (gamma) globulin"*; γ globulins are of five different types, the most commonly used one is IgG; other common ones are IgA and IgM. The basic structure of all types is the same (Figure 4.2). Each type is made up of two heavy and two light chains linked together by disulphide-bridges. The NH_2-terminal regions of the heavy and light chains are highly variable and constitute recognition sites specific to epitopes. Its molecular weight is usually 1.5×10^5 with a sedimentation coefficient of 6.5 S. Its electrophoretic mobility is very small. IgA and IgM are globulins with higher molecular weights and sedimentation coefficients than IgG. Production of IgG may be homogeneous or heterogeneous depending on the molecular structure of the antigen. Structures of antigens even in pure forms are not homogeneous and produce heterogeneous antibodies that are commonly called "Polyclonal Antibodies". Polyclonal antibodies are very useful for broad diagnostic tests, but for more specific detection "Monoclonal Antibodies" are preferred.

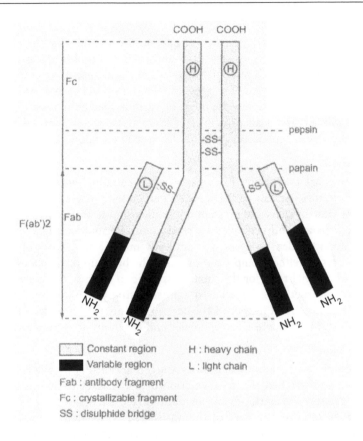

Figure 4.2 Structure of an immunoglobulin G (IgG) (Astier *et al* 2001; source: INRA 2001, France, with permission)

Production of monoclonal antibodies

Kohler and Milstein (1975) developed the "Hybridoma Technique" to produce monoclonal antibodies. In this technique spleen cells of a mouse previously immunized with antigens are fused with mouse *myeloma* cells. This fusion produces "*Hybridoma*" that inherits the traits from both the parents: specific antibody secretion from the spleen cells and the ability to grow continuously in cell-culture from the *myeloma* cells. *Antibodies* produced from these *hybridomas* are called "Monoclonal Antibodies" (*Mabs*). The specificity of these antibodies can be further improved by more cloning of the *hybridomas*. In the production of *Mabs*, one type of antibody reacts with only one epitope of a viral protein. If the epitope corresponds to a variable part of the protein, this method makes it possible to distinguish virus strains in which this epitope is present or not. If, on the contrary, the epitope corresponds to a *conserved motif* of the protein, the *Mab* will allow a very wide detection of all viruses (see Astier *et al* 2007).

Assay methods

Serological reactions rely on the unique nature of the way in which the *antigen* and *antibody* molecules fit together. The combining sites of the *antibody* molecules are complementary to the determinants on the surfaces of the homologous antigens. The specific relationship between an *antigen* and its *antiserum* leads to the development of various specific diagnostic tests or assays for viruses. The basic principles of these assays are: (a) Precipitation, (b) Agglutination, (c) Amplification and (d) Ultra-magnification.

Precipitation

Plant viruses are polyvalent. Each particle can combine with many antibody protein molecules; but antibodies are divalent. In the process of neutralization, polyvalent viral antigens and bivalent antibodies bind to form lattice-structured insoluble virus-antibody complex or precipitates. Formation of the precipitate depends upon the density of antigen-antibody mixture and purity of the virus antigen. Excess of either component prevents the build up of visible precipitate. There is an optimal ratio for the mixture specific to a virus. There are different types of assay methods for virus diagnosis using precipitation. The main methods are: (1) Tube Precipitin (2) Micro-precipitin, (3) Immuno-diffusion or Gel diffusion, (4) Immuno-electrophoresis and (5) Immuno-osmophoresis.

Tube precipitin

Various dilutions of antibody and virus antigen are mixed, incubated in small glass tubes and watched for the formation of the characteristic precipitate. A series of dilutions of antibody is normally mixed against a fixed amount of antigen to determine the titre of the antiserum. Elongated viruses produce bulky flocculent precipitates whereas isometric viruses produce dense granular precipitates. The amount of precipitate obtained from each mixture can be measured if quantitative estimation of the virus present is required. Uncombined virus antigens in the incubation mixture can also be estimated using analytical ultracentrifuge or density gradient centrifugation.

Micro-precipitin

This test is a modification of the tube precipitin test. In this test varying dilutions of antigen and antibody preparations are mixed. The quantity of the precipitate formed and the rate of its formation are recorded. A grid titration for this test may be set up within a Petri dish. This test may also be done by mixing a drop of antigen and antibody on a glass slide and observing the production of precipitate under a dark-field microscope. The evaporation of the mixture is minimized by covering it with a layer of mineral oil.

Immuno-diffusion or gel diffusion

In this test *antigen* and *antibody* diffuse towards each other in a gel and form a white band or line at a place where the two reactants meet and precipitate. Crude sap, clarified extract and highly purified *antigen* may be used in this test. The simplest form of this test is the "Ring Interference Test". When a little *antiserum* is put into a gel in a narrow glass or capillary tube and the virus preparation is layered on the top, both the *antigen* and the *antiserum* diffuse and precipitate in the form of a ring at their optimal ratio. This test is applicable only to those *antigens* that diffuse through agar gel. Gels that are used may be of noble agar, ion agar, bacto agar, agarose, etc. (Hampton *et al* 1990). There are many parameters such as size, relative concentration of the reactants, type of gel, electrolyte concentration in the gel, kind of buffer used, temperature, denaturing agent, etc. that influence the diffusion process and the formation of precipitation lines. The diffusion test can be performed with various modifications in a gel tube or plate. The most common one is double diffusion in plates and is known as "Ouchterlony two-dimensional or double diffusion". In this method, a layer of gel (agar or agarose) is made on a glass plate or dish. Wells are cut 2-4 mm apart in this layer and filled with the reactants. Various patterns of well size and arrangement can, however, be made in the gel for performing this test. The *antigens* and the *antibodies* diffuse together and precipitate. The position of the precipitate band at the optimal proportion depends upon the diffusion coefficients of the *antigen* and the *antibody* and is expressed by the equation: $Dag/Dab = Ag^2/Ab^2$, where Dag and Dab are diffusion coefficient of *antigen* and *antibody* respectively and Ag and Ab are distances of the precipitate bands from the source wells of the *antigens* and *antibody*. The diffusion coefficient of the γ globulin (IgG), the common antibody for all the viruses is 4.8×10^{-7} cm^2/ sec. Interpolating it to the equation and recording the distance of the precipitate bands from the source wells and optimal proportions of the reactants, the diffusion coefficients of the *antigens* or sizes are calculated.

When the reactants are put in the adjacent wells, a precipitation band is formed at right angles to the shortest line joining the wells; the band is usually straight when the diffusion coefficients of the reactants are identical otherwise the band curves around the wells containing the reactant of lower diffusion coefficient (Figure 4.3).

Immuno-diffusion techniques are useful for the diagnosis of isometric but not for filamentous viruses. Isometric particles (25-30 nm in diameter) when incubated at optimal proportions precipitate at 35-40% of the distance between the two wells. Larger particles precipitate near the *antigen* well. Protein subunits of the virus particles and low molecular plant proteins on the other hand precipitate near the *antisera* well. When *antigens* have no common determinant though the *antiserum* contains antibodies to both, the precipitate bands usually cross. When identical *antigens* react with their homologous *antiserum*,

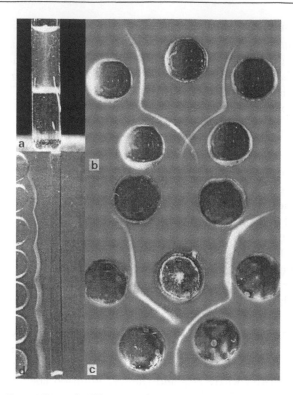

Figure 4.3 Formation of different patterns of precipitation in Gel Diffusion Test (Gibbs and Harrison 1976); reproduced by permission of Edward Arnold (Publishers) Ltd

the ends of the band usually join and do not cross. When the *antigens* have some common determinants and the *antiserum* is prepared against one of the *antigens*, a spur is formed where the precipitate band meets. The sensitivity of the immuno-diffusion tests may vary from 1-20 μg/ml of virus depending on the quality of the antiserum and the type of the medium. The Ouchterlony double diffusion method can be used to distinguish related, but distinct strains of a virus or even different but serologically related viruses. The disadvantages of this method include a lack of sensitivity in detecting viruses that occur in low concentration, the need to dissociate filamentous or rod-shaped viruses to allow them to diffuse through the gel matrix, and the need for a large quantity of antibodies.

Immuno-electrophoresis

This technique utilizes the combination of electrophoretic movement and agar diffusion of the virus particles. The reactants in gel and an electrical field differentially migrate. Virus *antigen* is put in wells on one side of the gel and electrophoresis is done from the cathode (–) to the anode (+) to separate the *antigens*. *Antiserum* is put in the troughs when immuno-diffusion occurs to precipitate the mixture at a specific distance.

Immuno-osmophoresis

This test is more or less similar to the electrophoresis test. It utilizes the property of the migration of *antigen* towards the anode (+) at pH 7.0 and that of *antibody* towards the cathode (–) due to endo-osmotic flow. Wells in gel are cut in pairs (1.4 cm apart) and align in the direction of the flow of the current. *Antigen* is put in the well closest to the cathode and the *antibody* is put in the well closest to the anode. The electrophoretic cell is placed on a black paper and the development of the precipitate is observed with white light through the agar bed. The *antigen* and the *antibody* are brought together by the electric current after the adjustment of appropriate conditions of pH, buffer and gel charge.

Agglutination

In the agglutination test, the antibody is coated on the surface of an inert carrier particle (e.g., red blood cells, latex, or *Staphylococcus aureus* cells), and a positive antigen-antibody reaction results in clumping/agglutination of the carrier particles that can be visualized by the naked eye or under a microscope. Agglutination tests are more sensitive than other precipitin tests and can be carried out with lower concentrations of reactants than are necessary for precipitin tests (Hughes and Ollennu 1993).

In plant virus diagnosis, two types of agglutination reactions are commonly used: Chloroplast agglutination and the Latex test. Chloroplast agglutination is a very simple method. A mixture of a drop of *antiserum* and a drop of the sap to be tested is placed on a glass slide. When the virus particles and the *antiserum* react together, the chloroplasts from the sap either co-precipitate or agglutinate. This test is possible only for elongated viruses in high concentration.

In passive agglutination tests sheep red blood cells or erythrocytes are sensitized with *antisera* or sap. When the sap or the *antisera* is mixed with the sensitized cells, haemagglutination takes place. In another passive agglutination test, sodium bentonite particles are sensitized with the *antibody*. Virus-containing sap causes flocculation of the bentonite particles.

In the Latex test, latex particles are sensitized by purified specific γ globulin. When homologous *antigen* is mixed with suspension of such particles, they aggregate (Figure 4.4). A modification of the latex test is the Protein A latex linked antiserum (PALLAS) test. It allows the use of low titre *antiserum* and overcomes the presence of inhibitory serum or plant cell components. Protein A (a bacterial protein) links the latex particles and the γ globulin.

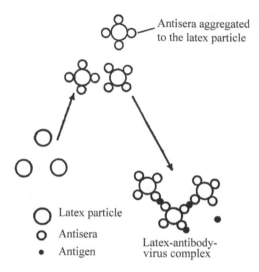

Figure 4.4   Diagrammatic representation of the aggregation of antigen and antibody in latex test

Amplification

There are several limitations in the common sero-diagnostic techniques of plant viruses. They can be performed only with the viruses that can be isolated and purified; they are time consuming and many samples cannot be tested at a time. Amplification methods, using labelled or conjugated antibody techniques, help to overcome these limitations.

The principle of *Amplification* is to amplify the *antigen-antibody* binding reaction. When a small quantity of *antibody* binds the viral *antigen*, the reaction in most cases remains invisible, but the signal-specific binding of *antibody* with the virus can be amplified to the threshold of visibility by labelling the *antibody* with enzyme, fluorescent, radioactive and chemiluminescent agents. Common amplification assay methods are: (1) Immuno-fluorescence assay, (2) Radio-immuno assay, (3) Enzyme-linked immunosorbent assay, (4) Dot immuno-binding assay, (5) Immuno-tissue printing assay and (6) Western Blot assay.

Immuno-fluorescence assay

In this method, the antibody is conjugated with a fluorescent dye (Fluorescein isothiocyanate, rhodamine, lissamine, etc.) and the *antigen antibody* complex is observed under a microscope. *Antisera* can also be conjugated with electrondense substances such as ferritin.

This technique can be applied to detect viruses both in plant and vector cells. It may be done directly or indirectly. In the direct method, *antibody* is directly conjugated with the dye and used to detect the *antigen*. In the indirect method, *antigen* and *antibody* are allowed to react. Then anti-globulin *antibody-*conjugate with the dye which is used to detect the *antigen-antibody* complex.

To detect the virus in the vector, the *antibody* after conjugating with the dye is injected into the vector and the tissues are examined under a fluorescent microscope.

Major limitation in the use of *antibody* conjugates is the non-specific binding of it with tissues and the presence of auto-fluorescent substances in many tissues.

Radio-immuno assay

In this method, the *antibody* is conjugated with radioactive substances (125_I, Hydrogen-^{3}H, etc.) and mixed with *antigen*. The mixture is then examined using a radioactivity measuring instrument.

Enzyme linked immuno-sorbent assay (ELISA)

Unlike precipitin and agglutination tests, ELISA is a solid phase heterogeneous immunoassay. It involves the immobilization of an *antigen* to a micro-titre plate made of polystyrene (inflexible rigid plates) or polyvinyl chloride (PVC, flexible plates) and their binding with the *antibody* conjugated to an enzyme. Enzymes normally used to prepare the conjugates are alkaline phosphatase (with p-nitrophenyl phosphate as substrate) or horseradish peroxidase (with tetramethyl benzydine as substrate). The reaction produces a yellow colour with the phosphatase and a blue colour with the peroxidase.

Immuno-specificity of the *antibody* and the catalytic activity of the enzyme are the basic principles of this assay. The *antigen-antibody* reaction is amplified by the enzymatic break down of the substrate, resulting in the visible coloured product, the concentration of which can be accurately measured by recording the absorbance data with a spectrophotometer at a wavelength of 405 nm; the absorbance may, in certain ranges of concentration, be proportional to the logarithm of the concentration. It is very sensitive and a large quantity of samples in the form of crude sap or purified preparation can be assayed within a very short time.

Many variations of ELISA have been developed that fall into two broad categories: Direct and Indirect procedures. They differ in the way the antigen-antibody complex is detected, but the underlying theory and the final results are the same.

In direct ELISA, a conjugate directly binds to the test *antigen*. The antibodies (usually as an immuno-γ-globulin or IgG fraction of the antiserum) bound to the well surface of the micro-titre plate capture the virus in the test sample. The virus is then detected by incubation with an antibody-enzyme conjugate followed by the addition of colour development reagents. The capturing and detecting antibodies can be the same or from different sources. Since the virus is sandwiched between the antibody molecules, this method is called the Double Antibody Sandwich (DAS) ELISA. In practice, this method is highly strain-specific and requires each detecting antibody to be conjugated to an enzyme.

In indirect ELISA, the conjugate binds the test *antigen* via another layer of *antibody* that is specifically bound with the test antigen. Indirect ELISA has many variations. One approach, known as Direct Antigen Coating (DAC), Antigen-Coated Plate (ACP) or Plate Trapped Antigen (PTA) ELISA allows the virus, in the absence of any specific virus trapping layer as in DAS-ELISA, to absorb on the plate surface by adding the test sample directly to the wells. In the second step, virus antibody (usually called primary antibody) is added either as IgG or crude antiserum. The primary antibody is then detected with antispecies antibodies (secondary or detecting antibody) conjugated to an enzyme, followed by the addition of colour development reagents. The detecting antibody binds specifically to the primary antibody since the former is produced against IgGs from the animal in which virus antibodies are raised. The disadvantages of this method are: competition between plant sap and virus particles for sites on the plate and high background reaction.

Another widely used approach is Triple Antibody Sandwich (TAS) ELISA. This is similar to DAS-ELISA, except that an additional step is involved before adding detecting antibody-enzyme conjugate. In this step, a monoclonal antibody (Mab), produced in another animal (usually mice) different from trapping antibody, is used. This Mab is then detected by adding an enzyme-conjugated species-specific antibody (e.g., rabbit anti-mouse IgG) that does not react with the trapping antibody, followed by colour development reagents. TAS-ELISA generally lead to a sensitivity less than 1 ng/ml, reaching some times 10 pg/ml, and allow for the detection of a large number of virus strains (Devergne *et al* 1981, Koenig 1981).

In the third, called Protein A Sandwich (PAS) ELISA, the micro-titre wells are usually coated with protein A before the addition of trapping antibody. The protein A keeps the subsequently added antibodies in a specific orientation which traps virus particles. This can often increase the sensitivity by increasing the proportion of appropriately aligned antibody molecules. The trapped virus is then detected by an additional aliquot of antibody (the same antibodies that were used for trapping) that in turn is detected by enzyme-conjugated protein A and subsequently colour development reagents. Thus, in this method the antibody-virus-antibody layers are sandwiched between two layers of protein A. As a result, different orientations of the IgG in the trapping and detecting layers of antibodies enable the protein A to conjugate and discriminate between them (Naidu and Hughes 1998) (Figure 4.5).

In indirect ELISA procedures, the virus is detected by using a heterologous antibody conjugate that is not virus-specific, but specific for the virus antibody

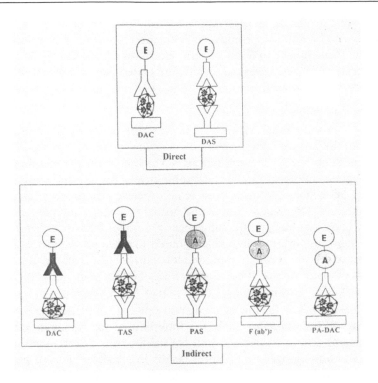

Figure 4.5 Diagrammatic representation of different versions of direct and indirect forms of ELISA. DAC = Direct antigen coated ELISA, DAS = Double antibody sandwich ELISA, TAS = Triple antibody sandwich ELISA, PAS = Protein A sandwich ELISA, PA-DAC = Protein A direct antigen coated ELISA, Fab-ELISA with antigen binding fragments of IgG. E = Enzyme. Black "Y" anti-antibody and white "Y" = Virus-specific antibody (Mandal 2003)

or primary antibody. As a result, a single antibody conjugate can be used in indirect assays to detect a wide range of viruses. Indirect ELISA is more useful for routine testing.

The first and foremost requirement of the ELISA test is the polyclonal or monoclonal *antibodies* for the test viruses depending upon the type of method to be used. Direct methods require purified γ-globulin; γ-globulin or conjugates for many viruses are commercially available. In routine testing with these conjugates, optimal dilution for each γ-globulin and conjugate preparation has to be done experimentally in a test plate. Concentrations normally should not exceed 10 μg per ml^{-1}; excess concentrations increase the intensity of non-specific reaction. The conjugate dilution may be of the order of 1/500 and 1/1000.

Most of the ELISA assays employ alkaline phosphatase or horseradish peroxidase. The antibody conjugates with these enzymes are commercially available. Penicillinase may be a useful alternative for alkaline phosphatase (Sudarshana and Reddy 1989). It is not widely used due to inherent difficulties in quantifying the results.

Dot immuno-binding assay (DIBA)

DIBA is also known as Dot-ELISA. In this test, *antigens* are immobilized directly to the nitrocellulose membrane or nylon membrane in a grid pattern (Hibi and Suito 1985). Test *antigen* extract is dotted on the membrane. The remaining membrane surface is blocked by incubation in a solution of a non-reactive heterologous protein to prevent non-specific adsorption of *antibody* to the membrane. After incubation in the primary *antibody* solution, unreacted *antibody* is rinsed away and the subsequent procedure follows as that in ELISA and the substrate forms an insoluble coloured material on the membrane. The procedure of this test is simple, rapid and less expensive. It is also used in routine diagnosis of many viruses. It can detect viruses both in plants and vectors.

This technique is similar to ELISA except that the plant extracts are spotted on to a membrane rather than using a microtitre plate as the solid support matrix. Unlike in ELISA, where a soluble substrate is used for colour development, a precipitating (chromogenic) substrate is used for virus detection in the DIBA. Hydrolysis of chromogenic substances results in a visible coloured precipitate at the reaction site on the membrane. Chemiluminescent substrates, that emit light upon hydrolysis, can also be used and the light signal detected with X-ray film as with radio-labelled probes (Leong *et al* 1986).

Immuno-tissue Printing Assay (ITPA)

This method is more or less similar to DIBA but differs from that in the initial test material. Instead of the *antigen* extract, simply cut surface of the test plant organ or an insect is pressed on the nylon membrane. Presence of the virus in the treated portion of the membrane is detected following the method as in DIBA. This method is also simple, does not require elaborate sample preparation or extraction, and provides information on the distribution of plant viruses in plant tissues. The possibility of using low-cost paper for blotting samples makes this technique useful for laboratories with limited resources (Makkouk and Kumari 2002).

The disadvantages of DIBA are possible interference of sap components with the subsequent diagnostic reactions. Sometimes the colour of the sap prevents weak positive reactions from being observed and the results cannot be readily quantified.

Western Blot (WB) Assay

Western Blotting combines the protein fractionating power of electrophoresis in gel with the sensitivity and specificity of solid-phase immunoassay. To begin with, proteins in the sap are fractionated by sodium dodecyl sulphate–polyacrylamide gel electrophoresis (SDS-PAGE) followed by electrophoretic transfer of the protein bands from the gel to the activated paper or nitrocellulose. Free binding sites on the membrane are blocked with heterologous proteins such as serum albumin. The membrane is then incubated with specific *antibody* followed by incubation with enzyme-labelled anti-immunoglobulin and enzyme substrate as done in ELISA.

This technique identifies the virus by two independent characters of its coat protein: molecular weight and serological specificity.

Ultra-magnification

Ultra-magnification of *antigen-antibody* reactants is done for precise determination of several physical characters of the viruses by employing electron microscopic observation. The most common method used in virus diagnosis is Immuno-sorbent Electron Microscopy or Serologically Specific Electron Microscopy.

Immuno-sorbent Electron Microscopy (ISEM) or Serologically Specific Electron Microscopy (SSEM)

This method is based on the two properties of the virus: serological reactivity with the *antiserum* and particle morphology or, in other words, observation of virus particles and visualization of antigen-antibody reaction. In ISEM procedure the support films of an electron microscope specimen grid are precoated with specific *antibody* for test virus. Grids are then floated on the solution of the sample virus. Homologous virus particles in the sap become attached to the antibody and also become concentrated. Successive fixing of the *antibody* and the *antigen* can be done interpolating the removal of salts and undesirable component by washing. *Antigen-antibody* mixture is then observed under an electron microscope after metal shadow-casting or negative staining. This method allows the detection of viruses in a very low concentration and comparative serological relations between virus strains.

There is another procedure called the "Decoration" technique, a modification of the conventional ISEM. In this technique virus particles, after being adsorbed onto the grid, are coated with virus-specific antibody. This produces a halo of IgG molecules around the virus particles clearly observable in negatively stained preparation. Matthews (1992) dealt with the merits and demerits of this method and its comparative advantages and disadvantages over ELISA.

3. Molecular or Nucleic Acid Based Diagnosis

The inherent disadvantage of serological methods is its dependence on the antigenic properties of the virus coat protein that represents only 10% of the total virus genome. Nucleic acid based detection methods, on the other hand, have the advantage that any region of a viral genome can be targeted to develop the diagnostic test. In addition, these methods can be applied where immuninological procedures have limited application in particular for the detection of viroids, satellite RNAs, viruses that lack particles, viruses that occur as extremely diverse serotypes and viruses that are poor immunogens or are difficult to purify. Moreover, these methods are extremely sensitive and can detect a virus at an extremely low level of concentration even up to "Pico" and "Femto" gram. Three types of technology have been developed so far, to detect viruses on the basis of its genomics. These are: (i) Nucleic Acid Hybridization, (ii) Nucleic Acid Amplification and (iii) Gel Electrophoresis (Ramchandran and Agarwal 2003).

Nucleic acid hybridization

The basic principle of this technique is the complement base pairing (A-T, G-C) found in any genome. The nucleic acid from plant material (DNA or RNA) or the target is irreversibly bound or immobilized onto a solid support like nitrocellulose or positively charged nylon-66 (NYTRAN) membrane and allowed to hybridize with a specific fragment or sequence or probe. The probe is normally a radioactively or chemically labelled nucleic acid of the virus genome suspected to be present in the sample. The target, in appropriate condition, will base pair with the probe sequence. After washing out the unspecific products the binding of the target and the probe is detected by auto-radiography or suitable chemical methods using biotin, acetylaminofluorene (AAF), fluocytosine, or digoxigenin.

Depending upon the nature of the target, the Nucleic Acid Hybridization Technique may be of three different types: Southern Blotting, Northern Blotting and Western Blotting. In Southern blotting the Target and the Probe are DNA; in Northern blotting on the other hand the Probe is DNA but the Target is RNA. In Western blotting the probe is antibody protein and the target is the protein.

The essential features of the hybridization technique are: (a) Isolation of the nucleic acid from the test sample, (b) Denaturation of the nucleic acid, (c) Application of the target nucleic acid on the membrane, (d) Production of radioactive labelled probe, (e) Hybridization of the probe with the target and (f) Recording of the reaction end point.

There are various methods for isolation of the nucleic acids, DNA or RNA. Purity of the nucleic acid is to be of a high order for this test. The membrane of choice is to be pretreated with de-ionized water or suitable buffer (e.g. $20 \times$ SSC) at saturation level and the nucleic acid sample is applied with a commercial applicator. Commercial membrane packs are also available to avoid the preparatory stages.

Depending upon the conditions, probes (cDNA, ribo-probes) may be prepared in different ways. There may be radioactive or non-radioactive labelled probes. Radiolabelled probes may be prepared by labelling with random nucleotide primers, T4 DNA polymerase or Nick translation. Non-radioactive probes are prepared by labelling the primers with suitable chemicals having affinity for nucleic acids and that can intercalate between the bases. Two types of labelling are normally used: Biotin labelling and Digoxigenin labelling. In Biotin labelling, ethidium bromide or psoralens are commonly used. When these are exposed to long wave ultraviolet radiation, they become attached to the nucleic acids. Using their conjugates with biotin, the nucleic acid can be labelled with biotin. These labelled products may act as probes and can be stored for more than a year at −80°C. In Digoxigenin (a steroid haptene) labelling, it is incorporated to deoxy-uridine–triphosphate (DIG-dUTP). Labelled nucleic acid probes are generated with Klenow polymerase by random-primed incorporation of DIG-dUTP. This probe is now commercially available as DNA labelling, RNA labelling and nucleic acid detection kits. The hybridization process is detected by using chemi-luminescent detection kits that are commercially available. Colourimetric detection system can also be done using alkaline phosphatase substrates NBT and X-phosphate at sites of hybridization directly on the membrane.

The test itself is quick and relatively easy to apply. The samples to be tested (4-50µl) are deposited on a nitrocellulose or nylon membrane. After baking at 80°C or fixation with UV, the membranes can be conserved for several weeks at 4°C or even at room temperature before being analyzed, which is very useful for field-work.

Nucleic acid amplification
This technique is essentially based on Polymerase Chain Reaction (PCR) developed by Mullis *et al* (1986) and Saiki *et al* (1988). It was adapted during the 1990s for the detection of plant viruses and constitutes a major breakthrough in the field of diagnosis (Henson and French 1993). The basic principles of this technique are: (i) selection of the primer, (ii) extraction of DNA from test samples, (iii) denaturation of DNA, (iv) annealing of primers, (v) polymerization, (vi) amplification and (vii) detection of target DNA sequence.

Primers are short segments of nucleic acid that attach the specific target segment of nucleic acid. It can be designed for its attachment to either end of the target sequence. After attachment, the primers initiate the polymerization process to produce new nucleic acid. DNA is extracted from the test samples by standard techniques. In the case of RNA viruses, RNA is converted to complementary DNA by reverse transcriptase. If the target sequence is double stranded, single-stranded DNA is produced by denaturation. The primers are then attached to the ends of the segments to be amplified by a process called "annealing". After the attachment, the polymerization is induced that elongates the developing nucleic acid chain between them. New dsDNA (old template strand and new complementary strand) is then split apart to yield two ssDNA strands. The processes of denaturation, annealing and polymerization are repeated in an automated thermal cycler and the amount of DNA is amplified. The amplified target DNA is detected by conventional DNA detection techniques. Depending upon the protocols adopted the transcription and amplification can be conducted separately or simultaneously in a single tube.

Preparation of the sample for this test is very critical as the extracts of certain plants may contain inhibitors of enzymatic reactions. Consequently, the extraction or purification protocol of nucleic acids is to be applied.

The sensitivity and specificity of PCR or RT-PCR tests depend on the choice of primers, temperatures of hybridization and elongation and number of cycles. If a specific test for a virus species or virus strain is to be conducted, the primers are chosen in a variable region of the genome. When several viruses of the same genus or family need to be detected, the primers are to be defined in a conserved region within the genus or the family.

PCR-based technique is highly sensitive and can detect one molecule of nucleic acid within 100,000 cells. Because of selective amplification it can pick out a minute amount of target nucleic acid within an ocean of extraneous cellular or other non-specific DNA. Therefore it can detect the virus at a very early stage of its infection. The most important quality of PCR is its high level of specificity that allows discrimination of sequences differing by single nucleotide mutations when carefully designed primers are used. This high specificity makes it a prime candidate for the development of tests for identification of specific genetic sequences and pathogens. Coupling of PCR with a preliminary reverse transcription step allows the amplification of RNA sequences in a cDNA form and is used to detect and identify RNA viruses. This method is known as RT-PCR. It increases the speed of detection and makes the construction of a probe and hybridization unnecessary. The major drawback of the technique is commonly used contamination that may often give false positive results. Sometimes, a single or a few mismatches between a primer sequence and a viral nucleic acid target sequence can lead to a negative reaction in PCR or RT-PCR. One way to overcome this difficulty is to use degenerate primers consisting of a mixture of closely related primers with only a few nucleotide differences at a certain position. Another way to improve primers annealing to the target sequence is to change the annealing temperature within the PCR cycle. An increase in annealing temperature provides a more specific detection, while a decrease broadens the potential of detection. Amplification products are normally visualized by observation of bands separated by gel electrophoresis and stained with ethidium bromide. However, the refinement of techniques for detection of the amplification products (amplicons) by colorimetry can lead to not only an increase in sensitivity but also a significant simplification of the analysis when dealing with numerous samples (Astier *et al* 2007).

Adaptation of PCR techniques to detect plant viruses started from the 1990s (Rybicki and Hughes 1990, Vunsl *et al* 1990). This technique is also adapted to the detection of viruses in their vectors. It was possible to detect viruses in aphids (Lopez-Moya *et al* 1992, Hadidi *et al* 1993, Levy and Hadidi 1994, Singh *et al* 1995, 1996, Conning 1996, Hu *et al* 1996), mealy bugs (Minafra and Hadidi 1994), whiteflies (Navot *et al* 1992, Deng *et al* 1994, Mehta *et al* 1994), thrips (Tsuda *et al* 1994), leafhoppers (Takahashi *et al* 1993) and nematodes (Van der Wilk *et al* 1994).

There are several modifications of the conventional PCR-based techniques. Some of these are: (a) IC-PCR/IC-RT-PCR, (b) Multiplex or m-PCR, (c) Real Time PCR, (d) Molecular beacons, (e) NASBA, (f) Differential Display, etc.

(a)   IC-RT-PCR

IC-RT-PCR is an immuno-capture PCR for detecting RNA viruses. The immuno-capture itself can be carried out in separate tubes, ELISA plates or directly in the Eppendorff tubes in which PCR reaction is to be performed (Wetzel *et al* 1992). The PCR amplification can also be performed directly in the ELISA plate that has been used in the immunocapture step (Nolasco *et al* 1993). IC-PCR can detect viruses that give a negative reaction to the simple ELISA test. Besides providing an efficient way to remove interfering plant substances IC-PCR allows the processing of large volumes of plant extract (100-200 µl, equivalent to 10-50 mg of starting plant material).

(b)   Multiplex PCR

Multiplex PCR (m-PCR) allows simultaneous amplification of several viruses or viroids using a mixture of specific primer pairs. Success with this technique depends upon the concentration of the reagents and primers that are needed to be determined. This strategy offers the advantage of dramatically reducing the diagnostic cost. Both simultaneous amplification of closely related viruses using a single polyvalent primer pair (Colinet *et al* 1993, Rush *et al* 1994) and simultaneous detection of unrelated viruses using a mixture of virus-specific primer pairs (Minarfa and Hadidi 1994, Smith and Van de Valde 1994) have been recorded. Bariana *et al* (1994) reported the development of an assay allowing the simultaneous detection of five seed-borne viruses from legume seeds using a mixture of nine primers.

(c)   Real-time PCR

Real-time PCR is an advanced type of m-PCR. It is a quantitative procedure that helps to detect accumulation of PCR products during the PCR reaction. Instruments for conducting this test are commercially available.

(d) Molecular beacons

Molecular beacons are single-stranded oligonucleotides having a stem-loop structure. The loop portion contains the sequence complementary to the target nucleic acid, whereas the stem is unrelated to the target and has a double-stranded structure. One arm of the stem is labelled with a fluorescent dye (fluorophore) and the other arm is labelled with a non-fluorescent quencher. In this state the probe does not produce fluorescence because of the presence of the quencher close to the fluorophore. However, when the probe hybridizes to its target sequence it undergoes a conformational change that separates the fluorophore and the quencher resulting in bright fluorescence. The generated fluorescence can be measured with the use of a fluorescence meter, allowing automated detection of the amplified sequences. These probes can be used in NASBA and Real-time PCR detection system.

(e) NASBA

NASBA (Nucleic acid sequence based amplification) technique amplifies nucleic acids without thermocycling and mRNA can be amplified in a dsDNA background. A molecular beacon is used during amplification to enable real time detection of the product. NASBA is normally applied to amplify single-stranded target RNA, producing RNA amplicons. It is an isothermal nucleic acid amplification process involving the simultaneous activity of three enzymes: reverse transcriptase, RNase H and T7 RNA polymerase. This technique utilizes two oligonucleotide primers in which the downstream antisense primer contains a highly conserved 5′ promoter sequence recognized by T7 RNA polymerase and provides a single-tube amplification format. Contaminating double stranded DNA, which is a problem in RT-PCR does not participate in the amplification procedure that obviates the need for rigorous RNA purification.

(f) Differential display

The course of normal cellular development as well as pathological changes that arise in diseases are normally driven by gene expression. Differential display allows rapid, accurate and sensitive detection of altered gene expression.

The Differential Display technology works by systematic amplification of the 3′ terminal portions of mRNAs and resolution of those fragments on a DNA sequencing gel. Using anchored primers designed to bind to the 5′ boundary of the PolyA tails for the reverse transcription, followed by PCR amplification with additional upstream primers of arbitrary sequences, mRNA subpopulations are visualized by denaturing polyacrylamide electrophoresis. This allows direct side-by-side comparison of most of the mRNAs between and among related cells. The differential display method is thus far unique in its potential to visualize all the expressed genes in an eukaryotic cell in a systematic and sequence dependent manner by using multiple primer combinations. This method enables the recovery of sequence information and the development of probes to isolate their cDNA and genomic DNA for further molecular and functional characterizations.

Gel electrophoresis

This method is commonly known as polyacrylamide gel electrophoresis or PAGE. Components of different molecular size show different electrophoretic mobility, when passed through an electric field in a polyacrylamide gel plate. Nucleic acids are extracted from infected and healthy samples and electrophorezed using a specific buffer. They are resolved in the gel by staining with ethidium bromide, toluidine blue or silver nitrate. Nucleic acids are identified by their molecular masses comparing that with standard samples of known nucleic acid masses.

A modified PAGE called R-PAGE is widely used to detect viroids. It involves a bi-directional PAGE. During the first run viroid molecules are separated from the plant nucleic acids. In the second run electrophoresis is performed at high temperature and low salt with current applied at reverse polarity. Resulting open circular RNA molecules migrate at a slower rate than any other RNA of similar molecular weight (Singh and Dhar 1998).

Among the serological techniques, agglutination test, SDS-immunodiffusion tests, DAS-ELISA, RIA, ISEM and decoration IEM are most specific and capable of differentiating distinct viruses and sometimes strains. Precipitation test, indirect ELISA and Western blotting have rather broad specificity and are suitable to detect distant serological relationships among distinct viruses and strains. SDS-immunodiffusion is the least sensitive serological technique followed by precipitin tests, whereas ELISA, RIA, IEM and Western blotting are highly sensitive (detection limit 1-10 ng virus ml^{-1}). Dot-blot immunoassay can detect viruses at amounts as low as 0.5 pg. Nucleic acid hybridization using random probes has a very narrow specificity and cannot detect distantly related strains of the same virus. Probes based on defined gene sequences of the viral genome appear to provide a better approach. This technique is highly sensitive and can detect viruses at pg amounts. PCR amplification of gene segments using degenerative primers based on conserved genome sequences seem to be a very powerful technique for the detection and identification of those viruses that are most difficult to diagnoze by other known procedures.

4. Microarrays for Rapid Identification

Schena *et al* (1995) initiated Microarray Technology for investigating variations in messenger RNA accumulation. As such the layout of the process typically involves gene specific DNA fragments bound to a solid support in an aspatially separated fashion. The RNA sample to be examined is typically fluorescently labelled in an enzymatic reaction and this fluorescent nucleic acid is hybridized to the array. The fluorescence allows the hybridization events on the solid support to be identified and the identity of the gene is deduced from the position on the array. DNA can be bound to the solid support in extremely small spots allowing the expression of many different genes to be investigated simultaneously in a highly parallel fashion. Literature to date deals with the advancement of this technology particularly on the numerous variations in array methodologies including different solid supports, methods for immobilization of capture probes, sources capture probes and approaches for their design, strategies for the labelling and if necessary, amplification of the target nucleic acid, different methodologies for hybridization, washing, scanning of arrays, and analysis of the resulting images and data (see Boonham *et al* 2007). Serological and molecular detection of different viruses are given in Appendix IV.2.

5. Biological Methods

Biological properties that have much relevance to the detection of viruses are symptoms, host range and vector relations. Viruses normally produce characteristic symptoms in susceptible hosts that differ from those produced by other pathogens like bacteria, fungi and nematodes. Symptoms may be of different types ranging from alteration of colour, necrosis and alteration in the nature of the growth and development. As for the host range, most viruses infect restricted number of plants. Specific vectors transmit most of the viruses and the virus-vector relationships are also characteristic of the concerned viruses. Vectors and the virus-vector relationships with reference to different viruses have been described in Chapter 5.

A. Symptoms

Bos (1963) described the symptoms of virus diseases in plants. Alteration of colour may develop in stems, leaves, flowers and fruits but changes in the colour of the leaves are more common. There are two broad types of changes or chlorosis that are found in leaves: *mosaics* and *yellows*. Depending upon the distribution of the chlorotic and green tissues in the lamina, mosaics may be of different types (true mosaic, streak or stripe mosaic, vein mosaic, vein banding mosaic, inter-veinal mosaic, flecking or spotting or speckling or stippling or dotting, mottling, etc.). Mosaic symptom-producing viruses are normally sap transmissible, aphid transmitted in the non-persistent manner. These viruses normally remain confined in the parenchyma tissues of the host.

In case of yellows, the green colour of the lamina uniformly changes to different colours depending upon the virus (blanching or bleaching in *hoja blanca*, yellowing in *aster yellows*, reddening or purpling in *yellow dwarf of oat* and *barley*, browning in *pea early browning*, bronzing in *tomato spotted wilt*, tip yellow in *pea tip yellows*, edge chlorosis in *strawberry edge yellowing*, vein chlorosis or vein yellowing in *vein yellowing of Lucerne*, vein clearing in *tungro*, etc.). These viruses are not sap transmissible, are transmitted in the semi-persistent or persistent manners and are usually located in the phloem tissues of the host.

Necrosis in some cases provides the opportunity to detect some viruses that produce local lesion in specific hosts known as *local lesion hosts*. Local lesion hosts of some viruses and the nature of lesions on them are given in Appendix IV.3.

Necrosis caused by some viruses may also be systemic and may cause rotting in certain areas of the host system (rotting of phloem tissues by *potato leaf roll, sugar beet curly top, barley yellow dwarf, citrus tristeza*; rotting of chlorenchyma tissue by *potato virus Y*; rotting in pith and cortical tissue by *tobacco rattle virus*). Visual symptoms of such rotting may be breaking of the petiole, blackening and drying of roots, shoot tip drying, bark splitting, girdling of trunks and branches, cankers, lesions, etc. Many viruses cause diagnostic rotting in potato tubers and stems (circular rotting due to *mosaic* and *leaf roll*).

Most viruses normally restrict the growth of plants and thus causing stunting or dwarfing. However some viruses may also cause excess growth of plants but these characteristic growth abnormalities mostly appear in leaf lamina and have diagnostic value. There may also be production of characteristic tumours in stems and roots or swelling in stem or shoot tips (histoid or organoid enation in *Fiji disease of sugarcane*, Aspermy virus in *Nicotiana rustica*; Tumour in *clover wound tumour*; Leaf narrowing in *Tobacco mosaic virus* in tomato; Fan leaf in *Fan leaf* disease of grape; Rugosity in *Rugose mosaic* in potato; Crinkling in *Crinkle leaf* disease of beans; Curling in *Leaf curl* diseases; Rolling in *Leaf roll* diseases).

Recording symptoms may not be a reliable approach for identification of most viruses owing to the inherent problems associated with symptom expression. Symptoms produced by most of the viruses are affected by the strain of the virus, environmental conditions and the hosts. A particular strain may produce different symptoms in one host under different environmental conditions, while on the other hand different strains of the same virus may induce similar symptoms in one host making their identification difficult. Moreover symptom phenotypes can be changed dramatically by single point mutations in the genome.

Symptoms at cellular level

Astier *et al* (2007) illustrated the symptoms at the cellular level that can be used for detection of some viruses simply with a microscope. Fragments of epidermis or sections taken freehand are stained to observe inclusion bodies specific to a given genus (*Potyvirus, Carlavirus, Comovirus, Tobamovirus*). As for example,

Potyviruses produce cylindrical, cytoplasmic-amorphous inclusions in epidermis and mesophyll cells, *Tobamovirus* produces virus crystal stocks, X bodies in those cells. Apart from the inclusion bodies, viruses induce cytopathological modifications that are sometimes characteristic of a virus family, genus or species. As for example, *Tobacco rattle tobravirus* (TRV) causes hypertrophy of mitochondria; *Turnip yellow mosaic tymovirus* (TYMV) causes aggregation of chloroplast; *Potato leaf roll polerovirus* causes accumulation of lipid or starch; *Cucumber mosaic cucumovirus* (CMV) forms cytoplasmic vesicles; some *Rhabdoviruses* cause proliferation of endoplasmic reticulum.

B. Host Range

Specific plant viruses are known to infect a unique combination of plant species, and this phenomenon is known as the *"Host Range"* of the virus, which has traditionally played an important role in the identification of some plant viruses of which the host range is very restricted. There are viruses that have a very wide host range such as *Tobacco rattle virus* (400 plant species belonging to 50 different plant families), *Cucumber mosaic virus* (470 plant species belonging to 67 different plant families), but *Watermelon mosaic virus 1* infects plants belonging only to the plant family Cucurbitaceae, whereas *Watermelon mosaic virus 2* infects plants belonging to the plant families Cucurbitaceae and Leguminosae.

Host range determination was the only approach available to virologists during the 1920s and 1930s and is still applied in the identification of new isolates in most laboratories around the world. But conflicting results may be obtained in tests done in different laboratories with the same virus isolate because of differences in environmental conditions, or use of different cultivars or genetic lines of the same plant species. Moreover test viruses may have overlapping host ranges and induce similar symptoms in some of these hosts. There is also the possibility of changes in the characteristics of the host ranges due to repeated passages through specific hosts or low virus concentrations or presence of inhibitors of infection in the original host. The host range approach has, however, been found successful in separating different viruses using differential hosts. Sarkar and Mukhopadhyay (1979) separated different cucurbit viruses by host reactions. Mukhopadhyay and Bandopadhyay (1984) separated different tungro viruses by using differential host reactions. The host range study still remains to be a very important low cost approach to differentiate strains and pathotypes of some viruses.

C. Grafting on Indicator Plants

Grafting is still in use to detect viruses in woody vegetative propagated plants. Grafting is an ancient horticultural practice. To conduct graft transmission of a virus disease, the scion is collected from the diseased plant and the rootstock is collected from the indicator test plant. When the diseased scion is grafted to the healthy rootstock, the resultant plant produces a symptom. It is also possible to transmit the virus to indicator herbaceous plants, as for example from grapevine to *Chenopodium amaranticolor* by means of approach grafts. The success of this method, however, depends upon the compatibility between scion and the stock and the perfection of their union.

Grafting is generally adopted in routine indexing of seedlings for viruses in many plantation crops, particularly citrus. In practice, both the scion and the rootstock seedlings require certification by legislation. Indicator plants for common citrus viruses are listed in Table 4.1.

Table 4.1 Indicator Plants of Some Citrus Viruses, All at 30-32°C (Aubert and Vullin 1998)

Virus	Indicator Plant	Incubation Time
Citrus leaf rugose	Mexican lime	5 mon
Citrus mosaic	Mandarin	5 mon
Concave gum	Sweet orange	5 mon
	Mandarin	5 mon
	Dweet tangor	5 mon
Impietratura	Sweet orange	2 yr
	Mandarin	2 yr
	Dweet tangor	2 yr
Leprosis	Sweet orange	1 yr
Ring spot	Sweet orange	4 mon
	Mandarin	4 mon
	Grapefruit	4 mon
	Chenopodium	4 months
Satsuma dwarf	Satsuma	6 mon
	Sesame	6 mon
Tatter leaf	Rusk citrange	1-2 yr
Tristeza	Mexican lime	6 mon
Vein enation	Mexican lime	1 yr
	Volkameriana	1 yr

Tuber-borne viruses are often tested by tuber grafting or grafting tissue from one tuber to another by core or plug grafting, also known as *"Grow on Test"*. A plug of tissue from a flat area of the infected tuber is removed by a small cork-borer. This plug is to be deep enough to contain the vascular tissue without containing any eye. A similar plug with an eye is removed from a healthy tuber. The virus-containing plug is then inserted into the hole of the healthy tuber. The grafted tuber portion is immersed in melted wax to prevent desiccation and rotting. The tuber is then planted and observed. A similar test is conducted to index seed potato in which samples are first sprouted, eyes are cut and planted in seed trays. Plants are allowed to grow at 12°C in 24 h light for 6 wk under aphid-free condition and assessed for infection.

Veroeven and Roenhorst (2000) proposed herbaceous test plants for the detection of quarantine viruses of potato. They reported reaction of nine plant species: *Chenopodium amaranticolor, C. quinoa, Nicotiana benthamiana, N. bigelovii, N. debneyi, N. hesperis, N. mersii, N. occidentalis* and *N. tabacum* to six non-European potato viruses (*Andean potato latent tymovirus, Andean potato mottle comovirus, eggplant mosaic tymovirus, tobacco ringspot nepovirus,* and an unidentified agent) and including all the seedborne viruses (*potato black ringspot nepovirus, potato T capillovirus, potato yellowing alfamovirus*), Different isolates of all these viruses were detected on *N. hesperis*-67-A as well as on *N. occidentalis* P1.

Indicator plants are useful for broad-spectrum detection of viruses. But it requires large numbers of test plants under controlled conditions and it takes a long time for the characteristic symptom expression.

References

Astier, S., Albouy, J., Maury, Y., Robaglia, C. and Licoq, H. 2007. Principles of Plant Virology: Genome, Pathogenicity, Virus Ecology. Science Publishers Inc., Enfield, USA, pp. 207-243

Aubert, B. and Vullin, G. 1998. Citrus nurseries and planting techniques. CIRAD, Montpellier, France. pp. 82

Bariana, H.S., Shannon, A.L., Chu, P.W.G. and Waterhouse, P.M. 1994. Detection of the seed-borne legume viruses in one sensitive multiplex polymerase chain reaction test. Phytopathology 84: 1201-1205

Bertolini, E., Olmos, A., Carmen Martinez, M., Gorris, M.T. and Cambra, M. 2001. Single step multiplex RT-PCR for simultaneous and colorimetric detection of six RNA viruses of olive trees. J. Virol. Methods 46: 33-41

Boonham, N., Walsch, K., Mumford, R.N.A. and Barker, I. 2000. Use of multiplex real-time PCR (TaqMan) for the detection of potato viruses. Bull. EPPO 30: 427-443

Boonham, N., Tomlinson, J. and Mumford, R. 2007. Microarrays for Rapid Identification of Plant Viruses. Annu. Rev. Phytopathol. 45: 307-328

Bos, L. 1963. Symptoms of virus diseases in plants. Institute for Phytopathological Research, Wageningen, The Netherlands

Bystricka, D., Lenz, O., Mraz, I., Dedic, P. and Sip, M. 2003. DNA microarray parallel detection of potato viruses. Acta Virologica 47: 41-44

Cai, H., Li, J.M., Xi, Zhi, Kang, L., Deng, K.J., Zhao, S.Y. and Li, C.B. 2003. Detecting rice stripe virus (RSV) in the small brown plant hopper (*Laodelphax striatellus*) with high specificity. J. Virol. Methods 112: 115-120

Dikova, B. and Hristova, D. 2002. Diagnostics of squash mosaic virus, zucchini yellow mosaic virus and cucumber mosaic virus in squash seeds. Bul. J. Agric. Sci. 8: 519-526

Chen, J., Zheng, H.Y., Antoniw, J.F., Adams, M.J., Chen, J.P. and Lin, L. 2004. Detection and classification of allexiviruses from garlic in China. Archv. Virol. 149: 435-445

Clover, G.R.G., Ratti, C. and Henry, C.M. 2001. Molecular characterization and detection of European isolates of soil-borne wheat mosaic virus. Plant Pathology 50: 761-767

Clover, G.R.G., Ratti, C., Rubies Autonel, C. and Henry, C.M. 2002. Detection of European isolates of oat mosaic virus. Euro. J. Plant Path. 108: 87-91

Colinet, D., Kummert, J., Lepoivre, P. and Semal, J. 1993. Identification of distinct potyviruses in mixedly infected sweet potato by the polymerase chain reaction with degenerate primers. Phytopathology 84: 65-69

Conning, E.S.G., Penrose, M.J., Barker, I. and Costes, D. 1996. Improved detection of barley yellow dwarf virus in single aphids using RT-PCR. J. Virol. Methods 56: 191-197

Delanoy, M., Salmon, M., Kumert, J., Friscon, E. and Lpoivre, P. 2003. Development of real-time PCR for the rapid detection of episomal Banana streak virus (BSV). Plant Dis. 87: 33-38

Deng, D., McGrath, P.F., Robinson, D.J. and Harrison, B.D. 1994. Detection and differentiation of whitefly-transmitted geminiviruses in plants and vector insects by the polymerase chain reaction with degenerate primers. Ann. Appl. Biol. 125: 327-336

Devergne, J.C., Cardin, L., Burckard, J. and van Regenmortel, M.H.V. 1981. Comparison of direct and indirect ELISA for detecting antigenically related cucumoviruses. J. Virol. Methods 3: 193-200

Dovas, C.I. and Katis, N.I. 2003a. A spot multiplex nested RT-PCR for the simultaneous and generic detection of viruses involved in the aetiology of grapevine leaf roll and rugose wood of grapevine. J. Virol. Methods 109: 217-226

Dovas, C.I. and Katis, N.I. 2003b. A spot nested RT-PCR method for the simultaneous detection of members of the Vitivirus and Foveavirus genera in grapevine. J. Virol. Methods 107: 99-106

Dovas, C.I., Katis, N.I. and Avgelis, A.D. 2002. Multiplex detection of criniviruses associated with epidemics of a yellowing disease of tomato in Greece. Plant Dis. 86: 1345-1349

Eun, A.J.C., Seoh, M.L. and Wong, S.M. 2000. Simultaneous quantitation of two orchid viruses by the TagMan R real-time RT-PCR. J. Virol. Methods 87: 151-160

Foissac, X., Svanella Dumas, L., Dulucq, M.J., Candresse, T., Genht, P. and Clark, M.F. 2001. Polyvalent detection of fruit tree tricho, capillo and foveaviruses by nested RT-PCR using degenerated and inosine containing primers (PDO RT-PCR). Acta Horticulturae No. 550 (Vol. 1): 37-43

Gibbs, A. and Harrison, B. 1976. Plant Virology. Edward Arnold, London, UK, p. 106

Hadidi, A., Montasser, M.S., Levy, L., Goth, R.N., Coaverse, R.H., Madkour, M.A. and Skrzeckowski, L.J. 1993. Detection of potato leafroll and strawberry mild yellow edge luteoviruses by reverse transcription-polymerase chain reaction amplification. Plant Dis. 77: 595-601

Haider, M.S., Bushra, M., Evans, A.A.F. and Markham, P.G. 2003. Dot blot hybridization and PCR-based detection of begomoviruses from the cotton growing regions of Punjab. Pakistan Mycopath. 1: 195-203

Hall, C.E. 1964. Electron microscopy: Principles and application to virus research. In: Plant Virology. M.K. Corbett and H.D. Sisler (eds.). University of Florida Press, Gainesville, Florida, USA, p. 255

Hampton, R., Ball, E. and de Boer, S. 1990. Serological methods for detection and identification of viral and bacterial plant pathogens. A Laboratory Manual. APS Press, St. Paul, MN, USA

Henson, J.M. and French, R. 1993. The polymerase chain reaction and plant disease diagnosis. Annu. Rev. Phytopathol. 31: 81-109

Hibi, I. and Suito, Y. 1985. A dot immunobinding assay for the detection of tobacco mosaic virus in infected tissues. J. Gen. Virol. 66: 1191-1194

Hourani, H. and Aboujawdah, Y. 2003. Immunodiagnosis of cucurbit yellow stunting disorder virus using polyclonal antibodies developed against recombinant coat protein. J. Plant Path. 85: 197-204

Hu, J.S., Wang, M., Sether, D., Xie, W. and Leonhardt, K.W. 1996. Use of polymerase chain reaction (PCR) to study transmission of banana bunchy top virus by the banana aphid (*Pentalonia nigronervosa*). Ann. Appl. Biol. 128: 55-64

Hughes, J.d'A. and Ollennu, L.A. 1993. The virobacterial agglutination test as a rapid means of detecting cocoa swollen shoot virus. Annals Appl. Biol. 122: 299-310

Ito, T., Ieki, H. and Ozaki, K. 2002. Simultaneous detection of six citrus viroids and apple stem grooving virus from citrus plants by multiplex reverse transcription polymerase chain reaction. J. Virol. Methods 106: 235-239

Ivars, P., Alonso, M., Borja, M. and Hernandez, C. 2004. Development of a non-radioactive dot-blot hybridization assay for the detection of Pelargonium flower break virus and Pelargonium line pattern virus. Euro. J. Plant Pathology 110: 275-283

James, D. 1999. A simple and reliable protocol for the detection of apple stem grooving virus by RT-PCR and in a multiplex PCR assay. J. Virol. Methods 83: 1-9

Jan, F.J., Wu, Z.B., Kuo, A.J., Zheng, Y.X., Chang, H.H., Su, C.C. and Yang, Y.S. 2003. Detection of Apple chlorotic leaf spot and apple stem grooving viruses by the method of reverse transcription polymerase chain reaction. Plant Path. Bull. 12: 10-18

Koenig, R. 1981. Indirect ELISA methods for the broad specificity detection of plant viruses. J. Gen. Virol. 55: 53-62

Kohler, G. and Milstein, C. 1975. Continuous cultures of fused cells secreting antibodies of predefined specificity. Nature 256: 495-497

Kong, B.H., Cai, H., Liy, J.Y. and Chen, H.R. 2002. Identification and detection of carnation mottle virus on carnation by RT-PCR. Plant Protection 28: 1

Kuroda, T., Urushibara, S., Takeda, I., Nakatani, F. and Suzuki, K. 2002. Multiplex reverse polymerase chain reaction for simultaneous detection of viruses in gentian. J. Gen. Plant Path. 68: 169-172

Kusano, N. and Ibi, A. 2003. Detection of Apple stem grooving virus from citrus plants by one-step immunocapture reverse transcription polymerase chain reaction (one step IC-RT-PCR). Kyushu Plant Protection Research 49: 50-55

Legarreta, G.G., Garcia, M.L., Costa, N. and Grau, O. 2000. A highly sensitive heminested RT-PCR assay for the detection of citrus psorosis virus targeted to a conserved region of the genome. J. Virol. Methods 84: 15-22

Lei, J.L., Lu, Y.P., Jin, D.D., Chen, S.X. and Chen, J.P. 2001. RT-PCR detection of rice ragged stunt virus in rice plant and insect vector. Acta Phytopathologica Sinica 31: 306-309

Leong, M.M.L., Milstein, C. and Pannell, R. 1986. Luminescent detection method for immunodot, western, and southern blots. J. Histochem. Cytochem. 34: 1645-1650

Levy, L. and Hadidi, A. 1994. A simple and rapid method for processing tissue infected with plum pox potyvirus for use with specific 3′ non-coding region RT-PCR assays. EPPO Bull. 24: 595-604

Lopez-Moya, J.J., Cubero, J., Lopez-Abella, D. and Diaz-Ruiz, J.R. 1992. Detection of cauliflower mosaic virus (CaMV) in single aphids by the polymerase chain reaction (PCR). J. Virol. Methods 37: 124-136

Makkouk, K.M. and Kumari, S.G. 2002. Low cost paper can be used in tissue-blot immunoassay for detection of cereal and legume viruses. Phytopathol. Medit. 41: 275-278

Mandal, B. 2003. Protein-based diagnostic assays in plant virology. In: Refresher course-cum-workshop. Advanced Centre for Plant Virology, Division of Plant Pathology. Indian Agricultural Research Institute, New Delhi, pp. 19-28

Marinho, V.L.V., Daniels, J., Kumert, J., Chandelier, A. and Le Poivre, P. 2003. RT-PCR ELISA for detection of apple stem grooving virus in apple trees. Fitopathologia Brasileira 28: 374-379

Maroon, C.J.M., Zavriew, S. and Hammond, J. 2002. PCR-based tests for the detection of tobamoviruses and carlaviruses. Acta Horticulturae No. 568: 117-122

Martin, D.P. and Rybicki, E.P. 1998. Microcomputer-based quantification of maize streak virus symptoms in *Zea mays*. Phytopathology 88: 422-427

Mason, G., Roggero, P. and Travella, L. 2003. Detection of tomato spotted wilt virus in its vector *Frankliniella occidentalis* by reverse transcription-polymerase chain reaction. J. Virol. Methods 109: 69-73

Mathews, R.E.F. 1992. Fundamentals of Plant Virology. Academic Press, Inc., San Diego, New York, pp. 45-59

Mehta, P., Wyman, J.A., Nakhla, M.K. and Maxwell, D.P. 1994. Polymerase chain reaction detection of viruliferous *Bemisia tabaci* (Homoptera-Aleyrodidae) with two tomato-infecting geminiviruses. J. Econ. Entomol. 87: 1285-1290

Menzel, W., Zahn, V. and Mai, S.S.E. 2003. Multiplex RT-PCR-ELISA compared with bioassay for the detection of four apple viruses. J. Virol. Methods 110: 153-157

Minafra, A. and Hadidi, A. 1994. Sensitive detection of grapevine virus A, B or leafroll associated III from viruliferous mealybugs and infected tissue by cDNA amplification. J. Virol. Methods 47: 175-188

Monger, W.A., Seal, S., Cotton, S. and Foster, G.D. 2001. Identification of different isolates of cassava brown streak virus and development of a diagnostic kit. Plant Pathology 50: 768-775

Monger, W.A., Seal, S., Cotton, S. and Foster, G.D. 2001. Identification of different isolates of cassava brown streak virus and development of a diagnostic kit. Plant Pathology 50: 768-775

Mukhopadhyay, S. and Bandopadhyay, B. 1984. Incidence of different strains of rice tungro virus in West Bengal. Int. J. Tropical Plant Diseases 2: 125-132

Muller, G., Fuentes, S., Salazar, L.F. and Ames, T. 2002. Detection of sweet potato chlorotic stunt crinivirus by non-radioactive Nucleic Acid Spot Hybridization (NASH) technique. Acta, Horticulturae No. 583: 129-133

Mullis, K.F., Faloona, F., Scharf, S., Saiki, R., Hora, G. and Erlich, H. 1986. Specific enzymatic amplification of DNA in vitro: the polymerase chain reaction. Cold Spring Harbor Symp. Quant. Biol. 51: 263-273

Mumford, R.A., Walsch, K., Barker, I. and Boonham, N. 2000. Detection of potato mop top virus and tobacco rattle virus using a multiplex real-time fluorescent reverse transcription polymerase chain reaction assay. Phytopathology 90: 448-453

Mumford, R.A., Skelton, A.L., Bomham, N., Posthuma, K.I., Kirby, M.J. and Adams, A.N. 2004a. The improved detection of strawberry crinkle virus using real-time RT-PCR (TauMan. Reg.). Acta Horticulturae 656: 81-86

Mumford, R., Skelton, A., Metcalfe, E., Walsh, K. and Boonham, N. 2004b. The reliable detection of Barley yellow and mild mosaic viruses using real-time PCR (TaqMan). J. Virol. Methods 117: 153-159

Naidu, R.A. and Hughes, Jd'A. 1998. Methods for the detection of plant virus diseases. In: Plant Virus Disease Control. A. Hadidi, R.K. Khearpal and H. Koganezawa (eds.). APC Press, St. Paul, Minnesota, USA, pp. 233-258

Navot, N., Zeiden, M., Pichersky, R. and Czosnek, H. 1992. Use of the polymerase chain reaction to amplify tomato yellow leaf curl virus DNA infected plants and viruliferous whiteflies. Phytopathlogy 82: 1119-1202

Nie, X. and Singh, R.P. 2001. A novel usage of random primer for multiplex RT-PCR detection of virus and viroid in aphids, leaves and tubers. J. Virol. Methods 91: 37-49

Niu, J.X., Liu, L.K., Zhu, J. and Quin, W.M. 2003. Study on the RT-PCR detection system pear vein yellows virus by using dsRNA as template. J. Fruit Sci. 20: 143-145

Nolasco, G., de Blas, C., Torres, V. and Ponz, F. 1993. A method combining immunocapture and PCR amplification in a microtiter plate for the detection of plant viruses and subviral pathogens. J. Virol. Methods 45: 201-218

Raboudi, F., Ben Moussa, A., Makni, H., Marrakchi, M. and Makrii, M. 2002. Serological detection of plant viruses in their aphid vectors and their host plants. Bull. EPPO 32: 495-498

Ramchandran, P. and Agarwal, J. 2003. Nucleic acid based diagnostic assays in plant virology. In: Refresher course-cum-workshop. Advanced Centre for Plant Virology, Division of Plant Pathology. Indian Agricultural Research Institute, New Delhi, pp. 29-40

Roberts, C.A., Dietzgen, R.G., Heelan, L.A. and MaClean, D.J. 2000. Real-time RT-PCR fluorescent detection of tomato spotted wilt virus. J. Virol. Methods 88: 1-8

Rubio, L., Janssen, D., Cuadro, I.M., Moreno, P. and Guerri, J. 2003. Rapid detection of cucumber vein yellowing virus by tissue-print hybridization with digoxigenin-labeled cDNA probe. J. Virol. Methods 114: 105-107

Rush, C.M., French, R. and Heidel, G.B. 1994. Differentiation of two closely related furoviruses using the polymerase chain reaction. Phytopathology 84: 1366-1369

Rybicki, B.P. and Hughes, F. 1990. Detection and typing of maize streak virus and other distantly related geminiviruses of grasses by polymerase chain reaction amplification of a conserved viral sequence. J. Gen. Virol. 71: 2519-2526

Saiki, R.K., Gelfand, G.H., Stoffel, S., Schraff, S.J., Higuchi, R., Horn, G.T., Mullis, K.B. and Ehrlicch, H.A. 1988. Primer-directed enzymatic amplification of DNA with a thermostable DNA polymerase. Science 239: 487-491

Salama, M.I., Abdel Gaffar, M.H., Kumari, S.G. and Sadik, A.S. 2003. Comparison between different ELISA procedures for detection of some *Vicia faba* viruses. Annal Agric. Science Cairo 48: 57-67

Salmon, M.A., Vendrame, M., Kummert, J. and Lepoivre, P. 2002. Rapid and homogeneous detection of apple stem pitting virus by RT-PCR and a fluorogenic 3' minor groove binder DNA probe. Euro. J. Plant Path. 108: 755-762

Sarkar, B.B. and Mukhopadhyay, S. 1979. Separation of nine anisometric cucurbit viruses by host reactions. Veg. Science 6: 142-144

Schena, M., Shalon, D., Davis, R.W. and Brown, P.O. 1995. Quantitative monitoring of gene expression patterns with a complementary DNA microarray. Science 220: 467-470

Sharman, M., Thomas, J.E. and Dietzgen, R.S. 2000. Development of a multiplex immunocapture PCR with colorimetric detection for viruses of banana. J. Virol. Methods 89: 75-88

Singh, R.P., Kurz, Z.P., Boiteau, G. and Bernard, G. 1995. Detection of potato leafroll virus in single aphids by the reverse transcription-polymerase chain reaction and its potential epidemiological application. J. Virol. Methods 55: 133-143

Singh, R.P., Kurz, J. and Boiteau, G. 1996. Detection of stylet-borne and circulative potato viruses in aphids by duplex reverse transcription-polymerase chain reaction. J. Virol. Methods 59: 189-196

Singh, R.P. and Dhar, A.K. 1998. Detection and management of plant viroids. In: Plant Virus Disease Control. A. Hadidi, R.K. Khetarpal and H. Koganezawa (eds.). APS Press, St Paul, Minnesota, USA, pp. 428-447

Smith, G.R. and Van de Velde, R. 1994. Detection of sugarcane mosaic virus in diseased sugarcane using the polymerase chain reaction. Plant Dis. 78: 557-561

Sudarshana, M.R. and Reddy, D.V.R. 1989. Penicillinase-based enzyme-linked immunosorbent assay for the detection of plant viruses. J. Virol. Methods 26: 45-52

Takahashi, Y., Tiongco, E.R., Cabauatan, P.Q., Koganezewa, H., Hibino, H. and Omura, T. 1993. Detection of tungro bacilliform virus by the polymerase chain reaction for assessing mild infection of plants and viruliferous vector leafhoppers. Phytopathology 83: 655-659

Thompson, J.R., Wetzel, S. and Jelkmann, W. 2004. Pentaplex RT-PCR for the simultaneous detection of four aphid borne viruses in combination with a plant mRNA specific internal control in *Frageria* spp. Acta Horticulturae 656: 51-56

Tsuda, S., Fujisawa, I., Hanada, K., Hidaka, S., Hogo, K., Kammeya-Iweko, M. and Tomaru, K. 1994. Detection of tomato spotted wilt virus SRNA in individual thrips by reverse transcription and polymerase chain reaction. Ann. Phytopathol. Soc. Japan 60: 99-103

Tzanetakis, I.E., Halgren, A.B., Keller, K.E., Hokanson, S.C., Maas, J.L., McCarthy, P.L. and Martin, R.R. 2004. Identification and detection of a virus associated with strawberry pallidosis disease. Plant Dis. 88: 383-390

Uhde, K., Kerschbaumer, R.J., Koening, R., Hirschl, S., Lamaire, O., Boonham, N., Roake, W. and Himmler, G. 2000. Improved detection of beet necrotic yellow vein virus in a DAS-ELISA by means of antibody single chain fragments (ScFv) which were selected to protease-stable epitopes from phage display libraries. Archiv. Virology 145: 179-185

Vacke, J. and Cibulka, R. 2000. Comparison of DAS-ELISA and enzyme amplified ELISA for detection of wheat dwarf virus in host plants and leafhopper vectors. Plant Protection Science 36: 41-45

Van der Wilk, F., Korsman, M. and Zom, F. 1994. Detection of tobacco rattle virus in nematodes by reverse transcription and polymerase chain reaction. Europ. J. Plant Pathol. 1: 109-122

Vaskova, D., Spak, J., Klerks, M.M., Schoen, C.D., Thopson, J.R. and Jelkmann, W. 2004. Real-time NASBA for detection of strawberry vein banding virus. Euro. J. Plant Pathol. 110: 213-221

Vercruysse, P., Gibbs, M., Tirry, L. and Hofte, M. 2000. TR-PCR using redundant primers to detect the three viruses associated with carrot motley dwarf virus. J. Virol. Methods 88: 153-161

Verma, Y., Sood, S., Ahlawat, Y.S., Khurana, S.M.P., Nic, X. and Singh, R.P. 2003. Evaluation of multiplex reverse transcription polymerase chain reaction (RT-PCR) for simultaneous detection of potato viruses and strains. Indian J. Biotechnology 2: 587-590

Veroeven, J.T.J. and Roenhorst, J.W. 2000. Herbaceous test plants for the detection of quarantine viruses of potato. Bull. OEPP 30 (3-4): 463-467

Vunsl, R., Rosner, A. and Stein, A. 1990. The use of the polymerase chain reaction (PCR) for the detection of bean yellow mosaic virus in gladiolus. Ann. Appl. Biol. 117: 561-569

Wang, R.Y. and Ghabrial, S.A. 2002. Effect of aphid behaviour on efficiency of transmission of soybean mosaic virus by the soybean-colonizing aphid *Aphis glycine*. Plant Dis. 86: 1260-1264

Wang, Z.H., Zhou, V.J., Fan, Y.J., Cheng, Z.B. and Zhang, W.H. 2002. Rapid methods for the detection of rice black streaked dwarf fiji virus in a single plant hopper. Journal of Shanghai Jiaotong University of Agricultural Science 20: 340-343

Wetzel, T., Candresse, T., Macquaire, G., Ravelonandro, M. and Dunez, J. 1992. A highly sensitive immunocapture polymerase chain reaction method for plum pox virus detection. J. Virol. Methods 39: 27-37

Whitefield, A.E., Cambell, L.R., Sherwood, J.L. and Ullman, D.E. 2003. Tissue blot immunoassay for detection of tomato spotted wilt virus in *Ranunculus asiaticus* and other ornamentals. Plant Dis. 87: 618-622

Xu, H., DeHaan, T.L. and de Boer, S.H. 2004. Detection and confirmation of potato mop top virus in potatoes produced in the United States and Canada. Plant Dis. 88: 363-367

5

Vectors of Viruses

In contrast to animals, plants are sessile organisms and are not significant transmitters of viruses except for some instances of pollen and seed transmission and movement of plants resulting from human intervention. Thus, the great majority of plant viruses are dependent for their spread (and therefore survival) upon efficient transmission from plant to plant by specific vector(s). Plant-to-plant spread ensures virus survival, often resulting in disease. It is therefore critical to understand this essential component of the virus life cycle and the life cycle of the vectors as well.

Plant viruses cannot enter the host on their own. They need a method of introduction that allows them to enter host tissues where they can multiply. One method, often used in experiments, but not of great importance in natural systems, is via mechanical damage but more important in nature is by a range of invertebrates and soil-borne fungal and plasmodiophorid species. These species that act in this way are known as 'vectors'. They penetrate unwounded plant cells, usually when feeding on them or parasitizing them, and can acquire viruses from an infected plant and may subsequently transfer the virus (inoculate) to susceptible, healthy plants. Different types of feeding behaviours are found in different invertebrate vector species depending upon their mouthparts that may be adapted for chewing, sucking (lacerating-sucking, piercing-sucking) and sponging-lapping. The vector retains the virus in an infective form between acquisition and inoculation. In the fungal or plasmodiophorid vectors the virus is contained either on the surface or within the vector. The relationship between vector and virus is often very specific and the result of complex biological interactions. This specificity operates both within and between vector groups. Thus, a virus transmitted by a member or members of one taxonomic group are not normally transmissible by a member or members of another taxonomic group. In most plant virus-vector interactions, the vector is just a carrier of the virus but, in a few cases, some vectors also support multiplication of the viruses and even transmit them to their progeny. The role of vectors in virus spread is of immense significance. Understanding their biology and phenology is essential if strategies to control the virus diseases they transmit are to be developed.

Not all invertebrates or soil-borne fungi and plasmodiophorids are vectors of viruses. Only a few phytophagous invertebrate families contain vectors but even within these families not all members are vectors, although in some cases quite a few have been tested in detail. The groups containing the vectors are arthropods (insects and mites), nematodes, fungi and protista. The most common vectors in the field are insects. Among the insects, aphids predominate in temperate regions whereas whiteflies and leafhoppers predominate in tropical and sub-tropical areas. Other vector groups are beetles (Coleoptera), plant-hoppers, nematodes, thrips, mites, mealy bugs, scales and fungi (including protists). Nault (1997) recorded the distribution of plant virus vectors within selected homopteran and coleopteran families and Dijkstra and Khan (2002) described fungal vectors. Despite much intense research, new records of vectors occur continuously. A new species of aphid (Gonzales *et al* 1998), four new species of mealy bugs (Gullan *et al* 2003, Granara de Willink *et al* 2004), seven new species of Mirids (Schwarz and Scudder 2003), a new species of *Aleurocanthus, Asialeyroides indica* Sundararaj and David (David and Jesudasan 2002), a new species of Dorylaimidae (Wasim and Araki 2003), a new species of the family Issidae (Gnezdilov *et al* 2003), a new species of coccid (Remadevi and Mathukrishnan 2002, Ulgenturk *et al* 2001, Hodgson and Millar 2002), and a new spider mite (Giorgini 2001) have been recently reported. Thus the area is still very active, both in identifying vector species and, as described below, in unravelling how transmission takes place. Incorporating the records of Matthews (1992), Ben-Dov (1993), Hull (2002) and the new identifications, a modified list has been compiled of the groups of invertebrates known to be vectors (Table 5.1).

A. VECTORS: MORPHOLOGY AND BIOLOGY

1. Aphids

Aphids belong to the taxonomic family Aphididae in the sub-order Homoptera, Order Hemiptera. These are soft, delicate, pear-shaped, small insects (body length normally around 2 mm). The characteristic features of an aphid are its pear-shaped structure, fairly long antennae, a pair of compound eyes, several jointed rostrum, two-segmented *tarsi* with paired claws, nine pairs of lateral spiracles and a pair of dorsi-lateral elongate *siphunculi*, sometimes known as *cornicles*, at the posterior end of the abdomen arising from the dorsal side of the fifth or sixth abdominal segment, that secrete a waxy fluid. In some species the *siphunculi* may be reduced to form a ring or may even be absent.

Table 5.1 Conventional Groups, Family and Common Names, Total Number of Known Species of Vectors And Number of Virus Diseases Transmitted by them

Group	Family	Common name	Number of known species	Number of vector species	Number of viruses transmitted
Insects	Aphididae	Aphid	>4000	192	370
	Cicadellidae	Leafhopper	15,000	49	38
	Chrysomelidae	Leaf beetle	37,000	48	30
	Coccinellidae	Lady bird beetle	4500	2	7
	Cucurlionidae	Weevil	50000	10	4
	Meloidae	Blister beetle	3000	1	1
	Fulgoidea (Superfamily)	Plant-hopper	19000	28	24
	Aleyrodidae	Whitefly	1200	3	114*
	Coccoidea (Superfamily)	Mealy bug & scales	> 6000	19	10
	Thysanoptera	Thrips	~5000	6	13
	Miridae	Bugs	6700	–	–
Mites	Eryophydidae	Leaf & bud mites	–	8	10
	Tetranychidae	Spider mite	>900	2	3
Nematodes Order: Dorylaimida		Plant parasitic nematodes	>2600	>33	35
	Longidoridae	–		>255	>14	–
	Trichodoridae	–		>50	>13	–
Fungi	Order: Chytridales	–	>1000		
	Family: Olpidiaceae	–	–	3	12
	Family: Synchitriaceae	–	–	1	1
Protista	Order: Plasmodiophorales Family: Plasmodiophoraceae	–	–	3	15

*Jones (2003)

Both winged and wingless forms are commonly found (Figures 5.1 and 5.2) depending upon the environmental conditions, including the presence or absence of the favoured hosts. All the nymphal forms are wingless. Normally wingless adults predominate in the population in favourable conditions. Wings are held vertically above the body. They feed on many

Figure 5.1 Wingless aphid (*Myzus persicae*)
Color image of this figure appears in the color plate section at the end of the book.

different host species and show great polymorphism under different conditions which can create confusion in their identification especially at the nymphal stages. But the numerous biological and morphological characteristics (structure of mouth parts, mode of feeding, searching behaviour for hosts, biological cycle alternating the apterous and winged generations, reproduction rate) combine to give them the outstanding vector capacities.

There is wide variation in their host preferences and virus transmission. Many species are involved in virus transmission and one of them, *Myzus persicae* (Sulzer) transmits more than 100 different viruses. Aphid mouthparts are well suited to acquire and inoculate viruses. The mouthparts consist mainly of two pairs of stylets held together by longitudinal ridges and grooves in their adjacent surface, a labium, a large slender but rigid organ, with a deeply concave anterior surface, which forms two channels of beak, a labrum, mandible and maxillae. The two pairs of stylets form a compact bundle or fascicle that slides in the groove of the labium and constitutes the piercing organ. This organ has two canals. Saliva is injected into the plant through one (the salivary canal), and through the other (the food canal) food is sucked up from the plant. In the aphid's head are food and salivary pumps that help to regulate the flow of saliva

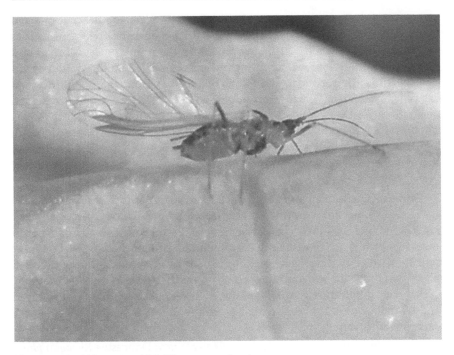

Figure 5.2 Winged aphid (*Myzus persicae*)
Color image of this figure appears in the color plate section at the end of the book.

and food. Aphids, when they land on a new host, first feed on parenchyma tissues and, if this probing or tasting action demonstrates that the plant is a suitable host, the stylets, supported by a salivary sheath, may then penetrate the phloem from where food is obtained. These actions and the aphid behaviour that controls them is critical to the transmission of plant viruses.

Biology and Life Cycle

The biology of aphids is very varied, flexible and unusual among insects. Depending upon the conditions, they can be *holocyclic* or *anholocyclic*. In a typical *holocyclic* life cycle, aphids reproduce both by sexual recombination, and by parthenogenesis. This type of biology is normally found in temperate countries where there is a marked cold period, usually with below zero centigrade temperatures during winter. In these areas, aphids over-winter on their primary hosts where eggs are sexually produced. These eggs hatch in the spring and give rise to apterous, viviparous, parthenogenetic females in the spring. This form is known as a *fundatrix,* and is characterized by well-developed sense organs, and whose legs and antennae are not well developed. *Fundatrices* reproduce parthenogenetically producing only apterous females. Parthenogenetic and viviparous progenies are produced by the *fundatrices* during the first one and two generations (*fundatrigenae*) and winged parthenogenetic viviparous females are produced during subsequent generations on the primary hosts. The winged forms migrate to the secondary hosts under favourable conditions. Individuals developed on the secondary hosts markedly differ from the *fundatrices* and are called *alienicolae*. These forms of parthenogenetic females undergo several generations on the secondary hosts throughout the summer. With the onset of shorter day lengths and cooler temperatures there is the production of a specialized form called *sexuparae*. These are either males or oviparous females that migrate to the primary hosts, mate and lay eggs and thus complete the life cycle.

Parthenogenetic, viviparous reproduction is the most obvious stage in the life cycle. Generation times are short and this enables populations to increase very rapidly and increases their chances of survival. Parthenogenetic reproduction is most rapid when the temperature is optimum and when the host is at its most nutritious stage of development. However, different aphids colonize different plant parts and do not multiply equally rapidly everywhere. For example, certain cereal aphids, e.g., *Sitobion avenae*, the English grain aphid, will feed and reproduce on most plant parts but rapidly moves to, and develops on the ears when they emerge.

Many aphids are *anholocyclic* [e.g. *Myzus ascalonicus* Doncaster, *Rhopalosiphum maidis* (Fitch)] either throughout their range or in particular regions where there is no sexual stage. It is not unusual for the species, particularly those that are damaging pests, to have both *anholocyclic* and *holocyclic* clones within the population. These alternative strategies enable aphid populations to take advantage of conditions that favour *anholocyclic* but also to have the *holocyclic* part of the population as security for survival if conditions deteriorate.

In tropical and sub-tropical countries, the aphids are mostly *anholocyclic* and feed on cool season crops as wingless parthenogenetic, viviparous forms. At the onset of hot conditions, when the temperature rises above 30°C, they become winged and migrate to cooler more favourable regions where they reproduce parthenogenetically.

Depending upon the species, an adult aphid may have a mean life span of 19-35 d. Adult females produce nymphs for 10-25 d. Numbers of *nymphs* produced per day, depend upon the species, environmental conditions and host quality. In optimal conditions (21°-32°C, relative humidity 65%) *nymphs* may molt three times and the *nymphal* period lasts for 5-6 d. Adult longevity, reproduction rate and nymphal period are influenced by morphology and colour (Jagadish and Jayaramaiah 2004).

The alternation of aphids between host plants provides a further variation in and flexibility of the life cycle. Such alteration appears to be a desirable strategy in regions where secondary hosts are short-lived and the aphids can exploit different plant species while they are nutritionally most favourable. The complexity and variation in the aphid life cycle makes the group facultatively opportunistic. Eastop (1983) reviewed the biology of aphids that transmit viruses.

2. Whiteflies

Whiteflies (Aleyrodids) are found in both tropical and sub-tropical regions. They have a wide host range and can reach a huge population in a short period under favourable condtions. Aleyrodids have seven segmented antennae, two *ocelli* and a pair of reniform compound eyes. Wings are of equal consistency, opaque, whitish, clouded or mottled with spots or bands. The *rostrum* is three segmented and the *tarsi* are two segmented, terminating in a pad-like *empodium* or spine between the claws. There is a pair of prominent *claspers* in male genitalia and an acute ovipositor in females.

The adults are very difficult to distinguish morphologically and they are usually separated on the basis of the characters of the fourth *instar* larva or pupa. Among the whiteflies, *Bemisia tabaci* (Gen) is the most common vector especially in tropical and sub-tropical regions (Figure 5.3). Its pupae undergo

Figure 5.3 Adult whitefly (*Bemisia tabaci*)
Color image of this figure appears in the color plate section at the end of the book.

considerable host-related and season-related variation that often creates confusion in their taxonomic identity. Females can be distinguished from the males in some species by a bigger and broader abdomen that does not taper at the posterior end as it does in males.

Biology and Life Cycle

Whiteflies are oviparous and the eggs are stalked. They normally reproduce sexually but parthenogenesis is also found in several species. Adult females normally lay eggs on the undersurface of the leaves. An individual female produces from 50 to 300 eggs depending upon the season and the host. The creamy white eggs produced may be both fertilized and unfertilized; unfertilized eggs develop into males while fertilized eggs produce both males and females. Eggs appear dark brown before hatching and usually about 75% hatch. The average incubation period is around 15 d. The larvae molt to produce 4 nymphal *instars* and the average nymphal period is around 24 d. In most cases first *instar* larvae are mobile. The second, third and fourth *instar* larvae have atrophied legs and are immobile. Fourth *instar* larvae stop feeding and adults develop inside them when the instars are called *pupal* cases. The adult emerges through a split on the dorsal side of the *pupal* case. Papillae or simple or compound pores of *nymphs* and pupae of many species often produce waxy secretions.

The speed of development of eggs and adults primarily depends upon temperature. A few eggs of *Bemisia tabaci* hatch at 10°C but the first *instar* nymph does not develop at this temperature. Equally there may be no hatching of the eggs at 36°C or above. However, completion of the life cycle is very rapid at temperatures between 27° and 30°C.

At 25.9°C temperature and 52.5% relative humidity, the duration of *puparia* destined to be male is normally 73.38 d, whereas in those destined to be female, it is 71.39 d. The longevity of male adults may be 6.5 d while that of female may be 7.65 d (Patel and Patel 2001). There may be several overlapping generations in a year depending upon the environmental condition and the host.

Verma *et al* (1990) found that accumulated degree-days had a profound influence on the lifecycle both under field and artificial conditions. It was estimated that for the developmental stages, 90.61-103.40 accumulated degree-days, above 10°C, are required for eggs to hatch and 303.28-336.80 for proper emergence of adults. The accumulated degree-days during the period of each generation ranged from 307.20 to 356.65.

Whitefly adults and immature instars are phloem-feeder with piercing-sucking mouthparts. Stylet penetration follows an intercellular course, becoming intracellular upon reaching the phloem. The adult stylet bundles enter the labial groove of the labium between the first and second segments and remains within the labium except during feeding (Rosell *et al* 1995). Stylet penetration requires physical force (Walker and Perring 1994). Alteration and change of the position of the head during feeding produce this force (Freeman *et al* 2001).

Johnson *et al* (2002) used the electrical penetration graph (EPG) technique to determine what part of stylet penetration behaviour by the whitefly vector is, inoculating *Lettuce chlorosis* (LCV), a semi-persistently transmitted *closterovirus*, to *Malva parviflora*. The virus was inoculated during the phloem phase of the stylet penetration behaviour.

Bemisia tabaci is a very complex species and its characteristics vary widely in different host species and geographical conditions particularly with respect to host-plant feeding preferences and virus transmission properties. The distinct population found in different hosts and locations are usually termed 'biotypes' A to T (Demichelis *et al* 2005), but there remains uncertainty over their degree of reproductive isolation (de Barro *et al* 2005). There are molecular differences between morphologically indistinguishable *B. tabaci* populations, and ribosomal and mitochondrial gene sequences of *B. tabaci* populations from around the world cluster into at least six phylogenetic groups, predominantly linked to their geographical region (de Barro *et al* 2005). *B. tabaci* adapts readily to new host plants and regions. Evidence suggests that *B. tabaci* is a highly dynamic species complex that is presently undergoing evolutionary change. Any control measure will exert selection pressure on the species complex, and may induce evolutionary change. Bosch, Jeger and Demon (October 2002-March 2005) in their study on the 'Evolution within *Bemicia tabaci* and associated begomoviruses: a strategic modelling approach', formulated a model and analyzed the evolution of *B. tabaci* (bio) types under the influence of the introduction of a new crop or variety; the use of pesticides; and the structure of the cropping system, comparing large-scale monocultures with intercropping.

Increased populations of *B. tabaci* are associated with a range of factors including conducive climatic conditions (Morales and Jone 2004), the spread of the more fecund B biotype (Perring 2001), the cultivation of particular crops or varieties (Morales and Anderson 2001; Varma and Malathi 2003) and virus infection of the host (Mayer *et al* 2002; Colvin *et al* 2004). The development of insecticide resistance in *B. tabaci* has also resulted in insecticide usage sometimes being counterproductive in reducing *B. tabaci* parasitoid populations, or favouring the selection of the more fecund insecticide-resistant B-biotype over other biotypes (Seal *et al* 2006).

Trialeurodes vaporariorum (Greenhouse whitefly, GHWF) has been emerging as a serious threat to vegetable and fruit production both in greenhouses and open fields. GHWF thrives on numerous crops as well as weed species particularly on plants in the families Asteraceae, Cucurbitaceae, Malvaceae and Solanaceae (Wintermentel 2004).

3. Leafhoppers

Leafhoppers belong to the Family Cicadellidae, Sub-order Homoptera, Order Hemiptera and Division Auchenorrhyncha. These small insects (body length around 3-4 mm) are often polymorphic, tangentially hop on the leaves of plants and suck juice by their *stylets* from the xylem, phloem or mesophyll cells depending upon the species. They are normally *brachypterous*, that is with short wings, nocturnal, and are most numerous in Asia, especially South East Asia and in Africa and Madagascar, although representatives are found very widely. Many are highly sensitive to temperature and relative humidity, and multiply, at a very rapid rate, on susceptible hosts during wet seasons with moderate temperature (20°-28°C) and multiplication slows down when conditions become dry and cold, particularly in sub-tropical regions.

Leafhoppers differ from other homopteran insects by the position of the *rostrum* that arises from the head in the members of the Auchenorrhyncha but in the hoppers, the vertex is smooth and convex, sometimes obtuse angled with compound eyes at the extremes. The antennae are thin and minute, bristle-like with two small basal segments. The forewings are non-transparent and variously coloured.

The identification of leafhoppers is based on size and colour, and the *aedeagus* is important in species differentiation. Another genital character used in taxonomy, is the numbers of spines on the *pygofer*. Leafhoppers are morphologically and biologically different from aphids but both the groups have similar feeding structures or mouthparts adapted to the piercing and sucking type of feeding.

There are 60 sub-families in Cicadellidae of which only 2 are known to contain virus vectors. The sub-family Agallinae feed on phloem or mesophyll cells of dicotyledonous herbaceous hosts whereas the members of the sub-family, Deltocephalinae, mostly feed on the xylem of monocotyledonous plants. Many species in the Deltocephalinae are vectors and transmit more than 70% of the leafhopper-borne viruses. They also have the ability to transmit other vascular-borne pathogens such as phytoplasma, spiroplasma, and bacteria. The transmission of viruses by leafhoppers is very specific.

Biology and Life Cycle

The life span of an adult leafhopper varies widely depending upon the species, host and the climate. It is normally between 16 and 53 d but the male life span is usually shorter than that of females. Females are sturdier than males and can be distinguished by more or less prominent ovipositors at the posterior end (Figure 5.4). After mating, adult females lay eggs on leaf blades, petioles, leaf sheaths and stems. The exact site depends on the species and the host. Several generations may be completed during crop growth. Generally temperature and host species are the dominating influences on the rate of completion of the life

Figure 5.4 Female rice green leafhopper (*Nephotettix virescens*)
Color image of this figure appears in the color plate section at the end of the book.

cycle. Nymphs usually *molt* four to five times before becoming adults. Some species may over-winter as fourth or fifth *instar* nymphs. The life cycle in some species may be completed within 21-35 d but can be as long as 35-59 d. Their normal preferred temperature range is 15° to 36°C.

The biology and life cycle of a leafhopper, rice green leafhopper [*Nephotettix virescens* (Distant)] both in field and laboratory conditions has been studied by Chakravarty *et al* (1979), Mukhopadhyay (1984), Sarkar (1988), and Das (1996). In many species nymphs emerge between 06.00 and 09.00 h. *Nymphal* emergence from the leaf tissue occurs with a bubble of leaf sap. As the sap dries neonate nymphs become active and start crawling. Newly hatched nymphs are transparent, sometimes yellowish and the eyes become conspicuous, reddish brown and oval. The average length of the first *instar* nymph is 0.69 mm. The colour of the eyes of the second *instar* nymph is superficially white but reddish underneath. There may be noticeable development of rudimentary wing pads along the posterior ends of ridges of meso- and meta-thorax and the average length is 1.09 mm. The colour of the third *instar* appears to be yellowish green.

At this stage the wing pads may reach up to the abdomen. The average length at this stage is 1.34 mm. The colour of the fourth and fifth *instar* nymphs is greenish yellow. The wing pad of the fourth *instar* nymphs may reach up to the fourth abdominal segment. The average length of this stage is 1.68 mm. The wing pad of the fifth *instar* may reach up to the ninth abdominal segment and the average length at this stage is 2.28 mm. Adults are normally recognizable by a pair of black dots on each of the apical end of the *tegmina* and the average length of an adult is 2.61 mm (Kiran *et al* 2003). Waquil *et al* (1999) studied the life cycle of corn leafhopper *Dabulus maidis* (DeLong and Welcott) under controlled conditions. The shortest incubation period was recorded at 26 and 29°C in which 70% hatching took place in 9 d. At 26.5°C nymphs went through four (76%) or 5 (18%) *instars* and reached the adult stage in 15.7 d (average 3.14 d per *instar*). Mean adult longevity was 51.4 d and the life cycle from eggs to adult averaged 26.3 d.

4. Plant Hoppers

Plant hoppers belong to the super-family Fulgoroidea, Division Auchenorryncha, sub-order Homoptera and Order Hemiptera. There are 20 families in this super-family of which the family Delphacidae contains most of the vector species. Members of this family normally feed on monocotyledonous plants primarily belonging to the family Poaceae.

Plant hoppers are morphologically and biologically different from leafhoppers but they have similar feeding structures. Planthoppers have a narrow rectangular *frons* between large compound eyes. Two basal segments of their antennae are distinctly thicker with distal segments reduced to a thin filament. Their fore- and hind-wings are usually hyaline or membranous, generally much less colourful than those of leafhoppers.

Plant hoppers normally occur in rice fields under tropical, subtropical and temperate conditions. Adults of several species are polymorphic particularly with respect to their wings. They exist in two forms *macropterous* (long winged) (Figure 5.5) and *brachypterous* (short winged). The *brachypterous* form is sedentary and when the crop is growing, this stage normally predominates.

Biology and Life Cycle

When individual *macropterous* forms arrive in rice fields they feed on the phloem of the young plants during which time, determined by temperature and food availability they complete *ovarial* development and the production of eggs. At times of emigration, the ovaries remain at the first stage whereas in the immigrants ovaries complete the developmental stages to produce mature eggs. Adult *macropterous* females lay eggs in leaf blades, petioles, leaf sheaths,

Figure 5.5 Macropterous brown plant hopper (*Nilaparvata lugens*). With permission from CABI, Wallingford, UK

Color image of this figure appears in the color plate section at the end of the book.

and stems depending upon the species and the host. In some species, under optimum environmental and host conditions, an adult may complete four generations within a month. Initially they produce *brachypterous* forms and later both *brachypterous* and *macropterous* forms with macropterous forms coming to dominate. *Brachypterous* forms have a greater fecundity than *macropterous* forms. The production of *macropterous* forms is linked to changes in the season, increases in the nymphal density and aging of the host plants. There is no over-wintering form.

5.   Treehoppers

These insects belong to the family Membracidae. They are characterized by the presence of two *ocelli*. Antennae in these hoppers lie in a depression beneath and in front of the eyes. They have no *stridulatory* organ. The *pronotum* extends backward over the abdomen and is ornamented with various spines, horns, keels, and bell-like knobs. Wings normally remain concealed by the *pronotum*. Membracids are normally very specific to hosts.

6.   Beetles

Beetle vectors belong to the family Chrysomelidae in the Order Coleoptera. These insects are oval, convex, brightly coloured and 3.5-12 mm long. The most characteristic feature of beetles is their forewings called *elytra*. These are thick, more or less hard, serving as a pair of convex shields to cover the hind wings and the abdomen from above. Hind wings, the real organs for flight, are thin and membranous.

The chrysomelid beetles or leaf beetles (Figure 5.6) usually feed and reproduce on a very limited range of hosts. Both larvae and adults have biting

Figure 5.6   Chrysomelid beetle
Color image of this figure appears in the color plate section at the end of the book.

and chewing type of mouthparts. They normally feed on parenchyma between vascular bundles leaving holes in the leaves. Many are very brightly coloured. All species are herbivorous, the adults chew leaves and flowers and the larvae feed externally or bore into stems, roots and leaves. Ladybird larvae are very active with well-developed legs. Fully-grown larvae pupate on foliage.

Most beetles reproduce sexually. Mate attraction and courtship involves sound production, pheromones and light display. The females of most beetles lay eggs, some species are ovoviviparous and the eggs hatch inside the reproductive tract of the female that then gives birth to larvae. Beetle larvae are very variable in appearance but all have toughened head capsules with forward facing biting mouthparts and short antennae having a maximum of four segments.

7. Thrips

Thrips that transmit plant viruses belong to the family Thripidae in the Order Thysanoptera meaning 'bristle wings'. These insects are very small, 1-2 mm long, slender and difficult to recognize. A characteristic feature of these insects is the absence of the right mandible. The left mandible is elongated and pointed and the maxillary *stylets* are elongated. Two maxillae fit together to form a tube with a sub-apical aperture. Their wings, when fully developed, are narrow and membranous with a fringe of long hairs around the margins. The antennae have six to nine segments and antennal segments three and four bear sense cones that are simple or forked. There may be a claw-like appendage. They feed by sucking the contents of the sub-epidermal cells of the host plants. Adults are winged and active. Larvae are largely inactive. Winged adults disperse over long distance by wind. They are polyphagus and feed in a variety of cell types and on various parts of the plant (Mound 1996).

Identification of thrips at the specific level is difficult due to the wide range of intra-specific variations in size and colour. The most widely known thrips species is *Thrips tabaci* Lindemann (Figure 5.7). It has a very wide host range and feeds on more than 140 species from over 40 families of plants and reproduces parthenogenetically.

Groves *et al* (2001) studied the over-wintering of *Frankliniella fusca* on winter annual weeds. The population was found to differ greatly among host plant species. The mean per plant population of *F. fusca* averaged 401, 162 and 10 thrips per plant on *Stellaria media*, *Scleranthus annus* and *Sonchus asper* during peak abundance in May. Adults collected from plant hosts were predominantly *brachypterous* throughout the winter and early spring, but *macropterous* forms predominated in late spring.

Figure 5.7 Thrips (*Thrips tabaci*)

Biology and Life Cycle

An adult *Thrips tabaci* normally lives around 20 d and there may be several generations in a year. Ohnishi *et al* (1999) studied the biology of *Thrips setosus* Moulton.

Most vector thrips species deposit their eggs into plant tissues and the eggs hatch after 2-3 d, depending on temperature and plant host. There are two feeding larval stages that are followed by two non-feeding pupal stages and an adult stage. The life cycle takes about 15-30 d from egg to adult depending on temperature (German *et al* 1992, Ullman *et al* 1992).

8. Mealy Bugs

Some members of the mealy bugs belonging to the family Pseudococcidae are vectors of viruses and feed on the phloem tissues of plants. The bodies of these insects appear to be dusted by white powdery material (Figure 5.8). They are elongate oval in shape with distinct segmentation. Antennae are present and legs are well-developed. However, they are not very mobile and usually move as a result of contacts between plants. *Nymphs* are more mobile than the adults and are often called crawlers.

Mealy bugs exhibit remarkable sexual dimorphism. Females are normally wingless, males are small and delicate, have a single pair of wings and have

Figure 5.8 Mealy bug

no mouthparts. Some mealy bugs may have three *nymphal* stages followed by a *pupal* stage. Lotfalizadeh *et al* (2000) reported the duration of the average incubation period and different stages (first, second and third larval *instar*) including *pupal* stage and total developmental time.

Biology and Life Cycle

At maturity, many species of mealy bugs secrete *ovisacs* in which eggs are deposited. In some species viviparous or ovoviviparous reproduction is also found. In those species producing eggs, fecundity may range from 152 to 356 per mature female. *Ovisacs* may lie entirely beneath the body, forming a pad, while in others they may be longer than the insects themselves. The incubation period may last up to 3-4 d. *Nymphal* stages for males and females may be four, and three instars, respectively. Completion of *nymphal* development of males and females normally takes 20 or 28 d respectively.

9. Leaf Bugs

Leaf bugs or plant bugs belong to the family Miridae. These are delicate small to medium-sized elongate insects. The antennae and rostrum of these insects are three-segmented. *Ocelli* may be absent; *tarsi* are three-segmented. Females have well-developed ovipositors. These bugs feed on plants by sucking sap through *stylets*.

Ash grey leaf bugs belong to the family Tingidae but these are with less lace-like sculpturing on the body than mirids.

Piesma quadratum Fieb. Heteroptera: Piesmatidae (beet leaf bug) has one generation in a year. Its occurrence is characterized by 2-3 periods when they are frequent every 9-11 y in Poland (Korez 2001).

10. Mites

Mites are very small, less than 1.0 mm in length, belonging to the Class Arachnida. Some species belonging to the families Eriophyidae and Tetranychidae are vectors (Figure 5.9). Eriophyid mites are worm-like animals and the adults usually measure around 0.2 mm in length. They have four legs with four rays of feathered claw. They have piercing and sucking mouthparts. The *stylets* are minute and feed on one or two layers of cells below the leaf surface of leaves, buds and other tender parts of plants. Populations may increase very rapidly in optimal conditions on susceptible hosts. Heavily infested leaves show stippling and bronzing, leaf drop and twig dieback. Mites have very limited mobility but they can be carried from plant to plant by wind and on plant debris.

Biology and Life Cycle

It takes 7-10 d to complete the lifecycle of eriophyid mites and adults normally live for 7-10 d at temperatures of 25°-30°C. Reproduction is mostly parthenogenetic. Males are normally rare.

Tetranychid mites lay eggs on the surface of leaves. The life cycle consists of six-legged larvae at the immobile stage, eight-legged ones at the mobile stage, *protonymph*, *deutonymph*, and adult males and females. In warm condition (30°C), the life cycle may be completed within 12 d and there may be 10-12 generations per year. They may not develop if the temperature is below 19°C.

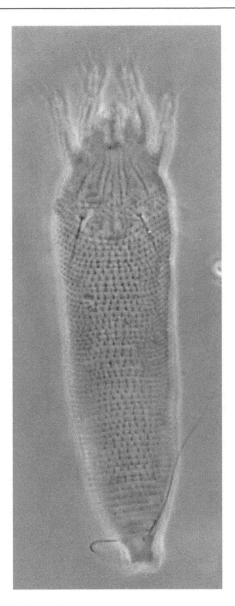

Figure 5.9 Mite (*Aceria* sp.)

In some species eggs are laid singly in a line on the upper surface of the leaf mostly along the midrib near the lateral veins. Freshly laid eggs are transparent, flat or disk-shaped measuring about 0.15 mm in diameter. They later become milky white and then dark brown. When newly hatched the larvae are yellow, slightly globular with a smooth dorsal surface measuring 0.17 mm in length and 0.12 mm in width. After emergence the *protonymph* moves on the upper surface of the leaf and remains active for a day, then enters the quiescent stage that lasts for 24 h. When the *deutonymph* emerges it resembles the *protonymph* except that it is larger and has genitalia. Marked sexual difference can be seen at the end of this stage. The female *deutonymph* has a broad and flattened *ophostoma* whereas that in the male is narrow and pointed. The mature *deutonymph* enters a quiescent phase before finally molting into an adult (Yadav *et al* 2003).

11. Fungi and Protista

Fungal vectors belong to the families Olpidiaceae and Synchitriaceae in the Order Chytridales and virus vectors in the Protists are in the family Plasmodiophoraceae in the Order Plasmophoridales. Chytridales contain *zoosporic* fungi, the form and structure of which are the simplest of all fungi. They have a vegetative body known as a *'thallus'* which is monocentric, and holocarpic without rhizoids. In the Olpidiaceae the vegetative body develops reproductive structures, sporangia or gametangia, to produce uninucleate zoospores bearing a single basal flagellum. There may be fusion of the gametes to produce a zygote that develops into resting sporangia that may survive in soil for a long period (Figure 5.10).

Species of the Synchitriaceae are endophytic with a holocarpic thallus that divides into several reproductive organs embedded in a common membrane forming a *sorus*. Sexual organs are either *sporangia* or *gametangia*. Sporangia are *inoperculate* and sexual reproduction is by the fusion of *isogametes* that results in the formation of resting sporangia.

Species of the Plasmodiophorales have a plasmodium and motile cells and the Plasmodiophoraceae produce resting spores, zoospores, plasmodia and zoosporangia. Resting spores germinate to release zoospores that penetrate into the host cells and ultimately develop into a plasmodium that cleaves and develops multinucleate portions, each of which becomes surrounded by a membrane and develops into a zoosporangium. Zoospores are released outside the host via exit papillae on the sporangium and through pores in the host cells.

12. Nematodes

Nematode vectors belong to the Order Dorylaimida and fall within two families Longidoridae and Trichodoridae. These are soil-inhabiting ecto-parasites that feed on roots. Two genera of Longidoridae (*Longidorus* and *Xiphinema*) and two genera of *Trichodoridae* (*Trichodorus* and *Paratrichodorus*) are known to

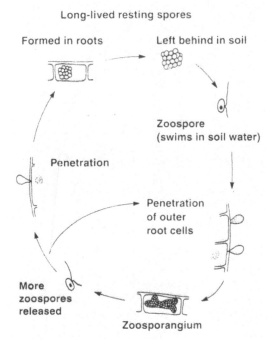

Long-lived resting spores

Formed in roots   Left behind in soil

Zoospore
(swims in soil water)

Penetration

Penetration
of outer
root cells

More
zoospores
released

Zoosporangium

Figure 5.10 Life cycle of a fungus vector (*Polymyxa graminis*) (Adams 2002)
With permission from the Rights Department, Elsevier LTD

be vectors (Figure 5.11). *Longidorus* is prevalent in temperate regions whereas *Xiphinema* is commonly found in the tropics. Adults of *Longidorus* and *Xiphinema* are 3 to 10 mm long. The mouthparts of all the vector nematodes are adapted for probing and consist of a single central *stylet*. The *stylets* are tubular and hollow; the distal part of the *stylet* is known as *odontostyle* and the proximal or basal region is the *odontophore*. The length of *odontostyle* and the *odontophore* together is around 200 μm in length. The *stylet* of the nematodes can penetrate to the xylem of plants. Tiefenbrunner *et al* (2002) developed a software tool as an aid to the identification of species of *Longidorus* Micoletzky 1922 (Nematoda: Dorylaimoidea).

Adults of *Trichodorus* are around 1.0 mm long. They have a *stylet* that is a ventrally curved solid tooth (around 50 μm long) known as *onchiostyle*. They feed on epidermal cells. The difference between *Trichodorus* and *Paratrichodorus* is apparent when they are fixed in heat-relaxed or acid fixatives when the cuticle of the fixed *Paratrichodorus* shows abnormally strong swelling. Nematodes may live from several months to several years. There may also be ecological intra- and inter-specific variants and some may also reproduce parthenogenetically (Chen *et al* 2004).

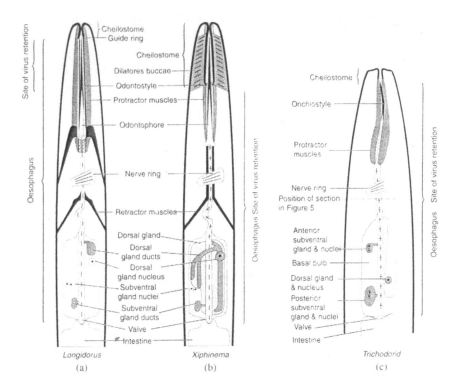

Figure 5.11 Schematic representation of the morphological structure of the anterior ends of *Longidorus*, *Xiphinema* and Trichodoridae showing the feeding apparatus, glands, and ducts in the oesophageal (pharyngeal in trichodoris) bulb, and site of the virus retention in (a) *Longidorus*; (b) *Xiphinema*; and (c) Trichodoridae vector nematodes (Macfarlane *et al* 2002). With permission from the Rights Department, Elsevier LTD

B. VECTORS: THEIR RELATION WITH VIRUSES

Vector transmission is a specific event in the virus life cycle. Virus-encoded determinants specifically interact with the vector, thereby facilitating virus transmission, and various plant virus utilize different but specific vectors to facilitate their spread. Diverse organisms particularly invertebrates are recognized as vectors for various plant viruses.

Vectors belong to 12 groups of invertebrates (Table 5.1). Of these, aphids transmit more viruses, 54% of those known, than all the other groups together. The second largest number of viruses is transmitted by whiteflies, 16-17% of those known, whereas all other groups transmit 5% or fewer. Also many aphid species are known to transmit viruses, whereas in other groups, very few species transmit.

Research on the relationships between plant viruses and their vectors has been influenced by where the earliest studies were done and the virus vector combinations studied. As knowledge has increased various descriptions have been used for these interactions.

Much of the early work was done on aphid transmission in Europe and North America and leaf and planthoppers in South-East Asia. This was a reflection of the relative importance of the vector groups, in these areas, on crops and plants of agricultural significance. In Europe and North America the crops were cereals, potatoes and sugarbeet and in South-East Asia rice. Most of the nomenclature now used was originally developed for aphid/virus interactions and tries to link the nature of the transmission process and the sites where viruses are in the vector. Such relationships and the terminology have been extended to other vector groups but this has not always been appropriate.

The first characteristics described were the length of time that aphids took to acquire and transmit virus and how long they retained the ability to transmit (Watson and Roberts 1939). The term used to describe the combination of these characters was 'persistence' and resulted in the separation of two basic groups of viruses: those that were capable of being acquired and transmitted in a time measured in a maximum of a few minutes, the 'non-persistently transmitted viruses', and those that required several hours or days to be transmitted, the 'persistently transmitted viruses'. In addition the persistently transmitted viruses are not lost when the vector molts as they are in non-persistent transmission. More recently it has become common to use the term to describe the virus, e.g. a non-persistent or persistent virus, but it is preferable that the term be used to describe the process.

A further characteristic of the two main types of transmission is the location of the virus in the host and their ability to be transmitted mechanically. The non-persistently transmitted viruses are nearly always located in the superficial tissues of the plant, the first two or three layers of parenchyma under the epidermis. Because of this location they can also be transmitted mechanically, as this damages the cells in which the virus can multiply and allows the virus to enter. The persistently transmitted viruses are usually restricted to the vascular tissue, nearly always the phloem, and cannot be mechanically transmitted. Subsequent work by Sylvester (1956) confirmed the distinction but also identified another type of transmission with properties somewhere between the non-persistently and persistently transmitted types and this was described as semi-persistent transmission.

The work also tried to identify the sites where the virus was present in the vector and thus 'the inferred route of transport' . This resulted in the description of viruses as stylet-borne, approximately corresponding to the non-persistently transmitted viruses, and circulative (Kennedy *et al* 1962), corresponding to the

Table 5.2 Some Properties of Viruses having Different Virus-Vector Relationships

Property	Non-circulative		Circulative	
	Non-persistent	*Semi-persistent*	*Non-propagative*	*Propagative*
Pre-acquisition fasting	Enhances acquisition	No effect	No effect	No effect
Acquisition time	< 1.0 min	> 30 min	Hours/days	Hours/days
Latent period	Nil	Nil	Hours/days	Hours/days
Persistence in vectors	~1.0 h	> 100 h	Days	Life
Retention after molt	No	No	Yes	Yes
Transovarial transmission	No	No	No	Yes
Mechanical transmission	++	+	+	No

persistently transmitted viruses. Both classifications have their merits and are used in the current literature. However, the persistence of a virus is of greatest significance when dealing with disease epidemiology and 'the inferred method of transport' when trying to understand the mechanisms involved in virus transmission. The routes and mechanisms of transmission will be considered in detail when discussing each vector group.

Other categories have also been identified. A few plant viruses also multiply in their vectors. These viruses are all persistently transmitted and circulative within the vector and are also described as propagative. Propagative viruses are usually transmitted to the progeny, whether asexually or sexually produced, but for some, the evidence for this is absent although they do appear to multiply in the parent vector. As evidence increases as to the location of viruses within their vectors other terms have been introduced such as foregut-borne (Nault and Ammar 1989). However, for very few viruses has any evidence been produced for their location in vectors (Plumb 2002) and for most an assumption of location is made based on other common characteristics. The readily observed, or measurable, properties associated with the various categories of transmission are summarized in Table 5.2.

Major parameters involved in specificity and virus-vector relationship are the properties of the vector, particularly the structure of the mouth-parts, quality of the saliva they produce at the time of feeding, their feeding behaviour and, ultimately, the recognition mechanism present in the vector for the concerned virus. Host-related factors also contribute to this process. Properties of the viruses that contribute to the relationship are the coat proteins and the expression of viral genes, depending upon the host-vector-virus combination.

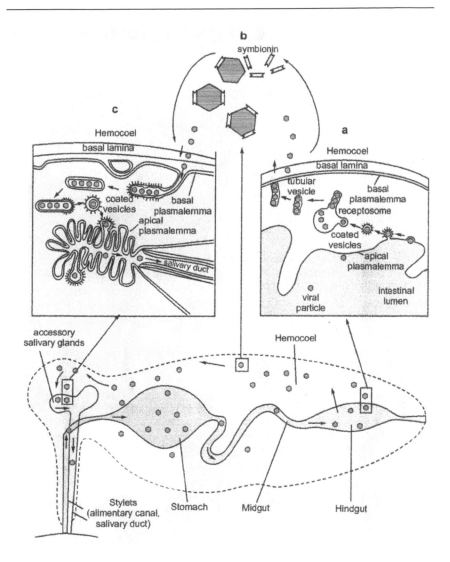

Figure 5.12 Movement of the circulative non-propagative virus in the body of an aphid vector and its intracellular transport. The virus after its ingestion follows the alimentary tract up to the intestine (midgut and hindgut). After crossing the intestinal epithelium, the virus reaches the hemocoel or general cavity where it interacts with symbionin and subsequently reaches the accessory salivary glands (Astier *et al* 2001, modified from Gildow 1982, 1985 and Gildow and Gray 1993; Source INRA 2001, France, with permission)

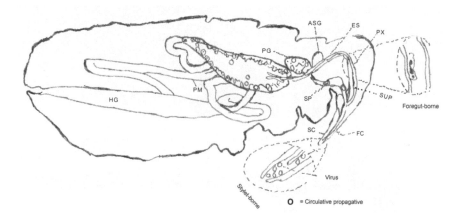

Figure 5.13 Models for movement of stylet-borne (non-persistent, non-circulative) and circulative non-propagative viruses in their vectors

FC = food canal, SC = salivary canal, SUP = foregut-borne sucking pump, PX = pharynx, AM = anterior midgut, PM = posterior midgut, ASG = accessory salivary gland, PG = principal salivary gland, SP = salivary pump, HG = hindgut (Hull 2002; Gray and Rochon 1999)

With permission from the Rights Department, Elsevier LTD

Conventional Relationship

The conventional relationship between vectors and viruses has already been described above. In the current literature viruses are usually differentiated into two broad groups: *foregut-borne* or *non-circulative* and *circulative*. The movement of circulative non-propagative virus in the body of the vector is given in Figure 5.12 and the movement of circulative propagative and foregut- borne viruses are given in Figure 5.13. The descriptive terminology for these transmission relationships was generally accepted and widely used among plant virologists and entomologists, but as more virus-vector interactions were studied in detail, more variations were found. Clearly these original terms could not describe all of the virus-vector relationships for Hemipteran transmitted plant viruses (Ng and Falk 2006).

Vectors become almost immediately infective when they acquire foregut-borne or non-persistent viruses. Infectivity is enhanced, and the retention period of the virus is increased, if the vectors are starved before acquisition. Acquisition and inoculation thresholds (transmission threshold) are only a few seconds or minutes. Prolonging acquisition and inoculation feeding for much more than a few minutes decreases vector infectivity.

Vectors retain semi-persistently transmitted viruses for several hours. Preacquisition fasting of vectors does not influence their infectivity and acquisition and inoculation thresholds are of several minutes. Prolonging inoculation feeding does not affect infectivity. These viruses are taken up by the *stylet* and probably retained in the food canal and foregut of the vector.

Vectors need to feed for at least 30 min both to acquire and inoculate *circulative* viruses. They also require a latent period, of 12 h or more, to become infective. Infection feeding involves ejection of virus in the saliva from the maxillary saliva canal. Viruses, after acquisition, are transported across vector membranes in the piercing-sucking insects. They enter the hemolymph during their passage along the digestive tract through midgut or hindgut and move from the hemolymph to the accessory salivary gland and pass into the saliva and are then ejected through the salivary duct into the host. In chewing insects the characteristics of transmission are different (see below) (Wang *et al* 1992). Ng and Falk (2006) illustrated the virus-vector interactions mediating non-persistent and semi-persistent transmission of plant viruses.

Circulative and propagative viruses multiply in both organisms. Propagative relationships include a long-term association of virus with the vector and the presence of the virus may have adverse effects on the insect hosts, particularly their longevity and fecundity.

B1. Relationship between Aphids and Viruses they Transmit

How aphids transmit viruses is closely linked to their behaviour. While in flight, aphids cannot distinguish and identify suitable host plants upon which they can feed and reproduce. On first landing on an unknown plant an aphid probes the leaf surface, penetrating both the epidermal cells and one or two layers of parenchyma, before sampling the cell contents. They do this by sucking cell contents into the food canal of the stylet. The cell contents are then tested to determine if the host is a suitable food source. This activity is known as 'probing' which may last for a few seconds or a minute or two, but rarely much longer. During a period of brief 'probing' on a virus-infected plant the epidermal cell plasma membranes can be punctured providing access of the stylets to cell contents and virions. Acquired virions can then be inoculated during subsequent probes on healthy plants. Inoculation assays monitored using electrical penetration graph analyses have confirmed that these brief shallow probes and cell membrane punctured by aphid stylets result in successful virus inoculation to plants (Collar *et al* 1997, Martin *et al* 1997, Powell *et al* 2006). Thus, even the plant is not a suitable host for the aphid, but is susceptible to the virus, transmission can result during aphid probing. Because the potential for non-persistent transmission depends on what might only be an ephemeral association of the host plant and aphid, it is often found that aphids can transmit virus to hosts on which they do not feed and on which they are rarely found (Harrington *et al* 1986) and a virus may have many vectors, although the relationship between each virus and vector is still intimate and biologically complex.

If the plant is not recognized as a food host the aphid may make another probe, on the same leaf, a different part of the same plant or move by walking or flying to a new plant and repeat the probing process until a suitable food source is located. If the aphid recognizes a suitable host it reinserts its stylets and, supported by a salivary sheath, they penetrate to the phloem from where feeding begins. Feeding can last for several hours or days and its length depends upon the continued suitability of the host and external conditions. Normally if aphids select and stay on a host plant, they are less likely to move and probe, thereby reducing the amount of virus transmission.

Once ended, feeding may start again on the same plant or the aphid may walk or fly to a new plant where the cycle or probing and feeding can begin again. Probing leads to the non-persistent transmission of viruses if the vector and the virus are compatible and the host is susceptible, whereas feeding can result in the persistent transmission of viruses if the appropriate criteria are fulfilled.

The earlier theories representing aphids as 'flying pins' has long since been superseded (Watson and Plumb 1972). To transmit a virus in a persistent manner an aphid has to feed on the plant host as the site of virus infection and multiplication is the phloem. Thus, potential virus vectors are very likely to be found on a host and generally viruses transmitted persistently have fewer vectors and a narrower host range than those transmitted non-persistently. It is also possible that a single aphid can transmit both non-persistently and persistently transmitted viruses to or from the same plant depending on its behaviour pattern.

Aphids transmit a wide range of viruses in 16 genera (Table 5.3) and show all forms of transmission and virus-vector relationships and the viruses transmitted by them differ in morphology, nucleic acid and genome structure. In some cases the presence of one virus being necessary for the transmission of another. For example *Bean leaf roll virus* (BLRV) helps the transmission of *Bean yellow vein banding virus* (BYVBV) by *Acyrthosiphon pisum*, and *Carrot red leaf vrus* (CRLV) is necessary for the transmission, by *Cavariella aegopodii*, of *Carrot mottle virus* (CMoV) Pirone and Megahed (1966).

Table 5.3   Viruses Transmitted by Aphids and their Transmission Characteristics

Virus	Particle shape and type of the nucleic acid	Vector species	Mode of transmission
1. Caulimovirus: *Cauliflower mosaic virus*	Isometric; dsRT-DNA	*Brevicoryne brassicae*	Semi-persistent

		(L.); *Myzus persicae* (Sulzer)	
(infects > 25 species)			
2. Nanovirus: *Subterranean clover stunt virus*	Isometric; ssDNA	*Acyrthosiphon pisum* (Harris); *Aphis craccivora* Koch; *Aphis fabae* Scop	Persistent: circulative; non-propagative
3. Babuvirus: *Banana bunchy top virus*	Isometric; ssDNA	*Pentalonia nigronervosa*	Persistent: circulative; non-propagative
4. Cytorhabdovirus: *Lettuce necrotic yellows virus*	Bacilliform; enveloped; ssRNA	*Hyperomyzus lactueae*	persistent; circulative; non-propagative
5. Cucumovirus: *Cucumber mosaic virus* (infects > 1200 species in 100 families)	Isometric; ssRNA	*Myzus persicae*; *Aphis gossypii*	Non-persistent
6. Alfamovirus: *Alfalfa mosaic virus*	Bacilliform; ssRNA	*M. persicae*; *A. gossypii*; *Acyrthosiphon kondoi*	Non-persistent
7. Fabavirus: *Broadbean wilt virus*	Isometric; ssRNA	*A. fabae*; *M. persicae*	Non-persistent
8. Potyvirus: *Potato virus Y*	Filamentous; ssRNA	*M. persicae*; *A. fabae*	Non-persistent
9. Macluravirus: *Maclura moaic virus*	Filamentous, ssRNAM	*M. persicae*	Non-persistent
10. Sequivirus: *Parsnip yellow fleck virus*	Isometric; ssRNA	*Cavariella aegopodii* (Scop.); *C. pastnacae* L.	Semi-persistent
11. Closterovirus: *Beet yellows virus*	Filamentous; ssRNA	*M. persicae*; *A. fabae*	Semi-persistent
12. Luteovirus: *Barley yellow dwarf virus* PAV	Isometric; ssRNA	*Ropalosihum maidis* (Fitch.); *R. padi*	Persistent; circulative; non-propagative
13. Polerovirus: *Potato leaf roll virus*	Isometric; ssRNA	*M. persicae;* *Macrosiphum euphorbae*	Persistent; circulative; non-propagative
14. Enamovirus: *Pea enation mosaic virus*	Isometric; ssRNA	*A. pisum*; *M. euphorbae*	Persistent; circulative; non-propagative
15. Carlavirus: *Carnation*	Filamentous;	*M. persicae*	Non-persistent

latent virus	ssRNA		
16. Umbravirus: *Carrot mottle virus*	Undefined; ssRNA	*C. aegopodii* (Scop); *M. persicae*	Persistent

Some viruses show bimodal transmission, i.e. with the characteristics of both non-persistent and semi-persistent transmission, depending upon the mode of feeding of the vectors, as for example *Pea seed-borne mosaic virus* (PSbMV). *Cauliflower mosaic virus* (CaMV) was also thought to have bimodal transmission but Palacios *et al* (2001) provided evidence for its semi-persistent transmission and it was found to be retained in the cuticle lining of the aphid stylet and/or anterior part of the alimentary tract but did not circulate through the aphid hemocoel.

(a) Non-persistent Transmission

Aphids transmit several virus genera in a non-persistent manner (Table 5.3). The most detailed studies of the transmission process have been made for the Potyviruses and Cucumoviruses transmitted by *Myzus persicae*, one of the most effective vectors.

What Determines Transmission

The importance of the capsid protein (CP) of the virus particles in the transmission process was demonstrated using cucumoviruses (Gera *et al* 1978, Chen and Francki 1990). It was also shown, for the CP of *Cucumber mosaic virus* (CMV) that an amino acid change from leucine to proline and from threonine to alanine determined transmissibility by *Aphis gossypii*; the alanine to threonine change was also found to be linked to transmission in another CMV isolate.

Studies of the transmission of potyviruses has been particularly rewarding in understanding the mechanisms involved in non-persistent transmission by aphids (Pirone and Perry 2002). Because non-persistently aphid transmitted viruses are usually readily transmitted mechanically this is often the preferred method for their maintenance, but can result in the loss of aphid transmissibility. In an isolate that was no longer transmitted by aphids, a mutation was noted in a conserved area of aspartine-alanine-glycine (DAG) near the N-terminal portion of the CP. Support for the suggestion that this change was linked to aphid transmission came from the work of Atreya *et al* (1990, 1991) with *tobacco vein mottling virus* (TVMV) but it was studies by Blanc *et al* (1997) that demonstrated that the reason for this change in transmissibility was the binding, or otherwise, of mutations in the CP to the helper component, which is vital to transmission. It was also found that changes in amino acids, especially the acidic glutamine or aspartine, or the basic lysine, in the region immediately following the DAG sequence abolishes transmission (Atreya *et al* 1995) and in the region immediately following DAG decreases transmissibility (Lopez-Moya *et al* 1999).

As indicated above, not only is the correct configuration of amino acids necessary in particular parts of the CP but there is also an essential requirement in the potyviruses for a virally encoded helper component (HC) (Kassanis and Govier 1971a and b; Govier and Kassanis 1974). The HC proved to have a subunit molecular mass of 58 kDa but in the functional form, a molecular mass of 100-150 kDa; the product is known as HC-Pro. Viruses showing 'Helper' strategy for transmission may be defined as HC-transcomplementation that have the consequences at the level of virus population genetics and evolution (Froissart *et al* 2002).

Cauliflower mosaic virus also requires a second component, in addition to the CP, for transmission (Lung and Pirone 1974, Blanc *et al* 1993). Leh *et al* (2001) demonstrated the requirement of two viral non-structural proteins, ORF II and ORF III, with products (P2 and P3) for transmission of CaMV by aphids. An interaction between a C-terminal domain of P2 and the N-terminal domain of P3 plays a crucial role in the formation of CaMV transmission-complex by serving as a bridge between P2 and the virus particle. Drucker *et al* (2002) determined that the viral protein P2 regulates the interactions between CaMV and its aphid vector that binds the virus to the aphid stylets. They suggested that P3 must be previously bound to virions so that the attachment of P2 will allow the aphid transmission. The P2-P3 complex exists in the absence of virions but it is non-functonal in transmission. Vectors cannot sequentially acquire P3 and virions. The evidence suggests that the transmissible complex P2:P3; virion forms primarily in aphids.

Burrows *et al* (2006) established the genetic control of aphid transmission of luteoviruses in the insects and the tissue-specific barriers to virus transmissons are not genetically linked. When sexual forms of two genotypes of the aphid *Schizaphis graminum*, one a vector, the other a non-vector of two viruses that cause barley yellow dwarf disease (BYDV-SGV) and cereal yellow dwarf virus-RPV, polerovirus were mated to generate F1 and F2 populations, it was indicated that the transmission phenotype is under genetic control and the parents are heterozygous for genes involved in transmission. Individual hybrid genotypes differed significantly in their ability to transmit each virus indicating that in addition to a major gene, minor genes can affect transmission of each virus independently. Gut and salivary gland associated transmission barriers were identified in the nonvector parent and some progeny, while other progeny possessed only a gut barrier or a salivary gland barrier. Hemolymph factors do not appear to be involved in determining the transmission phenotype.

Where in the vector is transmissibility determined?
Pirone and Perry (2002) have described the work leading to the present day understanding of how potyviruses are transmitted in a non-persistent manner.

The earlier theory (Govier and Kassanis, 1974) was that HC functioned as a bridge (see also above for CaMV) between virus and vector, with specific sites of attachment. Evidence in support of this theory came 20 yr later (Ammar 1994) using electron microscopy and immunogold labelling. Changes in two highly conserved regions of the HC-Pro, the KITC (lysine-isoleucine-threonine-cystine) motif and the PTK (proline-threonine-lysine) motif of TVMV were found to be associated with transmission (Atreya *et al* 1992, Atreya and Pirone 1993, Huet *et al* 1994)) and Pirone and Perry (2002) produced a diagrammatic illustration of how these components may link in the transmission process. When KITC was present HC was found in the food canal of 60% of aphids, whereas no HC was found in any aphids given access to a HC with the EITC motif (Thornbury *et al* 1990). Further results suggested that it was not the HC-CP binding but the stylet interaction that was disrupted (Pirone and Perry 2002). See also Ng and Falk (2006).

Inoculation

There is still uncertainty about how the virus is transferred from the stylets of one aphid to another plant. The ingestion-egestion theory of Harris (1977) would, potentially, involve all the viruses retained in the food canal up to the cibarial pump which helps to suck plant sap into the aphid's gut. If the pump's action is reversed during the probing process, when some sap has been taken up, the virus previously ingested with the sap would be egested into the plant. A more recent alternative hypothesis (Martin *et al* 1997) is based on the electrical changes during the host penetration of probing and feeding, and suggests that the transmissible virus is retained at the tip of the stylets, where the food and salivary canals are coincidental, and is inoculated into plants during salivation.

Vector specificity

As already noted non-persistently transmitted viruses may be transferred by many vector species. However, it is clear that specificity does exist. Pirone (1981) and Sako *et al* (1984) investigated the role of HC in specificity. Various combinations of purified viruses and HCs were tested using membrane feeding techniques and a range of transmission efficiencies, including non-transmission, were reported (Sako and Ogata 1981, Lecoq and Pitrat 1985, Flasinski and Cassidy 1998, Wang *et al* 1998, Pirone and Perry 2002). Homologous combinations of virus and HC always resulted in the most efficient transmission. The two hypotheses advanced to explain that the results were due to differences in the properties of the insect epicuticle within the food canal and the differential effects of saliva; a combination of these two is equally possible.

The cucumoviruses do not require a HC for their transmission, but changes in CP amino acids do affect transmission efficiency (Perry *et al* 1994, 1998) and it is possible that this is linked to direct binding of the CP to the vector mouthparts.

(b) Persistent Relationship

The persistent relationship between the aphids and the viruses they transmit is fundamentally different from that found between aphids and non-persistently transmitted viruses. The biological relationship between the virus and the vector is very specific as the viruses are to be acquired and circulated inside the body of the vectors and some also multiply in vector tissues. Two types of relationship may exist. Viruses may penetrate the vector tissues and produce reservoirs of virus that are sufficient for the aphids to remain viruliferous for some time and potentially for their life, or the viruses, after their entry into the vectors, infect them, multiply and spread to many tissues of the vectors. Whichever is the method of interaction of the viruses in the vectors the major issues in the persistent relationship are: Virus-vector-host recognition system or specificity" and "Circulation and movement of the virus inside the body of the vector and their release."

Specificity

The specificity of persistent viruses varies with the isolates (Gray 1999) and is dependent on factors related to both viruses and vector. The virus factor is principally the virus particle proteins. Rochow (1970) first drew attention to this possible explanation and showed that the capsid proteins of the viruses are the determinants of the specificity. When *S. avenae*, the vector of BYDV-MAV was fed on purified BYDV-MAV, treated with the MAV antiserum, transmissibility was abolished. However there was an apparent change in specificity when the vectors fed on virus mixtures. *R. padi* does not normally transmit BYDV-MAV but when this aphid was fed on BYDV-MAV antiserum treated BYDV-MAV and a CYDV-RPD mixture, *R. padi* transmitted both the viruses indicating the changes in the transmission determinants in BYDV-MAV in presence of CYDV-RPD. *R. padi* can also transmit BYDV-MAV when it is fed on a mixture of BYDV-MAV and BYDV-RMV (Rochow 1975). Changes in the transmission in these cases are probably due to transcapsidation of the non-transmissible RNA by the capsid protein of the transmissible isolate. Thus BYDV-MAV RNA is encapsidated by the protein of the other virus. Subsequently it was found that similar transcapsidation is common in several luteoviruses that facilitate the transmission of the RNA of the other luteoviruses, such as the helper dependent umbraviruses, luteovirus-associated RNAs or satellite RNAs (Falk and Tian 1999).

Assuming that different regions of the virus particles may be important in recognition events at different stages of the transmission process, the role of RT proteins have been investigated because BWYV or BYDV-PAV particles lacking RT proteins are not transmissible (Brault *et al* 1995, Chay *et al* 1996, Bruyere *et al* 1997). Some residues obtained by point mutation in the N-terminal part of the BWYV RT domain were found to be required for transmission (McGrath *et al* 1996, Brault *et al* 2000). This observation confirms the association of RT protein in virus transmission. However, the involvement of the N-terminal region of the RT protein is not universal. Jolly and Mayo (1994) found the transmissibility factor in the C-terminal part of the RT protein in PLRV and van den Heuvel *et al* (1993) observed the association of the coat protein of PLRV for its transmission.

Acquisition, circulation and movement

Acquisition of virus by the vector begins when infective sap is taken from the phloem sieve elements or companion cells of infected plants. Once the virus is inside the aphid it encounters several barriers to its circulation and movement. The first obstruction is the hindgut that has to be crossed if the virus is to accumulate in the hemocoel. However the ability to cross this barrier does not signify that transmission will occur as several non-transmissible luteoviruses can cross the hindgut and accumulate in the hemocoel. When *S. avenae*, a non-transmitter of CYMDV-RPV was fed on a plant infected with this virus, particles crossed the gut, passed through the gut lumen and accumulated in the hemocoel (Gildow 1993). However Rouzé-Jouan *et al* (2001) observed that certain clones of *M. persicae* that are inefficient vectors of PLRV-PAT became efficient transmitters when the virus was microinjected into the hemocoel of those clones and thus avoided the gut barrier.

As indicated above accumulation of virus particles in the hemocoel of an aphid does not always mean that the virus will be transmitted. *Macrosiphum euphorbiae*, an inefficient vector of PLRV, and *M. persicae* an efficient vector for the same virus, when fed on the same infected plant take up the same amount of virus but differ in their transmission ability (Tamada and Harrison 1981). The reason for such a difference is yet to be explained.

The differential interactions between the viruses and their vectors suggest that the transcellular movement of the virus particles inside the vector is very complex and involves possibly several different physico-chemical processes. Viruses normally enter through the aphid gut epithelium by receptor-mediated endocytosis that is initiated by the interaction of the virus coat protein with a cell receptor (Gildow 1993). Virus is recognized at the gut cell apical plasmalemma (APL), and becomes attached to the membrane. This initiates the process of the virus being engulfed into coated pits that bud off from the APL and form virus containing coated vesicles. These vesicles transport the virus to 'receptosomes' (large uncoated vesicles). Tubular vesicles are formed on the receptosomes and contain linear aggregates of virus. Garret *et al* (1996) observed the accumulation of PLRV in tubular vesicles and lysosomes in *M. persicae* fed on the PLRV-infected plants. This virus aggregates fuse with the basal plasmalemma (BPL) which allows the release of the virus from the cell. It then diffuses through the gut basal lamina into the hemocoel.

A protein produced by endo-symbiotic bacteria harboured in the cells lining the hemocoel may bind the virus particles. This binding may play an important role in preserving virus particle integrity and allowing them to survive in a transmissible state within the vector. It is an obligate, intracellular symbiotic bacterium of the genus *Buchnera* (for review see Baumann *et al* 1995; Douglas 1998). This bacterium is transmitted maternally and is located in specialized cells and is known as 'bacteriocytes' or 'mycetocytes'. *Buchnera* in the aphid vectors of luteoviruses produces a large quantity of a protein called 'symbionin' with a molecular mass of 63 kDa. van den Heuvel *et al* (1999) observed that a luteovirus-symbionin interaction is essential for the retention of the virus in the hemolymph. Treatment of *M. persicae* with tetracycline abolishes symbionin accumulation in the hemocoel and greatly decreases PLRV or BWYV transmission. Symbionin is also present in *A. pisum* and *R. padi* and binds to the particles of BWYV, CYDV-RPV, SbDV, PLRV, BLRV and BMYV. The mechanism of binding between the symbionin and the virus particle is not yet completely understood. Several authors have suggested a role for the RT protein in this linkage. Mutagenesis experiments showed that PLRV RT protein binds to the equatorial domain of symbionin (Hogenhout *et al* 1998, 2000). Symbionin-virus binding protects the luteoviruses from proteolytic degradation or attack by the insect immune system and is involved in the final stages of movement through the aphid as a virus particle-symbionin complex (see for review Reavy and Mayo 2002).

The virus particle-symbionin complex, after its formation, moves to the salivary glands and there undergoes selective association with the basal lamina of the accessory salivary glands (ASGs) (Gildow and Gray 1994) and penetrates into the plasmolemma of the ASGs (Peiffer *et al* 1997). Non-penetrating and non-transmissible virus particles are generally excluded from the basal lamina (Figure 5.12). Particles of CYDV-RPV or BYDV-PAV cannot penetrate the ASG's basal lamina when they were microinjected to the hemocoel of the non-vector *R. maidis*, and a Japanese isolate of SbDV was excluded from the basal lamina of the non-vector aphid *M. persicae* (Gildow *et al* 2000). Penetration of the basal lamina of the ASGs also is not sufficient for transmission to occur and in certain cases there may be a 'penetrating non-transmissible' interaction, where the virus particles may penetrate the outer half of the basal lamina but fail to penetrate the underlying plasmolemma as was found between BYDV-MAV and its non-vector *M. dirhodum*, and CYDV-RPV and *S. avenae*. In these latter cases, plasmalemma penetrability is the only determinant of aphid transmissibility (Gildow *et al* 2000). This determinant would appear to be a receptor molecule that recognizes some features of the surface of a transmissible virus particle. It is not known what part of the virus particle is involved in this recognition by the receptor molecules on the SAG basal lamina or plasmalemma but the RT protein domain may have a role in this regard, as mutants lacking RT protein are not transmissible, though the virus may accumulate in the hemocoel. It is, however, clear that cellular and molecular mechanisms are involved in the transmission of luteoviruses (Gray and Gildow 2003).

Gildow (1999) developed a model for transcellular transport of luteoviruses in aphid vectors. Recently Liu *et al* (2006) developed a simple wax-embedding method for isolation of aphid hemolymph for detection of luteoviruses in the hemocoel. This method uses warm melted wax to immobilize the aphid. Following the removal of the hind leg, the hemolymph is collected readily. Flushing with RNAs-free water allows for collection of sufficient hemolymph for RNA extraction from individual aphids and the presence of the BYDV was detected in the hemolymph of viruliferous *Rhopalosiphum padi* and *Acyrthosiphon pisum*. This technique provides a useful tool for studying molecular interactions between persistent and circulative plant viruses and their insect vectors.

B2. Relationship between the Leafhoppers, Plant Hoppers, Treehoppers and the Viruses they Transmit

Leafhoppers and plant hoppers do not transmit any virus non-persistently but do transmit three viruses in a semi-persistent manner, and all the other viruses are persistently transmitted. The viruses transmitted by leaf- and plant hoppers are given in Table 5.4.

Table 5.4 Viruses Transmitted by Leaf and Plant Hoppers and their Transmision Characteristics (Nault and Ammar 1989 with modification)

Virus, Type species: particle shape and nucleic acid	Important vector species	Transmission
1. **Curtovirus**: *Beet curly top virus* (isometric-geminate, ssDNA)	*Circulifer tenellus*	Persistent (circulative non-propagative)
2. **Cytorhabdovirus**: *Barley yellow striate mosaic virus* (bacilliform-enveloped, ss RNAs)	*Laodelphax striatellus*	Persistent (circulative propagative)
3. **Fijivirus**: *Sugarcane fiji disease* (polyhedral, dsRNA)	*Perkinsiella saccharicida*	Persistent (circulative non-propagative)
4. **Marafivirus**: *Maize rayado fino virus* (isometric, ssRNA)	*Dalbulus maidis*	Persistent (circulative non-propagative)
5. **Mastrevirus**: *Maize streak virus* (icosahedra-geminate, ssDNA)	*Cicadulina* spp.	Persistent (circulative non-propagative)
6. **Nucleorhabdovirus**: *Potato yellow dwarf virus* (bacilliform, enveloped, ssRNA)	*Acertgallia sanguinolenta*	Persistent (circulative propagative)

7. **Phytoreovirus**: *Wound tumour* virus (icosahedra, dsRNA)	*Agallia constricta, Agalliopsis novella*	Persistent (circulative propagative)
8. **Tenuivirus**: Rice stripe virus (bacilliform, dsDNA)	*Loadelphax striatelus*	Persistent (circulative propagative)
9. Rice tungro bacilliform virus (bacilliform, dsDNA)	*Nephotettix virescens, N. cincticeps, Recilia dorsalis*	Semi-persistent
10. **Waikavirus**: Rice tungro spherical virus (isometric, ssRNA)	*Nephotettix virescens*	Semi-persistent

(a) Semi-persistent Transmission of Viruses

The viruses transmitted semi-persistently by leafhoppers are the *badna virus, Rice tungro bacilliform virus* (RTBV), the *waika viruses* [*Maize chlorotic dwarf virus* (MCDV) and *Rice tungro spherical virus* (RTSV)]. Vectors for RTBV and RTSV are *Nephotettix* spp. and *Recilia dorsalis* (Motschulsky). The transmission efficiency of these species differs; the most efficient is *N. virescens* (Distant). MCDV is transmitted by six species of leafhoppers and the most efficient is *Graminella nigrifrons* (Forbes). The reasons for the differences in the efficiency of transmission of these viruses by different leafhopper species are not yet properly understood but host preference of the vectors certainly has a role.

Acquisition and inoculativity

The acquisition and inoculativity of all these viruses are more or less similar. They require a helper component either to acquire the virus or to become inoculative. Vectors can acquire RTSV or the mixture of RTSV and RTBV within 30 min of acquisition feeding (Inoue and Hirao 1981, Mukhopadhyay 1985). Longer acquisition access periods increase the efficiency of acquisition. The minimum acquisition period depends upon the time required by the vector stylet to reach the phloem tissues of the host where the viruses normally occur. Thus the feeding behaviour of the vectors on different varieties of the hosts and also the environmental conditions may affect acquisition (Mukhopadhyay 1975, Mukhopadhyay and Chattopadhyay 1975, Mukhopadhyay *et al* 1976, Ghosh and Mukhopadhyay 1978).

Nephotettix spp. can directly acquire RTSV from infected plants but RTBV requires a helper component for transmission. These leafhoppers cannot acquire RTBV from an infected plant unless they have previously acquired RTSV or the source plant has a mixed infection of RTSV and RTBV (Hibino 1989). When a leafhopper carrying RTSV is fed on RTSV antisera, its transmissibility of RTSV is lost but it retains its ability to acquire and transmit RTBV (Hibino and Cabautan 1987). This suggests that a helper component is required for acquisition of RTBV derived from the interaction between RTSV and the plant. The helper component is not required for acquisition and infectivity of RTSV.

The sex of the leafhoppers also affects acquisition and inoculativity; females are always more active transmitters than the males (Mukhopadhyay 1985). Dahal *et al* (1990) monitored the feeding behaviour of adult female *N. virescens* that had been starved for 2 h following a 4-d acquisition access period on plants infected with both RTSV and RTBV. They observed four waveform patterns indicating probing, salivation, phloem feeding and xylem feeding respectively and the transmission process was associated with phloem-feeding. The inoculativity of the leafhoppers was lost upon molting (Ling and Tiongco 1979). The localization of these viruses in the body of the vectors has yet to be properly delineated.

The acquisition of MCDV by *G. nigrifrons* and its inoculativity is more or less similar to that found with the RTSV and their *Nephotettix* vectors. *G. nigrifrons* acquires and inoculates the virus in 15 min and the vector can remain infective for 24 h to a few days depending upon the environmental conditions (Choudhury and Rozenkranz 1983).

Inoculation of this virus depends upon a helper component that is not strain-specific. Purified MCDV (WS strain) can be transmitted by the leafhopper to maize plants pre-infected by MCDV (MI strain) and vice versa. Viruliferous *G. nigrifrons* lost the ability to transmit MCDV-MI after feeding for 24 h on healthy plants but retained the ability to acquire and transmit purified MCDV-WS for 36 h, thus confirming the non-strain specific nature of the helper component (Creamer *et al* 1993).

Virus-like particles (VLP) were found embedded in a semi-opaque matrix attached to the cuticular lining of the oesophagus, cibarium, pre-cibarium and occasionally to the inner surface of the maxillary food canal of *G. nigrifrons*. The matrix and not the VLP is found in this vector after a 4-d feeding period following MCDV acquisition. The matrix in which the VLP is embedded may be the helper component required for transmission of MCDV (Ammar and Nault 1991).

(b) *Persistently Transmitted Non-propagative Viruses*

Leafhoppers and plant hoppers transmit viruses in both the circulative non-propagative and circulative propagative ways. Cicadellid leafhoppers transmit 11 viruses in the family Geminiviridae in the circulative non-propagative manner. Most are members of *mastrevirus* genus and one member belongs to the *curtovirus* genus. TPCTV is transmitted in this manner by a membracid treehopper. Acquisition, and the extent of transmission of these viruses, varies with different vector species.

Acquisition and transmission efficiency
Vectors acquire the viruses by sucking the sap from mesophyll and phloem. Important influences on acquisition and transmission are the localization of the

virus in the host tissues, the minimum acquisition access period (AAP), the minimum inoculation access period (IAP) and latency. These parameters vary with the virus vector combinations.

MSV, the type member of the mastreviruses, is localized in the mesophyll and phloem tissues of the infected maize plants. It is transmitted by *Cicadulina mbila*, *C. ghauri* and *C. arachidis* but their AAP, IAP and transmission efficiency differ greatly. AAPs for these vectors are 15 sec, 15 min and 1 h respectively and IAPs are 5 min, 1 h and 1 h (Ammar 1994, Asanzi *et al* 1995). With an increase in AAP the transmission efficiency of *C. mbila* for MSV increases significantly. Much greater transmission results when the AAP is 50 h than when it is 3 h. More virus particles are found in the leafhopper after 50 h AAP than after 3 h. But, with an increase of the post-AAP period, the virus content gradually decreases. Using a calibration curve obtained with purified MSV, a mean value of 0.36 ng of MSV/leafhopper is found after 3 d AAP but the value drops to 0.20 ng 14 d after the AAP, suggesting the non-propagative nature of the virus (Ammar and Nault 2002). Once acquired, this virus moves to the hemocoel via the filter chamber and anterior cells of the midgut by receptor-mediated endocytosis (Markham 1992) and the gut wall appears to be the first barrier for MSV transmission by *C. mbila*. Harrison (1985) observed that the latent period (LP) in this combination and in all geminiviruses is 4-19 h. Horn *et al* (1994) observed similar transmission characteristics of CCDV transmitted by the leafhopper *Orosius orientalis*.

Soto and Gilbertson (2003) determined the distribution and rate of movement of BMCTV. A 1.1 kb BMCTV DNA fragment was amplified from an adult leafhopper and from the organs involved in circulative transmission: digestive tract, hemolymph and salivary glands. The temporal distribution in the leafhopper was determined using insects given different AAPs (1-48 h) on BMCTV-infected shepherd's purse plants. The virus was detected in the digestive tract after all AAPs, in the hemolymph after 3 h or greater AAPs and in the salivary glands after 4 h or more AAPs. The amount of the virus in the hemolymph and the salivary glands increased with the length of AAP. The virus persisted up to 30 d in leafhoppers when given 3-d AAP on BMCTV-infected plants and maintained on maize plants.

In general, coat proteins of the viruses control the transmissibility by leafhoppers. Briddon *et al* (1990) demonstrated the role of the coat protein in vector specificity by gene replacement experiments. When the coat protein gene of the whitefly-transmitted ACMV was replaced by that of the leafhopper-transmitted BCTV, and injected into the leafhoppers, they transmitted ACMV.

The AAP and IAP of the treehopper *Micrutalis malleifera* for the virus TPCTV are both < 1 h and the LP is 24-48 h; retention of inoculativity is positively correlated with the length of the AAP (Simons 1962). Nymphs retain inoculativity after molting. Briddon *et al* (1996) indicated that the genome of TPCTV had features typical of both mastreviruses and curtoviruses. The coat protein was more similar to leafhopper-transmitted geminiviruses and also determines the vector specificity of this virus.

(c) Persistently Transmitted Propagative Viruses

Reo- and Rhabdo-viruses primarily infect animals but three genera of *Reoviruses* (*Phytoreovirus, Fijivirus* and *Oryzavirus*) and two genera of *Rhabdoviruses* (*Nucleorhabdovirus* and *Cytorhabdovirus*) infect plants. It may be that the genomes of these animal viruses had some chance mutations to produce mutants infective to invertebrates (insects) and plants. Similar chance mutations are apparent in insect adapted and plant adapted viruses. It appears that some Reoviruses are insect-adapted viruses but infect plants. 'Leafhopper virus A' (LAV), a member of the genus *Fijivirus* multiplies in its leafhopper vector *Cicadulina bipunctella bimaculata* Evan and transmits through the eggs of the leafhopper. But this virus does not multiply in its host plant in which it only transiently circulates (Ofori and Francki 1985). *Wound tumour virus* (WTV) and *Rice dwarf virus* (RDV) are *Phytoreoviruses* and multiply in their leafhopper vectors, and persist through generations, and also multiply in their host plants. Members of the *Nucleorhabdoviruses* and *Cytorhabdoviruses* also multiply in both their vectors and plant hosts. But in plants, these viruses remain restricted to the phloem whereas they multiply in many insect tissues. These differential sites of multiplication indicate that these viruses are more adaptable to vectors rather than to their plant hosts; the molecular basis of these differential relationships is yet to be properly understood.

The distinctive characteristic of the persistently transmitted propagative viruses is the multiplication of them inside the vectors before becoming viruliferous. Vectors suck the sap containing the viruses either from the mesophyll or phloem tissues of the infected plants. The viruses then cross the barrier of the gut cells of the vectors and accumulate in the hemocoel. In contrast to the persistently transmitted non-propagative viruses they gradually distribute themselves in different tissues and organs of the vectors and multiply in selected organs and tissues before entering the salivary glands. A protracted latent period is a consequence of this property. Nault (1994) indicated that the mean latent period of 13 propagative viruses from 4 genera was 368 +/- 41 h, compared to 23 +/- 4.1 h for 10 circulative non-propagative viruses from 3 genera.

Multiplication of viruses in vectors used to be studied through inheritance, by culturing infective insects on immune hosts and observing their transovarial transmission or by inoculating susceptible hosts by the infective progenies of the vectors. Another approach was mating infective females with virus-free males and testing the infectivity of the progenies. The virus content of infective insects used to be determined by immunological tests and later by demonstration of their presence by electronmicroscopy. Evidence for the multiplication of viruses in vectors was also obtained by crossing non-viruliferous and viruliferous insects. Another way of estimating the multiplication of viruses in vectors is by the serial injection method or bioassay. In this test the virus is passed serially by injecting one insect to another until the dilution exceeds the dilution endpoint of the initial inoculum (Banttari and Zeyen 1976, Sinha and Chiykowski 1969, Harris 1979, Nault 1997). Transmission microscopy of a thin section of viruliferous insects combined with immuno-labelling of virus particles or virus-encoded proteins can also provide evidence of virus multiplication in vectors (Shikata 1979, Ammar and Nault 1985). More direct evidence for virus multiplication in the vectors is obtained by quantitative serology using ELISA, dot blot, tissue blot or other assays that show an increase in virus titre in the vector after a brief AAP on a virus source (Gingery *et al* 1982, Falk *et al* 1987, Nault and Gordon 1988).

Transmission characteristics of the viruses belonging to different genera
The transmission characteristics of the viruses transmitted in a persistent and propagative manner follow more or less the same pattern, but their AAPs, IAPs and PPs differ.

Marafiviruses are transmitted by leafhoppers and have AAPs and IAPs that range from several minutes to several hours; LPs are at least 7 d. These viruses multiply in their vectors but no transovarial transmission has been reported (Banttari and Zeyen 1976, Gingery *et al* 1982, Gamez and Leon 1988). The efficiency of their transmission depends upon the vector population and Gamez and Leon (1988) observed that the efficiency of transmission is under genetic control. Selective breeding can rapidly increase the proportion of transmission and the rate of transmission can be reversed after a few generations of out-crossing. *Maize rayado fino virus* (MRFV) replicates in different organs of its vector insect *Dalbulus maidis* (De L. & W.) but it does not produce any cytopathic effects as are found for *Tenuivirus*-infected vectors.

Leafhoppers and planthoppers transmit most of the nucleorhabdoviruses and cytorhabdoviruses. The relationship between these viruses and their vectors has not been studied in detail. The AAP, IAP and LP of *Wheat American striate mosaic* (WASMV) < 1 min, < 15 min and < 3-66 d respectively while for *Rice transitory yellowing virus*, a strain of *Rice yellow stunt virus* (RYSV) they are 5-15 min, < 15 min and 3-66 d respectively (Lee 1967, Chen and Shikata 1971). Some plant rhabdoviruses are transmitted transovarially but at a very low rate (1-4%) in their leafhopper vectors (Sinha 1981).

Rhabdovirus particles are enveloped, bacilliform and with surface glycoproteins. These proteins are involved in the interaction with specific cellular receptors in the vectors analogous to those of animal infecting viruses. No genetic relationship between the rhabdoviruses and their vectors has been established and the loss of transmissibility is due to long-term propagation of viruses in plants.

The AAP and IAP for *Rice dwarf virus* (RDV) are only a few minutes. The LP for *Rice ragged stunt virus* (RRSV) is 2 d in the planthopper *Nilaparvata lugens* while that of *Pangola stunt virus* (PaSV) is between 7 and 14 d in *Sogatella furcifera*. Most of these viruses are transovarially transmitted in their vectors at a rate ranging from 0.2-100% depending upon the virus-vector combination.

Yan *et al* (1996) observed that one of the two outer capsid proteins of RDV is essential for the infection of insect cells. Tomaru *et al* (1997) cultured the virus in vector cell monolayers and observed the loss of transmissibility due to the loss of the P2 capsid protein. However, injecting the P2-free particles into the insects, the insects can regain the infectivity suggesting the retention of the ability of infection in the hemolymph (Omura *et al* 1998). There are also indications of the involvement of the P18 protein in virus-vector interactions. For WTV the loss of infectivity is due to the loss of both P2 and P5 proteins (Omura and Yan 1999) and for RRSV the viral spike protein, a 39-kDa protein, may be implicated in transmission. Zhou *et al* (1999) expressed this protein in bacteria and fed it to *Nilaparvata lugens* (Stål) before allowing them to feed on an infected host plant. This prefeeding made it unable to transmit the virus.

Planthoppers occur normally in two morphological forms, macropterous and brachypterous but Ridley *et al* (2004) found no significant difference in the AAP of *Fiji disease virus* transmitted by the two wing forms of the vectors, their life stages or sexes.

Tenuiviruses are normally transmitted by delphacid planthoppers. The AAP and IAP for these viruses range from 10 min to 4 h, and 30 sec to nearly an hour respectively. The LP for these viruses range from 3 to 36 d and the viruliferous vectors may remain inoculative for 84 d after acquisition. Most of the tenuiviruses show transovarial transmission (Nault and Ammar 1989, Falk and Tsai 1998) although *Rice grassy stunt virus* (RGSV) in its planthopper vector and *Maize stripe virus* (MSpV) in its leafhopper vector do not show any transovarial transmission (see review Ammar and Nault 2002).

Falk and Tsai (1998) demonstrated that there was a high degree of vector specificity and trans-ovarial transmission of tenuiviruses, except for RGSV. *Nymphs* are more efficient transmitters than adults, and female adults transmit Tenuiviruses more efficiently than adult males. The efficiency of transmission often differs with population or biotype and selective inbreeding colonies. The accumulation of MSpV in its plant hopper vector *Peregrinus maidis* (Ashmead) and its progeny by serology confirmed trans-ovarial transmission. These viruses multiply in different tissues of the insects (brain, digestive, respiratory and reproductive tracts, salivary glands, malphigian tubules, leg muscles, and fat bodies) indicating its insect-adaptive nature. It is often suggested that *tenuivirus* particles are associated with membranes containing virus-coded glycoproteins and that transmission involves membrane-based, receptor-mediated interactions. Tenuiviruses appear to have a complex genetic relationship with their host plants and the host insect. N protein and its mRNA of MSpV are equally expressed in plants and insects but the NCP and its mRNA are expressed only in plants (Falk and Tsai 1998).

B3. Relationship between the Whiteflies and the Viruses they Transmit

Four species of whitefly have been recorded as virus vectors; *Bemisia tabaci* (Genn.), *B. argentifolia*, *Trialeurodes vaporariorum* (West.), and *T. abutilonia* (Hold.). Of these, *B. tabaci* transmits most of the whitefly-transmitted viruses. Viruses are transmitted in non-persistent, semi-persistent and persistent manners. One member of the carlaviruses and two members of the ipomoviruses are non-persistent in their vectors; crimiviruses are semi-persistent whereas begomoviruses are normally persistent and non-propagative (Table 5.5).

Table 5.5 Viruses Transmitted by Whiteflies and their Transmission Characteristics (Valverde *et al* 2004, Brown and Czosnek 2002)

Virus: Type species (particle shape and nucleic acid)	Important vector species	Transmission
1. **Begomovirus**: *Bean golden yellow mosaic* (geminate, icosahedra, ssDNA)	*Bemisia tabaci*	Persistent (Circulative)
2. **Crinivirus**: *Lettuce infectious yellow virus* (filamentous, ssRNA)	*Bemisia tabaci*; *Trialeurodes vaporariorum*	Semi-persistent
3. **Ipomovirus**: *Sweet potato mild mottle virus* (filamentous, ssRNA)	*Bemisia tabaci*	Persistent (Circulative, non-propagative)
4. **Carlavirus**: *Cowpea mild mottle* (filamentous, ssRNA)	*Bemisia tabaci*	Non-persistent

(a) Non-persistently Transmitted Viruses

Ipomoviruses and carlaviruses that are transmitted by whiteflies in a non-persistent manner are *Cassava brown streak virus* (CBSV), *Sweet potato mild mottle virus* (SPMMV) and *Cowpea mild mottle virus* (CPMMV). Little information is available on the biological parameters, gene function or the cellular or molecular mechanisms of transmission of the ipomoviruses. Ingestion, transmission and retention periods for these viruses by their whitefly vectors are brief and are of minutes to hours. Transmission does not require any latent period.

It has been suggested that the capsid protein (31-34 K), located on a subgenomic RNA (1.3 kb) of the ipomoviruses may be the only viral-encoded protein essential for whitefly transmission (Brunt *et al* 1996).

(b) Semi-persistently Transmitted Viruses

Crimiviruses are phloem limited and depend on whiteflies or mealybugs for their transmission (Jelkmann *et al* 1997). *Bemisia tabaci* is the common vector for most of the crimiviruses but some viruses are transmitted by specific biotypes. *Cucurbit yellow stunting disorder virus* (CYSDV) is equally transmitted by the ubiquitous biotype B and the Spanish Q biotype of *B. tabaci* (Celix *et al* 1996, Livieratos *et al* 1998, Berdiales *et al* 1999). *Tomato chlorosis virus* (ToCV) is also transmitted by biotype Q (Navas-Castillo *et al* 2000) and A and B biotypes (Wisler *et al* 1998). *T. vaporariorum* is the only transmitter of *Beet pseudo yellows virus* (BPYV) (Coutts and Coffin 1996) and *Tomato infectious chlorosis* virus (TICV); also transmits ToCV (Wisler *et al* 1998) and *Strawberry pallidosis associated virus* (SpaV) (Wintermantel 2004).

Crimiviruses are acquired and transmitted within seconds or minutes, do not require a long latent period, and do not replicate in their vector. For example, BPYV is acquired by *T. vaporariorum* in 1 h and was transmissible less than 6 h later (Duffus 1965). Vectors retain the viruses for days or possibly weeks, and the viruses remain attached to a specific site in the foregut: transmissibility is lost after molting. *Lettuce infectious yellows virus* (LIYV) is known to encode two capsid proteins (CP and CPm). CPm occurs at one end of the particle and is involved in the transmission of this virus (Tian *et al* 1999). Johnson *et al* (2002) demonstrated by the electrical penetration graph technique that *closterovirus* inoculation by *Bemisia argentifolii* during the phloem phase of stylet penetration behaviour. Though the feeding behaviour of whiteflies was known but there was no method to estimate the presence of the virus in the natural *B. tabaci* population.

Ruiz *et al* (2002) made quantitation of CYSDV in its vector *B. tabaci* using digoxigenin-labelled cDNA CP and HSP70 probes and observed that HSP70 probes may be usueful to evaluate the virus content in natural vector population in commercial cucumber crops.

(c) Persistently Transmitted Viruses

Begomoviruses are transmitted exclusively by *B. tabaci* in a persistent, circulative manner. Virus acquisition requires at least several hours and there is a latent period of up to 24 h, depending on the virus-vector combination. Feeding behaviour of the whiteflies has a great influence on the acquisition process. The frequency of whitefly transmission of *Tomato yellow leafcurl virus-Spain* (TYLCV-ES) is strongly correlated with total duration of, and number of successful feeding episodes (Jiang *et al* 2000). *T. vaporariorum* is capable of ingesting begomoviruses but does not transmit them (Rosell *et al* 1999). Begomoviruses, once acquired, can persist for life of the vector. With an increase of AAP, virus transmission increases. In TYLCV transmission may increase from 10% to 20% after a 30 min AAP to 100% after 24-48 h, but the efficiency decreases with the age of the whitefly, and is correlated with a decrease in the amount of the virus detectable in the vector (Rubinstein and Czosnek 1997). During a single 24 h feeding event, females of *B. tabaci* can acquire a finite amount of virus, thought to be a saturated dose of an estimated 0.5×10^9 virions (Polston *et al* 1990) or 600 million genomes (Zeidan and Czosnek 1991).

After ingestion, begomoviruses are translocated from the digestive tract to the hemolymph, and from there, to the salivary glands. The LP varies with the virus-vector combination. In TYLCSV, the LP is 17 h (Caciagali *et al* 1995), but in SLCV, the virus can be detected in the saliva 8 h after the initiation of the AAP (Rosell *et al* 1999). TYLCV can be detected in the salivary gland 7 h after the beginning of the AAP but the whiteflies do not become inoculative until 1 h later (Ghanim *et al* 2001). In many combinations the virus particles may invade other tissues and survive for the life of the insect or even may multiply, as found for TYLCV. Replication of TYLCV in whitefly is associated with one or more of the prokaryotic organisms harboured by this insect, namely, the primary symbiont (Baumann *et al* 1993) or possibly by other endosymbiotic bacteria known to be associated with *B. tabaci* (Morin *et al* 2000).

Tomato yellow leaf curl virus is apparently trans-ovarially transmitted. It is propagative but no replication intermediates have yet been detected. Sexual transmission has been suggested (Ghanim and Czosnek 2000); Goldman and Czosnek (2002) confirmed trans-ovarial transmission. When *B. tabaci* eggs were bombarded with an infectious clone of TYLCV the viral DNA was detectable in all the developmental stages of the whitefly. Adult females developed from the bombarded eggs infected test plants with variable frequency. Viral DNA was also detected in the progeny of the whiteflies that developed from the bombarded eggs.

Pathways of the begomovirus in whiteflies
In whiteflies gut epithelial cells may serve as the first barrier to begomoviruses prior to their transmission. The non-transmissible AbMV may be blocked at the gut lumen before their passage to the hemocoel (Morin *et al* 2000). Contrastingly, non-vector *T. veporariorum* either do not permit the passage of the begomoviruses into the hemcoel (Rosell *et al* 1999), or the virions do not cross the gut membrane.

Begomovirus DNA can be detected in stylets, head, midgut, hemolymph, salivary glands, saliva, and honeydew of *B. tabaci* (Rosell *et al* 1999, Ghanim *et al* 2001). TYLCV can be detected in the head after only a 10 min AAP. The subsequent passage TYLCV from oesophagus to the midgut is rapid. The virion DNA is detectable approximately 30 min after it was found in the head. The movement of TYLCV from the midgut to the hemolymph is very quick. The virion reaches the hemolymph 30 min after it was first detected in the midgut, and only 90 min after the beginning of the acquisition feeding. It is detected in the salivary glands approximately 5.5 h after it was first detected in the hemolymph, 7 h after the beginning of the AAP, and approximately 1h before the whiteflies become inoculative. Observations made by different workers (Hunter *et al* 1998, Ghanim *et al* 2000) suggest that *begomovirus* particles may be selectively retained in the anterior hindgut and filter chamber from where they cross the digestive system lumen to the hemolymph.

Using *in situ* hybridization and immunology to study the localization of TYLCV in *B. tabaci* after a 24 h AAP on TYLCV-infected tomato plants, the virion was immunolocalized to the filter chamber and the proximal part of the descending midgut; immunogold label was associated with food in the lumen and the label was also associated with the microvilli and gut wall epithelial cells. *In situ* hybridization of sections of whitefly salivary glands revealed that the signal was associated with some, but not all, of the salvary gland cells (Brown and Czosnek 2002).

Specificity

Coat protein of begomoviruses determines their vector specifity. Transmission results strongly implicate the CP as the sole virus-encoded protein necessary for vector-mediated transmission. Exchanging the CP gene of whitefly-transmitted ACMV with that of the leafhopper-transmitted BCTV produced leafhopper-transmitted ACMV (Briddon *et al* 1990). A frameshift mutation in the CP gene of BGMV results in the loss of virion transmission by *B. tabaci* (Azzam *et al* 1994). Replacement of the CP of the non-transmissible AbMV mutant with a viable whitefly-transmissible SiGMV CP renders the mutant virus transmissible by *B. tabaci* (Hofer *et al* 1997). Norris *et al* (1998) showed that the region of the coat protein between amino acids 129 and 134 was essential for the correct assembly of the virus particle for whitefly transmission; full transmission is achieved by a third replacement at position 174 (Honle *et al* 2001).

Begomoviruses survive in the hemolymph of *B. tabaci* in the same way as persistently transmitted aphid-borne viruses. A GroEL homologue produced by endosymbiotic bacterium in *B. tabaci* has been implicated in the circulative transmission of TYCLV (Morin *et al* 1999, 2000). TYCLV particles display affinity for the *B. tabaci* GroEL homologue in a virus overlay assay. This homologue is detectable in the hemolymph but not in the digestive tract. The GroEL homologue interacts with TYCLV in the hemolymph as it does in aphids. This interaction is not specific to the vectors as non-vectors also harbour the same symbiont and show a similar reaction, but the reaction products have to move to the salivary glands if the vector is to become inoculative (for review see Brown and Czosnek 2002).

B4. Relationship of Thrips and Viruses Transmitted by them

Thrips normally transmit tospoviruses and ilarviruses in a persistent manner. Tospoviruses are transmitted by *Frankliniella occidentalis* Pergande, the western flower thrips and the ilarviruses by *Thrips tabaci* Lindemann. The vector acquires tospoviruses at its immature (larval) stage. Adults that develop from these immature stages are infective but cannot acquire the virus by feeding on infected plants. As larval development proceeds, competence in acquisition of tospoviruses decreases; the pupal stages do not feed. There may be variations in the efficiency of transmission by different species and between sexes. *F. occidentalis* is the most efficient vector and males are more efficient than females in transmitting the virus. Vectors may retain the virus throughout their life but may lose the virus if cultured in plants by successive transfer.

Pathway of the virus in the body of the vector

Most work has been done on the interactions between *Tomato spotted wilt virus* (TSWV) and *F. occidentalis*. The route of this virus through its thrips vector differ to that found for other persistently transmitted viruses. The virus has to cross several barriers to enter, and escape from, the salivary gland before the vector becomes infective. The virus first enters the midgut by crossing the apical membrane. The virus escapes from the midgut by crossing the basal plasmalemma and enters the salivary gland by crossing its basal plasmalemma. The virus ultimately escapes from the salivary gland by crossing the apical membrane of this gland. The presence of a basement membrane in the midgut and a basal lamina at the salivary gland adds additional complexity to each of these barriers. The virus does not move to the hemocoel from the gut cells.

Entry of the virus to the midgut takes place through a receptor-mediated mechanism. The glycoprotein spikes of the envelope of the particle are necessary for acquisition and this protein contains the binding site for a receptor in the vector's midgut. Bandla *et al* (1998) observed a 50 kDa thrips protein localized in the midgut gut of a larval thrip that may serve as the viral attachment protein for binding the thrips cellular receptors. The interaction between viral attachment proteins and cell receptors results in virus fusion with the cell membrane and immediate deposition of the virus core. Viral glycoproteins bind a cell receptor protein and are then often taken into the cell via receptosomes with N protein and viral RNA and polymerase. Kikkert *et al* (1998) observed a 94 kDa protein in *F. occidentalis* and *T. tabaci* bound to the G2 glycoprotein of TSWV. This protein is not found in the thrips midgut, but is abundant elsewhere in the thrips body. This protein is detected both in vector and non-vector insects. It has been suggested that the 94 kDa thrips protein may represent a receptor protein involved during circulation of the virus through the vector (Goldbach 1998).

The barrier to virus passage in adult *F. occidentalis* is primarily based on entry to the midgut (Nagata *et al* 1999), but in *T. setosus*, TSWV enters the adult thrips midgut epithelia and replicates (Ohnishi *et al* 1998). However, Nagata *et al* (1997) observed that the barriers operating *in vivo* are not found in cell cultures. It was possible to infect cell cultures from non-vectors *T. tabaci* and *F. occidentalis* as well by TSWV and the virus also replicated.

Immunofluorescent staining of *nymphs* at various times after virus acquisition showed the route of the virus through the vector. Infection is first found in the midgut-1 region 24 h post-acquisition. In the late larval stage the infection spreads to the circular and longitudinal midgut muscle tissues. In the adult stage the visceral muscle tissues of the midgut and the foregut become infected. Infection in the salivary glands and ligaments connecting the midgut with the salivary glands appears 72 h post-acquisition. The virus does not appear in the hemocoel or the midgut basal lamina. The virus reaches the salivary glands through the ligaments connecting the midgut-1 region and the salivary glands (Nagata *et al* 1999).

Kritzman *et al* (2002) determined the route of TSWV inside the thrip body in relation to transmission efficiency. First instar larvae of *F. occidentalis* and *T. tabaci* were allowed, immediately after hatching, to acquire virus from mechanically infected *Datura stramonium* plants for 24 h. The rate of transmission of adults was determined in inoculation access feeding test on *Emilia sonchifolia* leaf disks. Thrips tissues were analyzed for infection at 24 h intervals after the acquisition access period, and assayed by the whole-mount immuno-fluorescent staining technique. The virus was initially detected in the proximal midgut region in the larvae, then in the second and third midgut regions, foregut and salivary glands. The intensity of infection in various organs was positively related to the transmission efficiency. The virus must reach the salivary glands before thrips pupation to be transmitted by old second *instar* larvae and adults. The rate of virus replication in the midgut and the extent of virus migration from the midgut to the visceral muscle cells and salivary glands are crucial factors in the determination of vector competency (Nagata *et al* 2002).

Adult *F. fusca* (Hinds) and *F. occidentalis* are susceptible to infection of midgut cells by TSWV and subsequent virus replication, but adult thrips that feed on virus-infected plants do not transmit the virus. There may be a role of tissue behaviour in TSWV movement and infection from midgut muscle cells to salivary glands. de Assis Filho *et al* (2004) made a comprehensive study of the transmission of TSWV by thrips. Only larval thrips acquire virus and adults derived from such larvae transmit the virus. Nonviruliferous adults can ingest virus particles while feeding on TSWV-infected plants, but such adult thrips have not been shown to transmit TSWV. Thrips 1, 5, 10 and 20 d after adult emergence (DAE) fed on TSWV-infected plants acquired TSWV with virus replication and accumulation occurring in both epithelial and muscle cells of

F. fusca (tobacco thrips—TT) and *F. occidentalis* (western flower thrips—WFT) adults. WFT acquired TSWV more efficiently than TT. There was no significant effect of insect age on TSWV acquisition by TT. In contrast, acquisition by adult WFT at 1 and 5 DAE was higher than acquisition at 10 and 20 DAE. The susceptibility of adult TT and WFT to infection of midgut cells and TSWV and subsequent virus replication depends upon the movement of the virus by overcoming the tissue barrier to the movement.

Specificity

There are distinct levels of species and biotype specificity (Sakurai *et al* 2004). Nagata *et al* (2004) observed disparity in the competency of four thrips species (*F. occidentalis, F. schultzei, T. tabaci, T. palmi*) to transmit and replicate four Tospoviruses (*Tomato chlorotic spot virus* (TCSV), *Groundnut ring spot virus* (GRSV), TSWV, and *Chrysanthemum stem necrosis virus* (CSNV)). *F. occidentalis* transmitted all these viruses with different efficiencies but *F. schultzei* failed to transmit TSWV. No transmission of any of these viruses was found with *T. tabaci* or *T. palmi*. Viruses replicated in all the thrips tested and suggested involvement of barriers to virus circulation in the thrips' bodies required for transmission. The major barrier in this regard is the entry of the virus to the midgut of the vector that is mediated by the specific glycoproteins in the virus particles and compatible receptor protein in the vector hindgut. There are also other factors involved in the specificity as there may be entry and replication of the virus in some vectors but those vectors do not show any inoculability.

B5. Relationship between Beetles and the Viruses they Transmit

Four families of beetles normally transmit viruses: Chrysomelidae, Coccinellidae, Curculionidae and Meloidae. Most of the virus vectors are transmitted in the Chrysomelidae and Coccinellidae. Unlike other vector groups that have a piercing-sucking feeding type, beetles chew. They transmit viruses of the genera *Comovirus, Sobemovirus, Tymovirus, Carmovirus* and *Bromovirus* in a semi-persistent manner (Table 5.6).

Table 5.6 Viruses Transmitted by Beetles and their Transmission Characteristics (Hull 2002 with modification)

Virus: Type species (particle shape and nucleic acid)	Important vector species	Transmission
1. **Bromovirus**: *Brome mosaic virus; Cowpea chlorotic mottle virus* (icosahedra, ssRNA)	*Ceratoma trifurcata*	Semi-persistent
2. **Comovirus**: *Cowpea mosaic*	*Ootheca mutabilis*	Semi-persistent

virus (icosahedra, ssRNA)		
3. **Carmovirus**: *Carnation mottle virus; Turnip crinkle virus* (icosahedra, ssRNA)	*Phyllotreta* spp.	Semi-persistent
4. **Machlomovirus**: *Maize chlorotic mottle virus* (icosahedra, ssRNA)	*Diabrotica* spp.	Semi-persistent
5. **Sobemovirus**: *Southern bean mosaic virus* (icosahedra, ssRNA)	*Ceratoma trifurcata, Epilachna varivestis*	Semi-persistent
6. **Tymovirus**: *Turnip yellow mosaic virus* (icosahedra, ssRNA)	*Phyllotreta* spp.	Semi-persistent

Acquisition and transmission efficiency

Beetles can acquire the virus immediately after they start feeding and the efficiency of acquisition rises with the increasing consumption of virus-infected tissues. The virus concentration in the beetles gradually decline after acquisition (Ghabrial and Schultz 1983). Retention of the virus in the beetles is very variable and mostly depends on their feeding behaviour and on the virus-beetle combination. Beetles in diapause retain virus much longer than actively feeding beetles. Many viruses overwinter in diapausing beetles (Freitag 1956, Walters *et al* 1972). Starvation has little effect on transmissibility (Wang *et al* 1994). Viruses are usually lost from the beetles during the process of regurgitation.

There are controversies on the circulation of the viruses within the body of the beetles. Early researchers thought that the hemocoel was the site of the retention of the viruses (Freitag 1956, Slack and Scott 1971, Sanderlin 1973, Fulton and Scott 1974, Scott and Fulton 1978). Wang *et al* (1992) could not demonstrate the circulation of SBMV and BPMV in the hemocoel of the beetle vector *Epilachna varivestis* but SBMV was observed in the spotted cucumber beetle and bean leaf beetle but BPMV was not, indicating that virus movement from the digestive system to the circulatory system differs with the beetle-virus combination. Wang *et al* (1994) also found that BPMV remained confined to the gut lumen of the spotted cucumber beetle but that SBMV moved into the hemocoel of the same vector through the peritrophic membrane-lined midgut but not through the cuticle-lined foregut or hindgut. Wang *et al* (1994) found that the SBMV injected into *E. varivestis* or *Diabrotica undecimpunctata howardi*, did not move from the hemocoel through gnathal glands into the regurgitant. The exact processes of the movement of viruses within the beetles in different virus-beetle combination are still unclear.

Musser *et al* (2003) found a mutual beneficial relationship between the beetle (*Epilachna varivestis*) and the viruses it transmits. Beetles preferred and ingested more of the virus-infected bean leaf tissues than the non-infected ones. Beetle larvae fed on infected leaves weigh more than those fed on healthy plants.

Specificity

Transmission of viruses is very specific. A beetle can ingest several non-transmissible viruses that can move to the hemocoel and be deposited on the leaf surface by regurgitation during feeding, indicating the presence of a factor(s), or interactions, that determine whether a virus is transmissible that act or occur after the virus is deposited on the leaf surface. Transmissible viruses are not inactivated by proteases (Langham *et al* 1990) or by a high level of ribonuclease (Gergerich *et al* 1986) present in the virus-regurgitant mixture. It may be that the regurgitant selectively prevents infection of non-beetle transmissible viruses. However, the key factor preventing virus transmission by leaf-feeding beetles may be the immobilization of viruses within plants that prevents them from gaining access to the susceptible plant cells (Gergerich 2002).

B6. Relationship between MealyBug and the Viruses they Transmit

Mealy bugs (Coccoidea and Pseudococcoidea) have negligible mobility. They normally spread from plant to plant by contact or by crawling nymphs. Ants may also carry them from plant to plant and long-distance movement is possible by wind. They first received attention as vectors of *Cacao swollen shoot virus* (CSSV). Other viruses found to be transmissible by mealy bugs are: *Grapevine leafroll associated virus-3* (GLRaV-3), *Lettuce chlorosis virus* (LCV), *Pineapple mealybug wilt associated virus* (PMWaV1, PMWaV2), *Grapevine virus A* (GVA) and *Grapevine virus B* (GVB) (Table 5.7). Six species are known to transmit CSSV (Bigger 1981) but some species have transmitting and non-transmitting races (Posnette 1950).

Table 5.7 Viruses Transmitted by Mealy Bugs and their Transmission Characteristics (Hull, 2002 with modification)

Virus: Type species (particle shape and nucleic acid)	Important vector species	Transmission
1. **Badnavirus**: *Commelina yellow mottle virus* (bacilliform, dsDNA)	*Planococcus citri*	Semi-persistent
2. **Vitivirus**: *Grapevine virus A* (filamentous, ssRNA)	*Pseudococcus longispinus*	Semi-persistent
3. **Ampelovirus**: *Grapevine leafroll-associated virus 3* (filamentous, ssRNA)	*Planococcus ficus, P. citri*	Semi-persistent

Mealy bugs are phloem-feeders. *Pseudococcus njalensis* can acquire the virus within 20 min but they normally take at least 16 min to penetrate to the leaf tissues containing the virus. Once acquired virus persists in the mealy bug for less than 3 h (Posnette and Robertson 1950). Recent studies with other mealy bug-transmitted viruses showed very poor rates of transmission, even after 72 h acquisition access periods, with the retention of the inoculativity for 1 h (Cabaleiro and Segura 1997a, b; Peterson and Charles 1997), although Harris (1983) reported more acquisition of the virus by the nymphs and longer retention period.

The relationship between the CSSV and its mealy bug vector appears to be non-persistent as the virus is carried on or near the stylets (Hull 2002). The pink pineapple mealy bug, *Dysmicoccus brevipes*, and the grey pineapple mealy bug, *D. neobrevipes*, transmit PMWaV. The second and third instars have been found to be more effective in acquiring the virus than first instars and gravid females (Sether *et al* 1998). For GLRaV-3, *Planococcus citri* can retain the virus for approximately 24 h suggesting a semi-persistent relationship (Cabaleiro and Segura 1997a, b).

B7. Relationship between Bugs and the Viruses they Transmit

Mirid bugs (Miridae) feed by a mirid bug *Cyrtopeltis nicotianae* transmits *Velvet tobacco mottle* (VTMoV) and *Southern bean mosaic virus* (SBMV) and some other sobemviruses. The transmission process shows similarities to both non-circulative and circulative viruses. The acquisition threshold is less than a min and tranmission increases with up to 30 min acquisition access periods and 1-2 h inoculation access times; infectivity is retained for up to 10 d and passes through molts (Gibbs and Randles 1988).

VTMoV was detected in the gut, hemolymph and faeces of its vector *C. nicotianae* but not in the salivary glands; infective virus was found in faeces 6 d after acquisition (Gibbs and Randles 1990). The virus-vector association found in mirid-transmitted viruses appears to be distinct as non-infective mirids can inoculate plants from infectious sap deposits on the upper epidermis, and an ingestion-egestion mechanism may operate independently of the salivary gland. This may also explain the increase in transmission rates with increasing acquisition and inoculation times (Gibbs and Randles 1991).

The Piesmid bug *Piesma quadratum* (Fieb.) transmits *Beet leaf curl virus* (BLCV). The relationship between the virus and the vector is thought to be of a persistent propagative type but there is no evidence for its transmission through eggs (Proeseler 1980).

B8. Relationship between Mites and the Viruses they Transmit

Members of the families Eriophyidae and Tetranychidae transmit several viruses (Table 5.8). The most studied mite vector is *Aceria tulipae* (Keifer) that transmits *Wheat streak mosaic virus* (WSMV). The acquisition period can be as short as 15-30 min (Slykhuis 1955, 1973). The mite retains infectivity through molts and may remain infective (at least in glasshouse conditions) for 6-9 d after removal from the infected plant. On a virus-immune host held at 3°C, they may remain infective for more than 2 mon. The mites normally become infective as nymphs and not as adults (Slykhuis 1955, Orlob 1966). The particles of WSMV were found in the midgut, body cavity and salivary glands of the vector (Paliwal (1980), but Mahmood *et al* (1997) failed to detect WSMV in *A. tosichella* using immunofluorescence and dot-immunobinding assays.

Table 5.8 Viruses Transmitted by Mites and their Transmission Characteristics (Toros 1983 with modification)

Virus: Type species (shape of the particle, and nucleic acid)	Important vector species	Transmission
1. **Rymovirus**: *Ryegrass mosaic virus* (filamentous, ssRNA)	*Abacarus hystrix*	Semi-persistent
2. **Tritimovirus**: *Wheat streak mosaic virus* (filamentous, ssRNA)	*Aceria tosichella*	Semi-persistent
3. **Allexivirus**: *Shallot virus X* (filamentous, ssRNA)	*Aceria tulipae*	Semi-persistent
4. **Nepovirus**: *Tobacco ring spot virus; Black currant reversion-associated virus* (icosahedra, ssRNA)	*Cecidophyopsis ribis*	Semi-persistent
5. **Cytorhabdovirus**: *Barley yellow streak mosaic virus* (filamentous-enveloped, ssRNA)	*Petrobia lateens*	Persistent (Circulative; propagative)

In *Peach mosaic virus* (PMV) the minimum acquisition access period is 3 d and transmission increases if it is increased to 5 d; the minimum inoculation access period is 3-6 h and the relationship between the virus and the vector may be semi-persistent (Gispert *et al* 1998).

Eriophyid mites normally transmit *rymoviruses*. Schubert and Rabenstein (1995) could not detect the DAG motif in the coat protein in a virus of barley transmitted by mites.

Several members of the family Tetranychidae may acquire a number of viruses that accumulate in the body of the mites but remain non-inoculative (Orlob and Takahashi 1971), however the brown wheat mite, *Petrobia lateens*, transmits *Barley yellow striate mosaic virus* (BYSMV) (Robertson and Carroll 1988). Nymphs readily acquire this virus and both nymphs and adults can efficiently transmit the virus. There is indirect evidence that the virus is transovarially-transmitted (Robertson and Carroll 1988, Smidansky and Carroll 1996), but the virus is lost when the viruliferous mite is fed on a non-host plant for the virus (Hull 2002).

Larvae of False spider mite *Brevipalus phoenicis obovatus* transmit *Citrus leprosis virus* Chagas *et al* (1984) but the relationship is not clearly understood.

B9. Relationship between Nematodes and the Viruses they Transmit

Soil-inhabiting, plant-parasitic, ectoparasitic nematodes belonging to the families Longidoridae (order Dorylaimida) and Trichodoridae (Order Trilonchida) are involved in the transmission of plant viruses. 8 *Longidorus*, 1 *Paralongidorus* and 9 *Xiphinema* spp. of Longidoridae are vectors of 12 of the 38 putative members of the genus *Nepovirus*. Also 8 *Paratrichodorus* and 5 *Trichodorus* spp. of the family Trichodoridae are vectors of 3 viruses comprising the genus *Tobravirus* (Taylor and Brown 1997).

The mode of transmission of viruses by nematodes is similar to that recorded for insect vectors of non-circulative plant viruses. Nematode-transmitted viruses are retained for several weeks, and even up to several years, at specific sites in the feeding apparatus. Viruses do not replicate in the vector and are lost at the molting stage. Viruses transmitted by nematodes are given in Table 5.9.

Table 5.9 Viruses Transmitted by Nematodes and their Transmission Characteristics (Dijkstra and Khan 2002, MacFarlane *et al* 2002 with modification)

Virus type species (particle shape and nucleic acid)	Important vector species	Transmission
1. **Nepovirus:** *Tobacco ring spot virus* (icosahedra, ssRNA)	*Xiphenema* spp.	Non-persistent
2. **Tobravirus:** *Tobacco rattle virus* (icosahedra, ssRNA)	*Paratrichodorus* spp., *Trichodorus* spp.	Non-persistent
3. **Sadwavirus:** *Satsuma dwarf virus*; *Strawberry latent ring spot virus* (icosahedra, ssRNA)	*Xiphenema diversicaudatum*	Semi-persistent

Acquisition and transmission

Nematodes acquire virus while feeding on an infected plant and release into an uninfected cell during subsequent feeding. The process involves ingestion, acquisition, adsorption, retention, release, transfer and establishment (Brown and Weischer 1998) (for review see Visser and Bol 1999). These processes may be influenced by the physiological responses of the nematodes during feeding and the production of oesophageal gland secretions. Ingestion may not depend upon any relatonship between the virus and the nematode but it does require a specific interaction between the nematode and the plant. In the acquisition phase the ingested virus particles are adsorbed to specific receptor sites in the feeding apparatus of the nematode. Karanastasi *et al* (2003) determined how *Trichodorus* acquire and transmit *tobravirus* particles by feeding them on *N. tabacum* seedlings on agar and examining in real time using video enhanced intereference light microscopy. The feeding cycle was found to have four distinct phases: (i) host exploration; (ii) cell exploration; (iii) cell sampling; and cell feeding followed by a quiescent period preceding the initiation phase and (iv) and an average of four cells, which had been perforated by the nematode onchiostyle, were immediately abandoned and thus remain fully functional, providing a suitable environment for the establishment of virus infection.

One of the principal nematode vectors is *X. americana* but its taxonomy is highly controversial (Lamberti *et al* 2000). Several of the species within the *X. americanum* group transmit 2-4 distinct nepoviruses (Hoy *et al* 1984, Griesbach and Maggenti 1990, Brown *et al* 1994, 1996a). These differences in transmission in the population may be due to the specificity of transmission of 'local' isolates of the viruses by 'local' populations of the vector nematodes (Brown *et al* 1993). Similar, specific relationships also exist between isolates of tobraviruses and the population of trichodorid nematodes (Brown and Ploeg 1989, Ploeg *et al* 1992, Vasssilakos *et al* 1998). A plant-derived factor may be involved with the specific association of the virus and vector. *X. index* can acquire GLFV from root cells of both grapevine and *Chenopodium quinoa* but the virus is subsequently transmitted only to the grapevine, indicating the absence of the specific factor required for the release of the virus (Trudgill and Brown 1980).

Tobraviruses are usually vectored by specific nematode species; most Tobraviruses are transmitted by only one nematode species. *Paratrichodorus* spp. shows similar specificity. Nepoviruses, on the other hand, are vectored by several different species of *Longidorus*. The ability of certain nematodes to transmit viruses may also be genetically controlled as Brown (1986) observed inheritance of the ability of *X. diversicaudatum* to transmit ArMV and SLRSV.

Retention and release

Once acquired viruses may persist in a transmissible form in starved *Longidorus* for upto 12 wk, in *Xiphinema* spp. for about a year and for much more than a year in *Trichodorus* (van Hoof 1970). Retention sites were located in the mouthparts by electronmicroscopy (see review Brown *et al* 1996b). Nepovirus particles are associated with the inner surface of the odontostyle of various species of *Longidorus* and with cuticular lining of the odontophore and oesophagus of *Xiphinema* species. *Tobravirus* particles are absorbed to the cuticular lining of the oesophagus.

Brown *et al* (1996a) found genetic determinants for the transmissibility of *Raspberry ringspot virus* (RRSV) and *Tomato black ring virus* (TBRV) encoded by RNA2. Robertson and Henry (1986) found the association of a carbohydrate-like material on the food canal walls of *X. diversicaudatum* transmitting *Arabis mosaic virus* (ArMV). Ploeg *et al* (1993) observed segregation of transmissibility with RNA2 by making reciprocal pseudo-recombinants between a nematode-transmissible and a non-transmissible isolate of *Tobacco rattle virus* (TRV). MacFarlane *et al* (1995) demonstrated the involvement of more than one of the RNA2 genes. *Pea early browning virus* (PEBV) has two non-structural genes of 40 kDa and 32 kDa. Mutations of these two genes and the C-terminal mobile region of the coat protein abolish the transmissibility (MacFarlane *et al* 1996). Hernandez *et al* (1997) observed that the mutation of the 40 kDa non-structural gene (PpK20), but not the 32 kDa, of a TRV isolate abolished transmission by *Paratrichodorus pachydermis*. Visser and Bol (1999) detected interactions between TRV-PpK20 coat protein and 32.8 and 40 kDa proteins. Interference of the deletion of 19 amino acids of the C terminal portion was observed with only CP-40 kDa interaction and not that with CP-32.8 kDa, but the deletion of 79 amino acids from the C-terminal affected both interactions. These observations suggest that non-structural proteins may act as a helper component for transmission (MacFarlane *et al* 2002). Vellois *et al* (2002a) confirmed the localization of tobravirus 2b non-structural helper protein that functions by interacting with a small flexible domain located at the CP, forming a bridge between the virus particle and the internal surface of the vector nematode feeding apparatus. Vellois *et al* (2002b) compared the 2b proteins of the highly transmissible and poorly transmissible isolates of PEBV (TPA56 and SP5 respectively) and identified two amino acid substitutes (G90s and G177R) that might be responsible for the poor transmission of the virus. Examinations of the 2b proteins predicted the presence of a coiled-coil domain in the central region of the protein important for the association of interacting proteins and mediate interaction of the 2b protein with the virus coat protein or with the vector nematode.

Andret-Link *et al* (2004) observed the involvement of viral determinants in the specific transmission of *Grapevine fan-leaf virus* (GFLV) by *X. index*. These determinants are located within 513 C-terminal residues of the RNA2-encoded polyprotein and the coat protein is the sole viral determinant for the specific spread of GFLV by *X. index* (see review MacFarlane *et al* 2002).

B10. Relationship between Fungi and Protista and the Viruses they Transmit

Various degrees of host specificity exist in these vectors. The wide host range isolates tend to be better vectors than do those with only a narrow host range. Two types of association occur between the viruses and their vectors. Vectors can acquire the viruses *in vitro* and the viruses are carried externally on the spores or they acquire the viruses only *in vivo* and they are carried within the cytoplasm (Cambell and Fry 1966).

Viruses transmitted by fungi and protista are given in Table 5.10.

Table 5.10 Viruses Transmitted by Fungi and Protista and their Transmission Characteristics (Dijkstra and Khan 2002, Yilmaz *et al* 2003)

Virus: Type species (particle shape and nucleic acid)	Important vector species	Transmission
1. **Benyvirus**: *Bean necrotic yellow vein virus* (rod-shaped, ssRNA)	*Polymyxa betae*	Internal
2. **Bymovirus**: *Barley yellow mosaic virus* (filamentous, ssRNA)	*Polymyxa graminis*	Internal
3. **Furovirus**: *Soil-borne wheat mosaic virus* (rod-shaped, ssRNA)	*Polymyxa graminis*	Internal
4. **Necrovirus**: *Tobacco necrosis virus A* (icosahedra, ssRNA)	*Olpidium brassicae*	External
5. **Aureusvirus**: *Pothos latent virus, Cucumber leaf spot virus* (icosahedra, ssRNA)	*Olpidium bornovenus*	External
6. **Varicosavirus**: *Lettuce big vein associated virus; Tobacco stunt virus* (Rod-shaped, ssRNA)	*Olpidium brassicae*	Internal
7. **Carmovirus**: *Carnation mottle virus, Melon necrotic spot virus* (icosahedra, ssRNA)	*Olpidium radicale*	External
8. **Pomovirus**: *Potato mop top virus* (rod-shaped, ssRNA)	*Spongospora subterranea*	Internal
9. **Pecluvirus**: *Peanut clump virus* (rod-shaped, ssRNA)	*Polymyxa graminis*	Internal

(a) Externally-borne Transmission

Two species of *Olpidium* transmit several viruses belonging to the Tombusviridae where the fungal zoospores carry the viruses externally. Fungal zoospores acquire the virus, or the binding of the virus to the zoospores occurs, outside the host plant (Campbell and Fry 1966, Dias 1970). When *O. brassicae* transmitting TNV was observed under an electron microscope virus particles were found to be adsorbed on to the zoospore plasmalemma and exonemal sheath. Similar adsorption was also observed with STNV and *O. brassicae* and with CNV on *O. radicale* (Temmink *et al* 1970, Stobbs *et al* 1982).

Other external sources of the virus may be from leaching from plant roots or from plant debris (Smith *et al* 1969). Virus may also be carried on or in the seed coat and thus be available for transmission by a vector occurring in soil. This is known as vector-assisted seed transmission but appears to be very rare (Campbell *et al* 1996).

Acquisition

Campbell *et al* (1995) demonstrated that 12 single sporangial isolates of *O. bornovanus* transmitted MNSV and CNV but their ability to transmit CLSV, SqNV and CSBV differed. Robbins *et al* (1999) showed that the specificity of transmission was at least partly related to the degree of binding of the zoospores and that the specific binding site may not be the same for all the viruses that are transmitted.

Involvement of virus coat protein

McLean *et al* (1994) observed the involvement of the coat protein in the uptake of the virus by the zoospore and Robbins *et al* (1997) identified one amino acid in the coat protein responsible for the recognition of the virus by the zoospore. Adams *et al* (2001) found that plant virus transmission by plasmodiophorid vectors is associated with distinctive transmembrane regions of virus-encoded proteins. Computer analysis of published sequences identified two complementary transmembrane domains in the coat protein read through domains of *benyaviruses, furoviruses* and *pomoviruses* and in the P2 proteins of *bymoviruses.* These viruses differ in genome organization but are all transmitted by plasmodiophorid vectors. The second domain is absent or disrupted in naturally occurring deletion mutants that cannot be transmitted. In a non-transmissible substitution mutant of *beet necrotic yellow vein virus* (BNYVV) the alignment of the helices is disrupted. From conserved patterns detected in the transmembrane helix sequence and calculated relative helix tilts, structural arrangements consistent with tight packaging of transmembrane helices were identified. These included ridge/groove arrangements between the two helices and strong electrostatic associations at the interfacial regions of the membrane.

These data strongly suggest that these transmembrane helices facilitate the movement of virus particles across the fungal membrane.

Previous studies show that specific region of the CNV capsid are involved in transmission and that transmission defects in several CNV transmission mutants are due to inefficient attachment of the virus to the zoospore surface. Kakanik *et al* (2003) showed that binding of CNV to zoospores is mediated by specific mannose and/or fructose containing oligosaccharides. Kakanik *et al* (2004) also proved that CNV undergoes conformational change upon zoospore binding. It was observed that transmission of CNV by *O. bornovanus* involves specific adsorption of the virus particles onto the zoospore plasmalemma prior to the infection of host roots by virus-bound zoospores. Limited trypsin digestion of CNV, following *in vitro* CNV/zoospore binding assay, results in the production of specific proteolytic digestion products under conditions where native CNV is resistant. This digestion pattern is similar to that of swollen CNV particles produced *in vitro* suggesting a conformational change upon zoospore binding. This change is essential for the transmissibility of the virus.

The results and analyses suggest that binding of viruses to their zoospores involves receptors on the zoospore plasmalemma that interact with virion coat protein. This is not controlled by a simple amino acid motif but depends upon features of the coat protein that are brought together by folding into the tertiary structure (Adams 2002).

(b) Internally-borne Transmission

When zoospores containing externally-borne viruses are air-dried or treated with acid or tri-sodium phosphate, the treated zoospores become non-viruliferous. Similar treatments with many other viruses do not affect their viruliferous nature. These latter viruses are thought to be internally borne within the zoospores. Some of these viruses belong to the genera *Bymovirus*, *Furovirus* and *Varicosavirus* and *O. brassicae* and three Plasmodiophoral species are the vectors. The model developed for this relationship is based on observations on *O. brassicae* and LBVV, *P. graminis* and SBWMV, and *P. betae* and BNYVV (Campbell 1996).

Abe and Tamada (1986) first put forward the evidence for the presence of the BNYVV virions inside immature zoospores and zoosporangia of *P. betae* by electron microscopy. The first unequivocal, proof of the presence of a virus inside the zoospores and zoosporangia was provided by using immunogold labelling of viruses (Chen *et al* 1991, Rysanek *et al* 1992, Dubois *et al* 1994, Peng *et al* 1998). Merz (1995) and Chen *et al* (1998) provided conclusive demonstrations of the presence of virions inside the zoosporangia of *Spongospora subterranea* by scanning electron microscopy.

Acquisition and transmission

Viruses internally borne in zoospores do not multiply in the vector, as the fungal isolates can be freed from the viruses (Campbell 1962, Tomlinson and Garrett 1964, Adams *et al* 1987, Abe and Tamada 1986). It is therefore assumed that the virus moves out of the fungal cytoplasm and multiplies in the host plant cells. Acquisition and transmission occur while the fungal vector is growing within the host cells and involves the passage of virions across the fungal plasmalemma. This movement takes place by pinocytosis (Chen *et al* 1991, Rysanek *et al* 1992).

Virus-encoded proteins are thought to be the regulator of acquisition and transmission of the internally-borne viruses. When multi-partite RNA genomes are propagated by repeated mechanical transmission some parts of these genomes are deleted with consequent effects on virus transmission. BNYVV isolates lacking RNA 3 and RNA 4 were poorly transmitted by *P. betae* (Lemaire *et al* 1988) and the absence of RNA 4 had the largest effect (Tamada and Abe 1989, Richards and Tamada 1982). Han *et al* (1996) found the loss of transmission by deleting the 3'-part of the coding region of the RNA 4. Though the single 32 kDa product of this RNA appears to have only a minor effect on symptom production or virus movement it may have a role in acquisition and transmission.

Two different deletions in the coat protein readthrough (CP-RT) domain of RNA 2 produce a 75 kDa protein located at one end of the virion (Haeberlé *et al* 1994) that has been found to be required for virion assembly as well as fungal transmission (Schmitt *et al* 1992). A deletion in the C-terminal half of the CP-RT protein completely inhibited fungus transmission. Further studies using alanine-scanning mutagenesis showed that substitution of KTER (at amino acid numbers 553-556) by ATAR completely prevented transmission by *P. betae* (Tamada *et al* 1996). One A and one B type isolate of BNYVV which has been passaged for more than 15 yr on manually inoculated *Chenopodium quinoa* in the laboratory were found to have small deletions in the KTER-encoding domain on RNA2 (Koenig 2000).

In the case of fungal transmission of the filamentous *Bymoviruses*, RNA components produce polyproteins. BaMMV RNA 2 has two functional products and is prone to deletions in the C-terminal region (Timpe and Kühne 1994, Jacobi *et al* 1995, Peerenboom *et al* 1996) and it has been shown that the UK-M deleted form cannot be transmitted by *P. graminis* (Adams *et al* 1988). The deletions affect the P 2 protein, but detailed studies are yet to be done on the relation between the virus-encoded proteins on the transmission of viruses by fungi.

Studies so far showed that the C-terminal parts of the CP-RT of the rod-shaped viruses (*Benyvirus, Furovirus, Pecluvirus* and *Pomovirus*) and the P 2 of members of the *Bymovirus* are implicated in the virus transmission by fungi and Protista, but viruses with similar genomic organizations and with the same vector do not have readily identifiable common amino acid sequence motifs associated with virus transmission. Recent computer predictions suggest that all the CP-RTs and P 2 proteins contain two regions that, not only show strong evidence of transmembrane activity, but also show evidence of compatibility between their amino acids, suggesting that they could be closely paired within a membrane and with the region between them aligned on the inside of the membrane (Diao *et al* 1999, Adams *et al* 2001). It is possible that these regions are involved in attachment to the zoosporangial plasmalemma, and help virus particles move between the cytoplasm of the plant host and that of the fungus (for review see Adams 2002).

References

Abe, H. and Tamada, T. 1986. Association of beet necrotic yellow vein virus with isolates of *Polymyxa betae* Keskin. Annals Phytopath. Soc. Japan 52: 235-247

Adams, M.J. 2002. Fungi. In: Advances in Botanical Research—Plant Virus Vector Interactions. Vol. 36. R.T. Plumb (ed). Academic Press, San Diego, USA. pp. 47-60

Adams, M.J., Jones, P. and Swaby, A.G. 1987. The effect of cultivar used as host for *Polymyxa graminis* on the multiplication and transmission of barley yellow mosaic virus (BaYMV). Annals Appl. Biol. 110: 321-327

Adams, M.J., Swaby, A.G. and Jones, P. 1988. Confirmation of the transmission of barley yellow mosaic virus (BaYMV) by the fungus *Polymyxa graminis*. Annals Appl. Biol. 112: 133-141

Adams, M.J., Antoniw, J.F. and Mullins, J.G.L. 2001. Plant virus transmission by plasmodiophorid fungi is associated with distinctive transmembrane regions of virus-encoded proteins. Arch. Virol. 146: 1139-1153

Ammar, E.D. 1994. Propagative transmission of plant and animal viruses by insects: factors affecting vector specificity and competence. Adv. Dis. Vector Res. 10: 289-331

Ammar, E.D. and Nault, L.R. 1985. Assembly and accumulation sites of maize mosaic virus in planthopper vector. Intervirology 24: 33-41

Ammar, E.D. and Nault, L.R. 1991. Maize chlorotic dwarf viruslike particles associated with the foregut in vector and nonvector leafhopper species. Phytopathology 81: 444-448

Ammar, E.D. and Nault, L.R. 2002. Virus transmission by leafhoppers, planthoppers and treehoppers (Auchenorrhyncha, Homoptera). In: Advances in Botanical Research, Vol. 36, R.T. Plumb (ed). Academic Press, San Diego, USA. pp. 141-167

Ammar, E.D., Jarlfors, U. and Pirone, T.P. 1994. Association of potyvirus helper component protein with virions and the cuticle lining the maxillary food canal and foregut of an aphid vector. Phytopathology 84: 1054-1060

Andret-Link, P., Schmitt-Keichinger, C., Demangeat, G., Komar, V. and Fuchs, M. 2004. The specific transmission of *Grapevine fan leaf virus* by its nematode vector *Xiphinema index* is solely determined by the viral coat protein. Virology 320: 12-22

Asanzi, M.C., Bosque-Perez, N.A., Nault, L.R., Gordon, D.T. and Thottappilly, G. 1995. Biology of *Cicadulina* species and transmission of maize streak virus. African Entomol. 3: 173-179

Atreya, C.D. and Pirone, T.P. 1993. Mutational analysis of the helper component-proteinase gene of a potyvirus: Effects of amino acid substitutions, deletions, and gene replacement on virulence and aphid transmissibility. Proc. Nat. Acad. Sci. USA 90: 11919-11923

Atreya, C.D., Raccah, B. and Pirone, T.P. 1990. A point mutation in the coat protein abolishes aphid transmissibility of a potyvirus. Virology 178: 161-165

Atreya, P.L., Atreya, C.D. and Pirone, T.P. 1991. Amino acid substitutions in the coat protein result in loss of insect transmissibility of a plant virus. Proc. Nat. Acad. Sci. USA 88: 7887-7891

Atreya, C.D., Atreya, P.L., Thornbury, D.W. and Pirone, T.P. 1992. Site-directed mutations in the potyvirus HC-PRO gene affect helper component activity, virus accumulation and symptom expression in infected tobacco plants. Virology 191: 106-111

Atreya, P.L., Lopez-Moya, J.J., Chu, M., Atreya, C.D. and Pirone, T.P. 1995. Mutation analysis of the coat protein N-terminal amino acids involved in potyvirus transmission by aphids. J. Gen. Virol. 76: 265-270

Azzam, O., Frazer, J., de la Rosa, D., Beaver, J.S., Ahlquist, P. and Maxwell, D.P. 1994. Whitefly transmission and efficient ssDNA accumulation of bean golden mosaic geminivirus require functional coat protein. Virology 204: 289-296

Bandla, M.D., Campbell, L.R., Ullman, D.E. and Sherwood, J.L. 1998. Interaction of tomato spotted wilt tospovirus (TSWV) glycoproteins with a thrips mid-gut protein, a potential cellular receptor for TSWV. Phytopathology 88: 98-104

Banttari, E.F. and Zeyen, R.J. 1976. Multiplication of oat blue dwarf virus in the aster leafhopper. Phytopathology 66: 89-90

Baumann, P., Munson, M.A., Lai, C.-Y., Clark, M.A., Baumann. L., Moran, N.A. and Campbell, B.C. 1993. Origin and properties of bacterial endosymbionts of aphids, whiteflies, and mealybugs. American Soc. Microbiol. News 59: 21-24

Baumann, P., Baumann, L., Lai, C.-Y., Rouhbakhsh, D., Moran, N.A. and Clark, M.A. 1995. Genetics, physiology, and evolutionary relationships of the genus *Buchnera*: intracellular symbionts of aphids. Annu. Rev. Microbiol. 49: 59-94

Ben-Dov, Y. 1993. A systematic catalogue of the soft scale insects of the world. Flora & Fauna Handbook No. 9. Sand Hill Crane Press Inc. Gainesville, Florida, USA.

Berdiales, B., Bernal, J.J., Saez, E., Woudt, B., Beitia, F. and Rodriguez-Cerezo, E. 1999. Occurrence of cucurbit yellow stunting disorder virus (CYSDV) and beet pseudo-yellows virus in cucurbit crops in Spain and transmission of CYSDV by two biotypes of *Bemisia tabaci*. Euro. J. Plant Pathol. 105: 211-215

Bigger, M. 1981. The relative abundance of the mealybug vectors (Homoptera, Coccidae and Pseudococcidae) of cocoa swollen shoot disease in Ghana. Bull. Entomol. Res. 71: 435-445

Blanc, S., Cerutti, C.H., Louis, C., Devauchelle, G. and Hull, R. 1993. Gene II product of an aphid non-transmissible isolate of cauliflower mosaic virus expressed in a baculovirus system possesses aphid transmission fsctor activity. Virology 192: 651-654

Blanc, S., Lopez-Moya, J.J., Wang, R.Y., Garcia-Lampasona, S., Thornbury, D.W. and Pirone, T.P. 1997. A specific interaction between coat protein and helper component correlates with aphid transmission of a potyvirus. Virology 231: 141-147

Brault, V., van den Heuvel, J.F.J.M., Verbeek, M., Ziegler-Graff, V., Reutenauer, A., Herrbach, E., Garaud, J.-C., Guilley, H., Richards, K. and Jonard, G. 1995. Aphid transmission of beet western yellows luteovirus requires the minor capsid read through protein P74. EMBO J 14: 650-659

Brault, V., Mutterer, J., Scheidecker, D., Simonis, N.T., Herrbach, E., Richards, K. and Ziegler-Graff, V. 2000. Effects of point mutations in the readthrough domain of the beet western yellows virus minor capsid protein on virus accumulation in plants and on transmission by aphids. J. Virol. 74: 1140-1148

Briddon, R.W., Pinner, M.S., Stanley, J. and Markham, P.G. 1990. Geminivirus coat protein gene replacement alters insect specificity. Virology 177: 85-94

Briddon, R.W., Belford, I.D., Tsai, J.H. and Markham, P.G. 1996. Analysis of the nucleotide sequence of the treehopper-transmitted geminivirus, tomato pseudo-curly top virus, suggests a recombinant origin. Virology 219: 387-394

Brown, D.J.F. 1986. Transmission of virus by the progeny of crosses between *Xiphinema diversicaudatum* from Italy and Scotland. Revue de Nematologie 9: 71-74

Brown, D.J.F. and Ploeg, A.T. 1989. Specificity of transmission of tobravirus by their *(Para) Trichodorus* nematodes. J. Nematology 21: 553

Brown, D.J.F. and Weischer, B. 1998. Specificity, exclusivity and complementarity in the transmission of plant viruses by plant parasitic nematodes: an annotated terminology. Fundamental Revue of Nematology 21: 1-11

Brown, D.J.F., Halbrendt, J.M., Robbins, R.T. and Vrain, T.C. 1993. Transmission of nepoviruses by *Xiphinema americanum*-group nematodes. J. Nematology 25: 349-354

Brown, D.J.F., Halbrendt, J.M., Jones, A.T., Vrain, T.C. and Robbins, R.T. 1994. Transmission of three North American nepoviruses by populations of distinct *Xiphinema americanum*-group species (Nematoda: Dorylaimida). Phytopathology 84: 646-649

Brown, D.J.F., Robbins, R.T., Zanzinger, D. and Wickizer, S. 1996a. Transmission of tobacco and tomato ring spot nepoviruses by *Xiphinema americanum*-group nematodes from 29 USA populations. Nematropica 26: 246

Brown, D.J.F., Trudgill, D.L. and Robertson, W.L. 1996b. Nepoviruses: Transmission by nematodes. In: The Plant Viruses. Vol. 5: B.D. Harrison and A.F. Murant (eds). Plenum Press, New York, USA. pp. 187-209

Brown, J.K. and Czosnek, H. 2002. Whitefly transmission of plant viruses. In: Advances, in Botanical Research, Plant Virus Vector Interactions. R.T. Plumb (ed). Academic Press, San Diego, USA. pp. 65-92.

Brunt, A.A., Crabtree, K., Dallwitz, M.J., Gibbs, A.J. and Watson, L. 1996. Viruses of Plants. CAB International, Wallingford, UK.

Bruyère, A., Brault, V., Ziegler-Graff, V., Simonis, M.T., van den Heuvel, J.F.J.M., Richards, K., Guilley, H., Jonard, G. and Herrbach, E. 1997. Effects of mutations in the beet western yellows virus readthrough protein on its expression and packaging and on virus accumulation, symptoms, and aphid transmission. Virology 230: 323-334

Burrows, M.E., Caillaud, M.C., Smith, D.M., Benson, E.C., Gildow, F.E. and Gray, S.M. 2006. Genetic regulation of polerovirus and luteovirus transmission in the aphid *Schizaphis graminum*. Phytopathology 96: 828-837

Cabaleiro, C. and Sagura, A. 1997a. Field transmission of grapevine leafroll associated virus 3 (GLRaV-3) by the mealybug *Planococcus citri*. Plant Dis. 81: 283-287

Cabaleiro, C. and Segura, A. 1997b. Some characteristics of the transmission of grapevine leafroll associated virus 3 by *Planococcus citri* Risso. Eur. J. Plant Path. 103: 373-378

Caciagali, P., Bosco, D. and Al-Bitar, L. 1995. Relationships of Sardinian isolate of tomato yellow leaf curl geminivirus with its whitefly vector *Bemisia tabaci* Genn. Euro. J. Plant Path. 101: 163-170

Campbell, R.N. 1962. Relationship between lettuce big-vein virus and its vector, *Olpidium brassicae*. Nature 195: 675-677

Campbell, R.N. 1996. Fungal transmission of plant viruses. Annu. Rev. Phytopathology 34: 87-108

Campbell, R.N. and Fry, P.R. 1966. The nature of the associations between *Olpidium brassicae* and lettuce big vein and tobacco necrosis viruses. Virology 29: 222-233

Campbell, R.N., Sim, S.T. and Lecoq, H. 1995. Virus transmission by host-specific strains of *Olpidium bornovanus* and *Olpidium brassicae*. Euro. J. Plant Path. 101: 273-282

Campbell, R.N., Wipf-Scheibel, C. and Lecoq, H. 1996. Vector-assisted seed transmission of melon necrotic spot virus in melon. Phytopathology 86: 1294-1298

Celix, A., Lopez-Sese, A., Almarza, N., Gomez-Guillamon, M.L. and Rodriguez-Cerezo, L. 1996. Characterization of cucurbit yellow stunting disorder virus, a *Bemisia tabaci*-transmitted closterovirus. Phytopathology 86: 1370-1376

Chagas, C.M., Rossetti, V. and Chiavegato, L.G. 1984. Transmission of leprosies by grafting. Proc. 9th Conf. IOCV S.M. Garnsey, L.W. Timmer and J.A. Dodds (eds). Univ. Calif. Riverside, CA., USA. pp. 215-217.

Chakravarty, S., Ghosh, A.B. and Mukhopadhyay, S. 1979. Biology of the rice green leafhopper *Nephotettix virescens*. Int. Rice Res. Newsl. 4(1): 16-17

Chay, C.A., Gunasinge, U.B., Dinesh-Kumar, S.P., Miller, W.A. and Gray, S.T. 1996. Aphid transmission and systemic plant infection determinants of barley yellow dwarf luteovirus-PAV are contained in the coat protein read-through domain and 17-kDa protein respectively. Virology 219: 57-65

Chen, B. and Francki, R.I.B. 1990. Cucumovirus transmission by the aphid *Myzus persicae* is determined solely by the viral coat protein. J. Gen. Virol. 71: 939-944

Chen, D.Y., Ni, H.F., Yen, J.H., Cheng, Y.H. and Tsay, T.T. 2004. Variability within *Xiphinema elongatum* population in Taiwan. Plant Path. Bull. 13: 45-60

Chen, J., Swaby, A.G., Adams, M.J. and Ruan, Y. 1991. Barley mild mosaic virus inside its fungal vector, *Polymyxa graminis*. Annals Appl. Biol. 118: 615-621

Chen, J., Wang, Z., Hong, Z., Collier, C.R. and Adams, M.J. 1998. Ultrastructural studies of resting spore development in *Polymyxa graminis*. Myco. Res. 102: 687-891

Chen, M.J. and Shikata, E. 1971. Morphology and intracellular localization of rice transitory yellowing virus. Virology 46: 786-796

Choudhury, M.M. and Rosenkranz, E. 1983. Vector relationships of *Graminella nigrifrons* and maize chlorotic dwarf virus. Phytopathology 73: 685-690

Collar, J.L., Avilla, C. and Fereres, A. 1997. New correlations between aphid stylar paths and nonpersistent virus transmission. Environ. Entomol. 26: 537-544

Colvin, J., Omongo, C.A., Maruthi, M.N., Otim-Nape, G.W. and Thresh, J.M. 2004. Dual begomovirus infections and high *Bemisia tabaci* populations: Two factors driving the spread of a cassava mosaic disease pandemic. Plant Pathol. 53: 577-584

Coutts, R.H.A. and Coffin, R.S. 1996. Beet pseudo-yellows virus is an authentic closterovirus. Virus Genes 13: 179-181

Creamer, R., Nault, L.R. and Gingery, R.E. 1993. Biological factors affecting leafhopper transmission of purified maize chlorotic dwarf virus. Entomologia Experimentalis et Applicata 67: 65-71

Dahal, G., Hibino, H. and Saxena, R.C. 1990. Further studies on green leafhopper (GLH) feeding modes and tungro transmission. Int. Rice Res. Newsl. 15: 31-32

Das, B.K. 1996. Bio-geography of rice green leafhoppers *Nephotettix* spp. (Homoptera: Cicadellidae) in West Bengal. PhD Thesis, Bidhan Chandra Krishi Viswavidyalaya, West Bengal, India.

David, B.V. and Jesudasan, R.W.A. 2002. A new species of *Aleurocanthus* collected from banana (*Musa* sp.) in Andaman and Nicobar Islands. Entomon. 27: 323-325

de Assis Filho, P.M., Decom, C.M., Sherwood, J.L. and de Assis Filho, F.M. 2004. Acquisition of tomato spotted wilt by adults of two thrips species. Phytopathology 94(4): 333-336

de Barro, P.J., Trueman, J.W.H. and Frohlich, D.R. 2005. *Bemisia argentifolii* is a race of *B. tabaci* (Hemiptera: Aleyrodidae). The molecular differentiation of *B. tabaci* populations around the world. Bull. Entomol. Res. 95: 1-11

Demichelis, S., Arna, C., Bosco, D., Marian, D. and Caciagli, P. 2005. Characterization of Biotype T of *Bemisia tabaci* associated with *Euphorbia characias* in Sicily. Phytoparasitica 33: 196-208

Diao, A., Chen, J., Gitton, F., Antoniw, J.F., Mullins, J., Hall, A.M. and Adams, M.J. 1999. Sequences of European wheat mosaic virus and oat golden stripe virus and genome analysis of the genus *Furovirus*. Virology 261: 331-339

Dias, H.F. 1970. The relationship between cucumber necrosis virus and its vector, *Olpidium cucurbitacearum*. Virology 42: 204-211

Dijkstra, J. and Khan, J.A. 2002. Virus transmission by fungal vector. In: Plant Viruses and Molecular Pathogens. J.A. Khan and J. Dijkstra (eds). CBS Publishers & Distributors, New Delhi, India. pp. 77-98

Drucker, M., Froissart, R., Herbrad, E., Uzest, M., Ravalleo, M., Esperandieu, P., Mani, J.C., Pugniere, M., Roquet, F., Fereres, A. and Blanc, S. 2002. Intracellular distribution of viral gene products regulates a complex mechanism of cauliflower mosaic virus acquisition by its aphid vector. Proc. Natl. Acad. Sci. USA 99: 2422-2427

Dubois, F., Sangwan, R.S. and Sangwan-Norreel, B.S. 1994. Immunogold labelling and electron-microscopic screening of beet necrotic yellow vein virus in the fungus *Polymyxa betae* infecting *Beta vulgaris* root cortical parenchyma cells. Intl. J. Plant Sci. 155: 545-552

Duffus, J.E. 1965. Beet pseudo-yellows virus, transmitted by the greenhouse whitefly (*Trialeurodes vaporariorum*). Phytopathology 55: 450-453

Douglas, A.E. 1998. Nutritional interactions in the insect-microbial symbioses: aphids and their symbiotic bacteria *Buchnera*. Annu. Rev. Entomol. 43: 17-37

Eastop, V.F. 1983. The biology of principal aphid virus vectors. In: Plant Virus Epidemiology. R.T. Plumb and J.M. Thresh (eds). Blackwell Scientific Publications, Oxford, UK. pp. 115-132.

Falk, B.W. and Tsai, J.H. 1998. Biology and molecular biology of viruses in the genus *Tenuivirus*. Annu. Rev. Phytopath. 36: 139-163

Falk, B.W. and Tian, T. 1999. Transcapsidation interactions and dependent aphid transmission among luteoviruses and luteovirus-associated RNAs. In: The Luteoviridae. H.G. Smith and H. Barker (eds). CAB International, Wallingford, UK. pp. 125-134.

Falk, B.W., Tsai, J.H. and Lommel, S.A. 1987. Differences in the levels of detection for the maize stripe virus capsid and major non-capsid proteins in plants and insect hosts. J. Gen. Virol. 68: 1801-1811

Flasinski, S. and Cassidy, B.G. 1998. Potyvirus aphid transmission requires helper component and homologous coat protein for maximal efficiency. Arch. Virol. 143: 2159-2172

Freeman, T.P., Buckner, J.S., Nelson, D.R., Chu, C. and Henneberry, T.J. 2001. The process of stylet penetration by the silver leaf whitefly, *Bemisia argentifolii* (Homoptera: Aleyrodidae) into host leaf tissue. Annals Entomol. Soc. America 94(5): 761-768

Freitag, J.H. 1956. Beetle transmission, host range, and properties of squash mosaic virus. Phytopathology 46: 73-81

Froissart, R., Michalakis, Y. and Blanc, S. 2002. Helper component-transcomplementation in the vector transmission of plant viruses. Phytopathology 92: 576-579

Fulton, J.P. and Scott, H.A. 1974. Virus vectoring efficiency of two species of leaf-feeding beetles. Proc. Am. Phytopath. Soc. 1: 159

Gamez, R. and Leon, R. 1988. Maize rayado fino and relative viruses. In: The Plant Viruses, Vol. 3, R. Koenig (ed.). Plenum. New York, USA. pp. 213-233

Garret, A., Kerlan, C. and Thomas, D. 1996. Ultrastructural study of acquisition and retention of potato leafroll luteovirus in the alimentary canal of its aphid vector, *Myzus persicae* Sulz. Archiv. Virol. 141: 1279-1292

Gera, A., Loebenstein, G. and Raccah, B. 1978. Detection of cucumber mosaic virus in viruliferous aphids by enzyme-linked immunosorbent assay. Virology 86: 5442-5445

Gergerich, R.C. 2002. Beetles. In: Advances in Botanical Research. Vol. 36, R.T. Plumb (ed). Academic Press, San Diego, USA. pp. 101-112

Gergerich, R.C., Scott, H.A. and Fulton, J.P. 1986. Evidence that ribonuclease in beetle regurgitant determines the transmission of plant viruses. Phytopathology 76: 1122

German, T.L., Ullman, D.E. and Moyer, J.W. 1992. *Tospoviruses*: diagnosis, molecular biology, phylogeny, and vector relationships. Annu. Rev. Phytopath 30: 315-348

Ghabrial, S.A. and Schultz, F.J. 1983. Serological detection of bean pod mottle virus in bean leaf beetles. Phytopathology 73: 480-483

Ghanim, M. and Czosnek, H. 2000. Tomato yellow leaf curl geminivirus (TYLCV-Is) is transmitted among whiteflies (*Bemisia tabaci*) in a sex-related manner. J. Virol. 74: 4738-4745

Ghanim, M., Rosell, R.C., Campbell, L.R., Czosnek, H., Brown, J.K. and Ullman, D.E. 2000. Microscopic analysis of the digestive, salivary and reproductive organs of *Bemisia tabaci* (Gennadius) (Hemiptera: Aleyrodidae) Biotype B. J. Morphology 248: 22-40

Ghanim, M., Morin, S. and Czosnek, H. 2001. Rate of tomato yellow leaf curl virus (TYLCV-Is) translocation in the circulative transmission pathway in its vector, the whitefly *Bemisia tabaci*. Phytopathology 91: 188-196

Ghosh, A.B. and Mukhopadhyay, S. 1978. Varietal preferences of the green leafhoppers. Int. Rice Res. Newsl. 3(6): 13

Gibbs, K.S. and Randles, J.W. 1988. Studies on the transmission of velvet tobacco mottle virus by the mirid, *Cyrtopelti nicotianae*. Annals Appl. Biol. 112: 427-437

Gibbs, K.S. and Randles, J.W. 1990. Distribution of velvet tobacco mottle virus in its mirid vector and its relationship to transmissibility. Annals Appl. Biol. 116: 513-521

Gibbs, K.S. and Randles, J.W. 1991. Transmission of velvet tobacco mottle virus and related viruses by the mirid *Cyrtopeltis nicotianae*. In: Advances in Disease Vector Research. Vol. 7. K. Harris (ed). Springer Verlag, New York, USA. pp. 1-17

Gildow, F.E. 1982. Coated vesicle transport of luteovirus through the salivary gland of *Myzus persicae*. Phytopathology 72: 1289-1296

Gildow, F.E. 1985. Intracellular transport of barley yellow dwarf virus into the hoemocoel of the aphid vector *Ropalosiphum padi*. Phytopathology 75: 292-297

Gildow, F.E. 1993. Evidence for receptor-mediated endocytosis regulating luteovirus acquisition by aphids. Phytopathology 83: 270-277

Gildow, F.E. 1999. Luteovirus transmission and mechanisms regulating vector specificity. In: The Luteoviridae. H.G. Smith and H. Barkers (eds). CAB International, Wallingford, UK. pp. 88-112

Gildow, F.E. and Gray, S.M. 1993. The aphid salivary gland basal lamina as a selective barrier associated with a vector-specific transmission of barley yellow dwarf luteovirus. Phytopathology 83: 1293-1302

Gildow, F.E. and Gray, S.M. 1994. The aphid salivary gland basal lamina as a selective barrier associated with vector-specific transmission of barley yellow dwarf luteovirus. Phytopathology 83: 1293-1302

Gildow, F.E., Damsteegt, V.D. Stone, A.L., Smith, O.P. and Gray, S.M. 2000. Virus vector cell interactions regulating transmission specificity of soybean dwarf lutoviruses. J. Phytopathol. 148: 333-342

Gildow. F.E., Reavy, B., Mayo, M.A., Duncan, G.H., Woodford, T., Lamb, J.W. and Hay, R.T. 2000. Aphid acquisition and cellular transport of potato leaf roll virus-like particles lacking P5 readthrough. Phytopathology 90: 1153-1161

Gingery, R.E., Gordon, D.T. and Nault, L.R. 1982. Purification and properties of maize rayado fino virus from the United States. Phytopathology 72: 1313-1318

Giorgini, M. 2001. A new spider mite injurious to rocket crops in Capania, Italy. Informatore-Fitopatologica 51: 88-91

Gispert, C., Oldfield, G.N., Perring, T.M. and Creamer, R. 1998. Biology of the transmission of peach mosaic virus by *Eriophyes insidiosus* (Aceri: Eriophyidae). Plant Disease 82: 1371- 1374

Gnezdilov, V.M., Guglielmino, A. and D'-Ursov. 2003. A new genus and new species of the family Issidae (Homoptera: Cicadina) from West Mediterranean Region. Russian Entomol. J. 12: 183-185

Goldbach, R. 1998 Binding of tomato spotted wilt virus to a 94-kDa thrips protein. Phytopathology 81: 1087-1091

Goldman, V. and Czosnek, H. 2002. Whiteflies (*Bemisia tabaci*) issued from eggs bombarded with infectious DNA clones of Tomato yellow leaf curl virus (TYLCV) from Israel are able to infect tomato plants. Archiv. Virol. 147: 787-801

Gonzales, W.L., Fuentes-Contreras, E. and Niemeyer, H.M. 1998. A new aphid species (Hemiptera: Aphididae) found in Chile: *Sipha flava* (Forbes). Revista Chilena de Entomologia 25: 87-90

Govier, D.A. and Kassanis, B. 1974. A virus-induced component of plant sap needed when aphids acquire potato virus Y from purified preparations. Virology 61: 420-426

Granara de Willink, M.C. and Miller, D.R. 2004. Two new species of mealy bugs (Hemiptera: Coccoidea: Pseudococcidae) from Patagonia, Argentina. Proc. Ent. Soc. Washington 106: 140-158

Gray, S.M. 1999. Intraspecific variability of luteovirus transmission within aphid vector populations. In: The Luteoviridae. H.G. Smith and H. Barker (eds). CAB International, Wallingford, UK. pp. 119-123

Gray, S.M. and Rochon, D'.A. 1999. Vector transmission of plant viruses. In: Encyclopedia of Virology. A. Granoff and R.G Webster (eds), 2nd edit. Academic Press, San Diego, USA. pp. 1899-1910

Gray, S. and Gildow, F.E. 2003. Luteovirus-aphid interactions. Annu. Rev. Phytopath. 41: 539-566

Griesbach, J.A. and Maggenti, A.R. 1990. Vector capability of *Xiphinema americanum* in California. Calif. J. Nematology 21: 617-523

Groves, R.L., Walgenbach, J.F., Moyer, J.W. and Kennedy, G.G. 2001. Over-wintering of *Frankliniella fusca* (Thysanoptera: Thripidae) on winter annual weeds infected with *Tomato spotted wilt virus* and patterns of virus movement between susceptible weed hosts. Phytopathology 91(9): 891-899

Gullan, P.J., Downia, D.A. and Steffan, S.A. 2003. A new pest species of the mealy bug genus *Frisia fullaway* (Hemiptera: Pseudococcidae) from the United States. Ann. Ento. Soc. America 96: 723-737

Haeberlé, A.M., Stussigaraud, C., Schmitt, C., Garaud, J.C., Richards, K.E., Guilley, H. and Jonard, G. 1994. Detection by immunogold labeling of P 75 readthrough protein near an extremity of beet necrotic yellow vein virus-particles. Archiv. Virol. 134: 195-203

Han, C.G., Wang, D.Y., Yu, J.L., Li, D.W., Yang, L.L., Cai, Z.N. and Liu, Y. 1996. The localization of functional sequence on RNA 4 of beet necrotic yellow vein virus (BNYVV) related to fungus transmission by inoculation with infectious RNA 4 and its mutants transcripted *in vitro*. In: Proceedings of the Third Symposium of the International Working Group on Plant Viruses with Fungal Vectors. J.L. Sherwood and C.M. Rush (eds). American Society of Sugarbeet Technologists. Denver, CO. USA. pp. 17-20

Harrington, R., Katis, N. and Gibson, R.W. 1986. Field assessment of the relative importance of different species in the transmission of potato virus Y. Potato Research 29: 67-76

Harris, K.F. 1977. An ingestion-egestion hypothesis of non-circulative virus transmission by aphids. In: Aphids as Virus Vectors. K.F. Harris and K. Maramorosch (eds). Academic Press, New York, USA. pp. 165-220

Harris, K.F. 1979. Leafhoppers and aphids as biological vectors: vector-virus relationships. In: Leafhopper Vectors and Plant Disease Agents. K. Maramorosch and K.F. Harris (eds). Academic Press, New York, USA. pp. 217-308

Harris, K.F. 1983. Sternorrychous vectors of plant viruses: virus-vector interactions and transmission mechanisms. Adv. Virus Res. 28: 113-140

Harrison, B.D. 1985. Advances in geminivirus research. Annu. Rev. Phytopath. 23: 209-241

Hernandez, C., Visser, P.B., Brown, D.J.F. and Bol, J.F. 1997. Transmission of tobacco rattle virus isolate Ppk20 by its nematode vector requires one of the two non-structural genes in the viral RNA2. J. Gen. Virol. 78: 465-467

Hibino, H. 1989. Insect-borne viruses of rice. Adv. Dis. Vector Res. 6: 209-241

Hibino, H. and Cabauatan, P.Q. 1987. Infectivity neutralization of rice tungro associated viruses acquired by vector leafhoppers. Phytopathology 77: 473-476

Hodgson, C.J. and Millar, I.M. 2002. A new sub-family, two new genera and three new species of Aclerdidae (Hemiptera: Coccoidea) from Southern Africa with a phylogenetic analysis of relationships. Systematic Entomol. 27: 469-517

Hofer, P., Bedford, I.D., Markham, P.G., Jeske, H. and Frischmuth, T. 1997. Coat protein gene replacement results in whitefly transmission of an insect non-transmissible geminivirus isolate. Virology 236: 288-295

Hogenhout, S.A., van der Wilk, F., Verbeek, M., Goldbach, R.W. and van den Heuvel, J.F.M. 1998. Potato leafroll virus binds to the equatorial domain of the aphid endosymbiotic GroEL homologue. J. Virol. 72: 358-365

Hogenhout, S.A., van der Wilk, F., Verbeek, M., Goldbach, R.W. and van den Heuvel, J.F.M. 2000. Identifying the determinants in equatorial domain of *Buchnera* GroEL implicated in binding potato leafroll virus. J. Virol. 74: 4541-4548

Honle, M., Hofer, P., Bedford, I.D., Briddon, R.W., Markham, P.G. and Frischmuth, T. 2001. Exchange of two amino acids in the coat protein result in whitefly-transmission of a non-transmissible *Abutilon mosaic virus*. 3rd International geminivirus Symposium, Abstract p. 35. John Innes Centre, Norwich, UK

Horn, N.M., Reddy, S.V. and Reddy, D.V.R. 1994. Virus-vector relationships of chickpea chlorotic dwarf geminivirus and leafhopper *Orosius orientalis* (Homoptera: Cicadellidae). Annals Appl. Biol. 124: 441-450

Hoy, J.W., Mircetich, S.M. and Lownsbery, B.F. 1984. Differential transmission of prunus tomato ring spot virus strains by *Xiphinema californicum*. Phytopathology 74: 332-335

Huet, H., Gal-On, A., Meor, E., Lecoq, H. and Raccah, B. 1994. Mutations in helper component protease gene of zucchini yellow mosaic virus affect its ability to mediate aphid transmissibility. J. Gen. Virol. 75: 1402-1414

Hull, R. 2002. Matthews Plant Virology. Academic Press, San Diego, USA. pp. 250-252

Hunter, W.B., Hiebert, E., Webb, S.E., Tsai, H. and Polston, J.E. 1998. Location of the geminiviruses in the whitefly *Bemisia tabaci* (Homoptera: Aleyrodidae). Plant Dis. 82: 1147-1151

Inoue, H. and Hirao, J. 1981. Transmission of rica waika virus by green rice leafhoppers, *Nephotettix* spp. (Hemiptera: Cicadellidae). Bull. Kyushu Nat. Agric. Exp. Sta. 21: 509-552

Jacobi, V., Peerenboom, E., Schenk, P.M., Antoniw, J.F., Steinbiss, H.-H. and Adams, M.J. 1995. Cloning and sequence analysis of RNA-2 of a mechanically-transmitted UK-isolate of barley mild mosaic bymovirus (BaMMV). Virus Res. 37: 99-111

Jagadish, S.K. and Jayaramaiah, M. 2004. Biology of VFC tobacco aphid (Homoptera: Aphididae). J. Eco-Biol. 16: 93-97

Jelkmann, W., Fechtner, B. and Agranovsky, A.A. 1997. Complete genome structure and phylogenetic analysis of little cherry virus, a mealybug-transmissible closterovirus. J. Gen. Virol. 78: 2067-2071

Jiang, Y.X., De Blas, C., Barrios, L. and Fereres, A. 2000. Correlation between whitefly (Homoptera: Aleyrodidae) feeding behaviour and transmission of tomato yellow leaf curl virus. Ento. Soc. America 93: 573-579

Jolly, C.A. and Mayo, M.A. 1994. Changes in the amino acid sequence of the coat protein readthrough domain of potato leafroll luteovirus affect the formation of an epitope and aphid transmission. Virology 201: 182-185

Jones, D.R. 2003. Plant viruses transmitted by whiteflies. Euro. J. Plant Path. 109(3): 195-219

Jonson, D.D., Walker, G.P. and Creamer, R. 2002. Stylet penetration behaviour resulting in inoculation of a semipersistently transmitted closterovirus by the whitefly *Bemisia argentifolii*. Entomologia Experimentalis et Applicata 102(2): 115-123

Kakanik, R., Robbins, M. and Rochon, D. 2003. Evidence that binding of cucumber necrosis virus to vector zoospores involves recognition of oligosaccharides. J. Virol. 77(7): 3922-3928

Kakanik, R., Reade, R. and Rochon, D. 2004. Evidence that vector transmission of a plant virus requires conformational change in virus particles. J. Mol. Biol. 338: 507-517

Karanastasi, E., Wyss, U. and Brown, D.J.F. 2003. An in vitro examination of the feeding behaviour of *Paratrichodorus* (Nematoda: Trichodoridae), with comments on the ability of the nematode to acquire and transmit Tobravirus particles. Nematology 5(3): 421-434

Kassanis, B. and Govier, D.A. 1971a. New evidence on the mechanism of transmission of potato virus C and potato aucuba mosaic viruses. J. Gen. Virol. 10: 99-101

Kassanis, B. and Govier, D.A. 1971b. The role of the helper virus in aphid transmission of potato aucuba mosaic virus and potato virus C. J. Gen. Virol. 13: 221-228

Kennedy, J.S., Day, M.F. and Eastop, V.F. 1962. A Conspectus of Aphids as Vectors of Plant Viruses. CAB, London, UK.

Kikkert, M., Meurs, C., Van de Wetering, F., Dorfmuller, S., Peters, D., Kormelink, R. and Goldbach, R. 1998. Binding of tomato spotted wilt virus to a 94 kDa thrips protein. Phytopathology 88: 63-69

Kiran, K., Bhat, N.S. and Kumar, K. 2003 Developmental biology of leafhopper, *Amrasca biguttula biguttula* (Ishada) (Cicadellidae: Homoptera) on sunflower, *Helianthus annuus* (L.). Insect Environ. 9: 8-10

Koenig, R. 2000. Deletion in the KTER-encoding domain which is needed for *Polymyxa* transmission, in manually transmitted isolates of beet necrotic yellow vein benyvirus. Archiv. Virol. 145(1): 165-170

Korez, A. 2001. The forecasting possibilities of appearance of the pest creating gradations on the example of beet leaf bug (*Piesma quadratum* Fieb. Heteroptera: Piesmatidae). J. Plant Proc. Res. 41(2): 153-163

Kritzman, A., Gera, A., Raccah, B., van Lent, J.W.M., Peters, D. and van Lent, J.W.M. 2002. Tomato spotted wilt virus inside the thrips body in relation to transmission efficiency. Archiv. Virol. 147(11): 2143-2156

Lamberti, F., Molinari, S., Moens, M. and Brown, D.J.F. 2000. The *Xiphinema americanum* group. I. Putative species, their geographical occurrence and distribution, and regional polytomous identification keys for the group. Russian J. Nematology 8: 65-84

Langham, M.A.C., Gergerich, R.C. and Scott, H.A. 1990. Conversion of comovirus electrophoretic forms by leaf-feeding beetles. Phytopathology 80: 900-906

Lecoq, H. and Pitrat, M. 1985. Specificity of the helper-component mediated aphid transmission of three potyviruses infecting muskmelon. Phytopathology 75: 890-893

Lee, P.E. 1967. Morphology of wheat striate mosaic virus and its localizaton in infected plant cells. Virology 33: 84-94

Leh, V., Jacquot, E., Geldreich, A., Haas, M., Blanc, S., Keller, M. and Yot, A. 2001. Interaction between the open reading frame III product and the coat protein is required for transmission of cauliflower mosaic virus by aphids. J. Virol. 75(1): 100-106

Lemaire, O., Merdinoglu, D., Valentin, P., Putz, C., Ziegler-Graff, V., Guilley, H., Jonard, G. and Richards, K. 1988. Effect of beet necrotic yellow vein virus RNA composition on transmission by *Polymyxa betae*. Virology 162: 232-235

Ling, K.C. and Tiongco, E.R. 1979. Transmission of rice tungro at various temperatures: a transitory virus-vector interaction. In: Leafhopper Vectors and Plant Disease Agents. K. Maramorosch and K.F. Harris (eds). Academic Press, New York, USA. pp. 349-366

Liu, S.J., Bonning, B.C. and Miller, W.A. 2006. A simple wax-embedding method for isolation of aphid hemolymph for detection of luteoviruses in the hemocoel. J. Virol. Methods 132: 174-180

Livieratos, I.C., Katis, N. and Coutts, R.H.A. 1998. Differentiation between cucurbit yellow stunting disorder virus and b eet pseudo-yellows virus by a reverse transcription polymerase chain reaction assay. Plant Pathol. 47: 362-369

Lopez-Moya, J.J., Wang, R.Y. and Pirone, T.P. 1999. Context of coat protein DAG motif affects poyvirus transmissibility by aphids. J. Gen. Virol 80: 3281-3288

Lotfalizadeh, H., Hatami, B. and Khalaghani, J. 2000. Biology of *Exochomus quadripustulatus* (L.) (Col.: Coccinellidae) on cypress tree mealy bug, *Planococcus vovae* (Nasanov) (Hom.: Pseudococcidae) in Shiraz. J. Ent. Soc. Iran 20: 61-76

Lung, M.C.Y. and Pirone, T.P. 1974. Acquisition factor required for aphid transmission of purified cauliflower mosaic virus. Virology 60: 260-264

MacFarlane, S.A., Brown, D.J.F. and Bol, J.F. 1995. The transmission of nematodes of Tobraviruses is not determined exclusively by the viral coat protein. Eu. J. Plant Pathol. 101: 535-539

MacFarlane, S.A., Wallis, C.V. and Brown, D.J.F. 1996. Multiple virus genes involved in the nematode transmission of pea early browning virus. Virology 219: 417-422

MacFarlane, S.A., Neilson, R. and Brown, D.J.F. 2002 Nematodes. In: Advances in Botanical Research: Plant Virus Vector Interactions. Vol. 36, R.T. Plumb (ed). Academic Press, San Diego, USA. pp. 169-198

Mahmood, T., Hein, G.L. and French, R.C. 1997. Development of serological procedures for rapid and reliable detection of wheat streak mosaic virus in a single wheat curl mite. Plant Disease 81: 250-253

Markham, P.G. 1992. Transmission of maize streak virus by *Cicadulina* sp. XIXth Int. Cong. Entomol. 28 June-4th July. Beijing, China, p. 345.

Martin, B., Collar, L.J., Tjallingi, W.E. and Fereres, A. 1997. Intracellular ingestion and salivation by aphids may cause the acquisition and inoculation of non-persistently transmitted plant viruses. J. Gen. Virol. 78: 2701-2705

Matthews, R.E.F. 1992. Fundamentals of Plant Virology. Academic Press, Inc., San Diego, New York. USA

Mayer, R.T., Inbar, M., McKenzie, C.L., Shatters, R., Borowicz, V., Albrecht, U., Powell, C.A. and Doostdar, H. 2002. Multitrophic interactions of the silver leaf whitefly, host plants, competing herbivores, and phytopathogens. Arch. Insect Biochem. Physiol. 51: 151-169

McGrath, P.F., Lister, R.M. and Hunter, B.G. 1996. A domain of the readthrough protein of barley yellow dwarf virus (NY-RPV isolate) is essential for aphid transmission. Eu. J. Plant Path. 102: 671-679

McLean, M.A., Campbell, R.N., Hamilton, R.I. and Rochon, D.M. 1994. Involvement of the cucumber necrosis virus coat protein in the specificity of fungus transmission by *Olpidium bornovanus*. Virology 204: 840-842

Merz, U. 1995. PMTV-like particles inside resting spores of *Spongospora subterranea*. J. Phytopathol. 143: 731-733

Morales, F.J. and Anderson, P.K. 2001. The emergence and dissemination of whitefly transmitted geminiviruses in Latin America. Archiv. Virol. 146: 415-441

Morales, F.J. and Jones, P.G. 2004. The ecology and epidemiology of whitefly transmitted viruses in Latin America. Virus Res. 100: 57-65

Morin, S., Ghanim, M., Zeidan, M., Czosnek, H., Verbeek, M. and Van den Heuvel, J.F.J.M. 1999. A GroEL homologue from endosymbiotic bacteria of whitefly *Bemisia tabaci* is implicated in the circulative transmission of tomato yellow leaf curl virus. Virology 256: 75-84

Morin, S., Ghanim, M., Sobol, I. and Czosnek, H. 2000. The GroEL protein of the whitefly *Bemisia tabaci* interacts with the coat protein of transmissible and non-transmissible begomoviruses in the yeast two-hybrid system. Virology 276: 404-416

Mound, L. 1996. The thysanopteran vector species of tospoviruses. Acta Horticulturae 431: 298-309

Mukhopadhyay, S. 1975. Varietal preference of *Nephotettix* spp. under artificial and field conditions. Int. Rice Path. Newsl. 1/75: 17

Mukhopadhyay, S. 1984 Ecology of Rice Tungro Virus and its vectors. In: Virus Ecology. A. Misra and H. Polasa (eds). South Asia Publishers, New Delhi, India. pp.139-164

Mukhopadhyay, S. 1985.Virus diseases of rice. In: Vistas in Plant Pathology. A. Varma and J.P. Verma (eds). Malhotra Publishing House, New Delhi, India. pp. 111-119

Mukhopadhyay, S. and Chattopadhyay, K. 1975. Preferential feeding of green leafhoppers. Int. Rice. Comm. Newsl. 24(2): 76-80

Mukhopadhyay, S., Mohsin, M.D. and Ghosh, P.K. 1976. Behaviour of rice virus vectors under different cultivational conditions. Proc. Nat. Acad. Sci. India 46 (B) I & II: 71-76

Musser, R.O., Hum-Musser, S.M., Felton, G.W. and Gergerich, R.C. 2003 Increased larval growth and preference for virus infected leaves by the Mexican bean beetle, *Epilachna varivestis* Mulsant, a plant virus vector. J. Insect Behavior 16: 247-256

Nagata, T., Storms, M.M.H., Goldbach, R. and Peters, D. 1997. Multiplication of the tomato spotted wilt virus in primary cell cultures derived from two thrips species. Virus Research 49: 59-66

Nagata, T., Inoue-Nagata, A.K., Smid, H.M., Goldbach, R. and Peters, D. 1999. Tissue tropism related to vector competence of *Frankliniella occidentalis* for tomato spotted wilt tospovirus. J. Gen. Virol. 80: 507-515

Nagata, T., Inoue-Nagataka, A.K., van Lent, J., Goldbach, R. and Peters, D. 2002. Factors determining vector competence and specificity for transmission of *Tomato spotted wilt virus*. J. Gen. Virol. 83: 663-671

Nagata, T., Almeida, A.C.L., Reseneda, R.O. and de Avila, A.C. 2004 The competence of four thrips species to transport and replicate four tospoviruses. Plant Path. 53: 136-140

Nault, L.R. 1994. Transmission biology, vector specificity, and evolution of planthopper transmitted viruses. In: Planthoppers, their Ecology and Management. R.F. Denno and T.J. Perfect (eds). Chapman and Hall, New York, USA. pp. 429-448

Nault, L.R. 1997 Arthropod transmission of plant viruses: A new synthesis. Ann. Entomol. Soc. Am. 90: 521-541

Nault, L.R. and Gordon, D.T. 1988 Multiplication of maize stripe virus in *Peregrinus maidis*. Phytopathology 78: 991-995

Nault, L.R. and Ammar, E.D. 1989. Leafhopper and planthopper transmission of plant viruses. Annu. Rev. Entomol. 34: 503-529

Navas-Castillo, J., Camero, R., Bueno, M. and Moriones, E. 2000. Severe yellowing outbreaks in tomato in Spain associated with infections of tomato chlorosis virus. Plant Disease 84: 835-837

Ng, J.C.K. and Falk, B.W. 2006. Virus-Vector Interactions Mediating Nonpersistent and Semipersistent Transmission of Plant Viruses. Annu. Rev. Phytopath. 44: 183-212

Norris, E., Vaira, A.M., Caciagli, P., Masenga, V., Gronenborn, B. and Accotto, G.P. 1998. Amino acids in the capsid protein of tomato yellow leaf curl virus that are crucial for systemic infection, particle formation, and insect transmission. J. Virol. 72: 10050-10057

Ofori, F.A. and Francki, R.I.B. 1985 Transmission of Leafhopper A virus, vertically through the eggs and horizontally through maize in which it does not multiply. Virology 144: 152-157

Ohnishi, D., Hosokawa, D., Fujisawa, I. and Tsuda, S. 1998. Recent progress in Tospovirus and Thrips Research. Fourth International Symposium on Tospoviruses and Thrips in Floral and Vegetable Crops. Wageningen, The Netherlands

Ohnishi, J., Hosokawa, D., Murari, T. and Tsuda, S. 1999. A simple rearing system for *Thrips setosus* Moulton (Thysanoptera: Thripidae) using a leaf cage method for the transmission experiment of tomato spotted wilt tospovirus. Appl. Entomol. Zool. 34(4): 497-500

Omura, T. and Yan, J. 1999 Role of outer capsid proteins in transmission of phytoreovirus in insect vectors. Adv. Virus Res. 54: 15-43

Omura, T., Yan, J., Zhong, B., Wada, M., Zhu, Y., Tomaru, M., Maruama, W., Kikuchi, A., Watanaba, Y., Kimura, I. and Hibino, H. 1998. The P2 protein of rice dwarf phytoreovirus is required for adsorption of the virus to cells of the insect vector. J. Virol. 72: 9370-9373

Orlob, G. 1966. Feeding and transmission characteristics of *Aceria tulipae* Keifer as vector of wheat streak mosaic virus. Phytopathol. Z. 55: 218-238

Orlob, G.B. and Takahashi, Y. 1971. Location of plant viruses in two-spotted spider mite *Tetranychus urticae* Koch. Phytopathol. Z. 72: 21-28

Palacios, I., Blanc, S., Leite, S. and Fereres, A. 2001. Studies of feeding behaviour related to the acquisition of cauliflower mosaic virus by aphids. In: Aphids in a New Millennium. Proc. VIth Intl. Symp. on Aphids. Rennes, France. pp. 471-477

Paliwal, Y.C. 1980. Relationship of wheat streak mosaic and barley stripe mosaic viruses to vector and non-vector eriophyid mites. Archiv. Virology 63: 123-132

Patel, P.S. and Patel, G.M.2001. Biology of the citrus blackfly, *Aleurocanthus woglumi* Ashby (Homoptera: Aleyrodidae) on kagzi lime. Pest Management and Economic Zoology 9(2): 147-150

Peerenboom, E., Jacobi, V., Antoniw, J.F., Schlichter, U.H.A., Cartwright, E.J., Steinbiss, H.-H. and Adams, M.J. 1996. The complete nucleotide sequence of RNA-2 of a fungally-transmitted UK isolate of barley mild mosaic bymovirus (BaMMV) and identification of amino acid compositions possibly involved in fungus transmission. Virus Res. 40: 149-159

Peiffer, M.L., Gildow, F.E. and Gray, S.M. 1997. Two distinct mechanisms regulate luteovirus transmission efficiency and specificity at the aphid salivary gland. J. Gen. Virol. 78: 495-503

Peng, R.H., Han, C.G., Yang, L.L., Yu, J.L. and Liu, Y. 1998. Cytological localization of beet necrotic yellow vein virus transmitted by *Polymyxa betae*. Acta Phytopathologica Sinica 28: 257-261

Perring, T.M. 2001. The *Bemicia tabaci* species complex. Crop Protect. 20: 725-737

Perry, K.L., Zhang, L., Shintaku, M.H. and Palukaitis, P. 1994 Mapping determinants in cucumber mosaic virus for transmission by *Aphis gossypii*. Virology 205: 591-595

Perry, K.L., Zhang, L. and Palukaitis, P. 1998. Amino acid changes in the coat protein of cucumber mosaic virus differentially affect transmission by the aphids *Myzus persicae* and *Aphis gossypii*. Virology 242: 204-210

Petersen, C.L. and Charles, J.G. 1997. Transmission of grapevine leafroll-associated closteroviruses by *Pseudococcus longispinus* and *P. calceolariae*. Plant Pathology 46: 509-515

Pirone, T.P. 1981. Efficiency and selectivity of the helper component mediated aphid transmission of purified potyviruses. Phytopathology 71: 922-924

Pirone, T.P. and Megahed, S. 1966. Aphid transmissibility of some purified viruses and viral RNAs. Virology 30: 631-637

Pirone, T.P. and Perry, K.L. 2002. Aphids: Non-persistent Transmission. In: Advances in Botanical Research: Plant Virus Vector Interactions, Vol. 36, R.T. Plumb (ed.). Academic Press, San Diego, USA. pp. 1-19

Ploeg, A.T., Brown, D.J.F. and Robinson, D.J. 1992. The association between species of *Trichodorus* and *Paratrichodorus* vector nematodes and serotypes of tobacco rattle virus. Annals Appl. Biol. 121: 619-630

Ploeg, A.T., Robertson, D.J. and Brown, D.J.F. 1993. RNA2 of tobacco rattle virus encodes the determinants of transmissibility by trichodorid vector-nematodes. J. Gen. Virol. 74: 1463-1466

Plumb, R.T. (ed.). 2002. Advances in Botanical Research: Plant Virus Vector Interactions. Vol. 36, Academic Press, San Diego, USA

Polston, J.E., Al-Musa, A., Perring, T.M. and Dodds, J.A. 1990. Association of the nucleic acid of squash leaf curl geminivirus with the whitefly *Bemisia tabaci*. Phytopathology 80: 850-856

Posnette, A.F. 1950. Virus diseases of cacao in West Africa, VII. Virus transmission by different vector species. Annals Appl. Biol. 37: 378-384

Posnette, A.F. and Robertson, N.F. 1950. Virus disease of cacao in West Africa. VI. Vector investigations. Annals Appl. Biol. 37: 363-377

Proeseler, G. 1980. Piesmids. In: Vectors of Plant Pathogens. K.F. Harris and K. Maramorosch (eds). Academic Press, New York, USA. pp. 97-113

Powell, G., Tosh, C.R. and Hardie, J. 2006. Host-plant selection by aphids: behavioral, evolutionary, applied perspectives. Annu. Rev. Entomol. 51: 309-330

Reavy, B. and Mayo, M.A. 2002. Persistent Transmission of Luteoviruses by Aphids. In: Advances in Botanical Research: Plant Virus Vector Interactions, Vol. 36, R.T. Plumb (ed.). Academic Press, San Diego, USA. pp. 21-46

Remadevi, O.K. and Mathukrishnan, R. 2002 A note on infestation of a new Coccoid pest *Hemilecanium imbricans* (Gren) (Hemiptera: Coccidae) on *Swetenia macrophylla* King. Annals of Forestry 10: 359-360

Richards, K. and Tamada, T. 1992. Mapping functions of the multi-partite genome of beet necrotic yellow vein virus. Annu. Rev. Phytopath. 30: 291-313

Ridley, A., Croft, B., Dhileepan, K., Walter, G. and Hogarth, D.M. 2004. How does the sugarcane plant hopper take up Fiji disease virus? Conf. Australian Soc. Sugarcane Technologists, 4-7 May.

Robbins, M.A., Reade, R.D. and Rochon, M.D. 1997. A cucumber necrosis virus variant deficient in fungal transmissibility contains an altered coat protein shell domain. Virology 234: 138-146

Robbins, M.A., Kakani, K. and Rochon, D. 1999. Evidence that fungal zoospores contain specific receptors for transmission of cucumber necrosis virus. In: Proceedings of the Fourth Symposium of the International Working Group on Plant Viruses with Fungal Vectors. J.L. Sherwood and C.M. Rush (eds). American Society of Sugarbeet Technologists. Denver, CO, USA. pp. 101-104

Robertson, N.L. and Carroll, T.W. 1988. Virus-like particles and a Spider Mite intimately associated with a new disease of barley. Science 240: 1188-1190

Robertson, W.M. and Henry, C.E. 1986. An association of carbohydrates with particles of arabis mosaic virus retained within *Xiphinema diversicaudatum*. Annals Appl. Biol. 109: 299-305

Rochow, W.F. 1970. Barley yellow dwarf virus: phenotypic mixing and vector specificity. Science 167: 875-878

Rochow, W.F. 1975. Barley yellow dwarf: dependent virus transmission by *Rhopalosiphum maidis* from mixed infections. Phytopathology 65: 99-105

Rosell, R., Lichty, J.E. and Brown, J.K. 1995. Ultrastucture of the mouthparts of adult sweet potato whitefly, *Bemisia tabaci* (Gennadius) (Homoptera: Aleyrodidae). International Journal of Insect Morphology and Embryology 24: 297-306

Rosell, R.C., Torres-Jerez, I. and Brown, J.K. 1999. Tracing the geminivirus whitefly transmission pathway by polymerase chain reaction in whitefly extracts, saliva, hemolymph, and honeydew. Phytopathology 89: 239-246

Rouzé-Jouan, J., Terradot, L., Pasquer, F., Tanguy, S. and Ducray-Bourdin, D.G. 2001. The passage of potato leafroll virus through *Myzus persicae* gut membrane regulates transmission efficiency. J. Gen. Virol. 82: 17-23

Rubinstein, G. and Czosnek, H. 1997. Long-term association of tomato yellow leaf curl virus (TYLCV) with its whitefly vector *Bemisia tabaci*: effect on the insect transmission capacity, longevity and fecundity. J. Gen. Virol. 78: 2683-2689

Ruiz, L., Janssen, D., Velasco, L., Segundo, E. and Cuadrodo, I.M. 2002.Quantitation of cucurbit yellow stunting disorder virus in *Bemisia tabaci* (Genn.) using digoxigenin-labelled hybridization probes. J. Virol. Methods 101(1-2): 95-103

Rysanek, P., Stocky, G., Haeberlé, A.M. and Putz, C. 1992. Immunogold labeling of beet necrotic yellow vein virus particles inside its fungal vector, *Polymyxa betae*. K. Agronomie 12: 651-659

Sako, N. and Ogata, K. 1981. Different helper factors associated with aphid transmission of some potyviruses. Virology 112: 762-765

Sako, N., Yoshioka, K. and Eguchi, K. 1984. Mediation of helper component in aphid transmission of some potyviruses. Ann. Phytopathol. Soc., Japan 50: 515-521

Sakurai, T., Inoue, T. and Tsuda, S. 2004. Distinct efficiencies of Impatiens necrotic spot virus transmission by five thrips vector species (Thysanoptera: Thripidae) of tospoviruses in Japan. Appl. Entomol. Zoology 39(1): 71-78

Sanderlin, R.S. 1973. Survival of bean pod mottle and cowpea mosaic viruses in beetles following intrahemocoelic injections. Phytopathology 63: 259-261

Sarkar, T.K. 1988. Studies on the ecology of rice green leafhoppers *Nephotettix virescens* Distant and *Nephotettix nigropictus* Stal (Homoptera: Cicadellidae) in West Bengal. PhD Thesis, Bidhan Chandra Krishi Viswavidyalaya, West Bengal, India

Schmitt, C., Balmori, E., Guilley, H., Richards, K. and Jonard, G. 1992. *In vitro* mutagenesis of biologically active transcripts of beet necrotic yellow vein virus RNA 2: evidence that a domain of the 75 kDa readthrough protein is important for efficient virus assembly. Proc. Nat. Acad. Sci. (USA) 89: 5715-5719

Schubert, J. and Rabenstein, F. 1995. Sequence of the 3′-terminal region of RNA of a mite transmitted potyvirus from *Hordeum murinum* L. Euro. J. Plant Path. 101: 123-132

Schwartz, M.D. and Scudder, G.G.E. 2003 Seven new species of Miridae (Heteroptera) from British Columbia and Alaska and synonymy of *Adelphocoris superbus* (Uhler.). J. New York Ento. Soc. 111: 2-3, 65-95

Scott, H.A. and Fulton, J.P. 1978. Comparison of the relationships of southern bean mosaic virus and the cowpea strain of tobacco mosaic virus with bean leaf beetle. Virology 84: 207-209

Seal, S.E., van den Bosch, F. and Jeger, M.J. 2006. Factors influencing begomovirus evolution and their increasing global significance: Implications for sustainable control. Crit. Rev. Plant Sci. 25: 23-46

Sether, D.M., Ullman, D.E. and Hu, J.S. 1998. Transmission of pineapple mealy bug wilt-associated virus by two species of mealy bugs (*Dysmicoccus* spp.). Phytopathology 88: 1224-1234

Shikata, E. 1979. Cytopathological changes in leafhopper vectors of plant viruses. In: Leafhopper Vectors and Plant Disease Agents. K. Maramorosch and K.F. Harris (eds). Academic Press, New York, USA. pp. 309-325

Simons, J.N. 1962. The pseudo-curly top disease in south Florida. J. Econ. Entomol. 55: 358-363

Sinha, R.C. 1981. Vertical transmission of plant pathogens. In: Vectors of Disease Agents. J.J. Mckelvey, Jr., B.F. Eldridge and K. Maramorosch (eds). Praeger, New York, USA. pp. 119-121.

Sinha, R.C. and Chiykowski, L.N. 1969. Synthesis, distribution and some multiplication sites of wheat striate mosaic virus in a leafhopper vector. Virology 38: 679-684

Slack, S.A. and Scott, H.A. 1971. Hemolymph as a reservoir for the cowpea strain of southern bean mosaic virus in the bean leaf beetle. Phytopathology 61: 538-540

Slykhuis, J.T. 1955. *Aceria tulipae* Keifer in relation to the spread of wheat streak mosaic. Phytopathology 45: 116-128

Slykhuis, J.T. 1973. Viruses and Mites. In: Viruses and Invertebrates. A.J. Gibbs (ed). Elsevier, New York, USA. pp. 391-405

Smidansky, E.D. and Carroll, T.W. 1996. Factors influencing the outcome of barley yellow streak mosaic virus-brown wheat mite-interactions. Plant Disease 80: 186-193

Smith, P.R., Campbell, R.N. and Fry, P.R. 1969. Root discharge and soil survival of viruses. Phytopathology 59: 1678-1687

Soto, M.J. and Gilbertson, R.L. 2003. Distribution and rate of movement of the curtovirus beet curly top virus (family: Geminiviridae) in the beet leafhopper. Phytopathology 93(4): 478-484

Stobbs, L.W., Cross, G.W. and Manocha, M.S. 1982. Specificity and methods of transmission of cucumber necrosis virus by *Olpidium radicale* zoospores. Can. J. Plant Path. 4: 134-142

Sylvester, E.S. 1956. Aphid transmission of non-persistent plant viruses with special reference to the *Brassica nigra* virus by the green peach aphid. Hilgardia 3: 53-98

Tamada, T. and Harrison, B.D. 1981. Quantitative studies on the uptake and retention of potato leafroll virus by aphids in laboratory and field conditions. Annals Appl. Biol. 98: 261-276

Tamada, T. and Abe, H. 1989. Evidence that beet necrotic yellow vein virus RNA-4 is essential for efficient transmission by the fungus *Polymyxa betae*. J. Gen. Virol. 70: 3391-3398

Tamada, T., Schmitt, C., Saito, M., Guilley, H., Richards, K. and Jonard, G. 1996. High resolution analysis of the readthrough domain of beet necrotic yellow vein virus readthrough protein: a KTER motif is important for efficient transmission of the virus by *Polymyxa betae*. J. Gen. Virol. 77: 1359-1367

Taylor, C.E. and Brown, D.J.F. 1997. Nematode Vectors of Plant Viruses. CAB International. Wallingford, UK

Temmink, J.H.M., Campbell, R.N. and Smith, P.R. 1970. Specificity and site of in vitro acquisition of tobacco necrosis virus by zoospores of *Olpidium brassicae*. J. Gen. Virol. 9: 201-213

Thornbury, D.W., Patterson, C.A., Dessens, J.T. and Pirone, T.P. 1990. Comparative sequence of the helper component (HC) region of potato virus Y and a HC-defective strain, potato virus C. Virology 178: 573-578

Tian, T., Rubio, L., Yeh, H.-H., Crawford, B. and Falk, B.W. 1999. Lettuce infectious yellows virus: *in vitro* acquisition analysis using partially purified virions and the whitefly *Bemisia tabaci*. J. Gen. Virol. 80: 1111-1117

Tiefenbrunner, A., Tiefenbrunner, M., Tiefenbrunner, W. and Wahra, A. 2002. A software tool as an aid to the identification of species of *Longidorus* Micoletzky 1922 (Nematoda: Dorylaimoidea). Nematology 4(7): 845-852

Timpe, U. and Kühne, T. 1994. The complete nucleotide sequence of RNA 2 of barley mild mosaic virus (BaMMV). Euro. J. Plant Pathol. 100: 233-241

Tomaru, M., Maruyama, W., Kikuchi, A., Yan, J., Zhu, Y., Suzuki, N., Isogai, M., Oguma, Y., Kimura, I. and Omura, T. 1997. The loss of outer capsid protein P2 results in non-transmissibility by the insect vector of the rice dwarf phytoreovirus. J. Virol. 71: 8019-8023

Tomlinson, J.A. and Garrett, R.G. 1964. Studies on the lettuce big-vein virus and its vector *Olpidium brassicae* (Wor.) Dang. Annals Appl. Biol. 54: 45-61

Trudgill, D.L. and Brown, D.J.F. 1980. Effect of the bait plant on transmission of viruses by *Longidorus* and *Xiphinema* species. Annual Report of the Scottish Horticultural Research Institute for 1979. p. 128

Ulgenturk, S., Kaydan, M.B., Zeki, C. and Toros, S. 2001. *Rodococcus perornatus* (Cockerell & Parrott) (Homoptera: Coccida): A new record on the pest of oil roses. Turkeye- Entomology Dergisi 25: 127-132

Ullman, D.E., Cho, J.J., Mau, R.F.L., Hunter, W.B., Westcot, D.M. and Custer, D.M. 1992. Thrips-tomato spotted wilt virus interactions: morphological, behavioral and cellular components influencing thrips transmission. In: Advances in Disease Vector Research. Vol. 9. K.F. Harris (ed.). Springer-Verlag, New York, USA. pp. 195-240

Valverde, R.A., Sim, J.G. and Lotrakul, P. 2004. Whitefly transmission of sweet potato viruses. Virus Research 100: 123-128 van den Heuvel, J.F.J.M., Verbeek, M. and Peters, D. 1993. The relationship between aphid transmissibility of potato leafroll virus and surface epitopes of the viral capsid. Phytopathology 83: 1125-1129

van den Heuvel, J.F.J.M., Hogenhout, S.A. and van der Wilk, F. 1999. Recognition and receptors in virus transmission by arthropods. Trends in Microbiol. 7: 71-76

van Hoof, H.A. 1970. Some observations on retention of tobacco rattle virus in nematodes. Neth. J. Plant Pathol. 76: 329-330

Varma, A. and Malathi, V.G. 2003. Emerging geminivirus problems. A serious threat to crop production. Ann. Appl. Biol. 142: 145-164

Vassilakos, N., MacFarlane, S.A., Weischer, B. and Brown, D.J.F. 1998. Exclusivity and complementarity in the transmission of tobraviruses by their respective trichodorid vector nematodes. Mededelingen van de Faculteit Landbouwwetenschappen Rijksuniversiteit Gent 62: 713-720

Vellois, E., Duncan, G., Brown, D. and MacFarlane, S.A. 2002a. Immunogold localization of tobravirus 2b nematode transmission helper protein associated with virus particles. Virology 300: 118-124

Vallios, E., Brown, D.J.F. and MacFarlane, S.A. 2002b. Substitution of a single amino acid in the 2b protein of pea early browning virus affects nematode transmission. J. Gen. Virol. 83: 1770-1775

Verma, A.K., Ghatak, S.S. and Mukhopadhyay, S. 1990 Effect of temperature on development of whitefly (*Bemisia tabaci*) (Homoptera: Aleyrodidae) in West Bengal. Indian J. Agric. Sci. 60 332-336

Visser, P.B. and Bol, J.F. 1999. Non-structural proteins of tobacco rattle virus which have a role in nematode transmission: expression pattern and interaction with viral coat protein. J. Gen. Virol. 80: 3273-3280

Walker, G.C. and Perring, T.M. 1994. Feeding and oviposition behavior of whiteflies (Homoptera: Aleyrodidae) interpreted from AC electronic feeding monitor wave forms. Annal Entomol. Soc. America 87: 363-374

Walters, H.J., Lee, F.N. and Jackson, K.E. 1972. Overwintering of bean pod mottle virus in bean leaf beetles. Phytopathology 62: 808

Wang R.Y., Gergerich, R.C. and Kim, K.S. 1992. Non-circulative transmission of plant viruses by leaf-feeding beetles. Phytopathology 82: 946-950

Wang, R.Y., Gergerich, R.C. and Kim, K.S. 1994. The relationship between feeding and virus retention time in beetle transmission of plant viruses. Phytopathology 84: 995-998

Wang, R.Y., Powell, G., Hardie, R.J. and Pirone, T.P. 1998. Role of the helper component in vector specific transmission of potyviruses. J. Gen. Virol. 79: 1519-1524

Waquil, J.M., Viana, P.A., Cruz, I. and Santos, J.P. 1999. Biological aspects of the corn leafhopper *Dabulus maidis* (DeLong and Welcott) (Hemiptera: Cicadellidae). Anais da Sociedade Entomologica do Brasily 28: 413-420

Wasim, A. and Araki, M. 2003. New and known species of of the family Dorylaimidae (Nematoda: Dorylaimida) from Japan. Intl. J. Nematol. 13: 51

Watson, M.A. and Roberts, F.M. 1939. A comparison of the transmission of Hyoscyamus virus 3, potato virus Y and cucumber mosaic virus 1 by the vectors *Myzus persicae*

(Schulz.), *M. circumflexus* (Buckton), and *Macrosiphum gei* (Koch). Proc. Royal Soc., London, Series B 127: 543-576

Watson, M.A. and Plumb, R.T. 1972. Transmission of plant pathogenic viruses by aphids. Annual Review of Entomology 17: 425-452

Wintermantel, W.M. 2004. Emergence of Greenhouse Whitefly (*Trialeurodes vaprariorum*) Transmitted Criniviruses as Threats to Vegetable and Fruit Production in North America. APS*net* Feature Story, June 2004 (http:/www.apsnet.org/online/feature/whitefly/)

Wisler, G.C., Li, R.H., Liu, H.-Y., Lowry, D.S. and Duffus, J.E. 1998. Tomato chlorosis virus, a new whitefly-transmitted phloem-limited, bipartite closterovirus of tomato. Phytopathology 88: 402-409

Yadav, S.K., Sharma, A., Yadav, L.N. and Sharma, A. 2003. Biology of mite, *Eutetranychus orientalis* (Klein) at ambient temperature. Annals. Agri. Bio. Res. 8: 73-76

Yan, J., Tomaru, M., Takahashi, A., Kimura, I., Hibino, H. and Omura, T. 1996. P2 protein encoded by genome segment S2 of rice dwarf phytoreovirus is essential for virus infection. Virology 224: 539-541

Yilmaz, N.D.K., Yanar, Y., and Erkan, S. 2003. Rigid rod shaped viruses transmitted by fungi. Ondokuz Mayis Universitesi, Ziraat Facultesi Dergisi 18: 73-79

Zeidan, M. and Czosnek, H. 1991. Acquisition of tomato yellow leaf curl virus by the whitefly *Bemisia tabaci*. J. Gen. Virol. 72: 2607-2614

Zhou, G., Lu, X., Lu, H., Lei, J., Chen, S. and Gong, Z. 1999. Rice ragged stunt oryzavirus: role of the viral-spike protein in transmission by the insect vector. Annals Appl. Biol. 135: 573-578

6

Dispersal, Movement and Migration of Vectors

Most viruses are vector-borne and are thus spread through their vector's dispersal, movement and migration. The movement of vectors is normally an active process, such as crawling, walking or flying, but can also be a passive process when they are carried by wind or moved on plants or in soil. Active movement is a part of the vector's biology, but passive movement is an incidental process. Vectors have to move in search of food, a change of habitat, place for reproduction and to overcome adverse conditions. A few vectors may crawl in the field, for example aphids (Alykhin and Sewell 2003), but many virus vectors have special adaptations to aid dispersal, the most obvious of which are wings and they undergo morphological and physiological changes to adapt themselves for flying. The flying process also depends upon the physiology of the host, climatic and meteorological conditions.

Normally, dispersal of vectors is over relatively short distances, from plant to plant or field-to-field, and possibly habitat-to-habitat, and covers a distance of a few kilometers at the most and mainly within the surface boundary layers. However, movement may also be over much longer distances especially when the insect vectors are under the influence of atmospheric processes above the boundary layer. Vectors may migrate from one climatic region to another by their own flight activity and by wind field systems of the upper atmosphere. Knowledge of these aspects of vector dispersal is essential in planning and developing strategic operational tactics in the fields to prevent or control the spread of vector-borne virus diseases.

Flying is the most common method for dispersal. Insects can fly at all times of the day and night depending upon the species. A few species only fly during the day. Aphid flight is normally restricted to the daylight hours and by light and temperature thresholds. Many other small, wind-borne migrant pests are also day flyers. Others are nocturnal and some small bugs are crepuscular. Flight activity measures the dispersal of the flying vector insects.

A. DISPERSAL AND FLIGHT ACTIVITY

Many different methods have been used to determine the flight activity of insects. This activity can be observed using a range of different accessories such as sources of artificial (red) light or night vision equipment, low light or IR-sensitive video equipment, IR-detecting devices, ground-based and air-borne radar, night viewing telescopes. Flying insects can also be collected in a range of devices for subsequent identification and counting including aircraft trailing insect collecting nets, pilot balloon releases and aerial netting (Greenbank *et al* 1980, Riley *et al* 1992, Cooter 1993).

Lingren *et al* (1986) used night vision equipment (Figure 6.1) to study the reproductive biology and nocturnal behaviour of insects with reference to their flights, dispersal and migration. Knowledge of nocturnal activities is sometimes necessary for studying the migratory activity of source populations. Nocturnal observations and collections from emerging source populations along with radar determinations of flight activity and associated aerial and migratory fall-out collections offer a means of providing source material for determinations of physiological and behavioural differences in migratory and non-migratory individuals and populations.

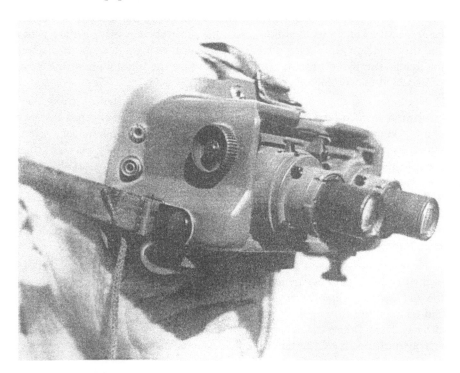

Figure 6.1 Night vision equipment (Lingren *et al* 1986). With permission from Springer Rights and Permission Department, Germany

Flight activity begins when the vector 'takes off' from the host. This may be at a particular time of the day or night and will depend on the species, the physiology of the insect, its host and meteorological conditions. It is the effort of the insect that makes it air-borne but its direction and the distance it travels is greatly influenced by the prevailing wind direction and speed. The majority of insects fly near the ground or at canopy height. Their density gradually decreases up to the boundary level. The insects that take off at dusk may get into the convective or advective up-draughts and may reach the inversion layer (normally 100-150 m above ground) through the temperature gradient and further displacement may occur through atmospheric circulation at different altitudes. The method of choice for monitoring insect dispersal depends upon its purpose.

The oldest, most widely used and relatively cheap method to record ground level flight activity is combined interception and attraction traps. Quantitative methods for estimation of the atmospheric flight activity are stationary light traps and suction traps. Suction traps can also be stationed at variable heights. Recently, radars are in use to estimate the flight activity at ground and higher altitudes. Nets are also used for aerial trapping, the simplest type of which is a suspended cone net that can be modified to different forms as required.

A1. Combined Interception and Attraction Traps

Common combined interception and attraction traps are sticky, water and light traps. These traps are normally used for population measurement of low-flying insects.

(a) Sticky trap
Many small insects including flies, bugs, and thrips are attracted by specific colours. Traps are devised according to the type of insects to be intercepted. Normally these traps are made of white or yellow rectangular galvanized sheets, plastic jars, plates or tin boxes suspended from or held by a support. A sticky adhesive is used for trapping the insects. Large screens consisting of a wooden lattice or series of boards are in use to measure movement at a range of heights of aphids and beetles. White and yellow coloured traps are effective to catch various flies and psyllids.

Glass plates are also used for sticky traps. Normally the upper surface of a plate (8 in. square) is coated with the adhesive and the underside painted with yellow (or any colour desired). Grease-coated microscope slides have been used to trap mites. These plates may be exposed vertically or horizontally but the influence of colour is less with vertical plates where most insects are caught by wind impaction. A good design, sampling at random from the passing air, is the cylinder sticky trap, which consists basically of a piece of plastic material covering a length of stove-pipe. If the wind speed is known, the catches of small insects on a white sticky trap of this type could be converted to a measure of the aerial density (Taylor 1962) but yellow traps catch more aphids than white or black (A'Brook 1973). Colour has the greatest effect at low wind speeds.

Sticky traps do not allow live trapping. A different method is needed where the infectivity of captured aphids is to be determined. Live vectors can be collected directly from vegetation, or by using collecting devices such as nets, trays or mobile suction traps. However, insects collected in this way are mostly those spreading viruses within crops rather than those bringing virus into plantings (Raccah 1983).

(b)   Water trap

Water traps are simply glass, plastic or metal bowls or trays filled with water to which a small quantity of detergent and a preservative (usually a little formalin) are added. The traps may be transparent or painted in various colours and can be placed at different heights. They are to be frequently attended or they may overflow in heavy rains or dry out in the sun. A model, however, is available with a reservoir for automatically maintaining a constant level (Adlerz 1971). The advantages of the water traps are: very low-cost, handy and the caught-insects remain in good condition for identification, as the catch can be easily separated by straining or individual insects picked out with pipettes or forceps. These traps not only record landing but also attracts aphids in flight. Irwin (1980) designed a modified water trap specifically to monitor aphid landing on soybean. The trap consists of a Perspex receptacle containing a horizontal, coloured (green) tile (HCT). The colour was chosen to resemble the characteristics of soybean vegetation based on reflection spectrophotometry. Instead of a preservative or sticky tanglefoot, Irwin and Goodman (1981) recommended the use of 50% ethylene glycol to facilitate collection and identification. A possible limitation of HCT is that the catches may be biased when the plants are small and separated by extensive areas of bare ground.

It is, however, very difficult to estimate any relationship between the wind speed and the catch in water traps.

(c) Light trap

C.B. Williams first designed a 'Light Trap' subsequently called the 'Rothamsted Light Trap' in 1948. It was used to record the distribution frequency of different insect species. The essential component of a light trap is its light source that may be 200-W clear glass tungsten filament lamp or 125-W clear glass mercury vapour (m.v) lamp. Both these traps have been extensively used in the Rothamsted Insect Survey in Britain. Traps with 125-W frosted glass m.v and 160-W frosted glass MBTU mixed tungsten and m. v lamps have been used in Africa and the USA. A 12-V battery operated spot lamp held parallel to the ground was also used in different parts of the USA to study flying insects

(Sparks *et al* 1985). A network of Coleman lantern light traps in New Brunswick was in operation from 1945 (Greenbank 1957). In India 100-W tungsten bulb traps have been used to study the flight activity of insects (Figure 6.2). The major limitation of such traps is in their use to trap only the photoperiod-sensitive noctuids. These traps may be used to study the effect of weather on insects, to develop and quantify the concept of insect diversity based on the

Figure 6.2 Rothamsted type light trap to catch noctuid insects (Courtesy: Professor P.S. Nath, BCKV, West Bengal)

observed species frequency distribution in the samples. The use of light trap catches for other types of population studies could not progress due the effect of moonlight on the quantum of the catches. Observations by most of the entomologists working in this field remained confined, primarily, to defining the effect of moonlight on light trap catches of different insects. A few of them were concerned in developing some 'correction factors' to make the catch data more useful, but the controversies continued.

Rothamsted light traps have been in operation at various sites throughout the world, as part of general studies into moth biodiversity. In Europe, this has involved sites in Denmark, Finland, France, and Ireland and also Aldabra, Iraq, Malaysia, Seychelles, Sulawesi and Tenerife (Woiwod and Harrington, 1994).

In India light traps have been used to study the ground level flight activity of an important group of plant virus vectors after verifying the suitability of the adjustment of moonlight interferences developed by Bowden (1973a, b). Utilizing the daily light trap data on leafhoppers *Nephotettix virescens* (Distant) and *N. nigropictus* (Stål) generated at the Department of Plant Pathology, Bidhan Chandra Krishi Viswavidyalaya (BCKV), West Bengal, over a period of several years, Bowden *et al* (1988) conducted phase group analysis of the catches. They first determined the time and duration of flight activity analyzing the hourly catches from sunset to sunrise, made on the dates during the suitable phase groups and recorded flight activity during 5 h after sunset and 2 h before sunrise. The phase group analysis showed the association between increasing amounts of moonlight in the early hours of the night and increased flight activity. There was a remarkable difference in the sensitivity to illumination and periodicity of flight found between *N. virescens* and *N. nigropictus*. Increased illumination heightened the activity in *N. nigropictus* by a factor of about 4 but only by about 50% for *N. virescens*. The flight activity of *N. virescens* was inhibited at very low illumination (0.02 lux). *N. nigropictus* on the other hand showed both lower (0.03 lux) and upper threshold (0.15-0.20) levels of illumination for its flight activity. The periodicity of flight activity of *N. virescens* was greatest during the period from four nights after new moon to about four nights after full moon. In the other half of the lunation the flight activity was much less confined to a short period following sunset. The periodicity of the flight activity of *N. nigropictus* remained confined between two threshold values. The greatest flight activity was found for a week between the first quarter and the full moon. During the other 3 wk of lunation, the flight activity was low and confined to a short period after sunset (Figure 6.3).

Mukhopadhyay (1991) analyzed the daily light trap catches of the leafhoppers collected from different locations during 1982-86 to determine the physical basis for the lunation induced variation in light trap (100-W tungsten lamp) catches. The phase groups of a lunation differentially affect the radius of a light

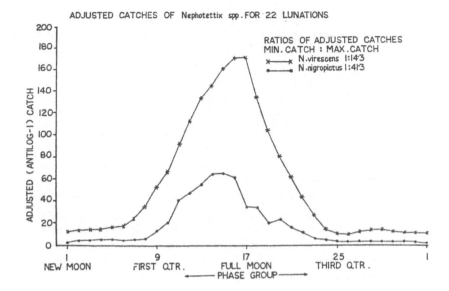

Figure 6.3 Flight activity of rice green leafhoppers during different lunations at 22°50.0′ N 80°20.0′ E (Bowden *et al* 1988)

trap depending on the illumination available in the particular phase group. The effective radius was found to be as far as 245.20 m in the phase groups 1-3 and 25-32 (during the new moon). The effective radius was as low as 16.72 m in phase group 17 (during the full moon). The geometric mean catches proportionately expanded with the increase in trap radius till the new moon and decreased with the drop in the trap radius till the full moon. Long range light trap data on the catch of leafhoppers was also useful to determine threshold minimum (15°C) and maximum temperature (35°C) for the flight activity of *N. virescens* and *N. nigropictus*. (Mukhopadhyay *et al* 1989).

Major limitations of studying the flight activity by light traps are: applicable only for night flying insects at ground level and the data are mostly empirical. To overcome these limitations, particularly to catch the insects continuously for 24 h with greater precision, another type of trap called the 'suction trap' was developed at Rothamsted.

A2. Suction Trap

C.G. Johnson and L.R. Taylor developed suction trap in 1955. The basic features are an electric fan that pulls air through a fine gauge cone that filters out the insects to collect them in a jar or cylinder at the base. The trap may be fitted with a segregating device that separates the catch according to predetermined time intervals and provides information, not only about the number of flying insects, but also the periodicity of their flight. If properly standardized, this trap can assess the aerial populations very precisely. The suction trap should be of a neutral colour to avoid any insect attraction or repulsion, and a consistent sample is obtained at all commonly occurring wind speeds.

The first suction trap began continuous operation at Rothamsted in 1964 after several trials to determine the best height at which samples should be taken. A height of 12.2 m with a sampling volume of 40-50 m^3/min of air was found to be the most economical. Sampling at this height excluded most of the insects flying locally but was low enough to sample adequately the densest layer of flying insects, particularly aphids that are important vectors for many viruses. Several other countries use this trap to monitor specific insects and their aerial dynamics. Most work is focussed on aphids and the countries of the European Economic Community established a 'Network' (AGRAPHID) to collaborate with each other and to develop appropriate strategies to control aphids and the viruses carried by them. To ensure uniformity and comparability of data from traps inled in different countries Macaulay (1988) described the design and construction of a typical 12 m suction trap (Figure 6.4). This trap has been widely adopted as a standard and the network of traps throughout Europe provides the most comprehensive standardized spatiotemporal database for any terrestrial invertebrate group. Woiwod *et al* (1984) developed a system for rapid collection, analysis and dissemination of aphid-monitoring data from suction traps. Subsequently more countries joined this effort with a view to assess the virus risks, climatic change impacts, biodiversity measurement, and dynamics of insecticide resistance (Harrington 1998). The number of suction traps operating in different countries is recorded by Harrington (1998) and is given in Table 6.1. In India, one 12 m Rothamsted type suction trap was locally fabricated at BCKV, West Bengal in 1997-98 to monitor the flight of different insects during different months and many aphids, leafhoppers, plant hoppers and other insects were found in November. Rothamsted Insect Survey (RIS) has successfully used suction trap data on infective aphid vectors to forecast Barley Yellow Dwarf Virus (BYDV) and other aphid-borne viruses particularly of wheat, barley, potato and sugar beet. To develop this forecasting system initially live trapping of migrating aphids was done using a suction trap of 50 mm diameter that samples air at 1.75 m above ground. The aphids were sorted and potential vectors were identified and indexed for viruses. The infectivity index of the viruses was determined and the risk of the appearance of the disease was calculated taking the RIS data into account (Plumb 1987, 2002).

Figure 6.4 Structure of a typical 12 m Rothamsted Suction Trap to catch aerial insects (Courtesy: Professor R.T. Plumb, IACR, Rothamsted, UK)

Table 6.1 Number and Operation of 12 m Suction trap to Study Insect Dynamics

Country	Number of traps	Frequency of emptying	Data usage
Belgium	2	Daily	Virus disease forecasting in seed potato production
Czech Republic	5	Daily	Weekly bulletin, aphid control decisions in a range of crops
Estonia	1	Twice/week	Forecasting aphids in cereals
Finland	1	Daily	Forecasting viruses
France	10	Daily	Monitoring vectors of viruses of cereals and sugar beet, control decision support in important crops, migration of apple aphids
Greece	5	Daily/Weekly	Forecasting aphids
Hungary	1	Daily	Forecasting PVY infection and haulm destruction
Italy	7	Daily	Aphid species composition and flight times, forecasting crop pests, risk assessment of CMV on tomato, non-persistent viruses of potato, BYDV in cereals
Kenya	4	Daily	Monitoring and forecasting cereal aphids
The Netherlands	3	Daily	Monitoring potato aphids, assessing best time for haulm destruction
Poland	1	Daily	Monitoring pests, forecasting aphids
Slovenia	2	Daily	Monitoring vectors of PVY and advice on control, haulm destruction times, planting dates for seed potatoes, BYDV vector dynamics
Spain	1	Daily	Aphid monitoring and advice on control
Sweden	9	3 × per week	Aphids and PVY, damage forecasting and control
Switzerland	2	Daily	Monitoring pests, date of potato haulm destruction, BYDV vector control
United Kingdom	15	Daily in season, weekly at other times	Population dynamics, forecasting aphid migration, aphid forms, decision support for BYDV control
United States/Florida	6	Weekly	Aphid flight activity, cereal aphid control, detection of exotic species
United States/Idaho	12	Weekly	Control of aphids, forecasting aphids in potatoes

A3. Aerial Netting

Riley *et al* (1991) developed the aerial netting method while working on the long-distance migration of brown plant hopper in China. In this method, flying insects are caught at a predetermined height by a net (0.4 mm mesh size with an entrance aperture of 0.64 m^2) attached to the tethering line of a helium-filled aerodynamically shaped balloon (kytoon) of gas capacity 11.3 m^3 (Figure 6.5) The kytoon is raised to and lowered from its operating altitude by a hand winch. The airflow through the net during the sampling interval is estimated from a wind-run meter hung below the kytoon and the sampling altitude is calculated from the length of the tethering line paid out and the elevation angle subtended by the kytoon at the tethering point. At the end of each sampling period, a radio-control device closes the net, the kytoon is winched down and a detachable bag containing the catch is removed from the net. Routinely the bag is placed in a container with a vapona (DDVP) strip to kill the insects that are later sorted, identified and counted. Live insects can also be indexed for viruses.

Aerial densities of the specific insect are calculated from the catch, volume of air samples or area of the net aperture and wind-run.

A4. Radar Systems

Insects produce radar echoes and G.W. Schaefer first used specifically modified radar in 1968 to study locust flight in Southern Sahara (Schaefer, 1969). Subsequently Microwave Radar techniques have been found to provide powerful tools for the study of insect movement in the upper air, both in daylight and after dusk (Riley 1974). A single radar allows the engineer to make precision estimates of a number of parameters related to air-borne insect movement including location, time and duration of take-off flight, density of insects at different altitudes, duration of flight activity, flight orientation, direction and velocity of movement and maximum detection range. Automated data collection and 'Signature analysis' equipment can categorize populations of insects on flight wing beat frequency, shape, orientation, air speed and densities. Ground-based radar can also work along with an aircraft based air-borne radar systems in a conjunctive way for more accurate information. The air-borne unit collects quantitative data on profiles, area densities, orientation, wing beat, frequencies and wind speeds at aircraft altitude. It may also sample plumules and layers of insects. There is an optical system (LIDAR) connected with the conjunctive radar system to quantify and differentiate species and activity of insects at ground level. There is also scope to archive the data and make them available for real time display.

Figure 6.5   Aerodynamically designed balloon (kytoon) for catching aerial insects from the atmosphere (Courtesy: Dr. J.R. Riley, NRI, Greenwich University, UK)

First generation radars are scanning type modified '*3.18 cm Marine Surveillance Systems*'. These systems were used during 1973, 1974 and 1975 in Canada to measure the effect of moth immigration on eggs populations in forests. In 1976 the capability of this type of radar was enlarged when a new air-borne radar (Solid State Marine Radar Decca RM 925) (Figure 6.6) was developed and employed in transects of the air-space to delineate source areas of the pests, flight duration, area of deposition and flight behaviour

Figure 6.6 Structure of a "Solid State Marine Radar". US Department of Agriculture, Entomological radar system (Greenbank *et al* 1980)

(Greenbank *et al* 1980). During the same period (1973, 1974, 1975), Riley and Reynolds (1979) studied migratory flight of grasshoppers in the middle of Niger area in Mali. They used two *3 cm pulse radar system*, employing a parabolic antenna rotating about a vertical axis in one of them to produce a conventional Plan Position Indicator (PPI) display for photography once per revolution. Radar 'signatures' were obtained by stopping the antenna rotation and recording the signals returned by individual targets that flew through the stationary radar beam. Subsequent frequency analysis of these signals provided the measure of the targets' wing beat frequencies. The second system provided information on 'body shape factor' and accurate measurements of the target heading distribution.

Conventional scanning entomological radars suffer from a serious limitation that their capability to identify insect species is very limited. During rotational scanning targets appear simply as anonymous dots on the radar screen and these radars cannot detect insects weighing more than 50 mg. To overcome this difficulty Riley (1992) developed a 'milli-metric radar' (8.8 mm wave length) that can monitor the passage of any insect of mass above about 1-2 mg, but the range at which these radars can detect individual insects is limited to a few hundred meters. Riley further envisaged novel methods of improving the target identification capacity of entomological radars and also to automate both their operation and the analysis of the radar data. He used a vertical-looking beam in which the plane of (linear) polarization is rotated. This system allows extraction of insect body mass and shape data as well as displacement velocity and orientation (Smith *et al* 1993). Based on these observations, Riley and Reynolds (1993, 1997) developed this new generation vertical looking (VLR) compact entomological radars (Figure 6.7) that permits continuous automatic monitoring of insect migration. This system can capture enormous data on individual migrants. A desktop computer can process the data very rapidly and present it in a form readily comprehensible by non-specialists, but these systems fail to track low-level flights. Riley *et al* (1996, 1999) also developed harmonic radar to record the flight trajectories of small low flying insects like bumble bees. Reynolds and Riley (1997) reviewed the radar studies on flight behaviour and migration of insect pests in developing countries.

Riley and his associates applied radar technology to study the migration of pests in different African and Asian countries but in the United States of America, the Department of Agriculture and Agricultural Research Service took up the atmospheric transport and pest migration in a comprehensive way in 1980

Figure 6.7 New generation vertical looking (VLR) compact entomological radar (Riley and Reynolds 1993). With permission from Rights and Licensing Department, Birkhäuser, Basel, Switzerland

by setting up a National Grid led by a team of radar engineers, meteorologists, entomologists and other relevant scientists for the development and application of a 'offensive strategy' to monitor atmospheric systems conducive to pest migration, and the migrating pests, in a regular way to disperse or kill them in the air before they landed on crops.

B. ATMOSPHERIC TRANSPORT AND MIGRATION OF VECTORS

Atmospheric transport and migration of insects have been extensively studied in the USA, Canada, and Australia. Transport determines the distance and direction of spread, influences the behaviour of insects on the ground, the behaviour of insects ascending through the surface boundary layer during the crepuscular activity period, the behaviour of insects that eventually exceed the surface boundary layer and their migration. Therefore atmospheric conditions need to be properly understood if precise knowledge of the migratory patterns of pests and virus vectors is to be determined.

1. Mechanics of Atmospheric Transport System

Insects may be active or passive fliers depending on their morphology and physiology. Insects that can bear stresses and strains and possess strong wings are active fliers whereas soft and fragile insects are mostly passive in their displacement. For both types of insects their atmospheric transport is determined by their ability to take off from their laminar boundary layers around their bodies into gusty winds. The take-off may be active or passive. Active take-off is triggered from outside by meteorological conditions when the insects become morphologically and physiologically ready for flying. There is always a temperature threshold only above which the take-off is accomplished. In case of passive take-off insects are propelled into the air by an outside force such as gusty wind, rain- drops, and electrostatic forces. Once in the air, the insects are then within convective or advective layers. In convective layers, vertical mixing of air takes place according to the temperature differences. During daytime the air temperature is usually lower than that in the ground and vegetation. Therefore, there will be a decrease of temperature upward and convective updroughts. On a cloudless night, on the other hand, especially with light winds, the ground and vegetation tend to be cooler than the air and heat flows downward in the atmosphere. Air temperature then decreases downward and temperature inversion occurs as much as 10°C in 100 m. Temperature gradients in the air can also be formed due to advection of warmer or cooler air from elsewhere. Such advection by the wind may be horizontal and vertical and depends upon strong winds with large gradients. Both convection and vertical advection may be responsible for the updroughts that carry insects into the atmosphere.

Atmospheric processes of displacement are the same for both active and passive fliers, but active fliers can modify their displacement by choosing where and when to fly. Active fliers may initiate flight, fly to latitudes of different wind speed or direction and at a heading that may differ from downwind to crosswind to upwind. Normally all passive insects move in the atmosphere much as balloons following the paths of the wind-fields, wind speed and the bearing. The place and time of the start and finish of their journey depends upon the insects and the path of the wind that carries them. Under certain circumstances many insects can cross their boundary layers and fly at heights of up to several kilometers above the ground. Strong winds at this level can lead to large displacements. An unbroken flight in wind of 20 m/sec would give a displacement of about 2000 km in a day. Such strong winds are common in cyclones of mid latitudes and in squall lines. Many insects can disperse on a global, synoptic or meso-scale of wind field and many insect species have been seen, or caught in flight, 1000 km or more over the ocean presumably displaced by global scale wind system such as the middle-latitude westerlies, the tropical easterlies (trade winds) and the monsoons. Displacement of insects can also take place within synoptic scale wind pattern particularly in the middle latitudes of the northern hemisphere that may carry the insects 1000 km or more. Such displacements often move polewards within warm winds blowing towards higher latitudes or blowing to the east of slow moving cyclones or the west of anticyclones and may last for a few days. Stronger displacements are possible in the stronger winds present in low-level jet streams. Meso-scale displacements of insects take place over tens of kilometers. It is likely that meso-scale wind fields particularly sea breezes, down drought squalls and rainstorms play important roles in the hour-to-hour changes in the spread of air-borne insects.

Weather changes at cold fronts, where cold air replaces warm, are also linked to the sudden movement of insects. Usually wind behind a front, or a squall line ahead of it, carries insects into a convergence. Such fronts may be tens of kilometers wide or hundreds or thousands of kilometers long across which the temperature may change 10°C or more. Such temperature changes, gustiness or lulls at many fronts may trigger off flights of insects.

Air-borne insects continue to be displaced by the atmosphere until they fall out, impinge or actively descend to the surface. The fall out rate depends on the imbalance of gravitational or convective forces. Heavier insects fall out at predictable densities within a moderate range of their sources. The ability of flying insects to alter their course makes the prediction of their movement much less accurate than of passive air-borne insects. Observations of the flight behaviour of air-borne insects will, however, increase the accuracy of predictive displacement models. Empirical parametrization of flight take-off, altitude, duration, vertical velocity, orientation and landing may enable transport/diffusion models to accurately predict displacements of various insect species. Pedgley (1980) reviewed the influence of weather on living things in the air and

Sparks *et al* (1985) described the atmospheric transport of biotic agents on a local scale. Long range migration of pests in the United States and Canada, with specific examples of occurrence and synoptic weather patterns conducive to migration, have been discussed by several workers (Sparks 1986). International efforts are now being made to correlate seasonal migration with wind fields at global, synoptic and meso-scales so as to adopt offensive strategies or appropriate predictions for pests/vectors management (Mukhopadhyay 1996).

2. Long Distance Migration

Extensive studies have been made in many countries on the long distance migration of pests but information on such migration of vectors of viruses is limited. The most well documented long distance migration of vectors is that of the brown plant hopper and aphids that are pests as well as vectors of different viruses. Other vectors are not strong fliers, or cannot fly at all, but those that do have wings can show passive migration through the wind field systems.

(a)   Long distance migration of brown plant hopper

Brown plant hopper [*Nilaparvata lugens* (Stål)] transmits *Rice grassy stunt virus* and *Rice ragged stunt virus*. Both the viruses are transmitted in a persistent manner. Brown plant hopper (BPH) itself is a serious pest in the absence of any viruses. This insect occurs in all Asian and Pacific countries. Adult hoppers occur as non-flying brachypterae with rudimentary wings that colonize rice where they feed and reproduce. These are sedentary and are the dominant form during the growing season. As soon as the crop matures, the active macropterae appear. They have well developed wings and great mobility. In tropical areas there may be several generations a year; there is no mechanism for the over-wintering of this insect. They survive through their migratory movement from one place to another depending up on the circulation pattern of the wind. Southwesterly winds associated with '*bai-u*' depressions moving north eastward from southern to central China, carry the brown plant hoppers from north eastward in China from April to July and to Korea and Japan in June and July. The return movement occurs in autumn. There is also wind-assisted return movement from Thailand, Laos and central Vietnam to tropical southern China as well. Once air-borne, synoptic weather patterns normally govern the hoppers' movement. Weather patterns, flight duration and temperature thresholds of flights determine the speed, direction and extent of displacement (Pender 1994). Pender (1994) also compiled laboratory and field observations on the migratory behaviour of BPH as determined by different workers (Table 6.2). Rosenberg and Magor (1983, 1987) used wind field trajectory models to determine the displacement of the brown plant hoppers. They showed that in November, under anti-cyclonic conditions over northeast India and Bangladesh, movements were generally towards the west. When eastward moving depressions were present

Table 6.2 Laboratory and Field Observations on the Migratory Behaviour of Brown Plant Hopper (*Nilaparvata lugens*) (Pender 1994)

Flight parameter	Method of observation	Value
1. Take-off time	Field studies	Dawn, dusk, 14-100 h, more in dusk 20-30 lux
	Laboratory studies	Dawn, dusk
2. Flight threshold	Field studies	Varies from 9°C – 16°C depending on the site, season and methodology
	Laboratory	10°C
3. Height of flight	Field studies	1.5 – 2 km in summer 1 – 1.5 km in spring 0.1 – 1 km in autumn 0.2 – 0.8 km in late autumn
4. Flight duration h	Tethered	4 – 11 h, 23.5 h, 23 h, 14 – 26 depending upon the site, season, time and height
	Trajectory analysis	30 h
5. Physiological status	Field studies	Females immature

over northern India, hopper movement was towards the east or northeast (Figure 6.8). Radar studies by Riley *et al* (1987, 1991, 1994) showed the predominance of short distance migration in the humid tropics in contrast to the temperate countries where long distance migration predominates. Riley *et al* (1995) made a direct demonstration of the mass movement of brown plant hoppers at altitudes over northeast India. Comparative aerial densities (numbers per 10^4 m^3) and flux values (numbers per m^2 per hour) taken at sites of different countries are given in Table 6.3.

Table 6.3 Comparison of Aerial Density (numbers per 10^4 m^3) and the Flux Values (numbers per m^2 per hour) of *Nilaparvata lugens* (Stål) Derived from Samples Collected at Sites of Different Countries (Reynolds *et al* 1999)

Country	Location	Year	Month	Height above the surface (m)	Mean density/ flux
India	Haringhata, West Bengal	1994	October	c. 150	11/12
	Patancheru, Hyderabad, AP	1985	Nov–Dec	70–130	0.62/1.5
Philippines	Marsiit, Laguna	1984	March–April	c. 75	2.7/5.5
China	Jiangpu, Jiangsu	1988	September	140–450	28/72
	Jiangpu, Jiangsu	1990	Aug–Sept	140–250	2.1–4.6
	Dongxiang, Jiangxi	1991	Sept–Oct	80–380	1.9–2.9
	Hubei, Jiangxi	1977–79	July	1500–2000	4.8
	Hunan, Anhui	1978–79	Sept–Oct	500–1000	4.6
		1978–79	Sept–Oct	100	4.4

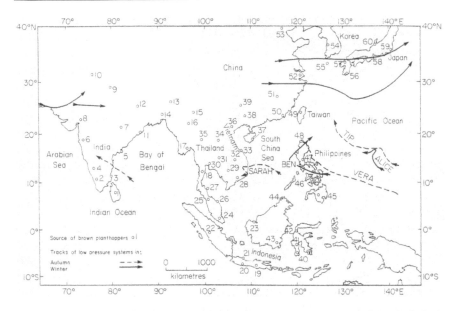

Figure 6.8 Sources of brown plant hoppers (indicated by numbered circles) and tracks of tropical and temperate low pressure systems during autumn (broken arrows) and winter (unbroken arrows) study periods. 'Alice', 'Sarah', 'Tip' and 'Vera' were typhoons and 'Ben' was a severe tropical storm (Rosenberg and Magor 1987). With permission from Wiley-Blackwell, UK.

Radar observations by Riley *et al* (1991) in China revealed that these insects flew up to altitudes corresponding to an air temperature of 16°C. As the air temperature varies in different geographical locations, their flying heights would vary in different countries. In West Bengal the flying height may be up to 1 km. Monitoring of migrant *N. lugens* in this region during April/May, provides a scope to estimate their build up in the kharif season. Studies during May at Haringhata showed the potential of this pest to invade crops at a density of 0.2 insects per m^2. The most interesting observation made in this study was migration of *Cyrtorhinus lividipennis*, a predator of the brown plant hopper more or less at the same altitude and same time as its prey. The presence of this predator makes it difficult to come to any conclusion on the prediction of their outbreaks and the viruses they carry.

(b) Long distance migration of aphid vectors

Plumb *et al* (1986), based on the results of the Rothamsted Insect Survey, recognized seasonal migration of aphids, in spring, summer and autumn. These data do not reflect long distance migration as the height of the trap used was 12.2 m, and stationed almost at the level of the surface wind (10 m). But aphids do migrate long distances as a result of different wind circulation systems in the upper atmosphere. Thresh (1983) reviewed the long-range dispersal of plant viruses by arthropod vectors. For aphid-borne viruses, he presented data on the migration of aphids carrying persistent and semi-persistent viruses (BYDV, PLRV, SCSV, GRV, and BYV), and also aphids carrying one non-persistent virus (MDMV). It has been postulated that infective aphids can be swept very rapidly by fast, low level jet-stream winds of up to 90-110 km h^{-1} blowing at an altitude of 300-1000 m, sometimes for several days. Infective aphids carried in this way can spread the viruses in fields where they land. Timing, altitude and prevailing wind during the flight also contribute to the distribution of viruses in the fields (Taylor 1986). Riley *et al* (1995) provided physical evidence for extensive nocturnal migrations of a number of insect taxa including aphids in North East India. Mass migration of the aphid species *Lipaphis erysimi* in this region was established by conducting trajectory analysis. It was estimated from the timing, altitude of flight, and the prevailing wind that the long flying individuals might have originated from sources between 100 and 300 km away to the north-east (Figure 6.9). Composition and density of different flying aphid species varied during different season or months. For example, during early March (1995), *Sitobion* population was high followed by *Ceratovacuna* sp, *Prociphilus* sp, *Rhopalosiphum rufiabdominalis*, *Aphis gossypii*, *Brachycaudus* sp and *Myzus persicae*. Among these aphids, *A. gossypii* and *M. persicae* transmit several viruses and may have epidemiological relevance for the spread of viruses in the summer crops of the catchments area. During October (1995) very large populations of only *R. rufiabdominalis* were found (Reynolds *et al* 1999). Redistribution of the density of the flying aphids at high altitudes also varies in different countries during different months and during different crop seasons (Table 6.4).

(c)   Long distance migration of leafhoppers

Long distance migration of leafhoppers particularly through upper wind field systems is a controversial issue as the leafhoppers are not strong fliers. It has been stated earlier that Riley *et al* (1995) and Reynolds *et al* (1999) studied wind-borne movement of insects in eastern India and reported occasional catches of some leafhoppers. As for example, 2 *N. virescens* and 11 *R. dorsalis* were caught during 15-17 March 1994 whereas 64 *N. virescens*, 1 *N. nigropictus* and 77 *R. dorsalis* were caught during 21-31 October and the respective catches during 14-24 November were 3, 2 and 99 respectively during the same year (Reynolds *et al* 1999). These catches reflect both short and long distance movement.

Most of the catch of *N. virescens* and *N. nigropictus*, however, were made during dusk or the post-dusk period indicating their flight for less than 1.5 h covering a short distance. This observation is comparable with the results of Bowden *et al* (1988) who analyzed the catches of *N. virescens* and *N. nigropictus* from light traps at Kalyani (West Bengal, India) and found that 85% of the flight activity occurred in the first 5 h after sunset.

Table 6.4 Comparisons of the Aerial Densities (numbers/10^4 m^3) of Aphids Flying at Altitudes in Different Countries (Reynolds *et al* 1999)

Location	Month	Time of day/night	Height (m)	Mean	Maximum
Day-time sampling					
Haringhata, West Bengal, India	October	*c.* 6.30–16.30	*c.* 150	47	110
Haringhata, West Bengal, India	November	*c.* 6.00–16.30	*c.* 150	6.3	17
East–Central Illinois, USA	May–August	varies between 6.40–14.35	7.5–1295	*c.* 20	113
Eastern Kansas, USA	May–July	13.00–17.00	610	1.0	3.8
Cardington, Bedfordshire, England	June–August	13.00–17.00	305	4.6	13.2
Night-time sampling					
Haringhata, West Bengal, India	October	*c.* 18.30–4.30	*c.* 150	35	202
Haringhata, West Bengal, India	November	*c.* 18.30–4.30	*c.* 150	8.7	27
Patancheru, Andhra Pradesh, India	November – December	varies between 18.00–6.15	70 – 130	1.7	8.8
Gezira, Sudan	October	*c.* 19.00–22.00	300 – 1200	*c.* 5	–
Eastern Kansas, USA	May–July	1.00–4.00	610	0.29	0.61
Cardington, Bedfordshire, England	June–August	1.00–4.00	305	0.19	0.94

A few of these leafhoppers, however, were caught after midnight, but their peak take-off time is at civil twilight or sunset (Perfect and Cook 1982, Bowden *et al* 1988). Therefore, insects caught at mid-night must have flown for about 7 h, indicating that their migration is from a long distance. Though the number of long distance migrants is very low it has much epidemiological significance, as the infective insects may be the primary source of spread in the field.

Mukhopadhyay and Mukhopadhyay (1997) also recorded the atmospheric transport of leafhoppers. They analyzed an off-time high catch of leafhoppers carried by cyclonic winds and deposited in a zone of low and variable winds (at a meteorological convergence) near the trapping site. Trajectory analysis indicated that these insects were brought from a north-eastern location 280 km away from the trapping site (Figure 6.10).

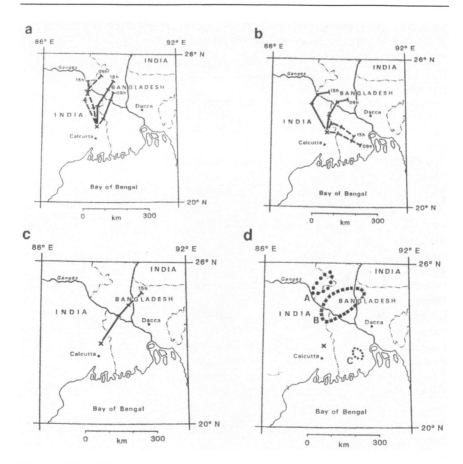

Figure 6.9 Wind-assisted migration of aphids in the tropics. a-c= back trajectories showing flight-paths of aphids migrating over the aerial netting site at Haringhata. d= putative source areas (Riley *et al* 1995)

C. DISPERSAL OF VECTORS OTHER THAN INSECTS

Common vectors other than insects are nematode, fungi and mite. Vector nematodes and fungi are mostly soil-borne and mites are mostly air-borne. Passive methods play a most important role in the dispersion of these vectors.

1. Role of soil

Vector nematodes do not have a resistant resting stage and survive adverse soil conditions by vertical movement through the soil profile. As soil becomes dry in summer or cold in winter, they move to the subsoil and return when conditions become favourable. The horizontal spread of nematodes in

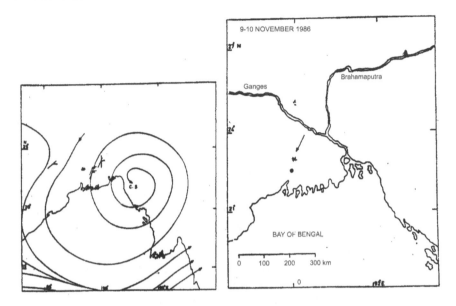

Figure 6.10 Atmospheric transport of rice green leafhopper (GLH) by cyclonic winds in the tropics (Mukhopadhyay and Mukhopadhyay 1997)

undisturbed soil is rather slow. It has been calculated that a population of *Xiphinema diversicaudatum* can invade uncultivated woodland at the rate of about 30 cm/year. Slow movement of the nematodes again depends upon the soil type particularly the pores through which they move. Their vertical movement occurs most frequently in light soil and favourable weather (Harrison 1964).

Fungal vectors prefer heavy soils where drainage is poor or rainfall is heavy. Their movement normally depends on the moving water.

2. Role of water

Water is the most important factor in the passive dispersal of soil-borne vectors. The soil must contain more than a critical amount of water for the transmission of a virus by a nematode and this amount depends on the pore structure of the soil. Fungus vectors also depend on soil water for mobility. Resting spores of the plasmodiphoroid vectors normally carry the viruses. Viruses can survive in these spores for many years. These spores or zoospores released by them cause localized spread and the local movement depends upon the soil water. Viruses that are carried externally by the zoospores of the fungal vectors probably survive for only a few hours. Virus that is free in soil may be picked up and transmitted by newly released zoospores. If the soil is moist, zoospores of a few vectors may retain the virus for at least 11 wk. Moreover, movement of the flagella of zoospores in water facilitates the dispersal of infected zoospores.

Water is the most efficient dispersion medium of soil-borne vectors. Water carrying the nematodes, fungal spores, infected soil and root fragment may move to a long distance through irrigation, drainage, rainfall and flood.

Most of the infectious virus probably moves in water while adsorbed on to organic and inorganic colloidal particles, especially clay. In this state, they would be substantially more resistant to inactivation than the free virus. In this form they may remain in rivers, lakes, ponds and all types of water-bodies from where they may be carried to distant locations.

3. Agricultural practices

All forms of agricultural practices contribute to the passive dispersion of soil-borne vectors including farm workers' foot wear, appliances and machinery contaminated with infested soil. Grass rats, farm animals like cows and donkeys can also carry viruses to short or long distances (Sarra and Peters 2003).

4. Seeds and pollens

Many viruses survive in seeds of the plants or in pollens that they produce. Dispersal of virus carrying seeds and pollens to short and long distances takes place in various ways. Wind and birds normally carry the seeds and pollinators carry pollen. Wind may also carry pollen. Dispersal of virus-containing weed seeds is the most effective means of spread of many nematode transmitted viruses. Seed and pollen transmission has considerable significance in epidemology (discussed in Chapter 7).

5. Human dispersal

Import of seeds/planting materials constitute a major strategy for the agricultural development of all the countries. Such importation makes significant contribution to local as well as global dispersal of many viruses and their vectors. For example, many viruses of potato and some of their vectors were brought to Europe from America and have since spread to many other countries through infected tubers. Large-scale trans-country dispersal of most of the citrus viruses and viroids takes place through export/import of the infected bud-sticks/rootstock seeds. Human dispersal has a big role in the epidemiology of viruses and will be discussed in Chapter 7.

References

A'Brook, J. 1973. Observations on different methods of aphid trapping. Ann. Appl. Biol. 74: 263-277

Adlerz, W.C. 1971. A reservoir-equipped Moericke Trap for collecting aphids. J. Econ. Ent. 64 (4): 966-967

Alykhin, A. and Sewell, G. 2003. On-soil movement and plant colonization by walking wingless morphs of three aphid species (Homoptera: Aphididae) in greenhouse arenas. Environ. Entomol. 32 (6): 1393-1398

Bowden, J. 1973a. The influence of moonlight on catches of insects in light-traps in Africa. Part I. The moon and the moonlight. Bull. Ent. Res. 63: 113-128

Bowden, J. 1973b. The significance of moonlight in photoperiodic responses of insects. Bull. Ent. Res. 62: 605-612

Bowden, J. 1981. The relationship between light-and suction-trap catches of *Chrysoperia carnea* (Stephens) (Neuroptera: Chrysopidae), and the adjustment of light-trap catches to allow for variation in moonlight. Bull. Ent. Res. 71: 621-629

Bowden, J. 1982. An analysis of factors affecting catches of insects in light-traps. Bull. Ent. Res. 72: 535-556

Bowden, J., Mukhopadhyay, S., Nath, P.S., Sarkar, T.K., Sarkar, S. and Mukhopadhyay, S. 1988. Analysis of light trap catches of *Nephotettix* species (Hemiptera: Cicadellidae) in West Bengal. Indian J. Agric. Sci. 58: 125-130

Cooter, R.J. 1993. The flight potential of insect pests and its estimation in the laboratory: Techniques, limitations and insights. The Central Association of Beekeepers. Essex, UK.

Greenbank, D.O. 1957. The role of climate and dispersal in the initiation of the outbreaks of the Spruce budworm in New Brunswick. Can. Entomol. 97: 1077-1089

Greenbank, D.O., Schaefer, G.W. and Rainey, R.C. 1980. Spruce budworm (Lepidoptera: Tortricidae) moth flight and dispersal: New understanding from canopy observations, radar and aircraft. Memoirs Entomol. Soc. Canada No. 110

Harrington, R. 1998. Workshop: Suction trapping. In: Aphids in Natural and Managed Ecosystems. J.M. Nieto Nafria and A.F.G. Dixon (eds). Leon, Spain, pp. 645-655

Harrison, B.D. 1964. The transmission of plant viruses in soil. In: Plant Virology. M.K. Corbett and H.D. Sisler (eds). University of Florida Press, Gainesville, Florida, USA. pp. 118-144

Irwin, M.E. 1980. Sampling aphids in soybean field. Sampling methods. In: Soybean Entomology. M. Kogan and D.C. Harzog (eds). Springer-Verlag, New York, USA. pp. 239-259

Irwin, M.E. and Goodman, R.M. 1981. Ecology and control of soybean mosaic virus. In: Plant Diseases and their Vectors: Ecology and Epidemiology. K. Maramorosch and K.F. Harris (eds). Academic Press, New York, USA. pp. 181-220

Joice, R.J.V. 1983. Aerial transport of pests and pests outbreaks. EPPO Bull. 13: 111-119

Lingren, P.D., Raulson, J.R., Henneberry, T.J. and Sparks, A.N. 1986. Night vision equipment, reproductive biology and nocturnal behavior: importance to studies of insect flight, dispersal, and migration. In: Insect Flight: Dispersal and Migration. W. Danthanarayana (ed). Springer-Verlag. Berlin-Heidelberg, Germany. pp. 253-264

Macaulay, E.D.M., Tatchell, G.M. and Taylor, L.R. 1988. The Rothamsted Insect Survey 12-meter suction trap. Bull. Ent. Res. 78: 12-129

Mukhopadhyay, S. 1991. Lunation-induced variations in catchment areas of light-traps to monitor rice green leafhoppers (*Nephotettix* species) in West Bengal. Indian J. Agric. Sci. 61: 337-340

Mukhopadhyay, S. 1996. Atmospheric Transport System and the Transport of Plant Virus Vectors in Eastern India. Indian J. Mycol. Pl. Pathol. 26: 154-161

Mukhopadhyay, S. and Ghosh, M.R. 1985. Use of traps for pest/vector research and control. Bidhan Chandra Krishi Viswavidyalaya, West Bengal, India

Mukhopadhyay, S. and Mukhopadhyay, S. 1997. Wind-field related light-trap catches of *Nephotettix* spp. (*N. virescens* Distant and *N. nigropictus* Stål), vectors of rice tungro virus in West Bengal. Mausam 48: 83-88

Mukhopadhyay, S., Mukhopadhyay, S. and Sarkar, T.K. 1989. Effect of meteorological factors on the incidence of rice green leafhoppers. Proc. Workshop on Agromet. Information for Planning and Operation in Agriculture with Particular Reference to Plant Protection, V. Krishnamurthy and G. Mathys (eds). WMO, Geneva, Switzerland. pp. 168-188.

Pedgley, D.E. 1980. Weather and air-borne organisms. WMO, No. 562

Pender, J. 1994. Migration of the brown plant hopper, *Nilaparvata lugens* (Stål.) with special reference to synoptic meteorology. Grana 33: 112-115

Perfect, T.J. and Cook, A.G. 1982. Diurnal periodicity of flight in some Delphacidae and Cicadellidae associated with rice. Ecol. Entomol. 7: 317-326

Plumb, R.T. 1987. Aphid trapping to forecast virus diseases. Span 30. 1 (87) 35-37

Plumb, R.T. 2002 Viruses of Poaceae: A Case History in Plant Pathology. Plant Pathology 51: 673-682

Plumb, R.T., Lennon, E.A. and Gutteridge, R.A. 1986. Forecasting barley yellow dwarf virus by monitoring vector populations and infectivity. In: Plant Virus Epidemics, Monitoring, Modelling and Predicting Outbreaks. G.D. MacLean, R.G. Garrett and W.G. Ruesink (eds). Academic Press. Sydney, Australia. pp. 387-398

Raccah, B. 1983. Monitoring insect vector populations and the detection of viruses in vectors. In: Plant Virus Epidemiology. R.T. Plumb and J.M. Thresh (eds). Blackwell Scientific Publications. Oxford, UK. pp. 147-157.

Reynolds, D.R. and Riley, J.R. 1997. Flight behaviour and migration of insect pests. Radar Studies in Developing Countries. Natural Resource Institute, The University of Greenwich, Bull. 71

Reynolds, D.R., Mukhopadhyay, S., Riley, J.R., Das, B.K., Nath, P.S. and Mandal, S.K. 1999. Seasonal variation in the windborne movement of insect pests over north east India. Int. J. Pest Management 45 (3): 195-205

Riley, J.R. 1974. Radar observations on individual desert locusts (*Schistocerca gregaria* Forsk.) Bull. Ent. Res. 64: 19-32

Riley, J.R. 1989. Orientation by high-flying insects at night: Observations and theories. Proc. Conf. Royal Inst. Navigation and Orientation and Navigation—Birds, Humans and other Animals. Cardiff, UK

Riley, J.R. 1992. A millimetric radar to study the flight of small insects. Electronics and Communication Engineering J. 40: 43-48

Riley, J.R. and Reynolds, D.R. 1979. Radar-based studies of the migratory flight of grasshoppers in the middle Niger area of Mali. Proc. R. Soc. Lond. B 240: 67-82

Riley, J.R. and Reynolds, D.R. 1993. Radar monitoring of locusts and other migratory insects. In: World Agriculture. A. Cartwright (ed). Sterling Publ. London, UK. pp. 51-53

Riley, J.R. and Reynolds, D.R. 1997. Vertical-looking radar as a means to improve forecasting and control of desert locusts. In: New Strategies in Locust Control.

S. Krall, R. Pavecling and D. Ba Diallo (eds). Birkhauser Verlag. Basal, Switzerland, pp. 47-54

Riley, J.R., Reynolds, D.R., and Farrow, R.A. 1987. The migration of *Nilaparvata lugens* (Stål) (Delphacidae) and other Hemiptera associated with rice during dry season in the Philippines: a case study using radar, visual observations, aerial tapping and ground trapping. Bull. Entomol. Res. 77: 148-169

Riley, J.R., Cheng, X.-N., Zhang, X.-X., Reynolds, D.R., Xu, G.-M., Smith, A.D., Cheng, J.-Y., Bao, A.-D. and Zhai, B.-P. 1991. The long distance migration of *Nilaparvata lugens* (Stål.) (Delphacidae) in China: radar observations of mass return flight in the autumn. Ecol. Entomol. 16: 471-489

Riley, J.R., Armes, N.J., Reynolds, D.R. and Smith, A.D. 1992. Nocturnal observations on the emergence and flight behaviour of the *Helicoverpa armigera* (Lepidoptera: Noctuidae) on the post-rainy season in central India. Bull. Ent. Res. 82: 243-246

Riley, J.R., Reynolds, D.R., Smith, A.D., Rosenberg, L.J., Cheng, X.-N., Zhang, X.-X., Xu, G.M., Cheng, J.Y., Bao, A.-D., Zhai, B.-P. and Wang, H.-K. 1994. Observations of the autumn migration of the *Nilaparvata lugens* (Homoptera: Delphacidae) and other pests in east-central China. Bull. Ent. Res. 84: 389-402

Riley, J.R., Reynolds, D.R., Mukhopadhyay, S., Ghosh, M. and Sarkar, T. 1995. Long-distance migration of aphids and other small insects in northern India. Eur. J. Entomol. 92: 639-653

Riley, J.R., Smith, A.D., Reynolds, D.R., Edwards, A.S., Osborne, J.L., Williams, I.H., Carreck, N.L. and Poppy, G.M. 1996. Tracking bees with harmonic radar. Nature 379: 29-30

Riley, J.R., Reynolds, D.R., Smith, A.D., Edwards, A.S., Osborne, J.L., Williams, I.H. and McCartney, H.A. 1999. Compensation for wind drift by bumble bees. Nature 400: 126

Rosenberg, L.J. and Magor, J.I. 1983. Flight duration of the brown plant hopper, *Nilaparvata lugens* (Homoptera: Delphacidae). Ecol. Entomol. 8: 341-350

Rosenberg, L.J. and Magor, J.I. 1987 Predicting wind-borne displacements of the brown plant hopper, *Nilaparvata lugens,* from synoptic weather data. I. Long distance displacements in the northeast monsoon. J. Anim. Ecol. 56: 39-51

Sarra, S. and Peters, D. 2003. Rice yellow mottle virus is transmitted by cows, donkeys, and grass rats in irrigated rice crops. Plant Disease 87 (7): 804-805

Schaefer, G.W. 1969. Radar studies of locusts, moth and migration in Sahara. Proc. R. Ent. Soc. Lond. C 34: 39-40

Smith, A.D., Riley, J.R. and Gregory, R.D. 1993. A method for routine monitoring of the aerial migration of insects by using a vertical-looking radar. Phil. Trans. R. Soc. Lond. B 340: 393-404

Southwood, T.R.E. 1978. Ecological Methods. The English Language Book Society and Chapman and Hall, London, UK. pp. 131-136

Sparks, A.N. 1986. Economic implications of long-range insect migration. In: Long-range Migration of Moths of Agronomic Importance to the Unites States and Canada: Specific Examples of Occurrence and Synoptic Weather Patterns Conducive to Migration. USDA, ARS-45. pp. 98-104

Sparks, A.N., Westbrook, J.K., Wolf, W.W., Pair, S.D. and Raulstor, J.R. 1985. Atmospheric Transport of Biotic Agents on a Local Scale. In: The Movement and Dispersal of Agriculturally Important Biotic Agents. D.R. Mackenzie, C.S. Barfield, G.G. Kennedy and R.D. Berer (eds). Claitor's Publishing Div. Inc. Baton Rouge LA, USA

Taylor, L.R. 1955. The standardization of airflow in the insect suction traps. Ann. Appl. Biol. 43: 390-408

Taylor, L.R. 1962. The efficiency of cylindrical sticky insect traps and suspended nets. Ann. Appl. Biol. 50: 681-685

Taylor, L.R. 1986. The distribution of virus diseases and the migrant vector aphid. In: Plant Virus Epidemics, Monitoring, Modelling and Predicting Outbreaks. G.D. McLean, R.G. Garrett and W.G. Ruesink (eds). Academic Press. Sydney, Australia, pp. 35-57

Taylor, L.R. 1989. Long Term Studies in Ecology: Approaches and Alternatives. Objective and experiment in long-term research. G.E. Likens (ed). Springer-Verlag, New York, USA. pp. 20-70

Thresh, J.M. 1983. The long-range dispersal of plant viruses by arthropod vectors. Phil. Trans. R. Soc. (London) (B) 302: 497-528

Williams, C.B. 1948. The Rothamsted Light Trap. Proc. Royal Entomol. Soc. London (A) 23: 80-85

Woiwod, I.P. and Harrington, R. 1994. Flying in the Face of Change: The Rothamsted Insect Survey. In: Long-term Research in Agricultural and Ecological Sciences. R.A. Leigh and A.E. Johnston (eds). CAB International, Wallingford, UK. pp. 321-342

Woiwod, I.P., Tatchell, G.M. and Barrett, A.M. 1984. A system for the rapid collection, analysis and dissemination of aphid-monitoring data from suction traps. Crop Protection 3: 273-288

7

Plant Virus Epidemiology and Ecology

A. INTRODUCTION

Epidemiology is the science of plant disease dynamics, their incidence and distribution and their control and prevention. An epidemic is the progress of a disease with time and space and depends on interactions between the pathogen, the host and the environment. If these events are known they may be used as structures and their effects may be modelled to predict, or forecast, epidemics particularly when the pathogen does not require a vector for transmission. However, vectors are normally required for the transmission of viruses and the relationships of the vectors with the hosts and the environment are different. So the structure of epidemics of virus diseases follows different and complex patterns. These patterns differ with the nature of the viruses, their relationship of the vectors, geographical locations and the weather/climate. Because of the complicated nature of these structures, an understanding of ecological concepts of epidemiology is also required for its proper understanding.

In the past epidemiology was mostly considered statistically. Statistical epidemiology primarily deals with the spread of disease under a particular set of conditions. An epidemic can be considered as, in a holistic way, taking all ecological components into consideration. Thresh (1987) emphasized the information on (i) *host populations* (longevity, propagation, mobility, growing season), (ii) *strategies of virus spread*, (iii) *disease progress curves*, and (iv) *virus strains and variation*. Hull (2002) outlined the following factors related to an epidemic: (i) *source of the virus*, (ii) *rate of movement and distribution within the host plants*, (iii) *severity of the disease*, (iv) *mutability and strain selection*, (v) *host range*, and (vi) *dispersal*. These outlines by Thresh and Hull included more or less similar parameters, but it is very difficult to quantify them in the field, particularly in tropical and sub-tropical humid or sub-humid climates with multi-culture systems where sources of the virus, availability of susceptible host and movement of vectors are very complex.

Most epidemiological studies have been made on arable crops of temperate countries planted in monocultures with dense, regular arrays of uniform genotypes in contrast to crops and viruses of the tropics. Plumb *et al* (2000) demonstrated the role of cropping systems and weeds in the perpetuation of the epidemics of different viruses in various crops in the tropics. The recent introduction of genetically engineered virus resistant crops throws a new challenge to the ecology and epidemiology of virus diseases as the genetic alterations may bring about changes both in the host and vector systems. The key factors for an epidemic are the viruses themselves, their change with time and space, perenation and infectivity, vector activity and climate that bring critical complexities to the events.

The International Society of Plant Pathologists (ISPP) constituted a separate committee 'The Plant Virus Epidemiology Committee' in 1981. This Committee periodically assesses the development of this most dynamic discipline aiming to bring together researchers working on the epidemiology and control of plant viruses in a global forum to address, from a world perspective, past, current and future thrusts. The activities of this Committee have introduced several new dimensions to the concept of the traditional epidemiology such as 'Molecular Epidemiology'. A reconsideration of plant virus epidemiology has also now become imperative because of the consequences of climate change on global agriculture, with ensuing changes in the cropping systems, host-vector-virus interactions, vector biology, and not least the introduction and movement of viruses and their vectors into new areas. Recently, advanced electronic technologies are being used to record the spread and assure bio-safety in some countries. Post-introduction mapping of plant virus spread can be done with GPS and GIS technologies that are currently practised by the USA National Plant Diagnostic Network, which provides a plant disease bio-security system now operating from five hubs in the USA. Gathering temporally and geo-spatially referenced diagnostic data is one of its roles. GIS is proving to be a powerful tool to provide maps that identify production areas with different degrees of risk for specific plant virus pathosystems.

B. NATURE OF VIRUSES AND THEIR EPIDEMIOLOGICAL RELEVANCE

According to the current record, there are 18 families and about 80 genera of viruses. Eighteen genera still remained to be assigned to specific families (Fauquet *et al* 2005). Epidemiology of the viruses belonging to these genera (vide Appendix I.6) differs widely. Some viruses are crop-adapted while others are wild plant adapted. There are also examples where the viruses are insect-adapted. Whatever their type, viruses are unable to move by themselves; most require vectors for their spread. Some, comparatively few, may multiply in their vectors, and propagate through their progeny, some may remain in the vectors till their death and others may remain in the vectors only for a short period of time. These relationships between the vectors and the viruses they transmit are generally controlled by the nature of the viruses at the molecular level but there may be influences of the products on host-virus interactions.

C. CONVENTIONAL EPIDEMIOLOGY

Epidemiology is an important component of Integrated Pest Management systems. It provides information on the prospect of the spread of a disease under a specific climatic and agro-meteorological condition. In any disease caused by a pathogen, other than a virus or allied one, an epidemic is the expression of the interaction between the pathogen, host and the environment; for viruses and an allied pathogen an additional interactor, the vectors play a very decisive role.

The epidemiological events start with the arrival of the virus in the field where susceptible crops are present. The arrival may be from a short or long distance by the vectors or through the seeds or planting materials. The virus may also be present in the soil. Once in the field, it multiplies and spreads following definite patterns depending upon the nature of the vector and agro-meteorological conditions.

(a) Arrival of the Virus in the Field

There are many ways for a virus to enter a virus-free environment. They may enter through seed, planting material or by vector. Insect vectors transport all types of viruses (non-persistent, semi-persistent and persistent) for short distances but only persistent viruses are transported from a long distance by migrating vectors (vide Chapter 6).

(i) Arrival through seed and pollen
Viruses may be present in the testa, endosperm, embryo and pollen. Irrespective of their place of localization, they infect the germinating seedlings. In the case of the embryo, the developing embryo becomes infected before fertilization by the viruses carried by the gametes or by direct invasion after fertilization. Both the processes may occur in certain host virus combinations but the pathways of the virus to the embryo are complex and involve a range of genetic and environmental interactions (Maule 2000). In direct embryo infection, the virus may move through the testa of the immature seed after fertilization and reach the micropylar region for embryo infection (Wang and Maule 1994).

There are a number of viruses that are transmitted through infected pollen. Self-pollination of infected plants may result in a higher percentage of infected seeds than when only one of the gametes comes from an infected individual. de Assis and Sherwood (2000), who made crosses between healthy and infected

Arabidopsis plants, observed that TYMV can invade the seeds from either the male or female parent but that invasion from the maternal tissue was the only way for the seed to be infected by TMV. Another way for the transfer of infection by the pollen is through the germ tubes growing from the pollen that may become infected by mechanical means and carry the virus to the ovule (Hamilton *et al* 1977). Insects may also be necessary for successful pollen transmission. Sdoodee and Teakle (1993) showed that pollen infected with TSV is carried both externally and internally in the thrips and is thought to infect through feeding wounds. In case of some soil-borne viruses, vectors may assist their seed-transmission as found in MNSV that is assisted by *Olpidium bornovanus* (Campbell *et al* 1996). Pollen transmission plays a significant role in the epidemiology of viruses in cross-pollinated trees (Mink 1993).

Seeds are a very effective means for the entry of the viruses in the fields. On germination infected seeds produce infected seedlings and the random foci of these primary sources of infection act as sources for the secondary spread by vectors. Cryptic viruses are exclusively seed transmitted whereas many other viruses are transmitted both through seeds and by vectors. Common seed transmitted viruses are given in Table 7.1.

Table 7.1 Common Seed/Pollen Transmitted Viruses (Fauquet *et al* 2005)

Virus	Mode
1. Badnavirus	May or may not be seed transmitted; transmission may be from pollen to the seed
2. Cavemovirus	Seed; Pollen
3. Hordeivirus	Seed; Pollen to the pollinated seed
4. Potexvirus	Seed; Pollen to the pollinated seed
5. Tobamovirus	Seed
6. Tobravirus	Seed; Pollen to the pollinated seed
7. Tymovirus	Seed
8. Tombusvirus	Seed; Pollen to the pollinated seed
9. Sobemovirus	Seed, Pollen
10. Nepovirus	Seed, Pollen; Pollen to the pollinated seed
11. Cucumovirus	Seed
12. Machlomovirus	Seed
13. Idaeovirus	Seed; Pollen to the pollinated seed
14. Enamovirus	Pollen to the pollinated seed
15. Trichovirus	Seed
16. Pecluvirus	Seed
17. Alfacryptovirus	Seed; Pollen to the pollinated seed
18. Betacryptovirus	Seed; Pollen to the pollinated seed
19. Ourmiavirus	Seed
20. Idaeovirus	Seed; Pollen to the pollinated seed
21. Furovirus	Seed

(ii)   Extent of infection in seeds

The extent of infection in seeds may range from 100% to 0%. Acquisition of the virus by seeds depends on many factors, including genomic determinants, properties of the host and time of infection. A wide range of host species may transmit several viruses through their seeds. The rate of seed transmission by different host species may also differ. McKirdy and Jones (1994) recorded that AMV is transmitted through seeds of *Melilotus indica, Stachys arvensis* and *Ornithopus cornessus* at rates of 10%, 2% and 0.1% respectively. Variation is also observed in different varieties of the same species. The transmission of BSMV through seeds of different varieties of barley differs from 0 to 75% (Carroll and Chapman 1970). Time of infection often affects the extent of seed infectivity. Normally the rate of infectivity is highest with the earliest invasion of the embryo by the virus from the mother plant (Ren *et al* 1997). In the case of the invasion of BSMV in barley however, a gradual increase of infectivity occurs with late invasion. Infectivity of seeds also varies with the age of the seeds. Some viruses are lost quite rapidly from seeds on storage, while others persist for years even when stored at a very low temperature. Frosheiser (1974) observed very little loss of AMV in infected alfalfa seeds after 5 y at –18°C or at room temperature.

Extent of seed transmission is however, controlled by the determinants in the virus. Wang *et al* (1997) observed that the removal of the 12-kDa gene from RNA1 of PEBV completely abolished seed transmissibility and concluded that this is involved in the infection of the gametic cells.

(iii)   Arrival through planting materials/bud-woods

Progeny derived from the vegetative propagation of virus-infected plants normally contains the virus. Such planting material used to establish a new crop is an important source of the virus in the field. Secondary spread takes place from this primary source of infection.

Tree crops are normally grown from grafts. If the stock or the scion contains a virus, the graft will also contain that virus (es). The infected graft when planted in the field acts as the primary source of infection. Budding/grafting of a healthy scion on infected rootstocks may also produce detrimental results. When healthy buds of *Burbank* and *Sultan* plums are grafted on to plum rootstocks infected with the vein-banding type of plum mosaic, the bud reacts with necrosis and dies (Chamberlain *et al* 1951).

Orchardists normally grow healthy rootstock seedlings and import bud-woods of desired quality. Infected bud-woods may sometimes become the source of the virus in that orchard and can spread in other local orchards. Importation of infected bud-wood that often remains symptomless is a serious concern in virus epidemiology as it spreads viruses from one country to another eluding routine diagnostics followed for quarantine certification.

(iv)   Arrival through vectors

Vectors normally bring viruses from different outside sources. These sources may be nearby infected crop fields, fields having other infected hosts particularly over-wintering weeds or shrubs or trees. Chances for transmission are greater for those viruses whose vectors have a wide host range. A virus may over-winter in various ways. As for example, yellowing viruses of sugarbeet in Europe, may over-winter in sugarbeet crops kept for a second year to produce seed; in other widely grown forms of *Beta vulgaris* such as fodder beet, mangold and beetroot or in spinach and in wild beet growing on foreshores around the coasts. Some of these viruses also occur in weeds such as *Capsella bursa-pastoris, Senecio vulgaris, Stellaria media* and *Chenopodium album*. Sugarbeet clamps can also provide suitable conditions for aphid vectors to over-winter and carry the viruses to the next year's crops under suitable environmental conditions.

In tropical countries, vectors frequently carry viruses from mature crops or infected weeds to near-by young crops in an overlapping sequence of cropping systems (Plumb *et al* 2000). In temperate countries, overlapping cropping may also contribute to the transfer of the viruses as, for example, in Canada where winter wheat is sown early in the autumn and germinates before the maturing crop of spring-sown wheat is harvested. Thus, mites carrying wheat streak mosaic virus are often blown from the maturing crop to the young crop. Volunteer wheat plant growing on summer-fallowed land may also play a role in such a carry over and spread of virus to the field (Slykhuis 1955).

Viruses may arrive in fields from near-by infected crops or weeds irrespective of the characteristics (non-persistent, semi-persistent, persistent) of their transmission but only persistent and semi-persistently transmitted viruses can arrive through vectors from a long distance away. Rosenberg and Magor (1983) mapped the long distance migration of *rice grassy stunt* and *rice ragged stunt* viruses by the vector *Nilaparvata lugens* (Figure 7.1). Thresh (1983) reviewed the dispersal of leafhopper-transmitted viruses particularly *beet curly top virus* transmitted by *Circulifer tenellus, sugarcane fiji virus* transmitted by *Perkinsiella saccharicida, maize streak virus* transmitted by *Cicadulina* spp. and *rice hoja blanca* transmitted by *Sogatodes oryzicola* and *S. cubanus*. Many persistent and semi-persistently transmitted viruses have the potential to arrive from long distances. Vectors of the common semi-persistent and persistent viruses are given in Table 7.2.

Many soil-borne viruses persist in soil even without the presence of the susceptible crop and infect such crops when they are planted, which may be many years later. For example, the nematode *Trichodorus pachydermus* maintains without the presence of a susceptible crop for more than a year and cereal mosaic viruses and *potato mop-top virus* persist in resting spores of *Polymyxa graminis* and *Spongospora subterranea* for many years.

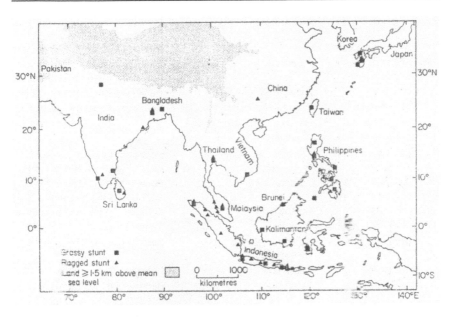

Figure 7.1 Distribution of rice grassy stunt and rice ragged stunt diseases. (Rosenberg and Magor 1983) (With permission from Wiley-Blackwell)

Table 7.2 Vectors of Some Common Semi-persistent and Persistently Transmitted Viruses (Fauquet *et al* 2005)

Virus	Vector	Nature
1. Fijivirus	Delphacidae (*Nilaparvata lugens*)	Persistent (trans-ovarial transmission)
2. Oryzavirus	Delphacidae (*Nilaparvata lugens*)	Persistent
3. Phytoreovirus	Cicadellidae (*Nephotettix cincticeps, N. nigropictus, Recilia dorsalis*)	Persistent (may or may not have trans-ovarial transmission)
4. Badnavirus	Aleyrodidae, Aphididae, Cicadellidae, Pseudococcidae	Persistent (not transmitted trans-ovarially), Semi-persistent
5. Caulimovirus	Aphididae	Semi-persistent, Non-persistent
6. Rice Tungro Virus	Cicadellidae (*Nephotettix virescens, N. nigropictus, N. cincticeps, N. malayanus, N. parvus, Recilia dorsalis*)	Semi-persistent
7. Begomovirus	Aleyrodidae	Persistent (may be trans-ovarial), Semi-persistent, Non-persistent
8. Curtovirus	Cicadellidae, Membracidae	Persistent (does not multiply in the vector), Semi-persistent

9. Mastrevirus	Cicadellidae	Persistent (does not multiply in the vector), Semi-persistent, Non-persistent
10. Nanovirus	Aphididae	Persistent (does not multiply in the vector), Semi-persistent
11. Carlavirus	Aleyrodidae	Semi-persistent, Non-persistent
12. Closterovirus	Aleyrodidae, Aphididae, Psyllidae, Pseudococcidae	Semi-persistent, Non-persistent
13. Ipomovirus	Aleyrodidae	—
14.Tenuivirus	Cicadellidae, Delphacidae	Persistent (multiplies in the vector, trans-ovarially or not)
15. Tospovirus	Thysanoptera	Persistent (multiplies in the vector), Semi-persistent
16. Tymovirus	Coleoptera	Semi-persistent, Non-persistent
17. Sequivirus	Aphididae	Semi-persistent
18. Sobemovirus	Aphididae, Cicadellidae, Coleoptera	Semi-persistent, Non-persistent
19. Polerovirus	Aphididae`	Persistent (non-propagative)
20. Luteovirus	Aphididae	Persistent (does not multiply in the vector)
21. Marafivirus	Cicadellidae	Persistent (multiplies in vector)
22. Comovirus	Coleoptera	Semi-persistent, Non-persistent
23. Cytorhabdovirus	Aphididae, Delphacidae	Persistent (multiplies in vector)
24. Nucleorhabdoirus	Aphididae, Cicadellidae	Persistent (multiplies in vector)
25. Umbravirus	Aphididae	Persistent
26. Enamovirus	Aphididae	Persistent (not multiplies in vector)
27. Trichovirus	Aphididae, Pseudococcidae, Eriophyidae	Semi-persistent
28. Waikavirus	Aphididae, Cycadellidae	Semi-persistent

(b)   Spread of Virus in Fields and Quantitative Epidemiology

The spread of plant virus diseases is the result of the interaction between a static and a dynamic factor; the host is static whereas the viruses carried by vectors are dynamic. The static factors that have received most attention are a few important temperate arable crops. These are mostly planted in monoculture and more or less synchronously in dense regular array of uniform genotypes. Thus any generalization on the mechanics of virus spread generated from such regional plant communities may be misleading elsewhere.

The spread of the virus in the field depends to a large extent upon its stability. Normally mechanically transmitted viruses are highly stable and are likely to survive and spread. Virus concentration in the infected tissues may be seasonal and the same host may differentially act as the source in different seasons. There are several other factors that affect spread. Strains that move slowly through the host plant from the point of infection are less likely to survive and spread. When a strain causes the death of the host plant, it is less likely to survive and cause further spread. The mechanics of the epidemiological work in temperate countries is the study of the determinants, dynamics and distribution of virus diseases in host populations. This includes a dimensional aspect of the factors determining the spread of a virus into a given situation. These dimensions are normally expressed in two different ways—spatial and temporal—and can be considered as quantitative epidemiology.

(i) Quantitative epidemiology

The concept of quantitative epidemology was first introduced by Vanderplank (1963) who suggested a formula to measure the progress of a disease which states: $X = X0^{ert}$, where 'X' = the proportion of the disease at one time; X0 = the amount of the critical inoculum; 'r' = average infection rate; 't' = time during which infection occurred. The value of 'e' is the base of the natural logarithm $e = (1 + 1/n)^n$.

The formula deals with an exponential function with the basic assumption that at any given time, the rate of disease increase is proportional to the amount of disease present at the moment, 'r' of this formula gives an overall measure of the rate at which an epidemic progresses, comparative 'r' can be derived by taking log and transposing

$$Log e^x = log e^{x0} + r t; r t = log ex - log e^{x0}; r = 1/t log e x/x0$$

When the disease reaches high intensity and the host becomes limiting, the formula is modified as: $r = 1/t_2 - t_1 \log e\ x_2 - (1 - x)/x_1 (1 - x)$

Vanderplank also provided mathematical treatments for the increase in incidence of virus infection with time resulting either from spread by the immigrating viruliferous vectors or from spread at the site. The spread in these two types of situations is calculated from compound and simple interest equations. Comparative rates of spread in space and time are determined by these formulae with respect to different viruses, crops, regions and seasons.

There may also be different mathematical equations for interpreting primary and secondary contaminations or spread. If the contamination is external in origin and occurs regularly, the progression of epidemics would be of 'simple interest' type (Vanderplanck 1963) or 'monocyclic' type (Zadoks and Schein 1979). In the secondary contamination or spread, the number of infected plants increases exponentially and epidemic would be of 'compound interest' (Vanderplanck 1963) or 'multicyclic' type (Zadox and Schein 1979) with a sigmoid shape.

There are also other transformations to describe plant virus epidemics (Madden and Campbell 1986, Nutter 1997) for better adjustment of the experimental data.

Following the above principles progress of epidemics in several virus diseases has been mathematically calculated and attempts have been made for appropriate modelling. Such modelling has been done mostly with synchronized crops, but Vanderplank's quantitative epidemiology was not maintained subsequently (Thresh 1983). Spatial and temporal patterns of virus spread were described and analyzed in a notable series of reviews by Thresh (Thresh 1974, 1976, 1983), and to some extent these patterns were related to the mobility characteristics of vectors.

(ii) Spatial distribution

In spatial distribution infected plants may be next to one another or randomly. This type of distribution is normally studied by gradient analysis which has been widely used. It involves recording the number, or proportion, of virus-infected plants as determined by symptoms. To analyze the gradient the cumulative incidence of disease is normally plotted against space. The gradient shows different types of steepness depending on the dispersal of the viruses from its source(s). The steepness is greatly influenced by the mobility of vectors. It can be very shallow for the leaf hopper-transmitted viruses, steeper for aphid borne viruses, and very steep for viruses with nematode vector. The gradient for non-persistently transmitted viruses may be steeper than those for persistent viruses with the same type of vector in the same crop. The spatial distribution is further influenced by wind direction and speed particularly when the vectors are aerially transported.

Zhang *et al* (2002) studied the spatial distribution patterns of tomato mosaic disease and its aphid vectors. The distribution pattern of the infected plants changed with time. The pattern was random at the beginning of the disease development followed by a clustering type of distribution and an even pattern towards the end of plant growth. Aphid vectors (*Myzus persicae*) always had a clustering type of distribution establishing the alatae as the most effective vector for the mechanical transmission of the virus.

(iii) Temporal distribution

Temporal distribution of diseases is recorded as progress of diseases with time. In this case the cumulative incidence of disease is plotted against time. Normally these curves tend to be sigmoid although within the same overall pattern there can be great differences between sites, seasons and disease in the onset, rate, duration and total amount of spread. After the first appearance of disease there is usually a rapid increase in the cumulative total infected plants. The rate of increase then declines as weather or other conditions become unfavourable or because progressively fewer healthy plants remain to be infected and multiple infections become increasingly important. Studies made so far on the progress of diseases indicate that the spread is more rapid in herbaceous annuals in comparison to that in woody perennials. The progress of diseases varies with respect to virus, vector, location, and weather conditions. Differential spread of the same virus transmitted by the same vector in the same crop in the different regions also occurs.

Lindblad and Sigvald (2004) studied the temporal spread of *wheat dwarf virus* and mature plant resistance in winter wheat in Sweden. The primary infection was as low as 5% but the infection rates increased up to 50% in summer. Infection rate ceased to increase as the crop reached the stage of stem elongation and the plant became resistant to infection at growth stage DC31 (when the first node is detectable). Groves *et al* (2003) studied the temporal dispersal of *Frankliniella fusca* and the spread of TSWV in different locations in Central and Eastern North Carolina, USA. The temporal dispersal patterns during spring normally differ between the locations. Regression analysis showed that the time of the first flight in spring also differ. The dispersal varies with height. The temporal patterns of the aerial capture at 0.1 and 1.0 m above the soil surface showed this difference. Fewer thrips are captured at 0.1 m, although thrips dispersal occurs earlier and over a greater time interval compared with that at 1.0 m. The temporal patterns of TSWV infection differ between locations in the spring whereas the patterns of virus occurrence are similar during the autumn. Patterns of *F. fusca* dispersal and subsequent TSWV occurrence are synchronous. The patterns of the vector dispersal and TSWV occurrence may be a useful indicator for identifying the time when susceptible crops will be at the highest risk of TSWV infection.

(iv)   Modelling, forecasting and prediction
Historically, modelling approaches to epidemiology is vector-based, dealing with empirical forecasting systems or simulation of vector population dynamics. Several workers of the temperate regions have modelled virus diseases of several economically important crops of those regions.

Werker *et al* (1998) made a detailed study on virus yellows diseases in England in relation to numbers of migrating *Myzus persicae*. A non-linear model was fitted to the data incorporating dual routes of infection: primary infection arising from winged immigrant aphids carrying the virus and secondary infection, arising from local dispersion of their wingless offspring transferring the virus from infected to uninfected plants. The model was fitted to the data assuming that: (a) the epidemic began when the first migrating aphid was recorded, (b) the rate of primary infection was dependent on the total number of aphids migrating up to the end of June, and (c) the rate of secondary infection was dependent on the proportion of the diseased plants in the crop. Good fits also depended on allowing the rates of the disease progress to decay with time to accommodate the effects of increasing host resistance to aphid feeding with plant age. This meant that early epidemics developed more rapidly than late epidemics, and this was a consistent observation. Given that the changes in the epidemiology of virus yellows over the years is associated with improvements in pest management practices, the model presented a useful extension to disease forecasting by providing predictions of the disease risks in the absence of pesticides.

Lu Jun *et al* (1999) made quantitative analysis of some epidemiological factors. The most important primary source of the *wheat rosette stunt virus* in spring wheat in northeast China was viruliferous nymphs of the vector *Laodelphax striatellus* that had over-wintered in underground wheat stubs, weed roots and soil cracks. It showed a linear relationship between the amount and the proportion of viruliferous vectors and the incidence of *wheat rosette stunt*. The equation was established by step-wise regression analysis. The disease incidence and the growth stage of wheat during infection were closely related. The most damaging inoculating period was before the 3-leaf stage and the disease index could reach 20-30%.

Dusi *et al* (2000) measured and modelled the spread of *beet mosaic virus* in sugarbeet in the Netherlands. Two plants in the centre of each plot were inoculated with BtMV using *Myzus persicae*. The spread of infection around these sources was monitored by inspecting the plants found on two diagonal transects through the centre of the plot. The spread was concentrated in patches around the inoculated plants and its rate was explained by vector pressure, as shown by regression analysis and mechanistic simulation model. This vector pressure was quantified using data obtained by catching aphids in a green water trap in the crop, catching aphids in a 12-m high suction trap at a distant location, and infection of the bait plants from adjacent virus source plants. The daily total aphids caught in a suction trap provided the best statistical explanation for the spread of the virus. The parameter 'r' describing the relationship between the vector pressure and the rate of disease progress was remarkably robust. This parameter varied, less than 10% between four experiments done at different sites in two years. The robustness of this parameter suggests that the spread of a potyvirus may be predicted on the basis of the initial infection date and vector abundance.

Madden *et al* (2000) used a continuous time and deterministic model to characterize a plant virus disease epidemic in relation to the virus transmission mechanism and the population dynamics of insect vectors. The model can be written as a set of linked differential equations for healthy (virus-free), latently infected, infectious and removal (post-infectious) plant categories and virus-free, latent and infective insects with parameters based on the transmission classes, vector population dynamics, immigration/emigration rates and virus-plant interactions. The rate of change of diseased plants is a function of the density of infective insects, the number of plants visited per time, and the probability of transmitting the virus per plant visit. The rate of change of infective insects is a function of the density of infectious plants, the number of plants visited per time by an insect, and the probability of acquiring the virus per plant visit. Numerical solutions of the differential equations were used to determine transitional and steady state levels of disease incidence (d*). d* was also determined directly from the model parameters. Clear differences were found in disease development in four transmission classes: non-persistently transmitted (NP), semi-persistently transmitted (SP), circulative persistently transmitted (CP) and persistently propagated (PP) with the highest disease incidence (d) for the SP and CP classes relative to the others, especially at low insect density when there was no insect migration or when vector status of emigrating insects was the same as that of immigrating ones. The PP and CP viruses were most affected by changes in insect mobility. When vector migration was explicitly considered, results depended on the fraction of infective insects in the immigration pool and the fraction of dying and emigrating vectors replaced by immigrants. The PP and CP viruses were most sensitive to changes in these factors. Based on model parameters, the basic reproductive number (RO) (number of new infected plants resulting from an infected plant introduced into a susceptible plant population) was derived and used to determine the steady-state level of disease incidence and an approximate exponential rate of the disease increase early in the epidemic.

Zhang *et al* (2000) prepared a general model of plant-virus disease infection incorporating vector aggregation. Power parameter can be regarded as a measure of the spatial aggregation of the vector or a coefficient of interference between the infection rates for plant-virus diseases that are of a more general form than the familiar bilinear inoculation and acquisition rate depending upon the context. When field data of *cassava mosaic virus* incidence in Uganda were examined, the vector (*Bemisia tabaci*) population density and disease incidence were found to be high, disease progress curves over the first six months from planting could not be explained using models with bilinear infection rates. Incorporation of new infection terms allowed the range of observed disease progress curve types to be described. New evidence of a mutually beneficial interaction between the viruses causing cassava mosaic disease and *B. tabaci* has shown that spatial aggregation of the vector is an inevitable consequence of infection, particularly with a severe virus strain or a sensitive host. Virus infection increases both vector fecundity and the density of vectors on diseased plants. This enhances disease spread by causing an increased emigration rate of infective vectors to other crops.

Webb *et al* (2000) modelled the effect of temperature on the development of *Polymyxa betae*, the vector of *beet necrotic yellow vein virus* (BNYVV) in UK.

The variation in inoculum build-up predicted when temperature data from a range of soil types were used in the model agreed with field observations where higher levels of infection were observed on sandy soils than on black fen peat soils. The difference is most distinct when daily maximum soil temperature values were used to drive the model rather than rolling 24-h average values.

Jeger *et al* (2004) made a critical assessment of quantitative approaches to plant virus epidemiology and the available general theoretical models. The general models can be used to answer strategic questions concerning plant virus epidemiology but need further simplification and/or refinement to be directed to specific diseases. These approaches were illustrated by reference to two examples of whitefly transmitted-diseases (African cassava mosaic disease and tomato leaf curl disease in India).

The parameters used in the model for *African cassava mosaic virus* were plant density, maximum planting rate, plant loss rate, harvesting rate, maximum vector abundance, maximum vector birth rate, vector mortality, infection rate, acquisition rate, infection cutting selection frequency and infected plant recovery rate (Holt *et al* 1997). Ranges of parameters and variables used in the model for Indian tomato leaf curl disease were crop turnover rate, infection rate, latent period, acquisition rate, vector loss rate, vector infectivity, vector arrival rate and scaled host abundance (Holt and Chancellor 1999).

D. ECOLOGICAL EPIDEMIOLOGY

Modelling approaches based on vector activity have several limitations. Recently, epidemiological models such as those used in human/animal epidemiology have been introduced in an attempt to characterize and analyze the population ecology of viral diseases. The theoretical bases for these models and their use in evaluating control strategies in terms of the interactions between host, virus and vector are considered here. Vector activity and behaviour, especially in relation to virus transmission, are important determinants of the rate and extent of epidemic development. Mukhopadhyay *et al* (1994) studied comparative epidemiology of some whitefly transmitted viruses (*mung bean yellow mosaic, bhendi (okra) yellow vein mosaic, tomato leaf curl*) in India particularly with respect to transmission characteristics, host range, spreading patterns and meteorological relations. Major transmission characteristics (minimum and optimum acquisition access, inoculation access, retention periods) of these viruses did not differ but their host range differed indicating their epidemiological differences. The spread of all these diseases was related to vector activity but the meteorological relations differed. Spread of mung bean yellow mosaic and bhendi (okra) yellow vein mosaic diseases and the vector population on theses crops were positively correlated to maximum temperature. For *tomato leaf curl*, on the other hand, the incidence of the disease and vector in winter crop were positively correlated not only to maximum temperature but also to relative humidity, indicating again the differential epidemiology. Isopathic contour maps of the spread of mung bean yellow mosaic disease showed vector activity dependent gradients.

In poly-culture systems in the tropics, alternate and weed hosts significantly contribute to the population build up of some viruses in fields. For example, different types of cucurbitaceous crops particularly pumpkin (*Cucurbita moschata* Poir), ridge gourd (*Luffa acutangula* Roxb.), pointed gourd (*Trichosanthes dioca* Roxb.), cucumber (*Cucumis sativus* L.) and bitter gourd (*Momordica charantia* L.) are grown on rotation throughout the year in eastern India. Amongst these, pumpkin, ridge gourd and bitter gourd are cultivated during different seasons but infected by a common virus called *Zucchini yellow mosaic virus*. This virus is also found in a perennial cucurbitaceous wild host called *Coccinia grandis* (L.) Voigt; in this plant this virus does not produce any prominent symptom. Similarly potato, brinjal (egg plant) and chilli are cultivated in the same or different seasons. *Potato virus Y* infects all of them. This virus is also found in two common weeds around the fields, called *Teramnus labilis* Sw. and *Datura metel* L. (Plumb *et al* 2000). Thus any attempt to study the epidemiology of any of these viruses in a particular crop during a particular season may be misleading.

Infectivity of the viruses also has epidemiological significance. Some viruses may infect many plants while the infection by others remains restricted to only a few. Some are adapted only to cultivated plants whereas some others are adapted to wild plants. *Cucumber mosaic virus* (CMV) can infect more than 1000 host species distributed in more than 85 families. Diversity of hosts gives a virus much greater opportunity to maintain and modify itself, and spread widely. Viruses that have perennial ornamental plants as hosts as well as other agricultural and horticultural species may attain global spread. There are some ornamentals and flowers that can carry several viruses. A perennial weed called *Plantago* may act as a reservoir of at least 26 viruses belonging to 19 different groups (Hammond 1982). The concentration of the viruses in the donor hosts and specific infectivity of them also contribute to any epidemic of a virus.

(a)   Weed hosts of different viruses and their epidemiological significance
Apablaza *et al* (2003) analyzed 211 weed samples in Chile collected from the fields of tomato, pepper, melon, watermelon, and squash that were virus infected. Detection was done by double antibody sandwich-ELISA and mechanical inoculation. The relative importance of tested weeds as sources in decreasing order was: *Datura stramonium, Amaranthus* spp., *Raphanus sativus, Chenopodium album, Galega officinalis, Conicum maculatum, Sonchus asper, Malva* spp., *Urtica urens, Bidens* spp., *Brassica campestris, Sorghum halepense* and *Solanum* spp. One to five of the following viruses were found infecting these weeds: *alfalfa mosaic virus* (AMV), *cucumber mosaic virus* (CMV), *tomato*

spotted wilt virus (TSWV), *potato virus Y* (PVY), *tomato mosaic virus* (ToMV), and *watermelon mosaic virus* (WMV-2).

Mehner *et al* (2003) while investigating the ecology of the *wheat dwarf virus* in Germany, found that self-sown cereals and grasses serve as the infection source. *Bromuis arvensis, B. commutatus, B. hordaceus, B. japonicus, B. sterilis,* and *Phalaris aurandinacea* were identified as new host plant of this virus.

Thomas (2002) recorded weeds predominating on potato and other annual crop fields and waste areas in the Columbia basin of the USA of which *Solanum sarrachoides* (Hairy Nightshade) was an important host of *potato leaf roll virus* (PLRV) and plays a significant role in the epidemiology of this virus.

Groves *et al* (2002) studied the role of weed hosts and tobacco thrips in the epidemiology of TSWV. Thrips (*Frankliniella fusca*) were monitored on 28 common perennial, biennial and annual plant species over two non-crop seasons at field locations across North Carolina, USA. *Sonchus asper, Stellaria media,* and *Taraxacum officinalis* consistently supported the largest populations of immature TSWV vector species. Perennial plant species (*Plantago rugelii, Taraxacum officinale*) were often locally abundant, and many annual species (*Cerastium vulgatum, Sonchus asper, Stellaria media*) were more widely distributed. Perennial species including *P. rugelii,* and *Rumex crispus* remained TSWV-infected for 2 y in small plot-field tests, where these species are locally abundant. They may serve as important and long-lasting TSWV inoculum sources. TSWV infection was documented by double antibody sandwich ELISA in 35 of 72 (49%) common perennial (N=10), biennial (N=4) and annual (N=21) plant species across 18 families. Estimated rates of TSWV infection were highest in *Cerastium vulgatum* (4.2%), *Lactuca scariola* (1.3%), *Mollungo verticillata* (4.3%), *Plantago rugelii* (3.4%), *Ranunculus sarodous* (3.6%), *Sonchus asper* (5.1%), *Stellaria media* (1.4%), and *Taraxacum officinale* (5.8%). Nine plant species were determined to be new hosts for TSWV infection including *Cardamine hirsuta, Eupatorium capillifolium, Geranium carolinianum, Mollugo verticillata, Gnaphelium purpureum, Linaria canadense, Pyrrhopappus carolianum, Raphanistrum* sp. and *Triodania perfoliata*. There is therefore a considerable potential of a number of common annual, biennial and perennial plant species acting as important reproductive sites for *F. fusca* and acquisition sources of TSWV for spread to susceptible crop.

Populations of tobacco thrips produced on TSWV-infected plants did not differ from that produced on healthy plants whereas populations vary greatly among host plant species. Adult *F. fusca* collected from plant hosts were predominantly brachypterous throughout the winter and early spring, but macropterous forms predominated in the late spring. Weed hosts varied in their ability to serve as over-wintering sources of TSWV inoculum. Following the initial infection by TSWV, 75% of *Scleranthus annuus* and *Stellaria media* retained infection over the winter and spring season, whereas on 17% of *Sonchus asper* plants remained infected throughout the same interval. Mortality of TSWV-infected *Sonchus asper* exceeded 25%, but mortality of infected *Stellaria media* and *Scleranthus annuus* did not exceed 8%. Dispersal of TSWV from the inoculum source extended to the limits of the experimental plot (>37 m). Significant directional patterns of TSWV spread to *R. sarodous* plots were detected in April and May, but not June, suggesting that over-wintering levels in an area can increase rapidly during the spring in susceptible weed hosts prior to planting of susceptible crops. The spread of TSWV among weeds in the spring serves to bridge the period when over-wintered inoculum sources decline and susceptible crops are planted (Groves *et al* 2001).

Gupta and Keshwal (2003) while studying the epidemiology of yellow mosaic disease of soybean, found *Paracalyx scubiosus* and *Corchorus olitorius* were reservoir hosts of *mung bean yellow mosaic virus*. Disease development was favoured when maximum temperature and relative humidity were 29.9-36.2°C and 62- 75% respectively. DI (Disease Index) was lower when soybean was alternated with mung bean. Cross inoculation revealed that YMD from mung bean or urad bean (*Vigna mungo*) was not directly transmitted to soybean but YMD from soybean was directly transmitted to French bean (*Phaseolus vulgaris*), *Alternenthera sessilis*, *Paracalyx scubiosus* and *Sida rhombiifolia* and *vice versa*.

Mojtahidi *et al* (2003) studied the weed hosts of *Paratrichodorus allius* and tobacco rattle virus in the Pacific Northwest of the USA. *P. allius* transmits corky ring spot disease of potato. Viruliferous nematodes multiplied on 24 out of 37 weed species evaluated indicating that these were suitable hosts of the vector. Out of 24 weeds, only 11 species were infected with *tobacco rattle virus*. The weeds that serve as hosts for both the virus and the vector are: *Kochia* (*Kochia scoparia* [*Bassia scoparia*]), *Lactuca serriola* (prickly lettuce), *Lamium amplexicaule* (Henbit), *Solanum triflorum* (Black nightshade), *Stellaria media* (Common chickweed) and *Sonchus oleraceous* (Annual sowthisle).

Weeds or wild hosts are also found to play a big role in the global distribution of different viruses.

(b)   Vector activity, host and environmental factors in epidemiology

Fiebig and Poehling (1998) observed that the epidemiology of aphid transmitted viruses is mainly determined by the feeding activity, the development (population density) and the dispersal behaviour of the vector population. Virus infections can alter host suitability of plants to favour vector development and spread. When interactions between *barley yellow dwarf virus* (PAV, MAV) and the vector *Sitobion avenae* were analyzed, non-preference of infected plants showing no symptoms was shown by alate and apterous aphids and preference was for plants showing symptoms. Video studies on probing behaviour showed a non-preference of *S. avenae* for PAV-infected but not for healthy wheat. Electronic monitoring of the probing and feeding behaviour of *S. avenae* on virus-infected plants indicated a reduction in host suitability after an infection. Virus induced effect on *S. avenae* was quite obvious after host plant acceptance. *S. avenae* numbers were decreased on virus-infected plants and indicated an antibiosis effects. Virus infection reduces the host suitability in terms of population growth on infected plants but development of increasing numbers of winged aphids accompanied by reduced host acceptance favours the dispersal of the vector population and the virus.

Irwin (1999) while studying the soybean mosaic virus pathosystem, illustrated the necessity of reasonable understanding of the virus transport by insect vector. Various aphids, with different flight activities, transmit this virus. The complex epidemiology involves vector movement over both landscape and eco-regional scales, and movement especially as it is influenced by atmospheric motion systems over these scales. The importance of conceptual, simulation and predictive models that take into consideration vector movement can not be overstated when dealing with a vector complex of this nature.

Effective movement of the vector within a crop field is related to prevailing meteorological conditions. Jovel *et al* (2000) studied the movement of *Bemisia tabaci* towards and within tomato plants and their relationship of climatic variables. During 26 d 0600-1800 h the immigration of adults (recorded by yellow sticky traps placed outside the perimeter of each plot) and movement between the plots (by counting the total number of adults present on marked plants (every 2 h) were determined and correlated with the speed and direction of the wind, temperature and relative humidity. Both immigration and repopulation of the plants were continuous throughout the day, with greatest activity in the morning and the repopulation depended only partially on immigration.

Wells *et al* (2002) studied the dynamics of spring tobacco thrips populations. Macropterous forms were found to be more abundant than brachypterous thrips but there was no difference in their ability to transmit TSWV. Macropterous forms were more capable of colonizing and subsequently transmitting TSWV to newly emerged crops. Brachypterous forms may help to perpetuate the disease cycle of TSWV over the winter, and by keeping inoculum in the population until temperature rises and the percentage of macropterous forms increase.

Kuchne (2004) studied the activity of aphids associated with lettuce and broccoli in Spain and their efficiency as vectors of *lettuce mosaic virus*. Green tile traps and Moericke yellow water trap were used to monitor aphid flights during the spring and autumn growing season in different locations. Different aphids were found on lettuce and broccoli at Madrid, Murcia and Navarra during spring. *Hyperomyzus lactuceae* and *Brachycaudus helichrysi* were common on lettuce at Madrid and Murcia. *Aphis fabae*, *Brachycaudus helichrysi*, and *H. lactuceae* were common on broccoli at Navarra. During autumn, *Hydaphis coriandri*, *Aphis spiraecola* were common at Madrid; *A. spiraecola* and *Myzus persicae* were common at Murcia; and *Therioaphis trifolii* was common at Navarra.

Nault *et al* (2003) studied seasonal patterns of adult thrips dispersal in eastern Virginia tomato fields. Populations of adult thrips (*Frankliniella fusca*), the only competent vector of TSWV in most of the mid-Atlantic US region, infesting tomato flowers and dispersing within tomato fields were monitored weekly. Seasonal patterns of *F. fusca* capture was dissimilar among tomato fields across all regions; yet more *F. fusca* were captured between mid-May and mid-June in all regions as compared with those captured between transplanting and mid-May each year. Despite the relatively low observed dispersal activity of *F. fusca* before mid-May, the threat of TSWV transmission warrants protection of the crop from the immigrating *F. fusca* from transplanting until the end of marketable fruit set.

Jurik *et al* (1987) demonstrated the effect of temperature on the retention of the non-persistent viruses like *bean yellow mosaic virus* by aphids. The temperatures tested were 4° and 21°C for *Acyrthosiphon pisum* and *Macrosiphum euphorbiae*, and 2, 4, 21 and 37°C for *Aphis craccivora*. Virus retention half-lives varied from 2.2 h (in *A. craccivora* at 21°C) to 6.6 h (in the same species at 4°C). Values for other species were intermediate.

Akhtar *et al* (2004) observed the influence of plant age on the spread of the virus (CLCuV) disease. All cotton mutant lines/varieties become increasingly resistant to CLCuV as plants aged. Expression of age-related resistance was more apparent during late growth stages. Maximum increase in % disease incidence occurred at 6 wk. There was significant correlation of whitefly levels with % disease incidence.

Latham *et al* (2004) conducted a field experiment to demonstrate the effect of time of infection on the spread of a disease. In this experiment lettuce seedlings from the nursery infested with lettuce big vein disease (LBVD) were inoculated with infested roots or left uninoculated before transplanting into subplots on land with no history of lettuce planting, disease progress followed a sigmoid curve with the former but an almost straight line with the latter. However, significant clustering of symptomatic plants was only in the subplots with the uninoculated plants. Leaf symptoms LBVD were more severe in lettuce infested later, whereas symptoms in those infested earlier were obvious initially but then became milder. When the leaf symptoms first appeared at 5-6 wk, there was a fresh weight loss of 14-15% for heads (all plants) and 39% for hearts (excluding plants without hearts). When leaf symptoms appeared 7 wk after transplanting, there were no significant yield losses for heads and hearts.

E. MOLECULAR ECOLOGY AND EPIDEMIOLOGY

Application of molecular techniques for the understanding of the viral genomes, their expression for disease development and spread is the most contemporary research in plant virology. Molecular techniques are used to study the complex patterns and processes of biological diversity at all levels of organization, from genes to organisms, populations, species, and ecosystems, but from a gene-centric perspective (Beebee and Rowe 2004) and lead to the development of molecular ecology. Studies on the molecular ecology of viruses lead to the crystallization of the concept of molecular ecology and epidemiology that focusses on the emergence of new viruses. Fargette *et al* (2006) recently reviewed the development of molecular ecology and emergence of new tropical viruses.

The path of molecular epidemiology of viruses however, was first indicated by the dependence of a large number of viruses on helper viruses, satellite viruses/RNAs, etc. for their infectivity (Table 7.3).

Table 7.3   Association of Different Genera of Viruses with Helper Virus/Factors for their Infectivity (Fauquet *et al* 2005)

Genus	Type of association
Fijivirus	Independent from functions during replication
Oryzavirus	Additional factor required for infectivity
Phytoreovirus	Additional factor required for infectivity
Badnavirus	Requires helper virus for vector transmission
Caulimovirus	Independent from functions during replication
Rice Tungrovirus	Requires helper virus (Rice Tungro spherical virus) for vector transmission
Cavemovirus	Independent from its functions during replication
Begomovirus	Independent from its functions during replication
Mastrevirus	Independent from its functions during replication
Bymovirus	Independent from its functions during replication
Carlavirus	Independent from its functions during replication
Closterovirus	Provide helper functions to dependent virus during replication; acts as helper for another virus
Hordeivirus	Independent from its function during replication
Varicosavirus	Independent from its functions during replication; Luteovirus required for vector transmission
Tymovirus	Independent from its functions during replication
Bromovirus	Coat protein specifically associated with RNA3′ terminal sequence
Ilarvirus	Coat protein specifically associated with 3′ terminal sequence to replicase recognition
Tombusvirus	Helper to dependent viruses; helper for a satellite RNA

(a)   Emergence of New Viruses and their Epidemiological Significance

Genetic RNA recombination is a major factor responsible for the emergence of new viral strains of species. Schnippenkoe *et al* (2001) tested this natural method leading to changes in the viral genome, and made forced recombination between distinct strains of *maize streak virus* (MSV): MSV-*kom* and MSV-*sat*. Heterodiametric agro-infectious constructs containing tandemly cloned mixtures of complete or partial MSV-*sat* and MSV-*kom* genomes were prepared. The pathogenicity of six predominantly MSV-*kom* like recombinants was tested in maize. While all were capable of producing a symptomatic infection in this host, none was more virulent than MSV-*kom* and only two were more virulent than MSV-*sat*. The most virulent recombinants were leafhopper transmitted to a range of differentially MSV-resistant maize, wheat and barley genotypes and both were found to have unique biological properties.

Garcia-Arenal *et al* (2000) studied the molecular epidemiology of *cucumber mosaic virus* and its satellite RNA in Spain. To characterize the genetic structure of CMV populations, 300 isolates, representing 17 outbreaks in different crops, regions and years, were compared. Genetic analyses of CMV isolates were done by ribonuclease protection assay of cRNA probes representing RNA1, RNA2 and the ORFs in RNA3. All isolates belonged to one of three genetic types: sub-group II and two types of sub-group I. The genetic structure of 17 sub-populations varied randomly, without correlation with location, year, or host plant species. Thus, CMV in Spain showed a meta-population structure with local extinction and random recolonization from local or distant virus reservoirs. Mixed infections and new genetic types were generated by re-assortment of genomic segments or by recombination. Heterologous genetic recombinations were not favoured. Thirty percent of CMV isolates supported *sat*-RNA. Molecular analyses of CMV-*sat* RNA isolates showed high genetic diversity, due to both accumulation of point mutations and recombinations. The CMV-*sat* RNA population has a genetic structure and dynamics different from those of its helper virus.

In contrast to the RNA viruses, it was initially thought that DNA viruses were very conservative and do not accumulate mutants to give rise to new strains or new viruses, but the discovery of the wide variations in begomoviruses rapidly changed this concept. Brown (2000) correlated the unprecedented upsurges in the population of whitefly (*Bemisia tabaci*) throughout the world and the emergence of new begomoviruses and recent epidemics, making this pathogen a major one, and development of the variants of the *B. tabaci*.

Major contributing factors for the emergence of and spread of new geminivirus diseases are the evolution of variants of the virus, the influence of the whitefly B biotype and the increase in the vector population. Variability arises through mutations, recombination and pseudo-recombination. Genomic recombination in geminiviruses occurs not only between the variants of the same virus but also between species and even between genera resulting in rapid diversification. Most virulent variants have developed through recombination of viral genomes such as those associated with cassava mosaic, cotton leaf curl, tomato leaf curl. Heterologous recombinants containing parts of the host genome and/or sequences from satellite-like molecules associated mono-partite begomoviruses provide unlimited evolutionary opportunities (Varma and Malathi 2003) and bring about complex epidemiological conditions. There may be additional virus-induced driving forces for begomovirus and other virus epidemics for altering the plant biochemistry so that infected plants increase vector fecundity and emit volatiles as vector attractants (Jiménez-Martinez *et al* 2004).

Molecular ecology has unravelled the emergence of several economically most important tropical viruses particularly *rice yellow mottle virus* (RYMV), cassava mosaic geminiviruses (CMGs), *maize streak virus* (MSV), *banana streak virus*, etc., during the last 10 y. The range of mechanisms involved included recombination and synergism between virus species, new vector biotypes, genome integration of the virus, host adaptation, and long distance dispersal. A complex chain of molecular and ecological events resulted in novel virus-vector-plant-environment interactions that led to the virus emergence (Fargette *et al* 2006). These discoveries also added a new dimension to virus ecology throughout the world.

F. EVOLUTIONARY EPIDEMIOLOGY

Evolutionary epidemiology is primarily based on the ecology of viruses that conventionally means distribution of viruses in different agro-ecosystems of the world. Rapid changes in the agro-ecosystems in the tropics with highly variable weather have made a breakthrough in the ecological and molecular epidemiology where rapid emergence of new viruses has been observed. This emergence is considered as a critical aspect of plant virus ecology in the tropics. Loebenstein and Thottappilly (2003) provided comprehensive and up-to-date information on the ecology of virus diseases of crops in tropical countries, but the primary focus was on the emerging infectious diseases that have increased in incidence, geographical distribution, or host range, changed pathogenesis and newly evolved, or discovered, pathogens (Frgette *et al* 2006).

From the ecological point of view, viruses may exist as components of host systems particularly at the points of origin of the concerned hosts where they may coexist without expressing recognizable symptoms showing compatible mutualism. This relationship changes with domestication, improvement and monoculture of those hosts as crops. At the points of origin epidemics may occur in susceptible hosts in favourable environments, particularly the agro-meteorological conditions.

Transportation of planting materials takes place from one country to another. Such transport has a different type of epidemiological potential. Apparently symptomless materials find a new climate, new biotic and abiotic stress including the prevalence of vectors. In favourable conditions, existing symptomless viruses may become aggressive and cause damaging epidemics. A more or less similar situation has been found with the spread of Swollen Shoot of Cacao in some countries in Africa where the mealy bug vector is prevalent, though the initial planting material brought from Latin America was apparently healthy. Therefore careful mapping of inter-country movement of planting materials and the occurrence of epidemics may demonstrate the evolution of virus epidemics in various crops and different countries.

Classic examples of coexistence of viruses and viroids are found in citrus where domestication and propagation separate the coexisting viruses and allow them to express as separate diseases. Citrus is a tropical to sub-tropical forest plant. Most of the species originated in a few South Asian countries. It has been domesticated, and transported to the Mediterranean and Caribbean regions from where it moved to the USA, Israel, Africa, and Australia for improvement and cultivation. Swingle and Weber in 1896 observed a symptom in plants grown from imported planting materials and recorded the disease as Psorosis. Subsequent research shows that 'Psorosis' is a complex of different viruses: Psorosis, infectious variegation, citrus ring-spot and viroids. Vegetative propagation separates these viruses and enables them to invade crops separately and show different patterns of epidemiology.

Epidemics are not always concerned with previously coexisting viruses. Viruses may occur in wild hosts expressing mild symptoms or without showing any symptom. When a new genotype of a crop is introduced where such hosts and vectors occur, viruses may be transported to these crops producing severe diseases. Insect viruses may also have a chance entry to plant systems causing severe diseases. This is the case with maize rayado fino and oat blue dwarf virus (Gamez and Leon 1988, Rivera and Gamez 1986, Kitazjima and Gamez 1983). Leafhopper A virus of *Cicadulina bipunctella bimaculata* even infects maize but does not multiply in that plant (Ofori and Francki 1985).

Nucleotide sequence diversity, genomic recombination and mutation of viruses occur naturally. When a crop is in a monoculture, resistant to normal strains of the viruses, or receives regular pesticide treatments for the control of vectors, the process of nucleotide sequence diversity and genomic recombination can adapt viruses to these conditions. The natural variation of viruses has been well studied with *Citrus Tristeza Virus* (CTV) particularly in the USA and Israel. Previously identified vectors of CTV were not common in the USA but variation in CTV made it transmissible by *Aphis gossypii* (Costa and Grant 1951, Wallace and Drake 1951, Dickson *et al* 1956). In Israel, after a sudden outbreak of the disease during 1970-1973, Raccah *et al* (1977) demonstrated the appearance of a new form of CTV. d'Orso *et al* (2000) recorded the uneven distribution of genomic RNA variants within the infected plants. They also observed that individual aphids transmit some of these variants and concluded that aphid transmission probably contributes to the changes observed in the CTV population. In CTV it is presumed that continuous genomic recombination is taking place and divergence of the nucleotide sequences are observed in different countries with time and space (Hilf *et al* 1999, Kong *et al* 2000). But the host passage may also contribute to the evolution of CTV genomes. Recently Sentandreu *et al* (2006) made evolutionary analysis of genetic variation in CTV after the host passage. Genetic variability was observed in two genes (p18 and p20) from two groups of CTV isolates. One group (isolates T385, T318, T317, and T305) was derived from a Spanish source by successive host passages while the other (isolates T388, and T390) was obtained after aphid transmission from a Japanese source. A total of 274 sequences were obtained for gene p18 and 451 for gene p20. In the corresponding phylogenetic trees, sequences derived from the severe isolates (T318, T305 and T388) clustered together and separately those from mild or moderate isolates (T385, T317 and T390), regardless of their geographic origin. Hierarchical analyses of molecular variance showed that upto 53% of the total genetic variability in p18 and upto 87% of the variation could be explained by differences in the pathogenicity features of the isolates and different selection forces had been acting between isolates and between genes. Recombination and transmission bottleneck did not play any significant role during p18 and p20 evolution whereas hosts can be an important evolutionary factor for CTV isolates.

Rice Tungro Virus (RTV) is probably unique in the history of plant virus epidemiology. It is now recognized that a mixture of two different viruses cause tungro disease. One is a ss RNA-containing spherical virus; the other is a ds DNA-containing bacilliform virus. The spherical virus normally coexists with *tall indica* genotypes of rice. In the 1960s, when *tall indica* was crossed with *dwarf indica*, it was assumed that the bacilliform virus was introduced into the system and started to cause severe disease in most of the varieties subsequently developed through breeding programmes, but how these two completely different viruses stabilized in a common system, resulting in the most severe and widespread disease of rice which still remains as an epidemiological enigma.

Wren *et al* (2006) put forward a novel concept on the ecology and evolution of viruses. For most of its history the ecology of viruses dealt with symptomatic diseases of crop plants. Little was known about the level of mixed virus infections or temporal patterns of virus accumulation in native plants and little is known about the potential for viruses in combination to cause disease in future. Understanding of the potential mutualistic interactions between viruses and their hosts is very limited. There is a huge gap in the understanding of viral diversity, evolution and ecology. A more objective view of virus populations in nature-affecting wild plants, symptomatically or asymptomatically, is necessary.

Virus-host interactions may be mediated by endophytes, mycorrhyzae or other symbionts. RNA silencing has dramatic but unexplored ecological implications. Small pieces of genetic information, when transferred from one organism to another have the potential to facilitate many interactions. Given their intimate genetic interactions with their hosts, viruses could, potentially, be prime drivers of evolutionary change. The degree to which viruses have determined micro-

and macro-evolutionary patterns in vascular plants is mostly unexplored. Comparing virus phylogenies to plant phylogenies may allow the development of a hypothesis about the nature of co-evolution and enable ecological studies to be pushed beyond the simple concepts of competition, mutualism, parasitism and predation.

Epidemiology is not a static process but a dynamic one that varies with change in the ecological, host, vector and virus systems. The genetics of viruses is mostly modulated by host response, transmission process and epidemic dynamics. Vector ecology, behaviour and genetic alteration also contribute to evolutionary epidemiology. The speed with which viruses evolve opens up possibilities for molecular epidemiology and tracking the virus strains through phylogenetic analysis. The rapidity of change in RNA viruses makes them useful experimental models for evolutionary epidemiology (Moya *et al* 2004).

Jeger *et al* (2006) illustrated the key features of evolutionary epidemiology particularly plant virus evolution, geographical sub-division and genetic variation, interactions, host resistance and pathogen virulence, molecular interpretations and evolution in natural plant and vector communities. The probability of resistant-breaking virus variants appearing and spreading depends on their fitness in the absence of virus resistance, the type of resistance encountered, and the number of resistance genes present (Harrison 2002, Hammond *et al* 1999, Tepfer 2002) assessed the epidemiological risks from the use of virus-resistant transgenic plants. Garracía-Arenal *et al* (2003) recorded that (i) high mutation rates are not necessarily adaptive, (ii) population of plant viruses are not highly variable, (iii) constraints in genetic variation in virus-encoded proteins are similar to those found in host plants and vector and (iv) genetic 'drift' may be important, as well as selection in shaping evolution.

Moya *et al* (2004) observed that the rapidity of the sequence change in RNA viruses make them ideal experimental models for the study of evolution and mutation rates. Several workers have studied their evolutionary changes (Malpica *et al* 2002, Kearney *et al* 1999, Fraile *et al* 1997). Sequence changes may take place both due to mutation and recombination. Aranda *et al* (1993) analyzed the contribution of mutation and recombination to the evolution of the satellite RNA of *Cucumber mosaic virus* (CMV). Fauquet *et al* (2005) explained the evolutionary divergence of the begomoviruses found in the Indian sub-continent and those causing similar diseases in other geographical regions by recombination following multiple infections with different begomovirus species. The novel recombinant phenotype may have selective advantage over the parental genotypes. Morilla *et al* (2004) explained the frequent occurrence of recombinants between two begomoviruses [*Tomato yellow leaf curl virus* (TYLCV) and *Tomato yellow leaf curl Sardinia virus* (TYLCSV)] due to co-infection of nuclei. Lin *et al* (2004) and Bonnet *et al* (2004) studied the contribution of reassortment in multipartite RNA or DNA in the evolution of population. Seal *et al* (2006) confirmed that both re-assortment and recombination have been important factors in the emergence of novel damaging begomoviruses.

Geographical sub-division also contributes to the evolution of viruses, and their vectors. These sub-divisions can substantially affect genetic variation provided that selection is sufficiently strong at a given locus (Kaplan *et al* 1991). Selection may act differently in the various patches comprising the spatial heterogeneity. Gravilets and Gibson (2002) modelled the situation in which a mutation may be advantageous in one patch but deleterious in another. Several workers (Padidam *et al* 1995, Harrison and Robinson 1999, Hameed and Robinson 2004, Simón *et al* 2003) conducted explicit studies on this aspect for plant viruses.

Recently several, seemingly new diseases and many virus strains with altered pathogenicity have been reported (Brown 1990, Polaston and Anderson 1997, Padidam *et al* 1999, Mansoor *et al* 2003a, b, Varma and Malathi 2003). The reasons for the emergence of new strains is related to more rapid evolutionary changes of the virus brought about by increased vector and virus population sizes, more polyphagous vector populations and the potential for rapid genetic changes. An excellent example of such changes is found with begomoviruses (Padidam *et al* 1999, Varma and Malathi 2003, García-Arenal and McDonald 2003, García-Arenal *et al* 2003, Seal *et al* 2006). Mathematical models have been used to study the effects of plant, plant virus and vector characteristics on the development of disease epidemics (Chan and Jeger 1994, Holt and Chancellor 1996, Jeger *et al* 1998, Madden *et al* 2000, Zhang *et al* 2000, Jeger *et al* 2002, Escriu *et al* 2003). These models, however, ignored the evolutionary dynamics of the plant virus. Considering this limitation Van den Bosch *et al* (2006) developed a model using within-cell multiplication to model the virus dynamics at the tissue level. Virus-host, virus-vector, vector-host and virus-vector interactions play a significant role in the evolutionary epidemiology. The spread and success of plant viruses include direct competition between viruses in host plants, direct competition within and between vectors, differences in transmission rates, and virus influences on vector behaviour and population dynamics (Power 1996). Opportunities for direct interaction within plant cells occur in many ways, between wild- type and mutant virus strains, between wild-type and defective interfering (DI) particles arising from deletion mutants, and with other viruses, where multiple particles are necessary for infection to occur (co-viruses). Several workers (de Zoeten and Skaf 2001, Mendez-Lozano *et al* 2003, Froissart *et al* 2004, Seal *et al* 2006) studied the production of 'co-viruses' and their implications to evolutionary epidemiology. There also occur molecular complexities underlying the virus epidemiology and evolution. Nanovirus-like DNA components associated with geminivirus diseases can adapt from whitefly to leafhopper transmission when co-inoculated with a leafhopper-transmitted virus (Saunders *et al* 2002). There are normally complex interactions between vector, host and vector dispersal patterns. Viral pathogens, often asymptomatic, play a significant evolutionary role in plant populations (Jeger *et al* 2006).

There may be adaptive evolution of pathogens in sympatric heterogeneous plant populations. Gudelj *et al* (2004) using adaptive dynamics, which assumes that sequential mutations induce small changes in the pathogen fitness, showed that evolutionary outcomes strongly depend on the shape of the trade-off curve between pathogen-transmission on sympatric hosts. The conditions under which the evolutionary branching of a monomorphic into a dimorphic population occurs, as well as the conditions that lead to the evolution of specialist (single host range) or generalist (multiple host range) pathogen population was determined.

G. ECOLOGICAL GENOMICS AND EPIDEMIOLOGY

The availability of new technologies and perspectives on biology, from computer simulation modelling to automated environmental sensing, and ecological genomics gives a new dimension to epidemiology. Epidemiological studies benefit from new genomics technologies in several ways. New diagnostic techniques enable the development of a 'community epidemiology'. New techniques also make it easier to extend genetic analyses of pathogens beyond virulence genes, by facilitating the study of the population structure and evolution of genes. Functional genomic analysis of pathogen virulence genes and host resistance and defense response genes enable better predictions in all respects (Garrett *et al* 2006).

Milgroom and Peever (2003) illustrated how epidemiology has benefitted from information about the population genetics of pathogens. Simultaneous studies on the change of gene frequencies within and among populations due to natural selection and gene flow and the growth and spread of pathogen populations, make it possible to track disease outbreaks (Zwankhuizen *et al* 1998), develop predictions about the sources of inoculum and pathogen life cycles (Cortesi *et al* 2000; Cortesi and Milgroom 2001), understand the evolution of virulence (Escriu *et al* 2000 a, b, 2003) and make predictions about the durability of resistance in crop genotypes (Escriu *et al* 2000 a, b). The developing synthesis of a functional genomics approach combined with a population and ecological perspective promises to lead to new avenues of research and understanding of plant-pathogen interactions. Evolutionary and functional genomics or EEFG (Feder and Mitchell-Olds 2003) has the goal of understanding ecological and evolutionary processes that maintain genotypes and phenotypes. Functional genomics, or the use of genomic technologies (e.g. microarrays) to find genes and polymorphisms that affect traits of interest and to characterize the mechanism underlying those effects, have been applied effectively in agricultural contexts. Genomics moves beyond simple sequence analysis to evaluate the function of particular DNA sequences through, for example, gene knockout mutants or gene activation mutations. These techniques have natural applications for the study of resistance and virulence, but might also be usefully applied in the study of other epidemiological features (Garrett *et al* 2006). Population genetics and population genomics can provide immense fundamental information on epidemiology. Models of population and communities may apply to systems of gene expression to provide 'population biology' of gene expression. Traditional epidemiology includes information about host species, pathogen and vector species, and environmental variables. The genomic status of the host and the pathogen communities can be incorporated to develop genomic epidemiology. Information about the soil metagenome may contribute to an understanding of disease suppressive soil that develops over time as microbial population respond to the buildup of pathogen population. Characterization of populations may include the composition of both qualitative features produced by different genotypes and quantitative features produced by different levels of gene expression in what may be the same genotype. The expression of plant genes in response to non-pathogenic microorganisms may be highly relevant to epidemiology. Microarray analysis can identify sets of coregulated elements (Maleck *et al* 2000, Chen *et al* 2002). New genetic information can be used to refine state transition models such as, the Susceptible-Infected (SI) model (Otten *et al* 2003) rather than modelling host individuals as simply 'susceptible and uninfected' or 'infected'. Genes related to pathogenicity, toxin production, and other epidemiological features could be used to more reliably measure genotypes in a population responsible for disease.

Models for population biology can be applied in the study of gene expression. Genes may be conceptualized to interact within a cell comparable to the way that species interact within an ecosystem (Mauricio 2005). Application of such a model is useful to study the interactions between defense response pathways. These pathways affect different pathogens and pests and also interact with each other (Glazebrook *et al* 2003). Depending on the response they may be complementary (van Wees *et al* 2000) or competitive (Spoel *et al* 2003).

An individual plant can be conceptualized as a population of cells or organs across which gene expression occurs. It is now possible to measure gene expression in individual plant cells (Kerk *et al* 2003, Nakazono *et al* 2003) so that spatial pattern of expressions through an individual host can be measured and modelled.

H. GLOBAL WARMING AND EPIDEMIOLOGY

Climatic change is occurring (Houghton *et al* 1996, 2001). During the past 100 years global-average surface temperatures have increased by approximately 0.6°C (Houghton *et al* 2001). The third IPCC report predicts that the global-average surface temperature will increase by a further 1.4-5.8°C by 2100 with atmospheric carbon dioxide (CO_2) concentrations expected to rise between 540 and 970 ppm over the same period (CCIRG 1991, 1996, Hulme and Jenkins 1998, Houghton *et al* 2001). Precipitation, ultra-violet B (UVB) penetration and extreme events (e.g. flooding, storminess and drought) are also predicted to increase, but there is less certainty of the magnitude of these changes.

Global change is a complex process that involves changes in atmospheric composition, climate and climate variability, land cover and land use. Multiple interrelated processes drive these changes. Projections of change in each component of global change have different levels of uncertainty; ranging from 'virtually certain' (e.g. concentration of greenhouse gases will continue to increase) to 'highly uncertain' (e.g. magnitude of climate change at regional or local scales) (Mahlman 1997). International and interdisciplinary research and communication is critical for tackling such an issue with so many interrelated causes and effects as global change. Proper understanding of the impact of global warming on the epidemiology of virus diseases and vectors needs the understanding of the physics of the climatic change and international approaches to deal with it, is an important necessity. These aspects however, have been discussed in Appendix VII.1.

(a) Global Change and Terrestrial Ecosystems (GCTE) Activity on Pests, Diseases and Weeds

Pests (arthropods, microbial pathogens, weeds) together cause average yield losses of *c.* 40% globally (Oerke *et al* 1994). Some pest species may also be useful as early indicators of global change because of their short generation times, high reproductive rates and efficient dispersal mechanisms (Sutherst *et al* 1996b). Pests may be affected by global change in many ways. For example, direct effects may include shifts in geographical distribution ('pest zones') due to changes in climate or land use, while indirect effects may result from changes in host-pest interactions due to altered plant physiology and/or morphology under increased CO_2 (Chakraborty *et al* 2000, Coakley *et al* 1999). The magnitude and direction of these potential effects cannot currently be predicted with great accuracy (Coakley and Scherm 1996, Coakley *et al* 1999, Scherm 2000). Thus GCTE Activity on these aspects was initiated with the goal of increasing the understanding of potential global change impacts on pests and developing a predictive capability for impact assessment and adaptation through international networks for research, data sharing and modelling (Sutherst *et al* 1996a). The organizational structure and objectives of this GCTE activity is given in Table 7.4.

Table 7.4 Organizational Structure and Objectives of Activity (Global Change: Impact on Pests, Diseases and Weeds) of the Global Change and Terrestrial Ecosystems (GCTE) (Sutherst *et al* 1996a)

Task	Overall goal	Core research projects
Pest distribution, dynamics and abundance under global change	Refine existing models and develop new models to predict the responses of populations of key arthropod pests to changes in land use, climate and atmospheric composition	Impact of global change on pests. Vulnerability of pests under global change. Adaptation of pest management strategies
Disease distribution, dynamics and severity under global change	Provide foundation for an analysis of impacts and develop a collaborative global network to facilitate comparative studies of modelling approaches and sharing of data	Disease zone mapping. Disease impact assessment and management as affected by global change
Weed distribution, dynamics and abundance under global change	Determine the effects of global change on the distribution, dynamics and competitive abilities of major weeds	Crop-weed competition under global change.A biographical approach to determine the extent of weed distribution and abundance under global change

(b) Impact on Virus and Vector Epidemiology

Changes in land use, host physiology, vector biology, dispersal and virus-vector, virus-host and vector-host interactions are all likely to be affected by climate change. Most of the literature is on insect pests. Scherm *et al* (2000) reviewed the global networking for assessment of impacts of global change on plant pests and Baker *et al* (2000) reviewed the role of climate mapping in predicting the potential geographical distribution of non-indigenous pests under current and future climates. Jeffery *et al* (2002) reviewed insect herbivory in global climate change research. Scherm and Coakley (2003) reviewed plant pathogens in a changing world and concerns about future crop productivity are highlighted by the results of a recent examination of yield trends in corn (maize) and soybean in the central United States, that documented a stronger than expected effect of gradual temperature change (Lobell and Asner 2003). Such long-term changes in crop productivity may be the results of direct effects of temperature or other global change elements on the plant, or they may occur indirectly via yield-reducing factors such as plant pathogens, insects and weeds.

There may also be a geographical shift in the prevalence of virus diseases (*rice stripe virus*) due to global warming that alters the synchronization between vector, small brown plant hopper (*Laodelphax striatellus* (Fallen)) (Hemiptera: Delphacidae) and the susceptible stage of the rice crop (Yamamura and Yokozawa 2002). Long-term data, however, are required to have a proper understanding of the impact of global warming on the incidence of virus diseases at regional levels and changed agro-climatic and agro-ecosystems.

Analysis of long-term data sets showed phenotypical responses to climate warming including changes in morphology, abundance, geographical distribution and life cycle timing. These analyses focussed on advances in spring phenology in temperate-zone species such as bud break or flowering in vascular plants (Fitter and Fitter 2002). Results of such phonological studies have provided strong evidence for a biological fingerprint of climate change. For example, a meta-analysis encompassing long-term records of 172 species of trees, shrubs, herbs, butterflies, birds, amphibians and fish revealed a significant average advancement in spring phenology by 2.3 d per decade (Parmesan and Yohe 2003). When geographical ranges of 99 species with detailed spatial records were examined in the same study, a significant range shift of 6.1 km per decade toward the poles was observed.

The Long Term Ecological Research (LTER) conducted for studying aphids by the Rothamsted Insect Survey enabled statistical relationships to be established between aphid phenology (e.g., annual time of first capture) and a range of land use variables. The time of first flight was strongly correlated with temperature for some species at some sites, with earlier first flights after higher winter temperature (Woiwod and Harrington 1994).

Within the United States, several LTERs are now being used to study pest populations. In a 12-y warming experiment at a site in the Rocky Mountains of Colorado, Roy *et al* (2004) documented an overall trend of increasing damage from herbivores and plant pathogens in plots with warmer temperatures and earlier snowmelt, although some pathogens and herbivores preferred the cooler plots.

Long-term data sets for pest phenology that are suitable for fingerprint analysis is rare, one of the exceptions is the continuous suction trap data at Rothamsted Research, UK. Harrington (2002) analyzed the data on some aphid vectors of viruses and predicted an advance in the timing of spring migration by 2 wk for every 1°C increase in winter temperature. It is anticipated that problems of BYDV in UK are likely to become more severe due to (1) changes in the cropping pattern (in particular increase in the area of maize); (2) increased survival of a vector species, *Rhopalosiphum maidis*, that prefers maize and requires warm winters; (3) increased prevalence of a BYDV strain transmitted efficiently by the vector species, and (4) increased efficiency of transmission of different strains of BYDV with increase of temperature in all cereals (Harrington 2003).

Global warming may also lead to the appearance of invasive pests. By the early 1990s, an estimated 239 species of non-indigenous plant pathogens had become established in the United States (National Research Council 2002). For several reasons, the acceleration of global change is likely to increase the success of invasive species, including non-indigenous plant pathogens. At first, most invasive plant pathogens possess 'weedy' life-history traits such as short generation times and excellent dispersal abilities; organisms sharing such traits are predicted to benefit from increased disturbance associate with global change (Duke and Mooney 1999, Simberloff 2000). Second, in the absence of the normal complement of antagonists and natural enemies present in their centres of origin (Torchin *et al* 2003), invasive species in new geographical ranges tend to be more limited by climate than by biotic interactions. Thus, climate warming may favour range expansion of non-indigenous species over that of resident species since the latter are more likely to be restricted by non-climatic factors (Scherm and Coakely 2003).

Reilly *et al* (2003) reviewed US agriculture and climate change and concluded that the risks from climatic change will more likely occur at regional levels, depending on changes in precipitation or changes in variability of climate, or stem from more complex climate-agriculture-environment interactions.

Scherm (2004) reviewed the predictability of the impacts of climate change on plant pathology and pest management. He emphasized the necessity to quantify the impacts of a changing climate on plant disease intensity and yield loss through retrospective analyses seeking to identify fingerprints related to climate change in long-term plant disease records, as well as the use of mathematical models to predict likely impacts on plant pathosystems in future.

Chakraborty (2005) reviewed the potential of climate change on plant-pathogen interactions highlighting a clear lack of knowledge for model host-pathogen interactions that must be met through baseline empirical studies. Models can be used to extrapolate, predict and validate potential impacts.

Wang (2005) examined the impact of greenhouse gas warming on soil moisture based on predictions of 15 global climate models. The models are consistent in predicting summer dryness and winter wetness in only part of the northern middle and high latitudes. Slightly over half of the models predict year-round wetness in central Eurasia and/or year-round dryness in Siberia and mid-latitude Northeast Asia. In the tropics and subtropics, a decrease of soil moisture is the dominant response. The models are especially consistent in predicting drier soil over Southwest North America, Central America, the Mediterranean, Australia, and South Africa in all seasons, and over much of the Amazon and West Africa in June-July-August (JJA) season and the Asian monsoon region in the December-January-February (DJF) season. Since the only major areas of future wetness predicted with a high level of model consistency are part of the northern middle and high latitudes during the non-growing season, it is suggested that greenhouse gas warming will cause a worldwide agricultural drought that may change land-use pattern and pests and diseases incidence as well.

Garrett *et al* (2006) reviewed the climate change effects on plant disease genomes to ecosystems. Global climatic change will affect plant disease in concert with other global change phenomena. There are potential effects of introductions of new species, in terms of new hosts that may boost pathogen inoculum levels, new vectors that may alter epidemic dynamics, and new pathogens themselves. Widespread changes in land-use patterns will alter the potential for populations of plants and plant pathogens to migrate through fragmented landscapes. If agricultural land-use decreases in temperate areas and expands in the tropics, policies in temperate areas may support restoration of natural areas, while the development of land-use policies in tropical areas will face related challenges for maintenance of agricultural productivity and plant biodiversity in a changing world.

Sutherst *et al* (2007) reviewed 'pests under global change' and concluded that some pests will be able to invade new areas and become increasingly problematic for the maintenance of biodiversity, the functioning of ecosystems and the profitability of crop production. Some pests that are already present but only occur in small areas, or at low densities may be able to exploit the changing conditions by spreading more widely and reaching damaging population densities. Specific studies on the effect of climate change on plant virus epidemiology are extremely limited. Yamamura and Yokozawa (2002) studied the epidemiological system of *rice stripe virus* (RSV) transmitted by the small brown plant hopper *L. striatellus* and observed that it is greatly influenced by the degree of synchronization between the vector occurrence and the susceptible stage of the crop. The susceptible stage of the virus infection is within several weeks after the transplantation of the crop. Assuming that rice seedlings are transplanted from May to June, a map was prepared in which the number of generations of the small brown plant hopper on 01 June was plotted by calculating the effective cumulative temperature. The influence of solar radiation was also considered in this calculation. The area located near the boundary of generation was judged as potentially vulnerable to disease prevalence. It was further calculated how the incidence of this disease will change after future global warming using the results of the Global Climatic Model experiments reported by the IPCC. The generation maps indicated that the major districts of rice production in Japan might be potentially vulnerable to MSV infection under future global warming.

Harrington (2003) estimated the effect of climate change on the *Barley yellow dwarf virus* (BYDV) and determined the probability of encountering two consecutive mild winters (December to February, mean temperature exceeding 4.5°C) with a wet summer in between (precipitation from June to August, exceeding 190 mm) in different regions of the UK according to climate predictions to the end of this century. Two mild winters with a wet summer in between provides the basis for increased risk of economically damaging virus spread. It was predicted that BYDV risk is to rise particularly in Northwest and Southwest England. In other parts, although the criterion of warm winters is met more often in the future, the summers are expected to be particularly dry, hence reducing risk by restricting the availability of suitable hosts for both vectors and viruses.

I. Epidemiology of Some Internationally Important Diseases

(a) Barley Yellow Dwarf Virus (BYDV)

Barley yellow dwarf virus is one of the most widely distributed and destructive viral diseases of small grains. Plumb 1983, 1992, 2002, Irwin and Thresh 1990 and Burges *et al* 1999 reviewed *Barley yellow dwarf virus* (BYDV) disease as a global problem. The disease of Poaceae, now known as barley yellow dwarf is likely to be a very ancient disease. Oswald and Houston (1951, 1952) first identified its causal agent as an aphid-transmitted virus. Five years after the first recognition of the virus aetiology, its incidence was reported from the United States, Canada, and the Netherlands. By 1963, infection had been reported widely in the United States, Canada, and Mexico, and has now been identified in most countries of the world.

(i) The Virus and its Strains
Barley yellow dwarf virus (BYDV) is a member of the genus *Luteovirus*. It is transmitted in a persistent, circulative, non-propagative manner by several aphid species. Different aphids have different host plant preferences and this will clearly affect disease epidemiology. Also crop growth stage will affect both aphid-feeding behaviour and host plant preference and thereby affecting the epidemiology.

Five strains of this virus were initially differentiated on the basis of transmission efficiency of different isolates by different aphid vectors. The strains were then designated as PAV, RPV, MAV, RMV and SGV in accordance to the specific vector species (Johnson and Rochow 1972). The efficiency of transmission (% of infection) of the strains by the vectors is given in Table 7.5.

Serological and nucleic hybridization studies clustered the strains into two groups, BYDV-1 (MAV, PAV, SGV) and BYDV-2 (RPV, RMV) (Rochow 1984, Gill and Chong 1979). Molecular studies, specifically, sequence analysis recognized these two groups as two different genera (Miller *et al* 2002) and subgroup I was retained as BYDV. The resistant gene *Yd2* found in cereals was effective only against subgroup I or BYDV (Herrera and Plumb 1991).

Table 7.5 The Efficiency of Transmission (% of Infection) of the Strains of BYDV by their respective Aphid Vectors (Rochow 1969, Johnson and Rochow 1972)

Strain	RP	MA	RM	SG
PAV	100	77	3	30
RPV	100	0	0	40
MAV	3	100	0	1
RMV	11	6	90	6
*PAV	100	72	0	31
*RPV	100	0	0	53
*MAV	6	97	0	3
*RMV	14	0	97	14
*SGV	6	0	0	75

RP = *Rhopalosiphum padi*, MA = *Macrosiphum (Sitobion) avenae*, RM = *Rhopalosiphum maidis*, SG = *Schizaphis graminum*

(ii)   Virus Sources

The sources of the virus may be local, distant, or both. Aphids acquire the virus by feeding on infected plants then migrate into the cereal crops. A green-bridge provided by grass, weeds, or cereal volunteers in the stubbles or the remains of the previous grass crop, often harbours aphids. Green-bridge aphids are favoured by warm, damp soil conditions and are not easily controlled by insecticides. Green-bridge aphids pose a special problem as they can transfer to the new crop and transmit BYDV before the risk of invasion by flying aphids bringing the virus from the sources outside the new crop. Kelley (1994) demonstrated that BYDV infection appears to generate advantages for rare sexually produced genotypes presumably through frequency dependent selection.

Wheat (*Triticum aestivum, T. durum*), barley (*Hordeum vulgare*), oats (*Avena sativa*), maize (*Zea mays*), rice (*Oryza sativa*), rye (*Secale cereale*) and sorghum (*Sorghum vulgare*), alfalfa, soybean, are all susceptible to BYDV and may act as the source of the virus. A large number of grasses—both annual and perennial—are alternate hosts of this virus. Perennial grasses such as couch, paspalum, perennial ryegrass, and kikuyu play an important role in allowing the virus to survive over summer. In northern and north-western Europe and New Zealand, perennial grasses are the principal virus source (Lindstan 1964, Smith and Wright 1964, Doodson 1967) and at least in the UK, appear to provide a reservoir of infection (Plumb 1977). Interactions between weather and vector biology probably determine which virus isolate is introduced to cereal crops in the spring and this may account for the regional differences in the frequency of occurrence of different isolates in cereals (Plumb 1974). Grasses also appear to be the predominant, often the only source of the virus for infection of autumn-sown cereals in the UK (Plumb 1983). In some areas of southern and western England, maize may act as a source and in France maize is the principal source of the virus carried by *R. padi* migrating during September to November (Dedryver and Robert 1981).

In areas of summer drought, irrigated grass or roadside verges may provide a local source of BYDV (Price 1970). *R. padi* can acquire the virus from barley roots (Orlob 1966) and in winter are often found feeding on cereals at or below soil level. *R. rufiabdominalis*, also a vector of BYDV, over-winters on the underground parts of wheat and barley and this has been suggested as being of potential epidemiological importance in North America (Jedlinski 1981). *Leersia oryzoides*, the only grass weed in rice in Italy is a host of *R. padi* and emerges before rice and acts as a focus for initial infection by migrant *R. padi* that can spread virus to the surrounding crop via apterous offspring of the initial migrants (Osler 1980). A perennial grass common reed (*Phalaris communis/P. australis*) acts as a source of BYDV-PAV in Turkey (Ilbag 2006). Other common grasses like *Echinochloa crus-galli*, *Setaria pumila* and *Phaleris canariensis* may also act as the source of BYVD-MAV in the Czech Republic (Pokorny 2006).

Successful over-wintering is not assured, as very cold conditions can kill eggs, as in some parts of Europe, while sexual forms of *R. padi* are produced, hosts (*Prunus padus*) on which eggs are often rare or absent (Plumb 1990). Therefore, some regions may be infested from sources in more favourable climates some distance away. In continental regions the northward progression of milder conditions in summer is often accompanied by weather conditions that favour aphid dispersal. Low-level jet winds have been implicated in long-distance dispersal of aphids transmitting BYDV in North America (Wallin *et al* 1967, 1971).

(iii)   Spread

Migration of aphids normally caused the primary infection, and the secondary infection within a crop is caused by the multiplication and spread of the aphid vectors. Where the population of infective migrants is large, as in regions where aphids can survive year round on Poaceae, primary infection may be extensive and secondary spread of little consequence. Where few migrants are infective, often where sexual generations are important, then primary infection may be of little consequence and most of the crop loss is due to the spread of the virus within the crop.

Patterns of dispersal within crops are also influenced by aphids' behaviour. In the UK, patches of infection are usually associated with autumn infection by *R. padi*, whereas spring infection by *S. avenae* usually results in single apparently random infections (Rochow *et al* 1965).

Three aphid species are the main vectors in most of Europe: *Rhopalosiphum padi* L., *Sitobion avenae* Fabricius and *Metopolophium dirhodum* Walker (Plumb and Johnstone 1995), but their transmission efficiencies differ. The virus

(RMV strain) is most efficiently vectored by *R. maidis* but can also be vectored with varying efficiency by *R. padi* and *Schizaphis graminum*. The transmission efficiency of *R. padi* increases if acquisition and inoculation feeding periods are allowed at higher temperatures. Under certain environmental conditions, *R. padi* can play a significant role in the epidemiology of BYDV-RMV. This may be especially significant in regions where maize is a major source of virus and aphids that can carry virus into an autumn-planted wheat crop (Lucio-Javaleta *et al* 2001). In Mediterranean type climatic conditions in Australia, spread of BYDV-PAV in autumn-sown wheat, appears to be related to the aphids immigrating early and remain active throughout the winter growing period (Thackray *et al* 2005).

Poaceae are important components of agricultural landscapes both as cultivated crops (e.g. wheat, barley, maize, oats, pasture grasses) and as volunteer, spontaneous species in non-cultivated areas. Only perennial species can be reservoirs of viruses and vectors during the whole year. To complete its life cycle, viral inoculum has to 'jump' several times within an agricultural landscape, from senescent to alternative hosts or the bridging crops (Irwin and Thresh 1990). In temperate climates, maize (Henry and Dedryver 1989, Irwin and Thresh 1990), cereal volunteers (Henry *et al* 1993, Masterman *et al* 1994) and some pasture grasses (Henry and Dedryver 1991) are important to carryover BYDV through summer, spontaneous grasses being of lesser importance (Irwin and Thresh 1990).

Considering just the virus, the host range of each BYDV strain is similar. However, in the field there are differences in which virus strains are prevalent in different hosts. For example, *R. maidis* does not tend to feed on oats and so RMV, a virus strain that relies on this vector, is not usually a problem in oats. Changes in the cultivars may have profound effects on epidemiology through effects on vectors and viruses (Harrington 2003).

(iv) Climate and Epidemiology

In the UK, general experience and some empirical evidence suggest that BYDV problems are greatest following two consecutive mild winters with a wet summer in between. The first mild winter leads to a high proportion of anholocyclic (that over-winter the active stage on grasses, including cereals, rather than as an egg on their primary host, *Prunus padus*) clones of *R. padi*, a major vector, in the following the year. The active forms are less tolerant of low temperature than are the eggs and a tendency can be seen for the proportion of anholocyclic forms in autumn to be related to temperature in the previous winter, with a higher proportion following a mild winter. If, in the meantime, the summer is wet, there is significant relationship between numbers of *R. padi* in autumn, and summer rainfall (Harrington 2003).

Fabre *et al* (2005) conducted studies on the effects of climate and land-use on the occurrence of viruliferous aphids and the epidemiology of BYDV in France.

Variations according to trap location were correlated with land use at the regional scale, annual variations being correlated with the climate of the year. The warmer the January to August period, the higher was the proportion of viruliferous aphids in the following autumn. The proportion of the viruliferous aphids also increased with the ratio of the area sown to small grain compared to maize, which accounted for the regional pattern observed.

Smyrnioudis *et al* (2000) studied the effect of drought stress and temperature on the dispersal of *R. padi* and the pattern of spread of BYDV within wheat plants and quantified the same in controlled environment chambers. Both drought and temperature have been found to have considerable importance in virus spread. Combinations of three different drought stress levels, unstressed, moderate and high stress level, and three different temperatures, 5 ± 1°C, 10 ± 1°C, and 15 ± 1°C were tested. With increased temperature there was an increase in the mean distance of visited plants from the point of release and in the number of plants visited and infected with BYDV. Drought stress had no effect on mean distance moved by aphids at any temperature or on plants infected with the virus at 10° and 5° C. A greater proportion of plants visited by aphids was infected with BYDV when plants were stressed than when not stressed. At 15°C a greater proportion of these plants were infected than at lower temperatures. There was no difference between treatments in the numbers of aphids present at the end of the experiment. Smyrniodis *et al* (2001) studied the efficiencies of seven clones of *R. padi* and five clones of *Sitobion avenae* (originating from Greece and the UK) as vectors of BYDV (PAV-like isolate) and evaluated at 5, 10, and 15°C using barley cv. Athenaida. When inoculation took place at 5 or 10°C, clones of *R. padi* differed in their ability to transmit. For *S. avenae*, there were no interclonal differences in the transmission efficiency at any of the temperatures.

Foster *et al* (2001) assessed the aphid abundance in autumn and incidence of PAV, MAV, and RPV isolates of BYDV the following spring in the UK. Fields were categorized on the basis of surrounding land use, field boundary type, aspect and shelter, soil, position, previous cropping and present husbandry. Results from multivariate analyses indicated that BYDV was more likely to occur in fields that were: early sown (P<0.01); close to the sea (P<0.01); in non-arable areas (P<0.01); east-facing (MAV only) (P<0.01); south-west-facing (PAV only) (P<0.05); or distant from arterial roads (P<0.05) than in fields without these features.

Northing *et al* (2004) used the internet for provision of user specific support for decisions for the control of aphid-borne virus. An internet-based decision support system (DSS) was developed to help growers assess the need to control BYDV vectors in autumn-sown cereals in the UK. The prototype DSS was incorporated into a dynamic web site that also provides comprehensive encyclopaedic information on the epidemiology of the virus, its vectors, and approved pesticides and regular updates on the latest information on the factors that may affect its spread. Using the latest internet-technology a two-tiered risk assessment (regional- and field-specific) is provided via an intuitive map-based interface. By pinpointing the position of the field on a regional map and providing some basic information, the users are able to generate a risk assessment that has greater relevance to their crop than the regional assessment summary.

Chaussalet *et al* (2001) made an attempt to estimate the rate of the spread of the BYDV by the nearest neighbour approach. The virus spread is described by the probability of a plant becoming infected, conditioning on the number of infected plants neighbouring it. This has the advantage that the influence of aphid movement can be incorporated into the definition of the probability of a neighbour becoming infected. The calculated simulations from the experimental data can be used to estimate the rate of spread of BYDV and also to perform a range of sensitivity analyses.

(b) Maize Streak Virus (MSV)

Maize streak virus (MSV) is a single-stranded DNA virus with circular genome and belonging to the genus *Mastrevirus* of the family Geminiviridae. The virus is transmitted by several species of *Cicadulina* China in a persistent, circulative, non-propagative manner.

MSV is widely distributed in sub-Saharan Africa, from Sudan to South Africa and Zimbabwe to Senegal, and also in the adjacent islands including Madagascar, Mauritius, Réunion, Sao Tomé and Principe (Storey 1936, Bock 1974, Rose 1978, Thottappilly *et al* 1993). It is also reported from Egypt (Ammar 1983) and Yemen. It is generally accepted that MSV is an endemic African virus that is restricted to the African continent and the neighbouring islands. This disease has been reviewed by Bosque-Pérez (2000) and Knoke *et al* (1992).

(i) Sources of the Virus

MSV infects a wide range of plant species in the Poaceae and occurs naturally in various indigenous African grasses (Rose 1978). Indigenous African crops like sorghum [*Sorghum bicolor* (L.) Moench], pearl millet [*Pennisetum americanum* (L.) K. Schum] and finger millet [*E. coracana* (L.) Gaertner], hardly get naturally infected in the fields (Bosque-Pérez and Buddenhagen 1999) but other introduced crops including rice (*Oryza sativa* L.), oats (*Avena sativa* L.), wheat (*Triticum aestivum* L.) and sugarcane are naturally infected in the fields (van Rensberg and Kuhn 1977, Rose 1978) and may act as the sources of the virus.

Weed grass hosts are usually considered to be sources of MSV and to present a potential threat to the susceptible crops depending upon the ecological condition of the area. As for example, virus isolates occurring in lowland areas in Nigeria are not readily transmissible to susceptible field maize (Mesfin *et al* 1992).

Survival of the grasses that harbour vectors and MSV isolates capable of infecting maize during the dry season (November-March) particularly in low-lying hydromorphic soils is critical for the ecology and epidemiology of MSV (Bosque-Pérez and Buddenhagen 1999). Perpetuation of MSV and its vectors during the dry season in West Africa appears to occur mostly in riverine areas and low-lying fields with residual moisture, and both maize and grass weeds occur. MSV outbreaks appear to occur only when favourable weather conditions allow vector survival and population build-up, and where maize-MSV infects grass hosts (Bosque-Pérez 2000). Mesfin *et al* (1992) observed that native annuals *S. barbata* and *B. lata*, and the introduced perennial *A. compressus* are weeds of importance in perpetuating epidemiologically competent maize strain of MSV in Nigeria. These weeds harbour MSV under field conditions in the humid forest zone and occur frequently in hydromophic areas. In Zimbabwe, the vectors are found in annual grass species like *Brachieria, Chloris, Cynodon, Digitaria, Eleusine, Paspalum, Pennisetum, Setaria* and *Sporobolus* (Rose 1978). In West Africa, the leafhoppers mostly occur on annuals: *Brachiaria deflexa* (Schumach) Hubbard, *B. lata, S. barbata, D. horizontalis* and *E. indica* (Mestin *et al* 1992; Asanzi *et al* 1994). *S. barbata* and *S. lata* harbour both virus and vectors. In certain parts of southern Nigeria, leafhoppers occur throughout the year, and streak symptoms often appear on annual grasses (Asanzi *et al* 1994).

(ii) Spread

MSV can be transmitted by nine species of *Cicadulina: C. arachidis* China, *C. bipunctata (bipunctella)* (Melicher), *C. ghauri* Debrowski, *C. lateens* Fennah, *C. mbila* (Naude), *C. niger* China, *C. parazeae* Ghauri, *C. similis* China (Bigirwa *et al* 1995; Bosque-Pérez 2000).

All individuals in a population may not be transmitters. Some members of a population of the vectors transmit the virus suggesting the transmissibility as a hereditary character (Asanzi *et al* 1995). All the transmitters are not equality efficient as transmitters. *C. mbila, C. storeyi* and *C. ghaurii* transmits more efficiently than *C. arachidis* (Asanzi *et al* 1995).

Spread of MSV depends on the environmental conditions, host plants and flight behaviour of the vectors. Vector population is low normally at the onset of the rainy season and then rises gradually as the host plants become abundant and succulent. Populations sharply decline in summer. Rose (1972) found a positive correlation between population density of *Cicadulina* species and rain during the preceding months.

The incidence of MSV varies according to the variety and planting date (Asanzi *et al* 1994, Bosque-Pérez *et al* 1998). Late season plantings (July-October) shows higher incidence than those planted in the early season (March-May) (Fajemisin *et al* 1976).

The nature of the epidemics differs with the season as the flight activity of the vector changes. Initial infection normally takes place through the movement of the infective insects from the nearby infected fields or grasslands. The overall disease incidence generally remains low and the foci tend to be scattered and increase slowly. With the change of the season new insects migrate that seldom settle for long and the population remains continuously changing. Two types of population are found, residential and migratory. The residential insects are larger in size and are short fliers. The migratory ones are short-bodied and can fly for a long period. The mean flight duration for short fliers is 15 s and can fly up to 4-12 m without the assistance of any wind. The duration and the extent of flight of the long fliers differ very widely. Some insects may travel 1.8 km down wind and some can travel 118 km (Thresh 1983).

The spread of MSV is critically dependent on the various *Cicadulina* species. Autrey and Ricaud (1983) suggested that three major factors influence the probability of the disease outbreaks: maize genotype, scrubland proximity, and cropping sequences. The ability of leafhoppers to produce many generations from four to five for *C. mbila*, *C. parazea*, and *C. storeyi* (Rose 1973) to eight or nine for *C. bipunctella* (Ammar 1977) and *C. chinai* (Ammar 1975), provides ample opportunities to accommodate climatic and plant growth variations and assures the transmission and dispersal of the virus to new hosts during a growing season.

The patterns of virus spread however, are dependent on the sexual composition of the vector population as well as their physiological state of arrival within a maize crop. Gravid females are less likely to fly; feeding on young wheat seedlings also inhibits flight. Rose (1972) found that males and non-gravid females from dry, mature wheat stems are most likely to fly. Short-bodied forms are also more likely to be long-distance fliers than are long-bodied forms. Females fly higher than males and *C. mbila* fly higher than *C. storeyi*. A steep infection gradient within 10 m of a field's edge is due largely to short-distance fliers and a more gradual gradient decline is due largely to long-distance fliers (Rose 1973).

The rate of spread is correlated with leafhopper population development and with the efficiency of vector transmission. Disease increase follows an arithmetical pattern when vector populations are low and become exponential when populations are high (Rose 1974). The environment may be the overriding factor in disease development when the availability of leafhopper vectors and susceptible plants are not limiting. Disease development is more likely when the rainfall distribution is uniform throughout a growing season than when rainfall is sporadic (Thresh 1986, Gorter 1953)

Disease development also may be influenced by the climatic diversity in the area of the crop production. MSV is widespread in the warmer parts of South Africa, but is seldom found at altitudes above 1,200 m (Storey 1926). The growing of maize, instead of the traditional pearl millet and sorghum in the drier savanna zones of northern Nigeria, where graminaceous grass weeds support leafhopper development, may be responsible for the high streak incidence (Okoth and Dabrowski 1987).

(c) Rice Tungro Disease (RTD)

Rice tungro disease is caused due to mutual infection of *Rice Tungro Bacilliform Virus* (RTBV), a double-stranded DNA virus belonging to the family Caulimoviridae and *Rice Tungro Spherical Virus* (RTSV), a single-stranded RNA virus belonging to the family Sequiviridae. Both the viruses are transmitted by a few leafhoppers in a semi-persistent manner. This disease occurs in most of the South and South East Asian countries, particularly, Thailand, Bangladesh, India, Malaysia, Philippines, Indonesia, Myanmar and Sri Lanka. The incidence of this disease has also been reported from Italy (Baldacci *et al* 1970) and China (Xia and Lin 1982). Global distribution of tungro disease coincides with the distribution of its important vectors *N. virescens* and *N. nigropictus*. The epidemiology of the disease has been reviewed by many workers (Ling *et al* 1983, Mukhopadhyay 1986, 1992, Azzam and Chancellor 2002, Hibino 2004).

(i) Sources of the Viruses

Rice tungro is a crop adaptive disease and it mainly infects and survives in rice plants and does not survive in seed or soil. Root debris and stubbles of infected plants may keep the pathogen for several days depending upon the variety and the season (Ling and Palomer 1966, Rao and John 1974, Tarafder and Mukhopadhyay 1979a, 1980a), but the chopped stubbles, when puddled in the field, failed to act as the source of the pathogen (Mukhopadhyay 1984). There are reports on the availability of the pathogen in some weeds (Rivera *et al* 1969, Mishra *et al* 1973, Tarafder and Mukhopadhyay 1979b, 1980b), but most of the weeds cannot keep the pathogen for a long time. Moreover many of them do not support the feeding/multiplication of the vectors. Thus the scope of stubbles and weeds for acting as the source of the disease is remote. In a mono-crop situation, infected stubbles may produce infected ratoons that may act as the source of the disease.

Wild rice and symptomless varieties have potentialities to keep the viruses separately or together. They may also support the feeding of vectors and therefore may act as the sources of the virus/viruses (Mukhopadhyay 1992). Though not common, viruliferous leafhoppers may migrate and act as the sources of the viruses.

Most of these studies are on the sources of the pathogen before the discovery of the involvement of two morphologically and serologically different viruses in the Tungro disease (Omura *et al* 1983, Cabauatan and Hibino 1988). This discovery opened a new vista for the understanding of the biological diversity of host ranges of these two different viruses and their symptomless hosts but subsequent research remained dominated by the molecular biology of the concerned viruses.

(ii) Spread

The spread of the RTD is directly dependent on the availability of viruliferous vectors or the presence of the sources of the viruses and the vectors. Ling (1975) estimated the relation between the proportion of plants infected in an area and size of the source of the pathogen. The spread was more when the source plants were scattered. A vector was found to spread the pathogen from a source plant to a maximum distance of 250 m. Theoretically a single infective vector can infect 288 plants (at a susceptible age and variety)/day but practically it is much lower (11-30 plants/day at 27°C and 40 plants/day at 40°C (Ling and Tiongco 1975).

The spread of RTD is a function of the flight activity of Rice Green Leafhoppers (GLH) the vectors of the viruses. The prevalence of the GLH and their flight activity again depends upon the climate, meteorological condition and cropping system. The climate and the meteorological conditions determine the biology, ecology and prevalence of the GLH. Tropical sub-humid climate favours the growth of the vectors but their biology is greatly influenced by the meteorological conditions particularly temperature. In tropical climates where the winter season is limited, vectors may remain abundant throughout the year subject to the availability of host plants whereas in the tropics and sub-tropics with a prolonged winter, the vectors over-winter as nymphs and the adults become scarce. The ecology of major leafhopper vectors has been studied in the Philippines, Indonesia, Malaysia (Cook and Perfect 1989, Bottenberg *et al* 1990) and India (Mukhopadhyay 1984, 1988, 1989, 1995).

In tropical or sub-tropical sub-humid irrigated rice crop field conditions with a prolonged winter (for example, West Bengal), the nymphs over-winters in summer rice seed beds/ratoons and the residential population develop in summer/pre-kharif rice that move to the seed-beds and transplanted fields of kharif rice where they multiply at a rapid rate and the population gradually increase till the maturity of the crop. The appearance of the peak population is often related to the occurrence of peak rainfall (Mukhopadhyay and Mukhopadhyay 1987).

The residential population shows flight activity particularly during favourable climatic and agro-meteorological conditions. This activity is influenced by the temperature regime between 20° and 34°C and moonlight. GLH is not a strong flier. They can hardly be caught 10 m above ground under normal conditions. They usually fly at the canopy height during the night usually up to 5 h after the civil twilight and 2 h before sun rise (Bowden *et al* 1988).

There may be occasional immigration of GLH from a short distance by active flying along the wind direction or by passive transportation from a long distance by the wind-field frontal system followed by climatic depression.

Cooter and Winder (1996) determined the flight potential of the vectors [*Nephotettix virescens* (Distant), *N. nigropictus* (Stål), *Recilia dorsalis* (Motschulsky)] using tethered flight techniques. Normally these insects do not tend to fly in response to stimulation or may fly for a very short time. A few leafhoppers, however, may fly for a long period even for almost 7 h that indicate the potential for long distance dispersal of the inoculum. When the leafhoppers are fed on mature plants, they show higher flight activity.

Chancellor *et al* (1996) studied the leafhopper ecology and rice tungro disease dynamics at the field conditions in the Philippines where favourable temperature conditions prevail throughout the year for the growth of these leafhoppers, but their abundance is seasonal because of the limited wet season and absence of the host during the dry period.

The flight activity of the three leafhoppers species [*Nephotettix virescens* (Distant), *N. nigropictus* (Stål), *Recilia dorsalis* (Motschulsky)] is closely associated with the rice cropping periods, crop growth stage and the season. *N. virescens* colonizes the rice plantings early, shows rapid population development and higher transmission efficiency. Early infected rice fields are randomly distributed and increasingly aggregate indicating the importance of secondary spread in disease dynamics. Vector population abundance is not directly related as the rapid disease spread may occur even when the population level of the vectors in the fields is very low. Presence of inoculum, favourable agro-meteorological conditions, varietal susceptibility and the susceptible crop age significantly contribute to the disease dynamics.

Asynchronous (grown in close succession with a virtual rice monoculture occupying a large tract of land) cultivation of rice promotes the spread of RTD as it gives greater variability in planting time between individual crops within a locality, greater opportunity for spread of the inoculum by movement of the vectors between the fields. In a particular cropping season, crops planted at a later date than average suffer a greater incidence of RTD. Several workers (Loevinsohn 1984, Suzuki *et al* 1992, Chancellor *et al* 1997) established this impact of asynchronous cultivation on the spread of the RTD in different locations.

Holt and Chancellor (1997) developed a mathematical model of the dynamics of a plant virus disease within a spatially referenced lattice of fields of a host crop in asynchronously planted cropping systems. This model is applicable to crops in continuous, contiguous cultivation such as tropical irrigated rice. Disease progress in each field is determined by incidence within the field itself as well as incidence in neighbouring fields, depending on the gradient of the disease spread. The frequency distribution of the planting dates may be assumed to follow a normal distribution and the variance of the planting date may be used as a measure of cropping asynchrony. Analysis of the model reveals that disease incidence within the lattice depends upon the infection efficiency, the slope of the dispersal gradient and the variance of the planting date. Disease endemically depends on planting date variance and the disease persists in the lattice if this variance exceeds a certain threshold. Above the threshold for persistence, the response of mean disease incidence to planting date variance is non-linear and the region of the greatest sensitivity is closest to the threshold. Thus the disease systems that show moderate rather than high cropping asynchrony are more likely to be influenced by changes in variance of the planting date. There are also relations between the spread and different agro-ecological conditions Cabunagan *et al* (2001).

The epidemiological studies so far have been based on infected plants as the source of the viruses. A consistent study was made in India for a period of 3 years (1997-2000) in a typical endemic situation through field surveillance on survival and spread of the two viruses. The field study was conducted in West Bengal (Chakda, 22°6/N, 88°32′E) (Bidhan Chandra Krishi Viswavidyalaya), and the detection study was conducted at BCKV and Delhi University. This study reveals some interesting features related to the survival and spread of tungro. The PCR-based technique developed at the Delhi University and the conventional indexing on T(N)1 test seedlings were used to detect the bacilliform virus (Dasgupta *et al* 1996) and tungro infection in plants and vectors during different crop seasons. Vectors were collected by standard sweeping technique and also using a 12.2 m suction trap (Rothamsted type). During the first year of the surveillance, boro-ratoons, kharif seedbeds, kharif standing crops and the vector, *N. virescens* were indexed by the conventional method and moderate infection was found in ratoons and standing crop. Strikingly, the vectors collected from all the situations showed moderate infectivity. Next year, boro-ratoons though reacted positively in indexing but did not react to PCR-based technique. But in highly susceptible varieties like *T(N)1* and *Swarna Masuri*, ratoons showed a positive reaction. Thus the ratoons of the moderately resistant varieties may carry only RTSV whereas highly susceptible varieties may carry both the viruses. When vectors were collected in July, indexing gives more prominent results than the PCR-based technique but the latter gives intense reactions with August collections. It indicates that RTBV may enter into the infection complex little later, but the reverse is found with *N. nigropictus*. Strong PCR-based reaction was observed in the July collection of this leafhopper but there is no response in August. It indicates that *N. nigropictus* may have a role in primary infection. In September both the plants and the vector show positive reaction whereas in October, only plants show positive reaction and not the vectors. In boro seedbeds (February), male and female *N. nigropictus* and *R. dorsalis* show positive reaction but does not respond to indexing, which may be due to very low concentration of the virus. Leaf samples when collected in May do not show positive reaction in PCR-based technique but positively respond to indexing indicating that the mature boro plants may contain only RTSV and not RTBV.

PCR-positive leafhoppers are more or less abundant in the light trap catches during April-May, but the number of the infective leafhoppers in the light trap catches gradually reduces in July and no infective vector could be caught in September-October although they are prevalent in the field. Migrating leafhoppers collected by the suction trap, though their number is very low, show strong positive reaction. Residential leafhoppers are normally low fliers. They hardly move above the canopy level (Chancellor *et al* 1997). Trapping them at 12.2 m suction trap in particular months of a year suggest their long distance migration. During the experimental period vectors could be collected only in 1997 (November-December) and 1998 (May-June). 2 *N. nigropictus* and 37 *R. dorsalis* were caught during 1997. 2 *N. nigropictus* and 2 *R. dorsalis* were caught during 1998. All these insects were strongly positive to the PCR tests. During December 1998, 4 *R. dorsalis* were collected that were PCR negative. Arrival of infective *N. nigropictus* and *R. dorsalis* in May-June, though apparently in an insignificant number, may have a role in initiating the primary sources of the virus in the field when *N. virescens* is scarce.

(*This study was conducted as part of a National Collaborative Project of the Department of Biotechnology, Government of India on 'Molecular characterization of Indian isolates of Rice Tungro Spherical and Rice Tungro Bacilliform Viruses and Development of Molecular Diagnostic Probes' between Delhi University, Indian Agricultural Research Institute, New Delhi and Bidhan Chandra Krishi Viswavidyalaya, West Bengal.*)

The enigma of the epidemiology of this disease caused by mutual effectiveness of two morphologically and immunologically different viruses still persists. The epidemiology so far studied is on the RTD. How the mixture works is yet to be known.

The crux of the epidemiology of the RTD may lie on the chance association of two contrasting viruses (RNA and DNA). It may be that *Indicas* and *Japonicas* at their points of origin coexisted either with the spherical or with the bacillary particles. Through extensive germplasm collections at the points of origin and extensive breeding for improved varieties, a chance may occur for the association of both types of particles in a cellular system of rice with genomic recognition between them. Critical study on evolutionary epidemiology can only throw light on this unresolved issue.

(d) Citrus Tristeza Virus (CTV)

Citrus tristeza virus (CTV) is a long, flexuous filament (longest among the known plant viruses), containing single stranded RNA. It belongs to the genus

Closterovirus and family Closteroviridae (Karasev *et al* 1995). It is transmitted in a semi-persistent manner by several species of aphids. Tristeza virus is the most economically important virus disease of citrus (Bar-Joseph *et al* 1989, Rocha-Peña *et al* 1995). Several authors have reviewed this disease (Bar-Joseph *et al* 1979, Bar-Joseph *et al* 1981, 1985, 1989, Davino *et al* 1986, Lee and Rocha-Peña 1992, Timmer *et al* 2003, Nolasco 2006). The losses caused by this virus have been estimated in several countries (Argentina > 10 million trees; Brazil > 6 million trees; California > 3 million trees). Cambra *et al* (2000) reported the loss of close to 40 million trees in Spain since 1935.

This disease known by various names (Quick decline, Seedling yellows, Stem-pitting, etc.) has been reported from South Africa, Java, Argentina, Brazil, California, New Zealand, Australia, West Africa, Ceylon, Hawaii and Spain. It has been speculated that tristeza originated in the orient or in other words from the points of origin of different citrus species and was distributed worldwide by the movement of citrus bud-woods and plants. South Africa is the first country to have imported the disease from the orient. From South Africa, the disease was probably distributed around the world to other citrus areas. This disease is now present in most of the major citrus areas of the world except in certain Mediterranean and Central American countries (Lee and Rocha-Peña 1992).

(i) Source and Spread

The virus does not spread through seeds. Infected plants and bud-woods are the primary sources of infection. Viruliferous vectors may also bring the virus from a short distance. There is no alternate or wild host for this virus.

Several aphid species transmit this virus. Principal vectors are brown citrus aphid *Toxoptera citricidus* (Kirk.), now called *T. citricida* Kirk. melon aphid *Aphis gossypii* (Glov.), and *Aphis citricola* Vander Goot. Five other vectors have also been reported. These are: spirea aphid *Aphis spiraecola*, Patch, black citrus aphid *Toxoptera aurantii* (Fonsc.), *Aphis craccivora* (Koch.), *Dactynotus jaceae* L. and *Myzus persicae* but not all are equally efficient in spreading the virus in the field. *T. citricidus* is the most efficient vector in the Southern Hemisphere. *A. gossypii* is the vector of tristeza in California whereas *A. spiraecola* and *T. aurantii* are vectors of tristeza in Florida. An aphid can acquire the virus after a feeding period of 24 h on an infected plant but the rate of transmission depends upon the strain of the virus and the variety of the citrus serving as the source of the inoculum. There is correlation between natural spread of the virus and flight activity of the vector. The main flight of *T. citricida* is closely correlated with the flush cycle of citrus, whereas the maximum number of alate *A. gossypii*, a more polyphagus aphid, is not.

Natural spread of the virus has been studied in different countries in various citrus crops. Vectors in some countries may be efficient transmitters whereas in other counties they may be inefficient transmitters as was the case in South

Africa and the USA respectively. Calavan *et al* (1980) found natural infection in Riverside, California as 2.4% to 6.3% during the 1950s which reached 50% after 8-12 y. Roistacher *et al* (1980) observed changes in the transmission efficiency of *A. gossypii* in California. Its efficiency to transmit both tristeza-quick decline and tristeza-seedling yellows reached at par with the efficiency of *T. citricida* found in other countries. This may be due to differences in the isolates. Of the 23 isolates tested, 100% efficiency was found with 13 isolates, 90% with 2 isolates, and an average efficiency between 21-28% with 4 isolates; and one isolate maintained the old transmissibility rate of 7%.

Catara and Davino (2006) observed that the main way for spreading of the disease in Sicily is the use of infected propagation material and *Aphis gossypii* was found to be the major vector.

Raccah *et al* (1977) in Israel, found segregation of tristeza virus isolates collected from a single sweet orange source plant, showing variation in the rates of transmission by *A. gossypii*. It is presumed that the donor and receptive host plants contribute significantly to the differences in the respective transmission efficiency. Hermoso de Mendoza *et al* (1988) found the lowest transmission efficiency ratios using lemon both as donor and receptor host, whereas highest efficiency was found with a sweet orange-sweet orange (scionic and rootstock combination) combination. In sweet orange-Mexican lime combination, the transmission efficiency of *A. gossypii* with respect to tristeza-seedling yellows isolate was 60% whereas that of tristeza-quick decline isolate was 90%. Kuno and Koizumi (1991) collected 20 subcultures of tristeza by feeding *T. citricida* on donor plants from 15 different original field sources. They also collected 15 cultures from the original field sources. When these subcultures and cultures were tested by double sandwich indirect ELISA using two monoclonal antibodies (MCA 13, 3DF1) and by double sandwich ELISA using polyclonal antibodies, 4 aphid transmitted subcultures derived from 2 field sources gave different reactions from their parent field sources. It indicates that there may be a changing dynamic state in the occurrence of strains and their transmissibility within a host system. This change is moving from a less transmissible to a more transmissible form.

Temperature also plays an important role in the transmission of tristeza. In Florida, *A. gossypii* transmit less efficiently during the summer than in winter. Bar-Joseph and Lobenstein (1973) experimenting with sweet orange seedlings grown at controlled temperature and using c. 100 aphids/plant, obtained transmission rates of 61% when source plants were kept at 22 ± 2°C, compared with only 12% for plants kept at 31°C. There is, however, variability in the sensitivity of tristeza isolates to heat (Roistacher *et al* 1974). Calavan *et al* (1972) computed infection gradient from two different years of virus indexing in a Valentine orange grove in California and quantified. The disease front was found to advance approximately 79 m in 2y. Tree spacing and row direction in relation to prevailing winds, differences in the trees size and their effect as wind barriers all affect vector movement. Close spacing within the row might facilitate spread by apterae and this will also influence the gradient.

The spread of tristeza is also receiving a new dimension with the increase of the geographical distribution of the vectors. The geographical area occupied by *T. citricida* has been expanding over the years. It has gradually been moving northward in South America (Geraud 1976). In Central America, *T. citricida* is found in Panama, Costa Rica, and El Salvador (Lastra *et al* 1991), Continental Portugal and Spain (Ilarco 2006). With the increase of the geographical distribution of the vector and expansion of citrus industry in different countries using susceptible combinations, the spread of tristeza is also on the rise. The threat is further aggravated by the development of more virulent strains. Strains of stem-pitting on grapefruit were found in South Africa that severely reduced fruit size and shortened tree life (Marais 1981, DaGraca *et al* 1982). In California, strains of tristeza-seedling yellows that cause severe stem-pitting on sweet orange were found to be spreading in the variety collection (Calavan *et al* 1980, Roistacher 1982). In Peru, severe strains of CTV were found that stem pit and reduce fruit size and vigour on Navel sweet orange scions (Roistacher 1988). In Israel, a very destructive decline strain of CTV has been found that often kills the tree before CTV can be diagnozed (Ben Zeev *et al* 1989, Bar-Joseph and Nitzan 1991). Genetic variability of CTV can also occur after host passage. The continued appearance of new strains and development of decline epidemics indicate that tristeza is constantly on the move and becoming economically important in more places worldwide.

Hardly any attempt has been made so far to model the incidence of CTV. Cambra *et al* (2000) studied the incidence and spread of CTV annually for 19 y in 2 adjacent orchards of 216 trees of Washington Navel (sweet orange) and 870 trees of *Clementina de Nules* (Clementine) and that the models differed with the plant species. The temporal progress of CTV in clementine followed the *Gompertz* mathematical model, whereas the logistic model represented the spread in sweet orange in a better way. Differences in the increase of incidence could be explained by a preferential alighting of aphid vectors on the clementines, associated with the number and availability of tender, succulent shoots.

Marroquin *et al* (2004) estimated the number of aphids carrying *citrus tristeza virus* that visit adult citrus trees and found that the infection rate was proportional to the number of aphids landing/tree. Nolasco (2006) observed that the epidemiology of *citrus tristeza virus* in the Mediterranean basin depends on the virus strains and the introduced vector *Toxoptera citricidus*.

(ii) Complexities of Tristeza

The complexity of tristeza is apparent from the wide variability in its expression in the fields from quick decline, dwarfing, stem-pitting to yellowing of seedlings. In nature all the components may remain together or may be separated during the process of propagation. The type of the disease normally depends on the source of the virus, composition of the complex, the kind of citrus tree into which the complex is introduced, and in the case of composite trees, the kinds of citrus used as scion and stock. Stem-pitting can severely injure scions of grapefruit, lime, and sweet orange directly, regardless of the rootstocks; seedling yellows induce stunting and yellowing at the seedling stage and tristeza causes bud-union failure of certain scions on sour orange rootstock. There also occur mild, moderate and severe strains of these three components. In mixed infection, one strain may not become fully systemic depending on the susceptibility of the host species. It is also apparent that in one set of conditions one strain may dominate, but if the conditions are changed, another strain may dominate. If there is a sub-dominant severe strain in a parent tree, this may become dominant in its daughter trees under different climatic conditions. The complexity has been further convoluted because of its continuous genomic recombination in nature and the appearance of new or more virulent strains. From 1980 to 1991, new or severe strains of tristeza stem-pitting virus were recorded in several countries notably South Africa, Japan, Brazil, Israel, Australia, and in the USA (California and Florida) causing destruction to mostly oranges (Calavan *et al* 1980, Miyakawa and Yamaguchi 1981, Da-Gracia *et al* 1982, Brlansky *et al* 1986, Roistacher 1988, Bar-Joseph and Nitzan 1991). These complexities have also increased the confusion of their epidemiology. The most important task for precision epidemiological understanding rests on the precise detection of the continuously evolved forms or strains of CTV. Niblett *et al* (2000) described the evolution of techniques developed to detect CTV and to differentiate the individual strains that may be applied for proper understanding of the epidemiology.

(e) Beet Curly Top Virus (BCTV)

Beet curly top virus is a single stranded DNA virus and belongs to the genus *Curtovirus* and family Geminiviridae. The virus is persistent circulative and normally transmitted by beet leafhopper *Circulifer tenellus*. Capsid protein of the virus is involved in systemic infection, virion formation and leaf hopper transmission (Soto *et al* 2005) .

BCTV is widespread throughout the western United States, southwestern Canada and Mexico. It appears to be endemic in the Mediterranean basin (Bennett 1979). Its occurrence has also been reported from South America (Bennett 1971), Australia (Thomas and Bower 1980), Turkey (Bennett and Tanrisaver 1957), Iran (Gibson 1967) and India (Sandhu and Bhati 1969).

In the Distribution Map of Plant Diseases (2001, Edition 5, Map 24), the following countries have been shown for the incidence of BCTV. Duffus (1983) and Mukhopadhyay (1992) have reviewed the virus and its epidemiology. The disease has been found to be a complex one because of the presence of several strains varying in virulence, host range and other properties and without any interference or cross protection phenomena. This virus also infects tomatoes, beans and other crops and a wide range of weed species.

(i) Sources of the Virus and Spread

Leafhopper vector acquires the virus by feeding on diseased plants for a period ranging from 1 to 20 min. The percentage of leafhoppers that acquired the virus increased from 3.3% in 1-min feeding to 28% in 30-min feedings. The increase continued from 44% in 1-h feeding to 76% in 4-h feeding. The total virus content and the ability to transmit may continue to increase over even longer periods (Bennett 1979).

As for the transmission of the virus, temperature has a profound influence. At high temperatures, the virus is occasionally transmitted within 4 to 6 h after acquisition (Bennett and Wallace 1938). The normal incubation period of the virus in the vector is 4 h, the time required for the virus to move into the alimentary tract of the insect, then passes through the tract wall into the haemolymph, then into the salivary glands for injecting into plants through saliva. It is circulative in the insect's body and the insect can keep the virus for more than 90 d.

The vector leafhopper has an extensively wide host range including several plant species from a large number of families. Among them, plants belonging to the families Cruciferae and Chenopodiaceae are most important for the survival and supporting reproduction. On the other hand, tomato, tobacco and bean get infected but do not support reproduction of the vector.

The major source of the virus in the field is the over-wintering beet plants in and near the breeding areas of the vector. Duffus (1982) described the epidemiology of the virus and the vector. The insect breeds readily on mustard (*Brassica* spp.), Russian thistle (*Salsola iberica*) and other weeds, and also on some crop plants, producing several generations during the summer. London rocket (*Sisymbrium irio*) is a common winter annual weed in southern New Mexico (USA) that can host both the virus and its vector and the weed can survive for the period of time needed to serve as the over-wintering host for beet leafhoppers (Ray *et al* 2006).

Harvesting and drying of crop and weed hosts in the autumn induce the leafhoppers to congregate on any remaining green summer annuals or perennials. As these plants dry, the leafhoppers move to breeding areas in the foothills and accumulate on perennials where favourable climate and environment prevail. Here they lay eggs and produce nymphs for some time. Egg-laying and the emergence of nymphs continues till February. In spring, it breeds to produce large number of winged adults that migrate to spring or summer crops and weeds. Migration usually involves sexually immature females for maximizing reproduction in the new environment. There may be long distance migration through a wind field transport system.

The severity of the curly top attack depends on the climatic factors affecting the weed hosts of the virus, the prevalence and severity of the virus and the reproductive capacity and migration of the leafhopper. The age of the plant at the time of infection contributes much to the infectivity and severity of the symptom (Wintermantel and Kaffka 2006).

In view of the economic significance of the crops and the virus, extensive studies have been made for the development of resistant varieties of sugarbeet and other susceptible economic crops.

Recently several resistant varieties have been released, as for example, Sugarbeet germplasm C931, CR11 and CZ25/2 301 that are resistant to beet curly top, beet chlorosis, beet yellow, beet necrotic yellow viruses (Lewellen 2006). A pinto bean variety (Quincy) has also been released as resistant to *bean common mosaic virus, beet curly top virus* and *bean common mosaic necrosis virus* (Hang *et al* 2006).

(f) Tomato Yellow Leaf Curl Virus (TYLCV)

Tomato yellow leaf curl virus (TYLCV) belongs the genus *Begomovirus* of the family Geminiviridae but has a single genomic component. Moriones and Navas-Castillo (2000) reviewed this disease.

The first report of the damage caused by this disease was from Israel in the late 1930s, later from the Middle-East countries in the 1960s. Subsequently this disease has become a threat to tomato production in many tropical and sub-tropical regions worldwide, particularly Africa, Europe, the Caribbean Islands and Central America (Cohen and Antignus 1994, Czosnek and Laterrot 1997). Recently, this disease has been reported from Japan (Kato *et al* 1998), Mexico (Ascencio-Ibañez *et al* 1999), Florida and Georgia in the United States of America (Momol *et al* 1999). It is a serious threat in Cuba (Martinez-Zobiaur *et al* 2006), Mexico (Brown and Idris (2006), China (Wu *et al* 2006), Italy (Parrella and Crescenzi 2005), Lebanon (Abou-Jawdah *et al* 1999), South Carolina, USA (Ling *et al* 2006) and the Dominican Republic (Salati *et al* 2002).

TYLCV has been studied in detail and differences were established in the collections of the isolates collected from different regions and countries. These isolates were clustered into eight groups representing eight recognized species for which CP sequences are available: TYLCV-Th (Thailand), TYLCV-Ch (China), TYLCV-Sar (Sardinia, Italy) TYLCV-Tz (Tanzania), TYLCV-Ng (Nigeria), TYLCV-SSA (Saudi Arabia), TYLCV-Ye (Yemen) and TYLCV-Is (Israel, Egypt,

Morocco, Portugal, Spain, the Dominican Republic, Cuba, France-Reunion, Jamaica, Japan, Iran) (Morione and Navas-Castillo 2000). This disease (TYLCV-Is) also causes severe damage to the common bean (*Phaseolus vulgaris*) in Spain (Sănchez-Campos *et al* 1999, Morilla *et al* 2005).

(i) Source of the Virus

So far only a few weeds, the common bean and pepper are known to act as the source of the virus depending upon the isolate and vector biotype. Natural infection of annual weed, *Datura stramonium, Solanum nigrum, Solanum lureum* and *Solanum luteum* and *Euphorbia* sp. by TYLCV-Sar was reported from Italy (Bosco *et al* 1993, Davino *et al* 1994, Sanchez-Campos *et al* 1999, Moreones and Navas Castellow 2000). In Israel, natural infection of TYLCV-Is was found in the perennial weed *C. acutum* and the annual weed *Malva parviflora* (Cohen *et al* 1988). Transmission of the virus from the source plant depends much upon the vector biotype. TYLCV-Sar can infect five weed species: *D. stramonium, S. nigrum, Brassica kaber, Capsella barsa-pasoris* and *Malva parviflora*. But the vector, *Bemisia tabaci* differentially transmits the virus from these weeds depending upon the biotype of *B. tabaci*. Biotypes B and Q can transmit the virus from tomato to *D. stramonium* and *S. nigrum* and vice versa. The transmission efficiency of both the biotypes from infected tomato to weeds does not significantly differ. The efficiency of transmission, however, from infected *S. nigrum* to tomato significantly differs between B- and Q-biotypes. The S-biotype cannot survive in tomato for a long enough time to acquire or transmit the concerned isolate. This biotype transmits the virus from *S. nigrum* to *S. nigrum* at a very low efficiency (Jiang *et al* 2004).

Recently *Capsicum* sp. has been identified as a symptomless host of TYLCV that can act as its reservoir (Polston *et al* 2006). They could infect 30 of 55 genotypes of *C. annuum*, 01 of 6 genotypes of *C. chinensis*, 01 of 02 genotypes of *C. baccatum* and the only genotype of *C. frutescens* tested but it failed to infect the one genotype of *C. pubescens* tested. Unlike TYLCV isolates, mild strain cannot infect *C. chinensis*. No host differences occur between TYLCV-Is and the isolate from Florida, USA. Virus infection in the field may be as high as 100%. Whiteflies can transmit the virus from the infected plants but not from fruits. Natural infection is also found in two other weeds, *Mercularis ambigua* and *S. luteum* that may act as reservoirs of the virus (Sanchez-Compos *et al* 2000).

Vectors may also act as the source of the virus. It persists in the vector and the vector (Biotype B) may keep TYLCV-Is at least for two generations (Ghanim *et al* 1998). The vectors may also keep the virus between individuals through copulation (Ghanim and Czosnek 2000).

(ii) Spread

So far known, the spread of the virus mostly depend upon the isolates and the vector biotype. Aggressive isolate and locally available biotype contributes to the widespread epidemic. Studies conducted in Spain in a region in which two biotypes of *B. tabaci* (B-, Q-) and two TYLCV species (TYLCV-Ser and TYLCV-Is) coexist. Initially TYLCV-Ser was common in this region (Norris *et al* 1994). After the appearance of TYLCV-Is, TYLCV-Ser was completely displaced in some areas due to preferential transmission by local biotypes and the use of the common bean in crop rotations that serves as a bridge crop for TYLCV-Is (Sánchez-Campos *et al* 1999). Moreover, TYLCV-Is can be maintained in biotype B-populations through copulation and transovarial transmission (Ghanim *et al* 1998, Ghanim and Czosnek 2000).

Spread of TYLCV is also related to the growth stage of the crop at the time of infection. The virus prevalence is normally highest at mid-stage (first flowering to first harvesting), followed by late (first harvesting to late harvesting) and early (transplanting to first flowering) in the tomato cultivars, BARI-T1 (Manik), BARI-T2 (Ratan), BARI-T4, BARI-T5, BARI-T6 (Apurba), BARI-T7 (Chaity), BARI-T11, and BARI-T12 (Rahman *et al* 2006).

TYLCV is a gradually emerging threat to tomato cultivation due to the evolution of new strains at a rapid rate. Van Den Bosch *et al* (2006) made a modelling approach to show the impact of resistance expressed through different mechanisms, on selection for plant virus strains with a higher multiplication rate using TYLCV. A model for the epidemiology of the plant-virus-vector system was combined with a model for the within-plant dynamics of the virus. Four types of resistance were defined, each expressing resistance through one or model parameters. The evolutionary stable strategy (ESS) approach was used to study the effect of resistance on the evolution of within- plant virus multiplication rate. Resistance expressed through reduced virus acquisition by the vector, and the resistance expressed through reduced inoculation of the plant, do not put a selection pressure on the virus to evolve towards a higher multiplication rate. Resistance expressed through within-plant virus titer symptom reducing resistance does put selection pressure on the virus to evolve towards a higher multiplication rate or more harmful types.

References

Abou-Jawdah, Y., Maalouf, R., Shebaro, W. and Soubra, K. 1999. Comparison of the reaction of tomato lines to infection by tomato yellow leaf curl begomovirus in Lebanon. Plant Pathology 48: 727-734

Akhtar, K.P., Hussain, M., Khan, A.I., Haq, M.A. and Iqbal, N.M. 2004. Influence of plant age, whitefly population and cultivar resistance on infection of cotton plants by cotton leaf curl virus (ClcuV). Field Crops Res. 86: 15-21

Alyokhin, A. and Sewell, G. 2003. On-soil movement and plant colonization by walking wingless morphs of three aphid species (Homoptera: Aphididae) in greenhouse arenas. Environmental Entomol. 32: 1393-1398

Ammar, E.D. 1975. Biology of the leafhopper *Cicadulina chinai* Ghauri (Homoptera, Cicadellidae) in Giza, Egypt. Z. Angew. Entomol. 79: 337

Ammar, E.D. 1977. Biology of *Cicdulina bipunctella zeae* China in Giza, Egypt. Dtsch. Entomol. Z. 24: 345

Ammar, E.D. 1983. Virus diseases of sugarcane and maize in Egypt. In: Proc. Intl. Maize Virus Disease Colloquium and Workshop. D.T. Gordon, J.K. Knoke, L.R. Nault and R.M. Ritter (eds). Ohio Agricultural Research and Development Center. Wooster, OH, USA. pp. 122-126

Anonymous. 2006. Global Change Update. The National Academics. National Academy of Sciences, 500 Fifth Street, NW, Washington DC, USA

Apablaza, G., Apablaza, J., Reyes, P. and Moya, E. 2003. Determination of viral diseases and insect vectors on weeds adjacent to fields of vegetable crops. Ciencia e Investigacion Agraria 30: 175-186

Aranda, M.A., Fraile, A. and Gracía-Arenal, F. 1993. Genetic variability and evolution of the satellite RNA of cucumber mosaic virus during natural epidemics. J. Virol. 67: 5896-5901

Asanzi, C.M., Bosque-Pérez, N.A., Budenhagen, I.W., Gordon, D.T. and Nault, L.R. 1994. Interactions among maize streak virus disease, leafhopper vector populations and maize cultivars in forest and savanna zones of Nigeria. Plant Pathol. 43: 145-157

Asanzi, M.C. 1991. Studies of epidemiology of maize streak virus and its Cycadulina leafhopper vectors in Nigeria. PhD Thesis. The Ohio State University, OH, USA

Asanzi, M.C., Bosque-Pérez, N.A., Nault, L.R., Nault, L.R., Gordon, D.T. and Thottapilly, G. 1995. Biology of *Cicadulina* spp. (Homoptera: Cicadellidae) and transmission of maize streak virus. African Entomol. 3: 173-179

Ascenio-Ibáñez, J.T., Diaz-Plaza, R., Médez-Lozano, J., Monsalve-Fonnegra, Z.I., Argüello-Astorga, G.R. and Rivera-Bustamante, R.F. 1999. First report of tomato yellow leaf curl geminivirus in Yucatán, Mexico. Plant Dis. 83: 1178

Autrey, L.J.C. and Ricaud, C. 1983. The Comparative Epidemiology of Two Diseases of Maize Caused by Leafhopper-Borne Viruses in Mauritius. In: Plant Virus Epidemiology, R.T. Plumb and J.M. Thresh (eds). Blackwell Scientific Publ., Oxford, UK. pp. 277-286

Azzam, O. and Chancellor, T.C.B. 2002. The biology, epidemiology and management of rice tungro disease in Asia. Plant Disease 86: 88-100

Baker, R.H.A., Sansford, C.E., Jarvis, C.H., Cannon, R.J.C., Macleod, A. and Walters, K.F.A. 2000. The role of climatic mapping in predicting the potential geographical distribution and non-indigenous pests under current and future climates. Agriculture, Ecosystems and Environment 82: 57-71

Baldacci, E., Assici, A., Bello, G. and Bella, L. 1970. Research and symptomatology and etiology of rice quallume (yellow). Riso 19: 3-9

Bar-Joseph, M. and Lobenstein, G. 1973. Effects os strain, source of plant and temperature on the transmissibility of citrus tristeza virus by the melon aphid. Phytopathology 63: 716-720

Bar-Joseph, M. and Nitzan, Y. 1991. The spread and distribution of a severe citrus tristeza virus in sour orange seedlings. In: Proc. 11th Conf. Int. Organ. Citrus Virol. R.H. Brlansky, R.F. Lee and L.W. Timmer (eds). IOCV, Riverside, CA, USA. pp. 162-165.

Bar-Joseph, M., Garnsey, S.M. and Gonsalves, D. 1979. The Closteroviruses: a distinct group of elongated plant viruses. Adv. Virus Res. 25: 93-168

Bar-Joseph, M., Roistacher, C.N., Garnsey, S.M. and Gumpf, D.J. 1981. A review on tristeza, an ongoing threat to citriculture. Proc. 1st Int. Soc. Citric. ISC, Okitsu, Japan. pp. 419-423

Bar-Joseph, M., Roistacher, C.N. and Garnsey, S.M. 1985. The epidemiology and control of citrus tristeza virus. In: Plant Virus Epidemiology. R.T. Plumb and J.M. Thresh (eds). Blackwell Scientific Publications, Oxford, UK. pp. 61-72

Bar-Joseph, M., Marcus, R. and Lee, R.F. 1989. The continuous challenge of citrus tristeza virus conrol. Ann. Rev. Phytopathol. 27: 291-316

Beebee, T. and Rowe, G. 2004. An Introduction to Molecular Ecology. Oxford Univ. Press, Oxford, UK. 346 pp.

Ben Zeev, I.S., Bar-Joseph, M., Nitzan, Y. and Marcus, R. 1989. A severe citrus tristeza virus isolate causing the collapse of trees of sour orange before virus is detectable throughout the canopy. Ann. Appl. Biol. 114: 292

Bennett, C.W. 1979. The curly top disease of sugar beet and other crops. Monograph No. 7. The Am. Phytopathol. Soc.. St Paul, Minnesota, USA

Bennett, C.W. and Wallace, H.E. 1938. Relation of the curly top virus to the vector *Eutettix tenellus*. J. Agric. Res. 56: 31

Bennett, C.W. and Tarnisever, A. 1957. Sugar beet curly top disease in Turkey. Plant. Dis. Reptr. 41: 721

Bigirwa, G., Gibson, R.W., Page, W.W., Hakiza, J.J., Kyetere, D.T., Kalule, T.M. and Baguma, S.D. 1995. A new maize disorder in Uganda caused by *Cicadulina niger*. In: Maize Research for Stress Environments. D.C. Jewell, S.R. Waddington, J.K. Ransom and K.V. Pixley (eds). Proceedings of the Fourth Eastern and Southern Africa Regional Maize Conference, Harare, Zimbabwe, CIMMYT, Mexico, DF. pp. 202-204.

Bock, K.R. 1974. Maize streak virus description of plant viruses No. 133. Commonwealth Mycological Institute/Association of Applied Biologists. Kew, UK, p. 4

Bonnet, J., Fraile, A., Sacristan, S., Malpica, J.M. and García-Arenal, F. 2004. Role of recombination in the evolution of natural population of *Cucumber mosaic virus*, a tripartite RNA plant virus. Virology 332: 359-368

Bosco, D., Caciagli, P. and Norris, E. 1993. Indagini epidemiologiche sul virus dell'accartocciamento fogliare giallo del pomodoro (TYLCV) in Italia. Inf. Fitopatol 11: 33-36

Bosque-Pérez, N.A. 2000. Eight decades of maize streak virus research. Virus Research 71: 107-121

Bosque-Pérez, N.A. and Buddenhagen, I.W. 1999. Biology of *Cicadulina* leafhoppers and epidemiology of maize streak virus disease in West Africa. S. Afr. J. Plant Soil 16: 50-55

Bosque-Pérez, N.A., Olojede, S.O. and Buddenhagen, I.W. 1998. Effect of maize streak virus disease on the growth and yield of maize as influenced by varietal resistance levels and plant age at the time of challenge. Euphytica 101: 307-317

Bottenberg, H., Litsinger, J.A., Barrion, A.T. and Kenmore, P.E. 1990. Presence of tungro vectors and their natural enemies in different rice habitats in Malaysia. Agric. Ecosyst. Environ. 31: 1-15

Bowden, J., Mukhopadhyay, S., Nath, P.S., Sarkar, T.K., Sarkar, S. and Mukhopadhyay, S. 1988. Analysis of light-trap catches of *Nephotettix* species (Hemiptera: Cycadellidae) in West Bengal. Indian J. Agric. Sci. 58: 125-130

Brlansky, R.H., Pelosi, R.R., Garnsey, S.M., Youtsey, C.O., Lee, R.F., Yokomi, R.K and Sonoda, R.M. 1986. Tristeza quick decline in South Florida. Proc. Fla. State Hort. Soc. 99: 66-69

Brown, J.K. 1990. An update on the whitefly-transmitted geminiviruses in the Americas and the Caribbean basin. FAO Plant Protection Bulletin 39: 5-17

Brown, J.K. 2000. Molecular markers for the identification and global tracking of whitefly vector-*Begomovirus* complexes. Virus Research 71: 233-260

Brown, J.K.M. and Hovmoller, M. 2002. Aerial dispersal of pathogens on the global and continental scales and its impacts on plant disease. Science 297: 537-541

Brown, J.K. and Idris, A.M. 2006. Introduction of the exotic monopartite Tomato yellow leaf curl virus into West Coast, Mexico. Plant Dis. 90: 1360

Burdon, J.J. 1987. Diseases and Population Biology. Cambridge University Press, New York, USA

Burges, A.J., Harington, R. and Plumb, R.T. 1999. Barley and cereal yellow dwarf virus epidemiology and control strategies. In: The Luteoviridae. H.G. Smith and H. Barker (eds). CABI, Wallingford, UK. pp. 249-279

Cabunagan, R.C., Castilla, N., Coloquio, E.L., Tiongco, E.R., Truong, X.H., Fernandes, J., Du, M.J., Zaragosa, B., Hozak, R.R., Savary, S. and Azzam, O. 2001. Synchrony of planting and proportions of susceptible varieties affect rice tungro disease epidemics in the Philippines. Crop Protection 20: 499-510

Cabauatan, P.Q. and Hibino, H. 1988. Isolation, Purification and Serology of Rice Tungro Bacilliform and Rice Tungro Spherical Viruses. Plant Dis. 72: 526

Calavan, E.C., Pratt, R.M., Lee, B.N., Hill, J.P. and Blue, R.L. 1972. Tristeza susceptibility of sweet orange on Troyer citrange rootstock. Proc. 5th Conf. IOCV. W.C. Price (ed). Univ. Florida Press, Gainesville, USA. pp. 146-153

Calavan, E.C., Harjung, M.K., Blue, R.L., Roistacher, C.N., Gumpf, D.J. and Moore, P.W. 1980. Natural spread of seedling yellows and sweet orange and grapefruit stem pitting tristeza viruses at the University of California, Riverside. In: Proc. 8th Conf. Int. Organ. Citrus Virol. E.C. Calavan, S.M. Garnsey and L.W. Timmer (eds). Univ. Calif. Press, Gainesville, USA. pp. 69-75

Cambra, M., Gorris, M.T., Marroquin, M.P., Román, A., Olmos, M.C., Martinez, A.H., de Mendoza, A., López, L. and Navarro, L. 2000. Incidence and epidemiology of *Citrus tristeza virus* in the Valencian Community of Spain. Virus Res. 71: 85-95

Campbell, R.N., Wipf-Scheibel, C. and Lacoq, H. 1996. Vector-assisted seed transmission of melon necrotic spot virus in melon. Phytopathology 86: 1294-1298

Caroll, T.W. and Chapman, S.R. 1970. Variation in embryo infection and seed transmission by barley stripe mosaic virus within and between two cultivars of barley. Phytopathology 60: 1079-1081

Catara, A. and Davino, N. 2006. Citrus tristeza virus in citrus in Sicily. Rivista di Frutticoltura e-di-Ortofloricoltura 68: 18: 20-22

CCIRG (Climate Change Impacts Review Group). 1991. The Potential Effects of Climate Change in the United Kingdom. HMSO, London, UK

CCIRG (Climate Change Impacts Review Group). 1996. Review of the Potential Effects of Climate Change in the United Kingdom. HMSO, London, UK

Chakraborty, S. 2005. Potential impact of climate change on plant-pathogen interactions. Aus. Plant Path. 34: 443-448

Chakraborty, S., von Tiedemann, A. and Teng, P.S. 2000. Climate change and air pollution: potential impact on plant diseases. Environmental Pollution 108: 317-326

Chamberlain, E.E., Atkinson, J.D. and Hunter, J.A. 1951. Plum mosaic, a virus disease of plums, peaches and apricots in New Zealand. N. Z. J. Sci. Technol. 32: 1-16

Chan, M.S. and Jeger, M.J. 1994. An analytical model of plant virus disease dynamics with rouging and replanting. J. Appl. Ecol. 31: 413-427

Chancellor, T.C.B., Heong, K.L., Cook, A.G. and Villareal, S. 1996. Leafhopper ecology and rice tungro disease dynamics. In: Rice Tungro Disease Epidemiology and Vector Ecology. T.C.B. Chancellor, P.S. Teng and K.L. Heong (eds). NRI, University of Greenwich, UK; IRRI. Manila, Philippines. pp. 10-35

Chancellor, T.C.B. Cook, A.G., Heong, K.L. and Villareal, S. 1997. The flight activity and infection of major leafhopper vectors (Hemiptera: Cicadellidae) of rice tungro viruses in an irrigated rice area in the Philippines. Bull. Ent. Res. 87: 247-258

Chaussalet, T.J., Mann, J.A., Perry, J.N. and Francos- Rodriguez, J.C. 2001. A nearest neighbor approach to the simulation of spread of barley yellow dwarf virus. Computers and Electronics in Agriculture 28: 51-65

Cheddadi, R., Yu, G., Guuiot, J., Harrison, S.P. and Prentice, I.C. 1996. Climate Dynamics 13: 1-9. doi: 10.1111j.1365-3059.2004.01044.x

Chen, W., Provart, N.J., Glazebrook, J., Katagiri, F., Chang, H.S., Eulgem, T., Mauch, F., Luan, S., Zou, G., Whitham, S.A., Budworth, P.R., Tao, Y., Xie, Z., Chen, X., Lam, S., Kreps, J.A., Harper, J.F., Si-Ammour, A., Mauch-Mani, B., Heinlein, M., Kobayashi, K., Hohn, T., Dangi, J.L., Wang, X. and Zhu, T. 2002. Expression profile matrix of *Arabidiopsis* transcription factor genes suggests their putative functions in response to environmental stresses. The Plant Cell 14: 559-574

Coakley, S.M. and Scherm, H. 1996.Plant disease in a changing global environment. Aspects of Applied Biology 45: 227-238

Coakley, S.M., Scherm, H. and Chakraborty, S. 1999. Climate change and plant disease management. Annu. Rev. Phytopathology 37: 399-426

Cohen, S. and Antignus, Y. 1994. Tomato yellow leaf curl virus (TYLCV), a whitefly-borne geminivirus of tomatoes. Adv. Dis. Vector Res. 10: 259-288

Cohen, S., Kern, J., Harpaz, I. and Ben-Joseph, R. 1988. Epidemiological studies of the tomato yellow leaf curl virus (TYLCV) in the Jordan Valley, Israel. Phytoparasitica 16: 259-270

Cook, A.G. and Perfect, T.J. 1989. Population dynamics of three leafhopper vectors of rice tungro viruses, *Nephotettix virescens* (Distant), *N. nigropictus* (Stål) and *Recilia dorsalis* (Motschulsky) (Hemiptera: Cycadellidae) in farmers' fields in the Philippines. Bull. Entomol. Res. 79: 437-451

Cooter, R.J. and Winder, D. 1996. The flight behavior and dispersal range of the leafhopper vectors of rice tungro disease. In: Rice Tungro Disease Epidemiology and Vector Ecology. T.C.B. Chancellor, P.S. Teng and K.L. Heong (eds). NRI, University of Greenwich, UK; IRRI. Manila, Philippines. pp. 1-9.

Cortesi, P. and Milgroom, M.G. 2001. Out-crossing and diversity of vegetative compatibility types in population of *Eutypa lata* from grapevines. J. Plant Pathology 83: 79-86

Cortesi, P., Fischer, M. and Milgroom, M.G. 2000. Identification and spread of *Fomitiporia punctata* associated with wood decay of grapevine showing symptoms of Esca. Phytopathology 90: 967-972

Costa, A.S. and Grant, T.J. 1951. Studies on the transmission of the tristeza virus by the vector *Aphis citricidus*. Phytopathology 41: 105

Creamer, R., Hubble, H. and Lewis, A. 2005. Curtovirus infection of Chile pepper in New Mexico. Plant Disease 89: 480-486

Czosnek, H. and Laterrot, H. 1997. A world-wide survey of tomato yellow leaf curl viruses. Arch. Virol. 142: 1391-1406

DaGraca, J.V., Marais, L.J. and Von Broembsen, L.A. 1982. Severe tristeza stem pitting in young grapefruit. Citrus and Subtropical Fruit J. Nov. 1982: 18

Dasgupta, I., Das, B.K., Nath, P.S., Mukhopadhyay, S., Niazi, F.R. and Varma, A. 1996. Detection of rice tungro bacilliform virus in field and greenhouse samples from India using polymerase chain reaction. J. Virol. Methods 58: 53-58

Davino, M. and Catara, A. 1986. La tristeza degli agrumi, Inf. Fitopathologico 36: 9

Davino, M., D'Urso, F., Areddia, R., Carbone, M. and Mauromicale, G. 1994. Investigations on the epidemiology of tomato yellow leaf curl virus (TYLCV) in Sicily. Petria 4: 151-160

de Assis, F.M. and Sherwood, J.L. 2000. Evaluation of seed transmission of turnip yellow mosaic virus and tobacco mosaic virus in *Arabidopsis thaliana*. Phytopathology 90: 1233-1238

De Zoeten, G.A. and Skaf, J.S. 2001. Pea enation mosaic and the vagaries of a plant virus. Adv. Virus Res. 57: 323-350

Dedryver, C.A. and Robert, Y. 1981. Ecological role of maize and cereal volunteers as reservoirs of Gramineae virus transmitting aphids. Proc. 3rd Conf. on Virus Diseases of Gramineae, Rothamsted Experimental Station, Herpenden, UK. pp. 61-66

de Orso, F., Ayllon, M.A., Rubio, L., Sambada, A., Mendoza, A.H., de Guerri, J., Moreno, P. and de Mandoza, A.H. 2000. Contribution of uneven distribution of genomic RNA variants of citrus tristeza virus (CTV) with the plant to changes in the viral population following aphid transmission. Plant Pathology 49: 288-294

Dickson, R.C., Johnson, M.M., Flock, R.A. and Laird, E.F. Jr. 1956. Flying aphid population in Southern California citrus groves and their relation to the transmission of the tristeza virus. Phytopathology 46: 204-210

Doodson, J.K. 1967. A survey of barley yellow dwarf virus in s. 24 perennial ryegrass in England and Wales, 1966. Plant Pathology 16: 42-45.

Duffus, J.E. 1983. Epidemiology and control of curly top diseases of sugar beet and other crops. In: Plant Virus Epidemiology. R.T. Plumb and J.M. Thresh (eds). Blackwell Scientific Publications, Oxford, UK. pp. 297-304

Duke, J.S. and Mooney, H.A. 1999. Does global change increase the success of biological invaders? Trends in Ecology and Evolution 14: 135-139

Dusi, A.N., Peters, D. and Werf, W. vander. 2000. Measuring and modeling the effects of inoculation date and aphid flights on the secondary spread of beet mosaic virus in sugar beet. Annals Appl. Biol. 136: 131-146

Emanuel, K. 2005. Increasing destructiveness of tropical cyclones over the past 30 years. Nature 436: 6866-688

Escriu, F., Fraile, A. and Garcia-Arenal, F. 2000a. Evolution of virulence in natural populations of the satellite RNA of Cucumber mosaic virus. Phytopathology 90: 480-485

Escriu, F., Perry, K.L. and Garcia-Arenal, F. 2000b . Transmissibility of Cucumber mosaic virus by *Aphis gossypii* correlates with viral accumulation and is affected by the presence of its satellite RNA. Phytopathology 90: 1068-1072

Escriu, F., Fraile, A. and García-Arenal, F. 2003. The evolution of virulence in a plant virus. Evolution 57: 755-765

Fabre, F., Plantegenest, M., Mieuzet, L., Dedryver, C.A., Leterrier, J.-L., and Jacquot, E. 2005. Effects of climate and land use on the occurrence of viruliferous aphids and the epidemiology of barley yellow dwarf disease. Agriculture, Ecosystems and Environment 106: 49-55

Faccini, N., Delogu, G., Stanca, A.M. and Vale, G. 2006. Barley: fungal and viral diseases in the Italian cultivation areas. Informatore Fitopatologico 56: 13-19

Fajemisin, J.M., Cook, G.E., Okusanya, F. and Shoyinka, S.A. 1976. Maize streak virus epiphytotic in Nigeria. Plant Dis. Rep. 60: 443-447

Fargette, D., Konaté, G., Fauquet, C., Muller, E., Peterschmitt, M. and Thresh, J.M. 2006. Molecular ecology and emergence of tropical plant viruses. Annu. Rev. Phytopathol. 44: 233-259

Fauquet, C.M., Sawyer, S., Idris, A.M. and Brown, J.K. 2005. Sequence analysis and classification of apparent recombinant begomoviruses infecting tomato in Nile and Mediterranean basins. Phytopathology 95: 549-555

Feder, M.E. and Mitchell-Olds, T. 2003. Evolutionary and ecological functional genomics. Nature Reviews Genetics 4: 651-665

Fiebig, M. and Poehling, H.M. 1998. Host-plant selection and population dynamics of the grain aphid *Sitobion avenae* (F.) on wheat infected with Barley yellow dwarf virus. In: Integrated Control of Cereal Crops—Meeting held at Lleida, Spain, 13-14 March 1997. Bull. OILB-SROP. 21: 51-62

Fitter, A.H. and Fitter, R.S.R. 2002. Rapid changes in flowering time in British plants. Science 296: 1689-1691

Foster, G., Blake, S., Tones, S., Barker, I., Harrington, R., Taylor, M., Walters, K., Northing, P. and Morgan, D. 2004. Decision support for BYDV control in the United Kingdom: can a regional forecast be made field specific? Aphids in a new millennium: Proc. 6th Int. Symp. on Aphids, September 2001, Rennes, France, pp. 287-291

Fraile, A., Escrui, F., Aranda, M.A., Malpica, J.M., Gibbs, A.J. and Gracía-Arenal, F. 1997. A century of tobacco mosaic virus evolution in an Australian population of *Nicotiana glauca*. J. Virol. 71: 8316-8320

Froissart, R., Wilke, C.O., Montville, R., Remold, S.K., Chao, L. and Turner, P.E. 2004. Co-infection weakens selection against epistatic mutations in RNA viruses. Genetics 168: 9-19

Frosheiser, F.I. 1974. Alfalfa mosaic virus transmission to seed through alfalfa gametes and longevity in alfalfa seed. Phytopathology 64: 102-105

Gamez, R. and Leon, P. 1988. Maize rayado fin and related viruses. In: The Plant Viruse. R. Koenig (ed). Plenum, New York, USA. pp. 213-233

García- Arenal, F. and McDonald, B.A. 2003. An analysis of the durability of the resistance to plant viruses. Phytopathology 93: 941-952

García-Arenal, F., Esciru, F., Aranda, M.A., Alanso-Prados, José, L., Malpica José, M. and Fraile, A. 2000. Molecular epidemiology of *Cucumber mosaic virus* and its satellite RNA. Virus Res. 71: 1-8

García-Arenal, F., Fraile, A. and Malpica, J.M. 2003.Variation and evolution of plant virus populations. International Microbiology 6: 225-232

Garrett, K.A., Hulbert, S.H., Leach, J.E. and Travers, S.E. 2006a. Ecological genomics and epidemiology. Euro. J. Plant Pathology 115: 35-51

Garrett, K.A., Dendy, S.P., Frank, E.E., Rouse, M.N. and Travers, S.E. 2006b. Climate Change: Effects on Plant Diseases: Genomes to Ecosystem. Annu. Rev. Phytopathology 44: 20. 20.21

Geraud, F. 1976. El afido Negro de los citricos, *Toxoptera citricida* Kirkaldy en Venezuela (Resumen), 1. Encuentro Venezolano and Entomologia. UCV Fac. Agronomia Inst. Zool. Agric. Maracay, Venezuela

Ghanim, M. and Czosnek, H. 2000. Tomato yellow leaf curl geminivirus (TYLCV-Is) is transmitted among whiteflies (*Bemisia tabaci*) in a sex-related manner. J. Virol. 74: 4735-4745

Ghanim, M., Morin, S., Zeidan, M. and Czosnek, H.G. 1998. Evidence for transovarial transmission of tomato yellow leaf curl virus by its vector, the whitefly *Bemisia tabaci*. Virology 240: 295-303

Gibson, K.E. 1967. Possible incidence of curly top in Iran—A new record. Plant Dis. Reptr. 51: 976

Gill, C.C. and Chong, J. 1979. Cytopathological Evidence for the Division of Barley Yellow Dwarf Virus Isolates into Two Subgroups. Virology 95: 59

Glazebrook, J., Chen, W., Etes, B., Chang, H.S., Nawrarath, C., Métraux, J.P., Zhu, T. and Katagiri, F. 2003. Topology of the network integrating salicylate and jasmonate signal transduction derived from global expression phenotyping. The Plant J. 34: 217-228

Gorter, G.J.M.A. 1953. Studies and the Spread and Control of the Streak Disease of Maize. Union S. Afr. Dep. Agric. For. Sci. Bull. 341: 20

Gravilets, S. and Gibson, N. 2002. Fixation probabilities in a spatially heterogeneous environment. Popul. Ecol. 44: 51-58

Groves, R.L., Walgenbach, J.F., Moyer, J.W. and Kennedy, G.G. 2001. Over-wintering of *Frankliniella fusca* (Thysanoptera: Thripidae) on winter annual weeds infected with tomato spotted wilt virus and patterns of virus movement between susceptible weed hosts. Phytopathology 91: 891-899

Groves, R.L., Walgenbach, J.F., Moyer, J.W. and Kennedy, G.G. 2002. The role of weed hosts and tobacco thrips, *Frankliniella fusca* in the epidemiology of tomato spotted wilt virus. Plant Disease 86: 573-582

Groves, R.L., Walgenbach, J.F., Moyer, J.W. and Kennedy, G.G. 2003. Seasonal dispersal patterns of *Frankliniella fusca* (Thysanoptera: Thripidae) and tomato spotted wilt virus occurrence in Central and Eastern North Carolina. J. Econ. Entomol. 96: 1-11

Gudelj, I., van den Bosch, F. and Gilligan, C.A. 2004. Transmission rates and adaptive evolution of pathogens in sympatric heterogeneous plant populations. Proc. R. Soc. Lond. B 271: 2187-2194

Gupta, K.N. and Keshwal, R.L. 2003. Epidemiological studies on yellow mosaic disease of soybean. Annals. Pl. Proc. Sci. 11: 324-328

Hameed, S. and Robinson, D.J. 2004. Begomoviruses from mung beans in Pakistan: Epitope profiles, DNA A sequences and phylogenetic relationships. Arch. Virol. 149: 809-819

Hamilton, R.I., Leung, E. and Nichols, C. 1977. Surface contamination of of pollen by plant viruses. Phytopathology 67: 395-399

Hammond, J. 1982. Plantago as a host of economically important viruses. Adv. Virus Res. 27: 103-140

Hammond, J., Lecoq, H. and Raccha, B. 1999. Epidemiological risks for mixed virus infections and transgenic plants expressing viral genes. Adv. Virus Res. 54: 189-314

Hang, A.N., Niklas, P.N., Silbernagel, M.J. and Hosfield, G.L. 2006. Registration of 'Quincy' pinto bean. Crop Science 46: 991

Harrington, R. 2002. Insect pests and global environmental change. In: Encyclopedia of Global Environmental Change Vol. 3. I. Dougglas (ed). Wiley, Chichester, UK. pp. 381-386

Harrington, R. 2003. Turning up the heat on pests and diseases: A case study for Barley yellow dwarf virus. Proc. British Crop Protection Council International Congress. Vol. 2, Glasgow, UK. pp. 1195-1200

Harrison, B.D. and Robinson, D.J. 1999. Natural genomic antigenic variation in whitefly-transmitted geminiviruses (begomoviruses). Annu. Rev. Phytopathol. 37: 369-398

Harrison, B.D. 2002. Virus variation in relation to resistance breaking in plants. Euphytica 124: 181-192

Henry, M. and Dedryver, C.A. 1989. Fluctuations in cereal aphid populations on maize (*Zea mays*) in Western France in relation to the epidemiology of barley yellow dwarf virus. J. Appl. Entomol. 107: 401-410

Henry, M. and Dedryver, C.A. 1991. Occurrence of barley yellow dwarf virus in pastures of western France. Plant Pathol. 40: 93-99

Henry, M., George, S., Arnold, G.M., Dedryver, C.A., Kendall, D.A., Robert, Y. and Smith, B.D. 1993. Occurrence of barley yellow dwarf virus (BYDV) isolates in different habitats in western France and south-west England. Ann. Appl. Biol. 123: 315-329

Hermoso de Mendoza, A., Bellester-Olmos, J.F. and Pina, J.A. 1988. Differences in transmission efficiency of *Citrus* tristeza virus by *Aphis gossypii* using sweet orange, mandarin, and lemon trees as donor or receptor host plants (1988). Proc 10th Conf. IOCV. L.W. Timmer, S.M. Garnsey and L. Navarro (eds). Univ. Calif., Riverside, CA, USA. pp. 62-64

Herrera, G.M. and Plumb, R.T. 1991. Effects of MAV-, PAV-, and RPV-like isolates of BYDV on spring and winter barley cultivars. Acta Phytopathologica Hungarica 26: 41-45

Hibino, H. 2004. Tungro virus disease and its management. Recent Research Development in Crop Science. Vol. 1, Part II: 273-293

Hilf, M.E., Karasev, A.V., Albiach-Marti, M.R., Dowson, W.O. and Garnsey, S.M. 1999. Two paths of sequence divergence in *Citrus* tristeza virus complex. Phytopathology 89: 336-342

Holt, J. and Chancellor, T.C.B. 1996. Simulation modeling of the spread of rice tungro virus disease: the potential for management by rouging. J. Appl. Ecol. 33: 927-936

Holt, J. and Chancellor, T.C.B. 1997. A model of plant virus disease epidemics in asynchronously-planted cropping systems. Plant Pathology 46: 490-501

Holt, J. and Chancellor, T.C.B. 1999. Modeling the spatio-temporal development of resistant varieties to reduce the incidence of rice tungro disease in a dynamic cropping system. Plant Pathology 48: 453-461

Holt, J., Jeger, M.J., Thresh, J.M. and Otim-Nape, G.W. 1997. A model of plant disease epidemiology with vector population dynamics applied to African cassava mosaic virus. J. Appl. Ecol. 34: 793-806

Houghton, J.T., Meiro-Fihlo, I.G., Callander, B.A, Harris, N., Kattenberg, A. and Maskell, K. 1996. Climate Change 1995: The Science of Climate Change. Cambridge University Press, Cambridge, UK

Houghton, J.T., Ding, Y., Griggs, D.J., Noguer, M., van der Linden, P.J., Xiaosu, D., Maskell, K. and Johnson, C.A. 2001. Climate Change 2001: The Scientific Basis. Cambridge University Press, Cambridge, UK.

Hull, R. 2002. Matthews's Plant Virology. Academic Press, London, UK.

Hulme, M. and Jenkins, G.J. 1998. Climate Change Scenarios for the United Kingdom: Scientific Report. UKCIP Technical Report No. 1. Climatic Research Unit, Norwich, UK

Ilbag, H. 2006. Common reed (*Phragmites communis*) is a natural host of important cereal viruses in the Trakya Region of Turkey. Phytoparasitica 34: 441-448

IPCC (Intergovernmental Panel on Climate Change). 2001. The Scientific Basis. Cambridge University Press, Cambridge, UK

Irwin, M.E. 1999. Implications of movement in developing and deploying integrated pest management strategies. In: Special Issue: Aerial dispersal of pests and pathogens. Agriculture and Forest Meteorology 97: 235-248

Irwin, M.E. and Thresh, J.M. 1990. Epidemiology of barley yellow dwarf virus: a case study in ecological complexity. Annu. Rev. Phytopathol. 28: 393-424

Jedlinski, H. 1981. Rice root aphid, *Ropalosiphum rufialominalis*, a vector of barley yellow dwarf virus in Illinois, and the disease complex. Plant Disease 65: 975-978

Jeffery, S., Bal, E., Gregory, J., Master, S., Hodkinson, Ian, D., Awmack, C., Benzemer, T.M., Brown, V.K., Butterfield, J., Buse, A., Coulson, J.C., Farrar, J., Good, J.E.G., Harrington, R., Hartley, S., Jones, T.H., Lindroth, R.L., Press, M.C., Symrnioudis, I., Watt, A.D. and Whittaker, J.B. 2002. Herbivory in global climate change research: direct effects of rising temperature on insect herbivores. Global Change Biology 8: 1-16

Jeger, M.J., van den Bosch, F., Madden, L.V. and Holt, J. 1998. A model for analyzing plant virus transmission—characteristics and epidemic development. IMA Journal of Mathematics Applied in Medicine and Biology 15: 1-18

Jeger, M.J., Dutmer, M.Y. and van den Bosch, F. 2002. Modeling plant virus epidemics in a combined field-nursery system. IMA Journal of Mathematics Applied in Medicine and Biology 19: 1-20

Jeger, M.J., Holt, J., van Den Bosch, F. and Madden, L.V. 2004. Epidemiology of insect-transmitted plant viruses: modeling disease dynamics and control interventions. Physiological Entomology 29(3): 291-304

Jeger, M.J., Seal, S.E. and van den Bosch, F. 2006. Evolutionary epidemiology of plant virus disease. Adv. Virus Res. 67: 163-203

Jiang, Y.X., Blas, C., de Bedford, I.D., Nombela, G. and Muniz, M. 2004. Effects of *Bemisia tabaci* biotype on transmission of tomato yellow leaf curl Sardinia virus (TYLCV-ES) between tomato common weeds. Spanish J. Agric. Res. 2(1): 115-119

Jiménez-Martinez, E.S., Bosque-Pérez, N.A., Berger, P.H., Zemetra, R.S., Ding, H. and Eigenbrode, S.D. 2004. Volatile cues influence the response of *Rhopalosiphum padi* (Homoptera: Aphididae) to barley yellow dwarf virus infected transgenic and untransformed wheat. Environ. Entomol. 33: 1207-1216

Johnson, R.A. and Rochow, W.F. 1972. An Isolate of Barley Yellow Dwarf Virus Transmitted specifically by *Schizaphis graminum*. Phytopathology 62: 921

Jovel, J., Hilje, L., Kleinn, C., Cartin, V. and Valver de, B. 2000. Daily movement of *Bemisia tabaci* in tomato plots in Turrialba, Costa Rica. Manejo Integrated de Plagas No. 55: 49-55

Jurik, M., Mucha, V. and Vorosova, T. 1987. The effects of and some other factors of retention of non-persistent viruses by aphids. Biologia Bratislava 42: 315-318

Kaplan, N., Hudson, R.R. and Iizuka, M. 1991. The coalescent process in models with selection, recombination and geographic subdivision. Genet. Res. 57: 83-91

Karsev, A.V., Boyko, V.P., Gowda, S., Nikolaeva, O.V., Hilf, M.E., Koonin, E.V., Niblett, C.L., Cline, K., Gumpf, D.J., Lee, R.F., Garnsey, S.M. and Dawson, W.O. 1995. Complete sequence of citrus tristeza virus RNA genome. Virology 208: 511-520

Kato, K., Onuki, M., Fuji, S. and Hanada, K. 1998. The first occurrence of tomato yellow leaf curl virus in tomato (*Lycopersicon esculentum* Mill.) in Japan. Ann. Phytopathol. Soc. Jpn. 64: 552-559

Kazinczi, G., Horvath, J. and Takacs, A. 2001. Role of weeds in the epidemiology of viruses. Proc. 5th Slovenian Conf. Plant Protection, Cetez, ob Savi, Slovania 6-8 March Zhornik Predavanj in refeatov 5: 222-226

Kearney, C.M., Thomson, M.J. and Roland, K.E. 1999. Genome evolution of tobacco mosaic virus populations during long-term passaging in a diverse range of hosts. Arch. Virol. 144: 1513-1526

Keldish, M., Pomazkov, Y., Arushanova, E., Chervyakov, O. and Hadidi, A. 1998. Vectors as epidemiological factors in widening the host range for viruses of fruit trees and small fruit crops. Acta Horticulturae No. 472: 147-152

Kelley, S.E. 1994. Viral pathogens and the advantage of sex in the perennial grass *Anthoxanthum audoratum*. Phil. Trans. R. Soc. Lond. B 346: 295-302

Kerk, N.M., Ceserani, T., Tausta, S.L., Sussex, I.M. and Nelson, T.M. 2003. Laser capture micro-dissection of cells from plant tissues. Plant Physiology 132: 27-37

Kitazima, E. M. and Gamez, R. 1983. Electron microscopy of maize rayado fina virus in the internal organs of the leaf hopper vector. Intervirology 19: 129-134

Knoke, J.K., Gingery, R.E. and Louie, R. 1992. Maize dwarf mosaic, Maize chlorotic dwarf, and Maize streak. In: Plant Disease of International Importance. Vol. 1, U.S. Singh, A.N. Mukhopadhyay, J. Kumar and H.S. Chaube (eds). Prentice Hall, Englewood Cliff, New Jersey, USA. pp. 235-281

Kong, P., Rubio, L., Polek, M., Falk, B.W. and Kong, P. 2000. Population structure and genetic diversity within California *Citrus* tristeza virus (CTV) isolates. Virus Genes 21: 139-145

Kuchne, T. 2004. Activity of aphids associated with lettuce and broccoli in Spain and their efficiency as vectors of the lettuce mosaic virus. Virus Research 100: 83-88

Kuno, T. and Koizumi M. 1991. Separation of Citrus tristeza virus (CTV) serotypes through aphid transmission. Proc. 11th Conf. IOCV. R.H. Brlansky, R.F. Lee and L.W. Timmer (eds). Univ. Calif. Riverside, CA. USA. pp. 82-85

Kumari, S.G., Muharram, I., Makkouk, K.M., Al-Anasi, A., El-Pasha, R., Al-Motwkel, W.A. and Kassem A.H. 2006. Identification of viral diseases affecting barley and bread wheat crops in Yemen. Australian Plant Pathology 35: 563-568.

Latham, L.J., Jones, R.A.C. and McKirdy, S.J. 2004. Lettuce big vein disease: Sources, pattern of spread and losses. Aus. J. Agric. Res. 55: 125-130

Lastra, R., Meneses, R., Still, P.E. and Niblett, C.L. 1991. The citrus virus situation in Central America. In: Proc. 11th Conf. Int. Organ. Citrus Virol. R.H. Brlansky, R.F. Lee and L.W. Trimmer (eds). IOCV. Riverside, CA, USA. pp. 166-169

Lee, R.F. and Rocha-Peña, M.A. 1992. Citrus tristeza virus. In: Plant Disease of International Importance, Vol. 3. J. Kumar, H.C. Chaube, U.S. Singh and A.N. Mukhopadhyay (eds). Prentice Hall. Englewood Cliff, New Jersey, USA. pp. 226-249

Lewellen, R.T. 2006. Registration of C931, C941, CR 11, and CZ25/2 self- fertile, genetic-male-sterile facilitated, random-mated, sugar beet germplasm populations. Crop Science 46: 1412-1414

Lin, H.X., Rubio, L., Smythe, A.B and Falk, B.W. 2004. Molecular population genetics of *Cucumber mosaic virus* in California: Evidence for founder effects and re-assortment. J. Virol. 78: 6666-6675

Lindblad, M. and Sigvald, R. 2004. Temporal spread of wheat dellarf virus and mature plant resistance in winter wheat. Crop Protection 23: 229-234

Lindstan, K. 1964. Investigations on the occurrence and heterogeneity of barley yellow dwarf virus in Sweden. Lantbrukshögskolans Annaler 30: 581-600

Ling, K.C. 1972. Rice virus diseases. International Rice Research Institute, Los Banos, Philippines

Ling, K.C. 1975. Experimental Epidemiology of Rice Tungro Disease II. Effect of Virus Source on Disease Incidence. Philipp. Phytopathol. 11: 21

Ling, K.S., Simmons A.M., Hassell, R.L., Keinath, A.P. and Polston, J.E. 2006. First Report of *Tomato Yellow Leaf Curl Virus* in South Carolina. Plant Disease 90: 379

Ling, K.C. and Palomar, M.K. 1966. Studies on rice plants infected with the tungro virus at different ages. Philippine Agriculturist 50: 165-177

Ling, K.C. and Tionco, E.R. 1975. Effect of Temperature on the Transmission of Rice Tungro Virus by *Nephotettix virescens*. Philipp. Phytopathol. 11: 46

Ling, K.C., Tiongco, E.R. and Floes, Z.M. 1983. Epidemiological studies of rice tungro. In: Plant Virus Epidemiology, R.T. Plumb and J.M. Thresh (eds). Blackwell Scientific Publications, Oxford, UK. pp. 249-258

Lobell, D.B. and Asner, G.P. 2003. Climate and management contributions to recent trends in US agricultural yields. Science 299: 1032

Loebenstein, G. and Thottappilly, G. 2003. Virus and Virus-like Diseases of Major Crops in Developing Countries. Kluwer, Dordrecht, The Netherlands

Loevinsolin, M.E. 1984. The ecology and control of rice pests in relation to the intensity and synchrony of cultivation. PhD thesis, University of London, London, UK.

Lu, J., Liu, X.M., Zhang, C.Q. and Zhang, M.G. 1999. Studies on the epidemc of wheat rosette stunt in spring wheat in northeast China. China Acta Phytophyacica Sinica 26: 45-49

Lucio-Zavaleta, E., Smith, D.M. and Gray, S.M. 2001. Variation in transmission efficiency among Barley yellow dwarf virus-RMV isolates and clones of the normally inefficient aphid vector, *Rhopalosiphum padi*. Phytopathology 91: 792-796

Madden, L.V. and Campbell, C.L. 1986. Description of virus disease epidemics in time and space. In: Plant Virus Epidemics. G.P. McLean, R.G. Garrett and W.G. Ruensink (eds.). Academic Press, Orlando, Florida, USA. pp. 273-293

Madden, L.V., Jeger, M.J. and van den Bosch, F. 2000. A theoretical assessment of the effects of vector-virus-transmission mechanism on plant virus disease epidemics. Phytopathology 90: 576-594

Mahlman, J.D. 1997. Uncertainties in projections of human-caused climate warming. Science 278: 1416-1417

Maleck, K., Levine, A., Eulgem, T., Morgan, A., Schmid, J., Lawton, K.A., Dang, J.L. and Dietrich, R.A. 2000. The transcriptome of *Arabidopsis thaliana* during systemic acquired resistance. Nature Genetics 26: 403-410

Malpica, J.M., Fraile, A., Moreno, I., Obies, C.I., Drake, J.W. and Gracía-Arenal, F. 2002. The rate and character of spontaneous mutation in an RNA virus. Genetics 162: 1505-1511

Mansoor, S., Briddon, R.W., Zafar, Y. and Stanley, J. 2003a. Geminivirus disease complexes: an emerging threat. Trends in Pl. Sc. 8: 128-134

Mansoor, S., Amin, I., Iram, S., Hussain, M., Zafar, Y., Malik, K.A. and Briddon, R.W. 2003b. Breakdown of resistance in cotton to cotton leaf curl disease in Pakistan. Plant Pathology 52: 784

Marais, L.J. 1981. Trisreza stem pitting. The problem facing the South African industry. Is it caused by new strains? Citrus and Subtropical Fruit J., March 1981: 11

Marroquin, C., Olmos, A., Gorris, M.T., Bertolini, E., Martinez, M.C., Carbonell, E.A., Hermoso-de-Mendoza, A. and Cambra, M. 2004. Estimation of the number of aphids carrying Citrus tristeza virus that visit adult citrus trees. Virus Res. 100: 101-108

Martinez-Zubiaur, Y., Muniz-Martin, Y. and Quinones-Pantoja, M. 2006. A new begomovirus infecting pepper plants in Cuba. Plant Pathology 55: 817

Masterman, A.J., Holmes, S.J. and Foster, G.N. 1994. Transmission of barley yellow dwarf virus by cereal aphids collected from different habitats on cereal farms. Plant Pathol. 43: 612-620

Maule, A.J. 2000. Virus transmission seeds. In: Encyclopedia of Plant Pathology. O.C. Maloy and T.D. Murray (eds). John Wiley and Sons, NJ, USA, pp. 1168-1170

Mauricio, R. 2005. Can ecology help genomics: the genome as ecosystem? Genetica 123: 205-209

McKirdy, S.J. and Jones, R.A.C. 1994. Infection of alternate hosts associated with annual medics (*Medicago* spp.) by alfalfa mosaic virus and its persistence between growing seasons. Aust. J. Agric. Res. 45: 1413-1426

Mehner, S., Manurung, B., Guntzig, M., Habekuss, A., Witsack, W. and Fuchs, E. 2003. Investigations into the ecology of the wheat dwarf virus (WDV) in Sxony-Anhalt, Germany. Zeitschrift-fur-Pflanzenkr ankheiten und Pflanzenschutz 110: 313-323

Méndez-Lozano, J., Torres-Pacheco, I., Fauquet, C.M. and Rivera- Bustamante, R.F. 2003. Interactions between geminiviruses in a naturally occurring mixture: Pepper huastecovirus and Pepper golden mosaic virus. Phytopathology 93: 270-277

Mesfin, T., Bosque-Pérez, N.A., Buddenhagen, I.W., Tottappilly, G. and Olojede, S.O. 1992. Studies of maize streak virus isolates from grass and cereal hosts in Nigeria. Plant Dis. 76: 789-795

Milgroom, M.G. and Peever, T.L. 2003. Population biology of plant pathogens: the synthesis of plant disease epidemiology and population genetics. Plant Disease 87: 608-617

Miller, W.A., Liu, S. and Bockett, R. 2002. Barley yellow dwarf: Luteoviridae or Tombusviridae. Molecular Plant Pathology 3: 177-183

Mink, G.I. 1993. Pollen- and seed-transmitted viruses and viroids. Annu. Rev. Phytopathol. 31: 375-402

Mishra, M.D., Ghosh, A., Niazi, F.R., Basu, A.N. and Raychaudhuri, S.P. 1973. The Role of Germinaceous Weeds in the Perpetuation of Rice Tungro Virus. J. Indian Bot. Soc. 52: 176

Miyakawa, T. and Yamaguchi, Y. (1981). Tristeza stem pitting in *Citrus* diseases in Japan. Plant Protection Association. Komagome, Toshima, Tokyo, Japan

Mojtahadi, H., Boydston, R.A., Thomas, P.E., Crosslin, J.M., Santo, G.S., Riga, E. and Anderson, T.L. 2003. Weed hosts of *Paratrichodorus allius* and tobacco rattle virus in the Pacific Northwest. Am. J. Potato Res. 80: 379-365

Momol, M.T., Simone, G.W., Dankers, W., Sprenkel, R.K., Olson, S.M., Mimol, E.A., Polston, J.E. and Hiebert, E. 1999. First report of tomato yellow leaf curl virus in tomato in South Georgia. Plant Dis. 83: 487

Morilla, G., Krenz, B., Jesloe, H., Bejarano, E.R. and Wege, C. 2004. Tête à tête of Tomato yellow leaf curl virus and Tomato yellow leaf curl Sardinia virus in single nuclei. J. Virol. 78: 10715-10723

Morilla, G., Janssen, D., Garcia-Andres, S., Moriones, E., Cuadrodo, I.M. and Bejarano, E.R. 2005. Pepper (*Capsicum annuum*) is a dead-end host for Tomato yellow leafcurl virus. Phytopathology 95: 1089-1097

Morione, E. and Navas-Castillo, J. 2000. Tomato yellow leaf curl virus, an emerging virus complex causing epidemics worldwide. Virus Res. 71: 123-134

Moya, A., Holmes, E.C. and Gonzalez-Candelas, F. 2004. The population genetics and evolutionary epidemiology of RNA viruses. Nature Rev. Microbiol. 2: 279-288

Mukhopadhyay, A.N. 1992. Curly top of sugar beet. In: Plant Disease of International Importance, Vol. 4. A.N. Mukhopadhyay, J. Kumar, H.C. Chaube and U.S. Singh (eds). Prentice Hall, Englewood Cliff, New Jersey, USA. pp. 103-137

Mukhopadhyay, S. 1984. Ecology of rice tungro virus and its vectors. In: Ecology of Plant Viruses. A. Mishra and H. Polasa (eds). South East Asian Publishers, New Delhi, India

Mukhopadhyay, S. 1986. Virus Diseases of Rice in India. In: Vistas in Plant Pathology. A. Varma and J.P. Verma (eds). Malhotra Publishing House, New Delhi, India. pp. 111-122

Mukhopadhyay, S. 1992. Rice Tungro. In: Plant Disease of International Importance, Vol. 1. U.S. Singh, A.N. Mukhopadhyay, J. Kumar and H.C. Chaube (eds). Prentice Hall, Englewood Cliff, New Jersey, USA. pp. 186-200

Mukhopadhyay, S. and Mukhopadhyay, S. 1987. Lag correlation between the peak monsoon rains and peak appearance of rice green leafhoppers in West Bengal. Proc. Indian Nat. Sci. Acad. B 83(2): 189-191

Mukhopadhyay, S., Nath, P.S., Sarkar, T.K., Sarkar, S. and Mukhopadhyay, S. 1988. Interrelationship between rainfall and rice green leafhopper population in West Bengal. Indian J. Agric. Sci. 58(1): 34-38

Mukhopadhyay, S., Mukhopadhyay, S. and Sarkar, T.K. 1989. Effect of some meteorological factors on the incidence of rice green leafhopper. In: Meteorology and Plant Protection. V. Krishnamurthy and G. Mathys (eds). WMO, Geneva, Switzerland. pp. 168-188

Mukhopadhyay, S., Nath, P.S., Das, S., Basu, D. and Verma, A.K. 1994. Comparative Epidemiology and Cultural Control of Some Whitefly Transmitted Viruses in West Bengal. In: Virology in the Tropics. N. Rishi, K.L. Ahuja and B.P. Singh (eds). Malhotra Publishing House, New Delhi, India. pp. 285-304

Mukhopadhyay, S., Chowdhury, A.K., De, B.K, Nath, P.S. and Sarkar, T.K. 1995. Dynamics of rice green leafhoppers in relation to spread of Rice Tungro Disease, Proceedings of the Second Agricultural Science Congress. M.S. Swaminathan, N.S. Randhawa, T.N. Ananthakrishnan and P. Narain (eds). National Academy of Agricultural Sciences, New Delhi, India. pp. 208- 221

Nakazono, M., Qiu, F., Borsuk, L.A. and Schnable, P.S. 2003. Laser capture micro-dissection, a tool for the global analysis of gene expression in specific plant cell types; identification of genes expressed differentially in epidermal cells or vascular tissues of maize. The Plant Cell 15: 583-596

National Research Council. 2002. Predicting Invasions of Non-indigenous Plants and Plant Pests. National Academy Press. Washington, DC. USA

Nault, B.A., Speese, J.I.I.I., Jolly, D. and Groves, R.L. 2003. Seasonal patterns of adult thrips dispersal and implementations for management in eastern Virginia tomato fields. Crop Protection 22: 505-512

Nibblet, C.L., Genc, H., Cevik, B., Halbert, S., Brown, L., Nolasko, G., Bonacalza, B., Manjunath, K.L., Febres, V.J., Pappu, H.R. and Lee, R.F. 2000. Progress of strain differentiation of *Citrus tristeza virus* and its application to the epidemiology of citrus tristeza disease. Virus Res. 71: 97-106

Nolasco, G. 2006. Citrus tristeza epidemiology in the Mediterranean basin: a changing scenario? Bull.OILB/SROP 29: 3

Noris, E., Hidalgo, E., Accotto, G.P. and Moriones, E. 1994. High similarity among the tomato yellow leaf curl virus isolates from the West Mediterranean Basin: the nucleotide sequence of an infectious from Spain. Archiv. Virol. 135: 165-170

Nutter, F.W. 1997. Quantifying the temporal dynamics of plant virus epidemics: a review. Crop Protect. 16: 603-618

Oberforster, M., Ruckenbauer, P., Rabb, F., Kern, R., Buchgraber, K. and Schaumberger, A. (eds.). 2003. Yellow dwarf virus Austrian winter cereals variety reaction and counter strategies. Bericht uber die Arbeitstagung 2002 der Vereinigung der Pflanzenzuchter und Saatgutkaufleute Osterreichs, 26, bis 28 November 2002, Gumpenstein, pp. 99-106

Oerke, E.C., Dehne, H.W., Schönbeck, F. and Weber, A. 1994. Crop Production and Crop Protection: Estimated Losses in Major Food and Cash Crops. Elsevier, Amsterdam, The Netherlands

Ofori, F.A. and Francki, R.I.B. 1985. Transmission of leaf hopper A virus, vertically through the eggs and horizontally through maize in which it does not multiply. Virology 114: 152-157

Okoth, V.A.O. and Dabrowski, Z.T. 1987. Population Density, Species Composition and Infectivity of Maize Streak Virus (MSV) of *Cicadullina* spp. Leafhoppers in Some Ecological Zones of Nigeria. Acta Oecol. Appl. 3: 191

Omura, T., Saito, Y., Usugi, T. and Hibino, H. 1983. Purification and serology of rice tungro spherical and rice tungro bacilliform viruses. Ann. Phytopath. Soc. Japan 49: 73-76

Orlob, G.B. 1966. The role of subterranean aphids in the epidemiology of barley yellow dwarf virus. Entomologia Experimentalis et Applicata 9: 85-94

Osler, R. 1980. II Giallume del Riso: Attuali Consocenze sul Ciclo di Infeziole del virus, con Partcolare Attenzione alle Piante *Oryza sativa* e *Leersia oryzoides* e all Afiade *Ropalosiphum padi. Risco* 29: 217

Oswald, J.W., and Houston, B.R. 1952. Barley Yellow Dwarf, a virus disease of barley, wheat and oats readily transmitted by Four Species of Aphids. Phytopathology 42: 15

Oswald, J.W. and Houston, B.R. 1951. A new disease of cereals transmitted by aphids. Plant Dis. Reptr. 35: 471-475

Otten, W., Filipe, J.A.N., Bailey, D.J. and Gilligan, C.A. 2003. Quantification and analysis of transmission rates for soil-borne epidemics. Ecology 84: 3232-3239

Padidam, M., Beachy, R.N. and Fauquet, C.M. 1995. Classification and identification of geminviruses using sequence comparison. J. Gen. Virol. 76: 249-263

Padidam, M., Sawyer, S. and Farquet, C.M. 1999. Possible emergence of new geminiviruses by frequent recombination. Virology 265: 218-225

Parrella, G. and Crescenzi, A. 2005. The present status of tomato viruses in Italy. Acta Horticulturae 695: 37-42

Parmesan, C. and Yohe, G. 2003. A globally coherent fingerprint of climate change impacts across natural systems. Nature 421: 37-42

Petit, J.R., Jouze, J., Raynaud, D., Barkow, N.I., Barnola, J.M. *et al.* 1999. Climate and atmospheric history of the past 420,000 years from from the Vostok Ice Core, Antarctica. Nature 399: 429-436. doi: 10.1038/ 20859

Plumb, R.T. 1974. Properties and isolates of barley yellow dwarf virus. Annals. Appl. Biol. 77: 87-91

Plumb, R.T. 1977. Grass as a Reservoir of Cereal Viruses. Annales de Phytopathoogiel. 9: 361-364

Plumb, R.T. 1983. Barley yellow dwarf virus— a global problem. In: Plant Virus Epidemiology. R.T. Plumb and J.M. Thresh (eds). Blackwell Scientific Publications, Oxford, UK. pp. 185-198

Plumb, R.T. 1986. A Rational Approach to the Control of Barley Yellow Dwarf Virus. J. R. Agric. Soc. Engl. 147: 195

Plumb, R.T. 1990. The Epidemiology of Barley Yellow Dwarf in Europe. In: World Perspectives on *Barley Yellow Dwarf.* P.A. Burnett (ed). CIMMYT, Mexico

Plumb, R.T. 1992. Barley yellow dwarf. In: Plant Disease of International Importance, Vol. 1. U.S. Singh, A.N. Mukhopadhyay, J. Kumar and H.C. Chaube (eds). Prentice Hall. Englewood Cliff, New Jersey, USA. pp. 41-79

Plumb, R.T. and Johnstone, G.R. 1995. Cultural, chemical, and biological methods for the control of barley yellow dwarf. In: Barley Yellow Dwarf, 40 Years of Progress. J.C. D'Arcy, and P.A. Burnett (eds.) APS Press, St Paul, MN, USA. pp. 307-319

Plumb, R.T. 2002. Viruses of Poaceae: a case history in plant pathology. Plant Pathology 51: 673-682

Plumb, R.T., Mukhopadhyay, S. and Jones, P. (eds). 2000. Viruses of crops, weeds in Eastern India, IACR-Rothamsted, Hertfordshire, United Kingdom and Bidhan Chandra Krishi Viswavidyalaya, West Bengal, India

Pokorny, R. 2006. Occurrence of viruses of the family Luteoviridae on maize and some annual weed grasses in the Czech Republic. Cereal Research Communications 34 (2/3): 1087-1092

Polaston, J.E. and Anderson, P.K. 1997. The emergence of whitefly-transmitted geminiviruses in tomato in the western hemisphere. Plant Dis. 81: 1358-1369

Polston, J.E., Cohen, L., Sherwood, T.A., Ben-Joseph, R. and Lapidot, M. 2006. Capsicum species: symptomless hosts and reservoirs of tomato yellow leaf curl virus. Phytopathology 96: 447-452

Power, A. 1996. Competition between viruses in a complex plant-pathogen system. Ecology 77: 1004-1010

Price, R.D. 1970. Stunted patches and deadheads in Victorian cereal crops. Technical Publication, Department of Agriculture, Victoria 23

Raccah. B., Bar-Joseph, M. and Lobenstein, G. 1977. The role of aphid vectors and variation in virus isolates in the epidemiology of tristeza disease. Contribution from the Volcani Center, Agri. Res. Organization, 273-E

Rahman, A.H.M.A., Akanda, A.M. and Alam, A.K.M.A. 2006. Relationship of whitefly population build up with the spread of TYLCV on eight tomato varieties. Journal of Agriculture and Rural Development, Gazipur 4: 67-74

Rao Prasada, R.D.V.J. and John, V.T. 1974. Alternate Hosts of Rice Tungro Virus and its Vectors. Plant Dis. Rept. 62: 955

Ray, J., Schroeder, J., Creamer, R. and Murray, L. 2006. Planting dates affects phenology of London rocket (*Sisymbrium iris*) and interaction with beet leafhopper (*Circulifer tenellus*). Weed Science 54: 127-132

Reilly, J., Tubiello, F., McCarl, B., Abler, D., Darwin, R., Fuglie, K., Hollinger, S., Izaurralde, C., Jagtap, S., Jones, J., Mearns, L., Ojima, D., Paul, E., Paustian, K., Riha, S., Rosenberg, N. and Rosenzweig, C. 2003. US agriculture and climate change: New Results. Climate Change 57: 43-69

Ren, R., Pfeiffer, T.W. and Ghabrial, S.A. 1997. Soybean mosaic virus incidence level and infection times: interaction effects on soybean crop. Sci. 37: 1706-1711

Rivera, C.T., Ling, K.C. and Ou, S.H. 1969. Suspect Host Range of Rice Tungro Virus. Philipp. Phytopathol. 5: 16 Riverside, CA, USA.

Rivera, C. and Gamez, R. 1986. Multiplication of maize rayado fino virus in the leaf hopper vector *Dalbulus maedis*. Intervirology 25: 76-82

Rocha-Peña, M.A., Lee, R.F., Lastra, R., Niblett, C.L., Ochoa-Corona, F.M., Garnsey, S.M. and Yokomi, R.K. 1995. Citrus tristeza virus and its aphid vector *Toxoptera citricida* threats to citrus production in the Caribbean and Central and North America. Plant Dis. 79: 437-445

Rochow, W.F. 1969. Biological properties of four isolates of barley yellow dwarf virus. Phytopathology 59: 1589-1587

Rochow, W.F. 1984. In: Bumett, P.A. (ed.), Barley Yellow Dwarf. Proc. Workshop (CIMMYT): 204. Mexico.

Rochow, W.F., Jedlinski, H., Coon, B.F. and Murphy, H.C. 1965. Variation in Barley Yellow Dwarf of Oats in Nature. Plant Dis. Rep. 49: 692

Roistacher, C.N. 1982. A blueprint for disaster, Part 3: The destructive potential for seedling yellows. Citrograph 67: 48

Roistacher, C.N. 1988. Observation of the decline of sweet orange trees in coastal Peru caused by stem pitting tristeza. FAO Plant Prot. Bull. 36: 19

Roistacher, C.N., Blue, R.L., Nauer, E.M. and Calavan, E.C. 1974. Suppression of tristeza virus symptoms in Mexican lime seedlings grown at warm temperature. Plant Dis. Reporter 58: 757-760

Roistacher, C.N., Nauer, E.M., Kishaba, A. and Calavan E.C. 1980. Transmission of citrus tristeza virus by *Aphis gossypii* reflecting changes in virus transmissibility in California. Proc. 8th Conf. IOCV. E.C. Calavan, S.M. Garsney and L.W. Timmer (eds). Univ. Calif. Riverside, CA, USA. pp. 76-82

Rose, D.J.W. 1972. Times and sizes of dispersal flight by *Cicadulina* spp. vector of maize streak disease. J. Anim. Ecol. 41: 494-506

Rose, D.J.W. 1973a. Distance flown by *Cicadullina* spp. (Hem. Cycadellidae) in relation to distribution of Maize Streak Disease in Rhodesia. Bull. Entomol. Res. 62: 497

Rose, D.J.W. 1973b. Field studies in Rhodesia on *Cycadullina* spp. (Hem. Cicadellidae), vectors of Maize Streak Disease. Bull. Entomol. Res. 62: 477

Rose, D.J.W. 1974. The epidemiology of Maize Streak Disease in Relation to Population Densities of *Cicadullina* spp. Ann. Appl. Biol. 76: 199

Rose, D.J.W. 1978. Epidemiology of maize streak disease. Annu. Rev. Entomol. 23: 259-282

Rosenberg, L.J. and Magor, J.I. 1983. A technique for examining the long-distance spread of plant virus diseases transmitted by the brown plant hopper, *Nilaparvata lugens* (Homoptera: Delphacidae), and other wind-borne insect vectors. In: Plant Virus Epidemiology. R.T. Plumb and J.M. Thresh (eds). Blackwell Scientific Publications, Oxford, UK. pp. 229-238

Rosenberg, L.J. and Magor, J.I. 1987. Predicting windborne displacements of the brown planthoppers, *Nilaparvata lugens* from synoptic weather data. I. Long-distance displacements in the north-east monsoon. J. Anim. Ecol. 56: 39-51

Roy, B.A., Gusewell, A. and Harte, J. 2004. Response of plant pathogens and herbivores to warming experiments. Ecology 85: 2570-2581

Sabin, A.L. and Pisias, N.G. 1996. Sea surface temperature changes in the Northern Pacific Ocean during the past 20,000 years and their relationship to climate change in Northwestern North America. Quaternary Research 46: 48-61

Salati, R., Nahkla, M.K., Rojas, M.R., Guzman, P., Jaquer, J., Maxwell, D.P. and Gilbertson, R.L. 2002. Tomato yellow leaf curl virus in the Dominican Republic: characterization of an infectious clone, virus monitoring in whiteflies, and identification of reservoir host. Phytopathology 92: 487-496

Sanchez-Campos, S., Navas-Castillo, J., Camero, R., Soria, C., Diaz, J.A. and Moreones, E. 1999. Displacement to Tomato yellow leaf curl virus (TYLCV)-sr by TYLCV-Is in tomato epidemics in Spain. Phytopathology 89: 1038-1043

Sandhu, S.S. and Bhati, D.S. 1969. The incidence of sugar beet disease in Punjab. J. Res. Ludhiana 6: 924

Sánchez-Compos, S., Navas-Castillo, J., Monci, F., Diaz, J.A. and Morione, E. 2000. *Mercurialis ambigua* and *Solanum luteum*: two newly discovered natural hosts of tomato yellow leaf curl gemini viruses. Eur. J. Plant Pathol. 106: 391-394

Saunders, K., Bedford, I.D. and Stanley, J. 2002. Adaptation from whitefly to leafhopper transmission of an autonomously replicating nanovirus-like DNA component associated with Ageratum yellow vein disease. J. Gen. Virol. 83: 907-913

Scherm, H. 2004. Climate change: can we predict the impacts on plant pathology and pest management? Can. J. Plant Pathol. 26: 267-273

Scherm, H. and Coakley, S.M. 2003. Plant pathogens in a changed world. Aus. Plant Pathology 32: 157-165

Scherm, H., Sutherst, R.W., Harrington, R. and Ingram, J.S.I. 2000. Global networking for assessment of impacts of global change on plant pests. Environmental Pollution 108: 333-341

Schnippenkoe, W.H., Martin, D.P., Willment, J.A. and Rybicki, E.D. 2001. Forced recombination between distinct strains of maize streak virus. J. Gen. Virol. 82(12): 3081-3090

Sdoodee, R. and Teakle, D.S. 1993. Studies on the mechanism of transmission of pollen-associated tobacco streak ilarvirus by *Thrips tabaci*. Plant Pathology 42: 88-92

Seal, S.E., Vandenbosch, S. and Jeger, M.J. 2006. Factors influencing begomovirus evolution and their increasing global significance: implications for sustainable control. Crit. Rev. Plant Sci. 25: 23-46

Sentandreu, V., Castro, J.A., Ayllon, M.A., Rubio, L., Guerri, J., Gonzles-Candelas, F., Moreno, P. and Moya, A. 2006. Evolutionery analysis of genetic variation observed in *Citrus tristera virus* (CTV) after host passage. Archiv. Virol. 151: 875-894

Simberloff, D. 2000. Global climate change and introduced species in the United States forests. Science of the Total Environment 262: 253-261

Simón, B., Cenis, J.L., Beitia, F., Khalid, S., Moreno, I., Fraile, A. and García-Arenal, F. 2003. Genetic structure of field populations of begomoviruses and of their vector *Bemisia tabaci* in Pakistan. Phytopathology 93: 1422-1429

Slykhuis, J.T. 1955. *Aceria tulipae* Keifer (Acarina: Eriophydae) in relation to the spread of wheat streak mosaic. Phytopathology 45: 116-128

Smith, H.C. and Wright, G.M. 1964. Barley yellow dwarf virus on wheat in New Zealand. New Zealand Wheat Review 9: 60-79

Smyrnioudis, I.N., Harrington, R., Katis, N. and Clark, S.J. 2000. The effect of drought stress and temperature on spread of barley yellow dwarf virus (BYDV). Agriculture and Forest Entomology 2: 161-166

Smyrnioudis, I.N., Harrington, R., Hall, M., Katis, N. and Clark, S.J. 2001. The effect of temperature on variation in transmission of a PAV-like isolates by clones of *Rhopalosiphum padi* and five clones of *Sitobion avenae*. Euro. J. Plant Pathology 107: 167-173

Soto, M.J., Chen, L.F., Seo, Y.S. and Gilberton, R.L. 2005. Identification of regions of beet mild curly top virus (family Geminiviridae) capsid protein involved in systemic infection, virion formation and leafhopper transmission. Virology 341: 257-270

Souchez, R., Petit, J.R., Jouzel, J., de Angelis, M. and Tison, J.L. 2003. Reassessing Lake Vostok's behavior from existing and new ice core data. Earth and Planetary Science Letters 217: 163-170

Spoel, S.H., Koormneef, A., Claessens, S.M.C., Korzelius, J.P., van Peir, J.A., Mueller, M.J., Buchala, A.J., Métraux, J.P., Brown, R., Kazan, K., van Loon, L.C., Dong, X. and Pieterse, C.M.J. 2003. NPRI modulates crosstalk between salicylate- and jasmonate-dependent defense pathways through a novel function in the cytosol. The Plant Cell 15: 760-770

Stone, O.M. 1968. The elimination of four viruses from carnation and sweet William by meristem-tip culture. Ann. Appl. Biol. 62: 119-122

Storey, H.H. 1926. Streak Disease of Maize. Farming S. Afr. 1: 183

Storey, H.H. 1936. Virus disease of east African plants. Virus streak disease of maize. East Afr. Agric. J 1: 471-475

Sutherst, B., Ingram, J. and Steffen, W. (eds). 1996a. GCTE Activity 3.2: Global Change Impacts on Pests, Diseases and Weeds. Implementation Plan (GCTE Report No.11), GCTE. Canberra, Australia

Sutherst, R.W., Yonow, T., Chakraborty, S., O'Donnell, C. and White, N. 1996b. A generic approach to defining impacts of climate change on pests, weeds and diseases in Australia. In: Greenhouse Coping with Climate Change. W.J. Bouma, G.I. Pearman and M.R. Manning (eds). CSIRO, Melbourne, Australia. pp. 281-307

Sutherst, R.W., Bajer, R.H.A., Coakley, S.M., Harrington, R., Kriticos, D.J. and Scherm, H. 2007. Pests Under Global Change— Meeting Your Future Landlord? In: Terrestrial Ecosystems in a Changing World. J.G. Canadell, D. Pataki and L. Pitelka (eds). The IGBP Series. Springer-Verlag, Berlin, Heidelberg, Germany. pp. 211-226

Swingle, W.T. and Weber, H.J. 1896. The principal diseases of citrus fruits in Florida. US Dept. Agric. Div. Physio and Path Bull: 12

Tarafder, P. and Mukhopadhyay, S. 1979a. Potential of Stubbles in Spreading Tungro in West Bengal, India. Int. Rice Res. Newslt. 4: 18

Tarafder, P. and Mukhopadhyay, S. 1979b. Potential of Weeds to Spread Tungro in West Bengal, India. Int. Rice Res Newslt. 4: 11

Tarafder, P. and Mukhopadhyay, S. 1980a. Further Studies on the Potential of Stubbles in Spreading Tungro in West Bengal, India. Int. Rice Res. Newslt. 5: 12

Tarafder, P. and Mukhopadhyay, S. 1980b. Further Studies on the Potential of Weeds to Spread Tungro in West Bengal, India. Int. Rice Res. Newslt 5: 10

Tepfer, M. 2002. Risk assessment of virus-resistant transgenic plants. Annu. Rev. Phytopathol. 40: 467-491

Thackray, D.J., Ward, L.T., Thomas-Carroll, M.L. and Jones, R.A.C. 2005. Role of winter-active aphids spreading barley yellow dwarf virus in decreasing wheat yields in a Mediterranean type environment. Aus. J. Agric. Res. 56: 1089-1099

Thomas, P.E. 2002. First report of *Solanum sarrachoides* (Hairy Nightshade) as an important host of potato leaf roll virus. Plant Disease 86: 559

Thomas, J.E. and Bowyer, J.W 1980. Properties of tobacco yellow dwarf and bean summer death viruses. Phytopathology 70: 214-217

Thresh, J.M. 1974. Temporal patterns of virus spread. Annu. Rev. Phytopathology 12: 111-126

Thresh, J.M. 1976. Gradients of plant virus diseases. Annals Appl. Biol. 82: 381-406

Thresh, J.M. 1983. Progress curve of plant virus disease. Applied Biol. 8: 1-85

Thresh, J.M. 1986. Plant Virus Disease Forecasting. In: Plant Virus Epidemics: Monitoring, Modeling and Predicting Outbreaks. G.D. McLean, R.G. Garrett and W.G. Ruesink (eds). North Ryde, NSW; Academic Press, Australia.

Thresh, J.M. 1987. The population dynamics of plant virus diseases. In: Populations of Plant Pathogens: their dynamics and genetics. Blackwell Scientific Publications, Oxford, UK. pp. 136-148

Timmer, L.W., Garnsey, S.M., Broadbent, P. and Ploetz, R.C. 2003. Diseases of citrus. Diseases of Tropical Fruit Crops: 163-195

Torchin, M.E., Lafferty, K.D., Dobson, A.P., McKenzie, V.J. and Kuris, A.M. 2003. Introduced species and their missing parasites. Nature 421: 628-629

Tottappilly, G., Bosque-Pérez, N.A. and Rossel, H.W. 1993. Viruses and virus diseases of maize in tropical Africa. Plant Pathol. 42: 494-509

van den Bosch, F., Akudibilah, G., Seal, S. and Jeger, M.J. 2006. Host resistance and the evolutionary response of plant viruses. J. Appl. Ecol. 43: 506-516

Vanderplank, J.E. 1963. Plant Diseases: Epidemics and Control. Academic Press, London, UK

van Rensberg, G.D.J. and Kuhn, H.C. 1977. Maize streak disease. South African Department of Agricultural Technology and Service. Maize Series Leaflet No. E3

van Wees, S.C.M., de Swart, E.A.M., van Pelt, J.A., van Loon, L.C. and Pieterse, C.M.J. 2000. Enhancement of induced disease resistance by simultaneous activation of salicylate- and jasmonate-dependent defense pathways in *Arabidopsis thaliana*. Proc. Nat. Acad. Sci. USA 97: 8711-8716

Varma, A. and Malathi, V.G. 2003. Emerging geminivirus problems: a serious threat to crop production. Annals Appl. Biol. 142: 145-164

Vilau, F. 2004. Infestation of cereals by some pests and barley yellow dwarf virus. Analele Institutului de Carcetari pentru Cereale si Plante Tehnice, Fundulea 71: 287-294

Walket, D.G.A. and Cooper, J. 1976. Heat inactivation of cucumber mosaic virus in cultured tissues of *Stellaria medica*. Ann. Appl. Biol. 84: 425-428

Wallace, J.M. 1957. Virus strain interference in relation to symptoms of psorosis disease of citrus. Hilgardia 27: 223-246

Wallace, J.M. and Drake, R.J. 1951. Recent developments in studies of quick decline and related diseases. Phytopathology 41: 705-793

Wallin, J.R. and Loonan, D.V. 1971. Low Level Jet Winds, Aphid Vectors, Local Weather and Barley Yellow Dwarf Virus Outbreaks. Phytopathology 61: 1068

Wallin, J.R., Peters, D. and Johnson, L.C. 1967. Low Level Jet Winds, Early Cereal Aphid and Barley Yellow Dwarf Virus Detection in Iowa. Plant Dis. Rep. 51: 527

Wang, D. and Maule, A.J. 1994. A model for seed transmission of a plant virus: genetic and structural analysis of pea embryo invasion by pea seed-borne mosaic virus. Plant Cell 6: 777-787

Wang, D., MacFarlane, S.A. and Maule, J.A. 1997. Viral determinants of pea early browning seed transmission in pea. Virology 234: 112-117

Wang, G. 2005. Agricultural drought in a future climate: results from 15 global climate models participating in the IPCC 4th Assessment. Climate Dynamics 25: 739-753

Webb, C.R., Gilligan, C.A. and Asher, M.J.C. 2000. Modeling the effect of temperature on the development of *Polymyxa betae*. Plant Pathology 49: 600-607

Webster, J.S., Bravo, C., Holland, G.J., Curry, A. and Chang, H.R. 2005. Changes in tropical cyclone number, duration, and intensity in a warming environment. Science 309: 1844-1846

Weiker, A.R., Dewar, A.M. and Harrington, R. 1998. Modeling the incidence of virus yellows in sugar beet in the UK in relation to number of migrating *Myzus persicae*. J. Appl. Ecol. 35: 811-818

Wells, M.L., Culbreath, A.K., Todd, J.W., Csinos, A.S., Mandel, B. and McPherson, R.M. 2002. Dynamics of spring tobacco thrips (Thysanoptera: Thripidae) populations: implications for tomato spotted wilt virus management. Environ. Entomol. 31: 1282-1290

Wintermantel, W.M. and Kaffka, S.R. 2006. Sugar beet performance with curly top is related to virus accumulation and age of infection. Plant Disease 90: 657-662

Woiwod, I.R. and Harrington, R. 1994. Flying in the face of change. The Rothamsted Insect Survey. In: Long Term Experiments in Agricultural and Ecological Sciences. R.A. Leigh and A.E. Johnston (ed.). CABI International, Wallingford, UK. pp. 321-342

Wren, J.D., Roossinck, M.I., Nelson, R.S., Scheets, K., Palmer, M.W., Marilyn, J. and Melchers, U. 2006. Plant Virus Biodiversity and Ecology. PloS Biol. 4(3): e80

Wu, J.B., Dai, F.M. and Zhou, X.P. 2006. First report of Tomato yellow leaf curl virus in China. Plant Dis. 90: 1359

Xia, L.H. and Lin, J.Y. 1982. The occurrence of rice tungro disease (spherical virus) in China. J. Fufian Agr. College 3: 15-23

Yamamura, K. and Yokozawa, M. 2002. Prediction of a geographical shift in the prevalence of rice stripe virus disease transmitted by the small brown plant hopper, *Laodelphax striatellus* (Fallen) (Hemiptera: Delphacidae) under global warming. Appl. Entomology and Zoology 37: 181-190

Zadoks, J.C. and Schein, R.D. 1979. Epidemiology and Plant Disease Management. Oxford Univ. Press, London, UK

Zhang, W.M. and Wang, Z.Z. 2002. Study on the spatial distribution patterns of tomato mosaic disease and its aphid vectors. J. South China Agric. Univ. 23: 23-26

Zhang, X.-S., Holt, J. and Colvin, J. 2000a. A general model of plant-virus disease infection incorporating vector aggregation. Plant Pathology 49: 435-444

Zhang, X.-S., Holt, J. and Colvin, J. 2000b. Mathematical models of host plant infection by helper-dependent virus complexes: why are helper viruses always a-virulent? Phytopathology 90: 85-93

Zwankhuizen, M.J., Govers, F. and Zadoks, J.C. 1998. Development of potato late blight epidemics: disease foci, disease gradients, and infection sources. Phytopathology 88: 754-763

8

Management: Strategies and Tactics

Recently most research has concentrated on the production of virus-resistant transgenic plants as the best way to keep crops free from viruses, avoid or minimize the use of pesticides and increase the yield and quality of the crop. The concept of 'Management', linking knowledge of virus ecology and epidemiology has been neglected or overlooked. The production and deployment of transgenic plants has much to offer, but is a laboratory process that only has its value when used in field-gown crops. Whereas the development of a management process necessitates long-range field studies and is a dynamic process involving virus, host, vector, volunteer sources of the virus and climate. The incidence of a virus disease depends upon the interaction of these components and this is true whether the host is a transgenic plant produced in the laboratory or one conventionally bred and selected.

The International Committee for the Taxonomy of Viruses (ICTV) (Fauquet *et al* 2005) recognized a total of about 2000 species of viruses as of 2005 and 77% of these were initially isolated from cultivated plants (Wren *et al* 2006). Certain combination of conditions makes these viruses cause economic loss in one or more regions of the world. The primary objective of an agriculturist is to obtain cost-effective economic yields. On many occasions, more than one virus may invade the same crop and this makes management more complex. In addition other pathogens, microorganisms and invertebrates will affect the plant and may interact with any viruses that are present. Thus management of virus diseases necessitates a holistic approach to modulate all the relevant components and keep the incidence below the threshold level for economic damage. Economic globalization, competitive economies, demand-driven agriculture, swift change of cropping systems, the rapid increase of inter-country movement of planting materials, global warming, etc. are constantly changing disease incidences and damage in the world. This focusses the need for effective 'Integrated Management' and designing appropriate 'Cropping Systems'.

A. INTEGRATED PEST MANAGEMENT (IPM)

The concept of 'Integrated Pest Management' emerged during the 1960s as lessons were learnt from the problems of pest control associated with the increased and widespread use of pesticides. IPM takes into account the associated environment and the population dynamics of the pest species and utilizes all suitable techniques and methods in as compatible a manner as possible. The simple presence of a pest alone does not warrant any control measures, the use of which should be reserved for when the threat crosses the economic threshold. In pest management, the application of pesticides is to be considered as the last option after the application of all available suppressive factors.

There are six basic elements in this management system:

1. People
2. Knowledge and information
3. Monitoring and sampling
4. Decision making levels, forecasting
5. Methods, agents and materials
6. Well trained system devisers and pest managers.

Decisions and actions should be taken on the basis of an understanding of the complex factors involved in the epidemics, the biology of the crops or resources, the potentially damaging pests/diseases/vectors and their natural enemies and their interrelationships with the surrounding ecosystems. Monitoring and sampling is essential to record the changes occurring in the state of resources or crops, the weather and the number of pests and their natural enemies present, particularly to establish economic thresholds and operational tactics or decision-making. Computer models are now gaining prominence in this respect. Pest management has to receive special attention in virus disease management as vectors transmit most of the viruses. The success of crop-based IPM depends upon the judicious application of the following management options:

- Regulatory measures
- Host resistance
- Natural enemies
- Autocidal phenomena
- Cultural manipulation
- Mechanical and physical methods
- Safe use of chemical or biopesticides.

Regulatory practices include quarantine, sanitation and eradication programmes; host resistance includes selective breeding and genetic manipulation of the host, virus and the vector; this option seems to be the most ecologically sound technique provided that the resistance is sought in the germplasms present in the ecological zone in which they must function. The application of natural enemies or biological control is often considered as the first line of ecological defence. In nature, biological balancing mechanisms always operate. The concerned autochtonous-balancing agents may be manipulated to maintain the population of the pest/disease/vector at an acceptable limit. Alien, exotic, natural enemies may also be put in use through careful manipulations but care must be used to ensure that the ecosystem is not adversely affected. Autocidal management utilizes tactics that cause the pest/vector to contribute to the reduction of its own population. Cultural manipulation normally includes adjustment of planting time, regulation of inputs and their application, and cropping and harvesting manipulation. Chemical/botanical pesticide management includes application of them at action levels determined after due monitoring or judicious forecasting (see below).

The crop-based managements are then subjected to further refinement to overcome and avoid disciplinary myopia and to incorporate more ecological considerations. This approach is termed as 'Systems Approach', a scientific philosophy with a holistic view of agricultural systems in particular agro-climatic conditions.

The successful use of the management options depends to a large extent upon national and international organizational systems on plant protection organizations. There needs to be comprehensive knowledge of the distribution and phenology of pests/vectors and pathogens in each country so that suitable places for the cultivation/plantation of relevant crops under minimum pest/vector and disease pressure can be identified. Surveys and detection of pest/vectors and pathogens are strategically important for their management. Thus, accurate and sensitive plant virus diagnosis is a prerequisite for designing management practices.

Detection of Viruses and Vectors and its Relevance to Management

The range of different methods for virus diagnosis is considered in Chapters 4 and 5. Diagnostics are vital for phyto-sanitation and quarantine and there should be coordinated governmental, institutional and private policies and activities in each country in accordance with international guidelines, so that viruses and vectors can be correctly diagnozed and managed. There should also be adequate information networking and communication to provide relevant technical information to the agriculturists and planters to plan their crops and management practices. The agriculturists and planters on their part, however, should pursue the appropriate strategies and tactics to manage the virus diseases.

Strategies and Tactics

The *basic strategy* for an agriculturist/planter is to grow *virus-free seed/planting material* in a *virus-free land* preferably located in an area where disease/vector pressure is very low. The next *strategy is to cultivate resistant plants* and *control the arrival of the vectors and* keep their numbers below the economic threshold. *Cultural practices* also have strategic significance and may be utilized as *tactics*, as and when necessary. *Cross protection* may also be considered as a strategy for the cultivation of certain crops. In disease/vector prone areas, *the system approach* may be used to minimize disease/vector incidence. As global agriculture is now influenced by agreements under the World Trade Organization (WTO) internationally agreed *Sanitation and Phyto-sanitation (SPS)* is also to be used as a strategy.

Virus-free Seed/Planting Material and Certification

Many viruses are seed-borne or carried by planting materials (see Chapter 7). The use of *Certified Seeds/Planting Materials* is necessary to restrict the initial build-up of infection in the field/orchard and their use generally gives higher yields. By decreasing the initial inoculum load in the field the spread of the virus is restricted and yield loss prevented. Marketable lettuce yield was increased in the Salinas Valley area of California by cultivating *lettuce mosaic virus* (LMV)-free seed (Kimble *et al* 1975). Seed/planting material production and certification schemes now operate mainly through legislation. Regulatory authorities of the governments normally monitor the production chains and issue licences to the producers of disease-free seeds/planting materials and the producers have to follow international standards.

Detection of viruses is most important in the seed certification process and also to determine the acceptable limits or threshold values of infection. For this purpose, large quantities of seeds are tested following appropriate statistical and serological, or molecular methods. Cultivation of certified seeds always give good dividends, as for example, seed certification schemes against BSMV have been credited with avoiding millions of dollars in losses due to this disease in barley crops in some states of the USA (Carrol 1983). Legislative certification of seed potatoes is followed in most of the potato growing countries of the world. Potato seeds are usually produced from virus-free 'Foundation Seeds' supplied by the authorized institutes or agencies following special procedures or techniques in vector-free areas or during the period of the year when the vectors are usually absent. National seed certifying agency(ies) follow standard protocols for certifying the seeds. The certification process normally excludes primary sources of viruses in the potato fields. Similar certification is also done or has the scope for doing so, in many commercial crops particularly, sugarcane, banana and citrus. In many countries, legislative certification of the planting materials of these crops does operate. For citrus, rootstock and scion mother groves are maintained in vector-free conditions. These are registered by the appropriate authority and annually certified by the authorized certifying agency or institute. Orchardists normally raise rootstock seedlings from the seeds collected from their own registered and certified rootstock mother grove and buy certified virus-free bud-wood from registered nurserymen or institutes. There are also several methods for the production of virus-free clone or eliminating the virus from the infected clone.

Elimination from the Clone

Common methods for the elimination of viruses from infected clones are thermotherapy and tissue culture, or meristem tip culture, combined with thermotherapy and chemotherapy.

Thermotherapy

Thermotherapy to exclude viruses from seeds and planting materials has been used for many years and is now routinely used when planting many horticultural crops. Heat treatment of perennial plants to eliminate viruses has been reviewed by Mink *et al* (1998).

Thermotherapy may be done by treating the seeds/planting materials in hot water, hot air, moist air or aerated steam. Hot water treatment (50°C for 30 min before planting) is a common practice to cure dormant trees, bud-sticks and growing plants to make them free from viruses. *Vinca rosea* and *Nicotiana rustica* plants were cured from *aster yellows* by keeping them at 35°C for 2 wk. Potato tubers were freed from *potato leaf roll virus* by keeping them at 37°C in a moist atmosphere for 10 to 20 d. *Citrus tristeza virus* can be eliminated from bud-sticks by treating them at 35°/45°C for 14-35 d before grafting. Hiroyuki and Yamada (1980) recommended two methods of heat treatment to eliminate CTV from bud-sticks; either keeping the plants directly in a high temperature chamber for a long period and the other is keeping the plants at a high temperature for a shorter period, after preconditioning at a lower temperature. The actual method of temperature treatment depends upon the heat tolerance of each variety; preconditioning at a low temperature for several weeks is desirable. By alternating temperatures 45°/35°C for 5, 7, or 9 wk Tristeza-free trees are generally obtained. A regime of alternating high (40°C) and normal (20°C) was used to eliminate CCMV in cowpea (Lozoya-Saldaña and Dawson 1982), and for GFLV and ArMV in grapes (Monette 1986). Lukicheva and Mitrofanov (2002) cleaned up cherry and plum infected by *prunus necrotic spot virus* and *apple chlorotic leaf spot virus* by thermotherapy (37± 1°C) under a 16 h photoperiod. The temperature was decreased by 10°C during darkness. For cherries, PNRSV and ACLSV were successfully eliminated after 20 and 40 d respectively. For plums, PNRSV was successfully eliminated after 30 d of treatment.

The therapeutic effect of heat on viruses is not related to the thermal inactivation point of the respective infectious sap. Many viruses with high thermal inactivation points *in vitro* may be inactivated *in vivo* when infected plants are kept continuously at a temperature of around 37°C. *Tomato spotted wilt virus* is very labile *in vitro* and is inactivated at 45°C when kept for 10 min but survives in plants at 36°C, whereas *tomato bushy stunt virus* can withstand 80°C for 10 min *in vitro* becomes labile at 36°C and infected cuttings could be made virus-free by heat treatments for 3 wk.

Qu *et al* (2003) developed a method to eliminate latent viruses (*apple chlorotic leaf spot virus, apple stem grooving virus* and *apple stem pitting virus*) of apple trees. The plants were heat-treated at the 4-5-leaf stage. Potted plants were treated in a glass cabinet and others were treated in a shed. Heat treatment for 26 d at 37°C in the shed achieved elimination rates up to 72.3%. Heat-treating tube plantlets in an illumination cabinet for 30 d at 38°C provided 100% virus elimination with a survival rate of 13.3 to 60%.

Tissue Culture/Meristem Culture

Tissue culture is a biotechnological method for improvement and regeneration of plants. This technique has received commercial attention since 1970s. Basically it is the culture of tissues in artificial media. Plant cells possess a unique property called 'totipotency' or the property to produce a mature plant from its appropriate cell under suitable environments. This property was first observed by P.R. White in 1934 to produce tomato plants from their root-tips. Subsequently it was also demonstrated that tissues of root, shoot, stem, flower or leaf produce plants when incubated in proper media and environment reacted in this way. This technique has also received attention to produce virus-free plants in view of the general hypothesis that the rapidly dividing tissues do not contain any virus or they are immune to viruses. Moreover, while meristem tip-derived cultures normally retain the genetic characteristics of the mother plants this may not be so with cultures derived from other tissues. Facioli and Marani (1998) reviewed virus elimination by meristem-tip culture.

Meristem Immunity

Rapid growth occurs in meristem shoot tip and it is devoid of any virus. White (1934) was the first to hypothesize that the plant meristem could be free from viruses. He studied the virus distribution in plants by inoculating *in vitro* grown tomato roots with the *aucuba strain of tomato mosaic virus* (*ToMV*) He determined the presence of the virus in successive portions of plants by inoculating the extract from them in *Nicotiana glutinosa*. Morel (1948) was the first to obtain virus-free plantlets from diseased ones by carrying out the aseptic culture of meristems excised from shoot-tips and proposed the hypothesis of 'Meristem Immunity' towards viruses. Limasset and Cornuet (1949) demonstrated a differential distribution of viruses in systemically infected plants. They established that the virus concentration in buds was 1/150 of that in developing leaves using biological and serological methods of detection. As they did not find any virus in the meristem, they supported the hypothesis of 'Meristem Immunity' to viruses. In the same period, Holmes (1948) obtained *Dahlia* plants free from the tomato spotted wilt virus by grafting meristem tips onto healthy root-stocks. Later Morel and Martin (1952, 1955) produced healthy plants from some virus-infected cultivars of *Dahlia* and potato, by excision and growth of meristem in a sterile synthetic medium. Their results opened up a new possibility for controlling plant viruses by eliminating them through tissue culture of apical meristem.

Meristem-tip Culture

Meristem-tip culture consists of three major steps: meristem sampling, meristem culturing and development in the culture medium, transplanting of plantlets to pots/soil and subsequent indexing.

(i) Meristem sampling

Different names have been used to indicate the explant suitable for virus elimination. Most frequently used names are 'Shoot-tip', 'Tip', 'Meristem' and Meristem-tip' (Hollings 1965). The last two names indicate that the explant is made of meristematic dome and one or two leaf primordia (100 μm in diameter and 250 μm in length). 'Shoot-tip' and 'Tip' indicate part of the shoot apex including the meristem with several leaf primordia and adjacent differentiated stem tissue. This kind of explant is usually removed from the heat-treated plants and used for clonal propagation, if freedom from virus is not required.

Meristem-tip explants suitable for virus elimination are about 0.1 mm in diameter and 0.2 to 0.4 mm in length (apical dome plus one or two couples of leaf primordial). These explants are normally dissected under completely sterile conditions using a stereomicroscope fitted with a cool light illuminator or glass fibre lamps to avoid desiccation during dissection. Alternatively, dissection could be done in a Petridish, containing moist sterile filter paper to prevent drying. The tip however is always excised inside a laminar airflow system.

The first important goal in realizing meristemtip culture is to avoid contaminating the explant by surface pathogens. If the meristem-tips are collected after bud-breaking from the bud-sticks by placing them in sterile media, the chances of obtaining contamination-free meristem-tip is more than that obtained by placing in non-sterile media.

Meristem-tips from underground plant organs, such as tubers, rhizomes, bulbs or corms, from buds or from seeds should be dipped in 75% ethanol to remove trapped air, then sterilized with sodium hypochloride or calcium chloride solution for 10 min and rinsed 3 times in sterile water.

(ii) Meristem culturing and development

Morphogenetic potential of meristems for culturing depends mainly on meristem size, plant species, physiological conditions of the donor plants and meristem position on the plant.

1. *Meristem size.* The majority of workers generally use tips between 0.2 and 1.0 mm in length, consisting of meristematic dome and 2 or more leaf primordia. As the number of leaf primordia per unit length vary remarkably, it is convenient to specify the number of primordia excised for culture instead of specifying the length. Selection of explant dimension is very important, as it is to accomplish the development of complete plant and virus elimination.

2. *Plant species.* The regenerative potential of meristems is always related to the species and even to cultivars. When 0.2 mm meristem of carnation cultivars is cultured, 68 to 83% regeneration is usually obtained. In another cultivar, using even 0.4 mm meristems, only 8% regeneration is found (Goethals and van Hoof 1971, Kowalska 1974, Maia *et al* 1969).

   In a general way, higher regenerative capacity is found in herbaceous plants. Bulbous plants (iris, lily, gladiolus, freesia, etc.) show very slow meristem growth than that found in dicotyledonous plants and takes a long period (>6 mon) to become transferable to soil.

3. *Physiological condition of the mother plant.* The best physiological conditions of donor plants for meristem sampling coincide with the highest vegetative activity. So the full control of the factors affecting this activity is desirable which can be obtained by using phytotrons or conditioned greenhouses, where light and temperature can be regulated. When plants are not to be grown under regulated conditions, seasonal fluctuations, with special reference to light intensity and photo period are to be properly taken into account. Attention should be given not to sample from buds already induced to flowering particularly for ornamental species where flower induction occurs precociously. In carnations, meristems survive better when excised from plants in early spring and autumn. Shoots derived from winter culture, root more easily than those cultured in summer but the latter originate more frequently in virus-free plants (van Os H, 1964). For most potato varieties, meristems isolated in spring and early summer, root more easily than those taken later in the year (Mellor and Stace-Smith 1977, Quak 1977). In bulbs and corms meristems are best taken at the end of the dormancy. For trees, the growth of which is limited to a short period of time in the spring, meristem tips should be excised in this season and in winter the tips should be excised by breaking the dormancy. In *Prunus* spp., this is done by maintaining stem cuttings at 4°C for almost 6 mon before excising of the tips (Boxus *et al* 1977). In *Citrus*, growth normally takes

place in three seasons. Buds collected in autumn (September) perform better followed by those collected in spring (March-April).

4. *Meristem position on the plant.* Although, in principle, all plant meristems are suitable as starting material, both morphogenetic potential and virus content depend on their position on the stem and shoot, apical or axillary, and in the latter case on the shoot position on the stem. Meristems from basal buds of *Asparagus* spears develop into plant-lets more easily than meristems of buds positioned near the apex (Yang and Clore 1973); in gooseberry successful culture is obtained only from the apical bud (Jones and Vine 1969); rose meristem that comes from mid-part of stems develop faster and more easily (Bressan *et al* 1982); in carnation and *Chrysanthemum* meristem tips excised from apical buds give better results (Hollings and Stone 1968).

Factors Affecting Meristem Development

The development in culture of meristems sampled depends mainly on the characteristics of the culture media, light and temperature.

(i) Culture media

Tissue culture media generally consist of mineral salts, a carbon source of energy, vitamins, growth substances and other minor components. Meristem development primarily depends on the composition of hormones, as the meristems do not synthesize them. Different groups of hormones exert contrasting effects on meristem differentiation and each sample demands separate treatment. Shoot development normally depends on cytokinin-rich medium and auxin-rich medium is required to stimulate root formation.

Meristems usually grow well in nutrient media solidified with agar (0.6-0.8%) but in some cases liquid media is quite useful. Agar concentrations can also affect the meristem development. Use of Difco Bacto agar, alternative products like biogel P200 or alginate, Gelrite etc. often are used as they do not contain inorganic contaminants toxic to meristems. In some cases, liquid medium is also used that requires growth regulators at a lower concentration since their uptake by the tissues is faster than in agar. Use of liquid medium is suitable for rooting of woody species. Use of double phase technique (liquid above solid) is found to be suitable for some fruit trees (Maene and Debergh 1985). pH of the culture medium is an important variable which affect meristem development according to the plant species. Normal pH range used in tissue culture is 5.0-6.5. pH lower than 5.5 inhibits shoot rooting. Carnation meristems, cultured at pH 5.5 differentiated 59% of shoots but the differentiation is only 4% when cultured at pH 6.0. In case of tobacco cells, shoots differentiate at pH 5.9 whereas roots differentiate at pH 4.9 (Stone 1963). Use of buffers also needs to be standardized as they can give different results according to the medium composition.

Other important components of the media are mineral salts, carbon and energy source, vitamins, growth regulators, etc. Concentrations of all these components in the media need to be standardized for the development of the meristem of each species. There are, however, several standard media commonly used in tissue culture of meristems. These are White's medium, media of Murashige and Skoog (MS), Gamborg *et al* (B5), etc.

(ii)   Light

Light affects meristem organogenesis in a complex way through its various characteristics, involving exposure time (photo period), intensity and wavelength. Most of the time, 14-18 h of daylight with an intensity of 1-4 Klux of a cool white fluorescent lamp, either alone or implemented by enrich-red fluorescent light (Gro-lux) is adopted, even though a differentiated use of these parameters according to plant species and to type of wanted morphogenesis will improve, or even elicit, meristem development. It is normally assumed that photo period requirement for tissue culture resembles that of the whole plant. Photoperiod and light intensity are interdependent requirements and they are always referred together. Wavelength requirement is similar for all plant species, and supposedly culture tissues.

(iii)   Culture containers

Dimensions of the containers can influence the type of meristem development. *Cassava* meristems differentiate into plantlets when cultured in small tubes but produce callus when cultured in large containers (Kartha *et al* 1974). Most commonly used containers are glass tubes 10 × 120 mm, closed with caps permitting air exchanges. Sealing of culture vessels to avoid contamination, may inhibit effective gas exchanges and cause accumulation of various gases that could adversely affect morphogenesis.

Micro-grafting

Micro-grafting is a convenient technique to raise virus-free planting materials particularly in case of graft-propagated plants. In this case, the meristem (0.1-0.4 mm) excised from the infected cultivar, is aseptically grafted onto the vascular ring of a decapitated virus-free rootstock, such as a seedling or a shoot propagated in culture. This technique was first used to eliminate different viruses from *Citrus* species as an alternative to the use of nucellar seedlings to avoid juvenility (Murashige *et al* 1972, Navarro *et al* 1975, Russo and Starrantino 1973, Mukhopadhyay *et al* 1997). This technique has also been applied to several other fruit species such as *Prunus persica* (peach), *Prunus amygdalus* (almond), *Prunus salicina* (Japanese plum), *Prunus domestica* (Stanley plum), *Prunus insititia* (Pollizo plum), *Prunus armeniaca* (apricot), etc. with variable success.

Transplanting of Plantlets

After the development of satisfactory root system by the cultured shoots, they are transferred to well-drained soil or specialized` sand mixtures. At this step careful attention is to be given to remove the agar sticking on the root surface to avoid future contamination. The most important requirement in this operation is to keep the plants under very high humidity (90-100%) for at least 2 wk. Humidity should then be gradually decreased to adjust plants to normal conditions of the glasshouse. Plants forming resting organs, such as tubers, corms, rhizomes and bulbs, plantlets should be induced to produce these organs in culture.

Indexing

After transferring the regenerated plantlets from the culture medium to soil, they must be indexed to ensure that they are virus-free. Indexing should be done by more than one method. In addition to testing by conventional transmission and symptom development in indicator plants, sensitive serological methods such as ELISA, ISEM or nucleic acid hybridization/cDNA, cRNA, etc. should be used. In many cases, virus at a very low concentration may remain in the meristem, for which highly sensitive methods are required for detection. Moreover certain reverse transcribing viruses can be activated by stresses such as tissue culture and crossing (Harper 2002, Ndowora *et al* 1999, Hull 2000) indicating the necessity of robust diagnostic procedures.

Virus Elimination

Virus elimination by tissue culture/meristem-tip culture depends on meristem size, type of virus and virus strain, plant species and cultivar, physiological condition of the mother plants and meristem position on it. Among these factors meristem size is most important. Each plant-virus combination exists as a specific apical 'immunity' regarding a certain meristem size. Thus before starting a successful elimination programme, virus distribution in meristem tip is to be precisely determined.

Elimination by Tissue Culture

Culture of single cells or small clumps or cells from virus-infected plants may sometimes give rise to virus-free plants. For example, Preil *et al* (1982) eliminated two viruses from *Euphorbia pulcherima* by cell suspension culture followed by regeneration of plants *in vitro*. Toyoda *et al* (1985) made TMV-free tobacco plant by regenerating plants from shake culture of small pieces of infected tobacco tissues.

Meristem-tip Culture Coupled with Thermotherapy

To improve pathogen elimination, other techniques such as thermo- and chemotherapy coupled with meristem-tip culture can be used. Thermotherapy can be carried out on mother plants, before meristem excision or during the *in vitro* culture of meristem tips. The first type of application is less damaging and also has the advantage of allowing the use of somewhat larger meristems. This technique has been followed to eliminate viruses from many economic crops. In strawberry, 86% elimination of SMYEV is possible with heat treatment at 36°C for 6 wk when only 25% elimination is possible without such treatment (Mullin *et al* 1974). For *strawberry latent A virus* and *strawberry mottle virus* and mixed infection of the two, an increase in virus-free material produced from 17 to 71%, 35 to 100% and 0 to 17% respectively when the meristem-tips were heat treated for 5 wk at 38°C (Sobcykiewicz 1979). With this procedure several potato cultivars were freed from PVS and PVX (MacDonald 1973, Stace-Smith and Mellor 1968). High yield of PVS-free (93%) cv. Majestic plants were obtained from meristem tips sampled from plants heat treated at 35°C for 26 d. Direct heat treatment in tissue culture can also be effective in some cases. It was possible to eliminate CMV and AMV from *Nicotiana rustica* when treated at 30°C for 4-5 d or at 45°C for 9 d (Walkey 1976, Walkey and Cooper 1972). GLRV and the agent of grapevine necrosis were also similarly eliminated (Mur 1979). GFLV was eliminated from the culture of grapevine shoot apex by heat treatment at 35°C, while cultures held at normal temperature were all infected (Barlass *et al* 1982). The same temperature was used for 4 wk to free shoot-tip culture of sweet cherry from PNRSV (Snir and Stein 1985). Low temperature could also be utilized to free tissues from viruses before meristem sampling. PVA and PVY are eliminated from potato tissues grown at 5°C (Moskovets *et al* 1973). PSTVd is eliminated from potato by cooling at 5°C for 3-6 mon (Lizarraga *et al* 1980). Helliot *et al* (2002) cryo-preserved excised meristematic clumps of banana (cv. Williams-AAA, Cavendish subgroup) infected with CMV and BSV using PVS2 solution. The frequency of virus elimination in these plants for CMV and BSV was found to be 30% and 90% respectively. Wang *et al* (2003) successfully eliminated *grapevine virus A* (GVA) by cryo-preservation of *in vitro* grown shoot tips. The freezing step resulted in 97% of the GVA elimination.

Verma *et al* (2004) could eliminate CMV from *Chrysanthemum* by meristem tip culture. Seventy-two per cent of the plants were found to be negative to CMV. Criely (2003) produced BYMV and CMV-free gladiolus (cv. Morala) by meristem-tip culture, heat therapy and using a biovirus inhibitor. All the plants regenerated from the meristem were free from the viruses. In addition, hot water treatment of corms at 55°C could also eliminate these viruses from the plants. It was also possible to eliminate the viruses from the crops by spraying infected plants with 90 mg/l of *Asparagus adscendens* root extract.

When explants from infected plants were grown on medium supplemented with the virus inhibitor, even a concentration of 7.0 mg/l, could eliminate both the viruses. Balamuralikrishnan *et al* (2002) observed the effect of anti-viral chemo-therapeutants (ribavirin, and 8-azaguanine) applied to the medium for meristem-tip culture on the eliminated *sugarcane mosaic virus* from sugarcane (CoC 671). Ribavirin at 50 ppm eliminated the virus without any phytotoxicity. Williamson (2002) eliminated *raspberry bushy dwarf virus* from raspberry (cv. Gatineu) by *in vitro* meristem culture with or without thermotherapy (37 ± 1°C). Garcheva (2002) eliminated *plum pox virus* in plum (cvs. Kyustendiska Sinya and Valjevka) performing shoot culture. After the fourth subculture, 35.3% of obtained sub-clones of Kyustendiska and 33% of Valjevka were free from the virus. After 8 passages 88% of established sub-clones of Kvustendiska and 100% of Valjevka were PPV-free. James and Clark (2001) performed shoot culture of *Malus domestica* (*M. pumila*) infected with *apple stem grooving virus* (ASGV) and treated *in vitro* with a combination of quercetin and ribavirin (10 μg ml^{-1} each) for 9 to 12 wk. Over 3 y after the terminating treatment, the treated plants were normal, virus-free and slightly vigorous than the uncontrolled plants. Park-Insook *et al* (2002) produced gladiolus (Spic, Span, True love, Topaz), free from *broad bean wilt virus, bean yellow mosaic virus, cucumber mosaic virus, clover yellow vein virus* and *tobacco rattle virus* by corm tip culture. Webster (2000) eliminated *apple chlorotic leaf spot virus* from pear by *in vitro* thermotherapy and chemotherapy. Micro-propagated shoots of pear (cv. Pierre Corneille) were subjected to heat therapy at 36°C or/and chemotherapy using ribavirin (virazole) at different doses to eliminate the virus. ELISA did not show the presence of the virus in 78 and 80% of pear shoots treated with virazole in concentration of 25 and 50 mg/l respectively.

Yi *et al* (2001) made a comparative study on the different methods (direct meristem-tip culture, meristem-tip culture in combination with thermotherapy and meristem-tip culture in combination with ribavirin for elimination of carnation viruses. Meristem-tip (0.2 mm) culture in combination with thermotherapy showed best de-virus rate reaching 77.78%.

Virus-free materials thus produced are to be propagated and maintained under extreme phyto-sanitary conditions to preclude reinfection with periodic inspection for the virus and horticultural traits.

Risks of Tissue Culture

Tissue culture (TC) has become a convenient tool for commercial production of virus-free plants, but the process has several risk factors.

1. There are different types of explants and methods to raise TC plants. Some of these explants and methods may lead to variability particularly plants, raised from callus and somatic embryogenesis.

2. If the plantlets raised in media are repeatedly used for further proliferation, continuous exposure to the growth substances may induce genetic alterations of the parental type.

3. In many cases meristem tips may contain some viruses/viroids/ phytoplasmas even after thermo- and/or chemotherapy. Planting materials commercially produced by tissue culture are to be indexed before releasing them to the fields.

4. Tissue culture plants made free from one virus may be more susceptible to another virus as found in banana. Banana made free from *bunchy top virus*, becomes more susceptible to *banana mosaic virus*.

Cultivation of Resistant Plant

Cultivation of resistant plants is an ideal component for the management of virus diseases. Resistant crop cultivars may occur in nature but conventional breeding has developed most of the varieties that are cultivated. For example, many wheat cultivars have been developed and registered that are claimed to be resistant to *soil-borne wheat mosaic virus*. Other cultivars are also claimed to be resistant to *wheat spindle streak mosaic virus* (Bacon *et al* 2002, Sears *et al* 2001a, 2001b, Martin *et al* 2001a, Carver *et al* 2003). McPhee *et al* (2002) developed and registered 'cv. *Lifter*' green dry pea resistant to *pea enation mosaic virus* and *pea seed borne mosaic virus*. Hirai *et al* (2003) developed a new MNSV resistant melon rootstock variety '*Sorachi Dai Sako*. A sugar beet germplasm line, *CR09-1* resistant to *beet necrotic yellow vein virus* was developed, registered and released in 2001 (Lewellen 2002a, b) and sugarbeet germplasm lines *C67/2, C68/2, C78/3, C80/2, C869, C.869CMS* resistant to yellows and rhizomania (*beet necrotic yellows virus*) were registered and released in the USA (Lewellen 2004a, b). Similarly, there are several claims about the development of resistant cultivars.

There is however a real problem with resistance terminology. For example 'Tolerance', as the presence in the plant of the same concentration of virus as in susceptible plants, but the plant shows few symptoms or little damage — thus it is tolerant.

For judicious breeding, proper understanding of the resistance is required. A virus may become inactivated immediately after its entry into the host cell and thus get eliminated from the host's cellular system. In this type of reaction, the plant is called 'immune' or 'hypersensitive'. Immunity is the highest form of resistance in which the virus does not allow it to invade. In hypersensitive reactions that are more common than immunity in the field, the virus remains restricted around the point of infection by the collapse of plant tissue around the point of infection. In another form of resistance, the virus, after its entry into the host, multiplies in a limited manner without producing any symptoms. This type of infection is called 'latent infection' and the grade of resistance is termed 'tolerance'. The plasticity of resistance may be further extended to interactions leading to infection and production of symptoms but causing hardly any loss in yield or quality. Immunity and hypersensitivity seem at first sight attractive options, but in practice it is often disappointing. Viruses attacking the same host and designated as the same virus are not uniform and possess a wide range of flexibility to react with the host. There can also be pseudo-recombination of viral genomes and natural mutation. In a resistance-breeding programme, the plasticity of viruses is often overlooked, and desirable and durable results are often not obtained and viruses very often mutate to overcome the host resistance, depending upon the virus-host system.

From the conventional breeding point of view tolerance or toremicity (resistance operating to slow disease spread in the field) has both advantages and disadvantages. It allows the virus and the host to operate together without much effect on the desired end product and yield. However, tolerance is often environmentally determined and changes in conditions may lead to epidemics. Thus, this type of resistance alone as a control method has risks particularly in tropical countries. In addition, tolerant plants can act as source of viruses for other crops where epidemics may develop.

Resistance to viruses can be confused with resistance of hosts to the feeding of vectors. Most of the *rice tungro virus* resistant varieties bred in the past were, in fact, resistant to vectors. When the insects had adapted to feed on those varieties, epidemics reappeared. So the resistance or susceptibility of plants to vectors is also an important criterion in breeding for virus resistance.

Resistance to a virus may be due to a major or minor gene, dominant or recessive, monogenic or polygenic. Separate gene or gene pool/clusters, may be involved for resistance to different viruses, vectors, and other pathogens. All may affect the breeding strategy, as do the ploidy level of the crop, self-fertility or incompatibility necessitating embryo rescue or other *in vitro* techniques. The source of the gene and these factors will determine how many back-cross generations are required before the desired resistance can be introgressed into an agronomically adapted line. Overall the best and most durable results will be obtained when different resistance genes with effects on different stages of the viral life cycle are pyramided into a single cultivar.

Varieties bred in one place as resistant may not be resistant elsewhere because of virus variablility. Varieties of sugarbeet introduced in the USA as resistant to *curly top virus* became susceptible very quickly. Tomato (Pearl Harbor) was resistant to *tomato spotted wilt virus* in Hawaii, but not in North America. Potato seedlings hypersensitive to strains of PVY in Australia were not hypersensitive to some strains in the UK.

It is also necessary to take into account the co-evolution of host virus system in searching for resistance genes. Normally resistance is searched in the areas where both the host and the virus undergo simultaneous evolution. But the resistant genes can also be found where co-evolution of host and the pathogen does not occur. In this case, resistance seems to be accidental. Genetic balances of pathogens in this situation usually occur in plant systems other than the concerned crop to which they are known to occur. In some cases, resistance can be found in plant species that are geographically and genetically distant from the chronically infected plant species occurring in geographically different locations.

Mechanisms of virus resistance are normally of three different types: (i) Non-host resistance, (ii) Heritable resistance and (iii) acquired resistance. Resistance may be positive or negative. Positive resistance implies the presence of a mechanism that inhibits some phase of virus replication, as for example, in the hypersensitive reaction. It is primarily dominant or dosage dependent (i. e. homozygotes are more resistant than heterozygotes). Negative resistance implies the lack of a necessary component (e.g. a host factor) for the virus to complete the life cycle. These might include a protein that interacts with virus movement protein, or a compatible host-coded replicase subunit. Such resistance is typically recessive.

Acquired resistance results from exposure of susceptible plants to the virus, another pathogen, a chemical, or a cultural treatment, and is not heritable. Pathogen-derived resistance in transgenic plants is a special case of acquired resistance that would normally be heritable.

Occurrence of Resistance Genes in Crops to Different Viruses

Several genes have been identified in many countries that are resistant to different viruses. For example, the host gene *Tm 22* in tomato provides resistance to ToMV and TMV and the resistant tomatoes developed were effective in commercial plantings worldwide (Watterson 1993). Conventional breeding has been used successfully to develop resistant varieties in some cucurbits especially CMV resistant cucumber (Superak *et al* 1993, Munger 1993) and seeds of which are marketed as 'Marketmore' series in the USA.

Search for resistant genes against viruses has gradually increased and early records of some host genes for resistance to plant viruses are given in Table 8.1.

Table 8.1   Examples of host genes for resistance to plant viruses

Gene	Host species	Virus	Selected reference
Controlled by dominant genes			
N	*Nicotiana glutinosa*	TMV	Holmes (1929)
N′	*N. sylvestris*	TMV	Melchers *et al* (1966)
Zym[a]	*Cucurbita moschata*	ZYMV	Paris *et al* (1988)
Tm-2	*Lycopersicon esculentum*	TMV	Pilowski *et al* (1981)
Tm-2^2	*L. esculentum*	TMV	Cirulli and Alexander (1981)
Nx, Nb	*Solanum tuberosum*	PVX	Jones (1982)
By, By-2	*Phaseolus vulgaris*	BYMV	Schroeder and Provvidenti (1968)
Rsv$_2$, Rsv$_1$'	*Glycine max*	SbMV	Buzzell and Tu (1984)
Controlled by incompletely dominant genes			
Tm-1	*L. esculentum*	TMV	Fraser *et al* (1980)
L, L3, L^1	*Capsicum* spp.	TMV	Berzal-Herranz *et al* (1995), Dardick and Culver (1997), Dardick *et al* (1999)
Two genes	*Hordeum vulgare*	BSMV	Sisler and Timian (1956)
Multiple genes	*Vigna sinensis*	SCPMV	Hobbs *et al* (1987)
Apparently recessive genes			
By-3	*P. vulgaris*	BYMV	Provvidenti and Schroeder (1973)
Sw$_2$', Sw$_y$, Sw$_4$	*P. vulgaris, L. esculentum*	TSWV	Finlay (1953)

Abad *et al* (2000) identified resistance to CMV isolates from eastern Spain in accessions of the species *Lycopersicon hirsutum* (2 accessions), *L. chmielewskii* (1 accession), *L. pimpinellifolium* (1 accession), and *L. esculentum* (3 accessions), but in case of the severe *Fny*-CMV, only *L. chmielewskii- CHM-47* was found to be resistant to this severe strain. No accumulation of the virus was detected in the inoculated upper leaves of the plant. The resistance factor also did not interfere with replication or cell-to-cell movement. It may perhaps act at the early stage of the infection.

Daryono *et al* (2003) screened 40 melon cultivars collected from 17 Asian countries to locate resistance to an Indonesian isolate of CMV-B2 and found resistance only in 5 cultivars (Yamatauri, Miyamauri, Mawatauri, Sanuki-shirouri and Shinjong). A single dominant gene conferred this resistance and the gene was designated as '*Creb-2*'. Grube *et al* (2000) identified genes resistant to several isolates of CMV in *Capsicum frutescens* '*BG2814-6*' controlled by two major recessive genes. *C. annuum* '*Perennial*' is also resistant to CMV but there may be a mechanistic difference in the expression of resistance shown by these two species.

Dogimont *et al* (1997) identified that two independent complementary recessive genes ('*cab-1*' and '*cab-2T*') confer the resistance to *cucurbit aphid-borne yellows virus* (CABYV) in Indian line PI 124112 of *Cucumis melo*.

In potato, the pattern of resistance to CMV however, differed. Celebi *et al* (2003) studied the susceptibility of various potato genotypes to three different CMV strains, *pf*-CMV, *Fny*-CMV (subgroup I) and *A9*-CMV (subgroup II) and found systemic infection by one of these three strains. Normally potato cultivars are resistant to CMV at 24°C, but all become infected systemically when inoculated plants are grown at 30°C. Thus natural resistance that most of the potato crops express to CMV may be overcome under high-temperature growing conditions following infection and where resistance is found, it is strain specific.

Martin *et al* (2003) identified that the melon gene '*Vat*' confers resistance to non-persistent virus transmission (CMV) by aphids (*Myzus persicae, Aphis gossypii*). No relationship has been detected between resistance and the probing behaviour of aphid stylet tips but the resistance may be due to temporary prevention of the release of virus particles, the function conferred by the gene.

Parrella *et al* (2004) identified a gene (*Am*) in *Lycopersicum hirsutum* f. sp. *glabrum* (p1134417) that confers resistance to most strains of *alfalfa mosaic virus* including necrotic strains. '*Am*' gene acts by an inhibition of viral accumulation during the early events of the viral life cycle. The gene has been mapped genetically to the short arm of the tomato chromosome-6 that includes the '*R*'-genes *Mi* and '*cf-2/Cf-5*' and the quantitative resistance factors '*Ty-1*', '*01-1*' and '*Bw-5*'.

Rocha *et al* (2003) screened 24 cowpea white testa genotypes against *cowpea mosaic virus, cowpea aphid-borne mosaic virus* and *cucumber mosaic virus* and identified the broadest spectrum resistance in genotypes *TE-87-98-13 G* and *TE-87-98-9G-2* to all the 3 viruses while *IT 85F-2687* and *IT 86D-716* showed very high resistance to *cowpea aphid borne mosaic virus* associated with good seed yield. Nogueira *et al* (2001) observed that cultivar '*costelao*' has a dominant gene for susceptibility and the other parents have a recessive gene for resistance in Brazil.

Pico *et al* (2003) screened a collection of 46 Spanish cucumber land races, 4 melon cultivars and 1 accession of *Cucumis africanus* for resistance against cucumber vein yellowing virus (isolate from Almeria, closely related to that found in Israel) and found partial resistance displaying mild symptoms and significantly reduced virus accumulation.

Fedak *et al* (2001) identified resistance to *barley yellow dwarf virus* (BYDV) in wild relatives of wheat such as *Thinopyrum* species. There is no resistance to this virus in the primary gene pool of wheat and it was introgressed from this wild species for resistance to this virus. It was found in partial amphidiploids consisting of 42 wheat, 14 alien chromosomes obtained in hybrids between wheat and *T. intermedium* (*Elymus hispidus*) and *T. ponticum* (*E. elongatus*).

Huth (2003) identified BYDV tolerant 14 winter barley accessions from the German Gene Bank Gatersleben. This tolerance was not located on the '*Ryd 2*' gene. This gene together with '*Ryd 1*' was the tolerance loci.

Ordon *et al* (2004) mapped several recessive genes in barley for BYMV e.g. '*rym 4*', '*rym 9*', '*rym 11*', '*rym 13*', etc. and identified BYMV tolerance on chromosomes *2HL* and *3HL* of several accessions. Chelkowski *et al* (2003) while studying the major genes ('R' genes) against pathogenic viruses and aphids in *Hordeum vulgare* accessions, found the greatest resistance to *barley yellow mosaic* and *barley mild mosaic* viruses in genes 17 and 14 respectively. Werner *et al* (2003) identified '*Taihouku A*' derived from 'Taiwan' as resistant to all known members of *barley yellow mosaic virus* complex carrying a BaMMV resistance gene not allelic to '*rym 4*'/'*rym 5*', the current genetic base of European resistant cultivars. This resistance gene '*Taihoku A*' tentatively named '*rym 13*' is located in the telomeric region of chromosome *4HL*.

Okada *et al* (2003) found a Chinese land race of barley 'Mokusekko 3' is completely resistant to all strains of *barley yellow mosaic virus* (BaYMV) and *barley mild mosaic virus* (BaMMV) and known to have at least two resistant genes '*rym 1*' and '*rym 5*'. '*rym 1*' gene is completely resistant to BaYMV-I, -II, BaMMV- '*Ka 1*', '*Na 2*' and acceptable resistance to BaYMV-III.

McGrann and Adams (2004) identified 11 genes that confer resistance to barley (*Hordeum vulgaris*) 2 or more components of the *soil-borne barley mosaic virus* complex. Apart from the immunity conferred by the widely used gene '*rym 4*' little is known about their mode of action. Plants with the genes '*rym 3*' or '*rym 6*' were fully susceptible to the virus, whereas those with genes '*rym 1*', '*rym 2*', '*rym 5*' or '*rym 11*' appeared to be immune as BaMMV was never detected in any tissue type or was the virus transmitted from them to susceptible genotypes. Operation of translocation resistance in plants with '*rym 8*', '*rym 9*' or '*rym 10*' genes which is temperature sensitive in '*rym 8*' and '*rym 10*' and perhaps tissue specific in '*rym 9*' .

Chen *et al* (2003) found *Thinopyrum intermedium* (*Elymus hispidus*) (2n = 6x = 42; JJs Js SS) as a potentially useful source of resistance to wheat streak mosaic virus (WSMV) and its vectors the wheat curl mite-WCM (*Aceria tosichella*). Five amphidiploids (Zhong 1, Zhong 2, Zhong 3, Zhong 4 and

Zhong 5) derived from *Triticum aestivum* × *T. intermedium* crosses produced in China were screened for WSMV and WCM resistance. Zhong 1 and 2 had high levels of resistance to both WSMV and WCM while Zhong 3, 4 and 5 were resistant to WSMV, but were susceptible to WCM. Kanyuka *et al* (2004) screened 26 *Triticum monococcum* lines of diverse geographical origin for resistance to *soil-borne cereal mosaic virus* (SBCMV). The resistance to the virus and partial resistance to the vector was found only in one Bulgarian isolate.

Ndjiondjop *et al* (2001) identified high resistance in *Oryza glaberima* cv. Tog 5681 and *O. sativa* var. *indica* cv. Gidante to *rice yellow mottle virus* correlated with the failure of cell-to-cell movement.

Ogundiwin *et al* (2002) identified high level of resistance in *Vigna vexillata*, a wild *Vigna* species to *cowpea mottle virus*. Segregation patterns observed in the F2 and the back cross populations of all the crosses between two resistant *V. vexillata* line and three susceptible lines showed that the resistance is controlled by a single dominant gene and the level of resistance conferred by this gene is very high.

Mallor *et al* (2003) determined the mechanism resistance in melon (*Cucumis melo* L.) to *melon necrotic spot virus*. A single recessive gene called '*nsv*' controls this resistance but the genetic control in the progenies from the cross '*Doublin*' × '*ANC 42*' (a susceptible line) is conferred by two completely dominant genes '*Melon necrotic resistance 1*' ('*Mnr1*') and '*Melon necrotic resistance 2*' ('*Mnr2*').

Bosland (2000) determined the sources of *curly top virus* resistance in *Capsicum*. Resistance (0-25% devoid of any symptom) was found in 3 accessions of each of *C. annuum*, ('Burpee chiltepin', 'NumMex Bailley Piquin' and 'NuMex Twilight'), *C. frutescens* ('USDA-Grif-9322' from Costa Rica, 'PI 241675' from Ecuador and Tabasca) and 1 accession of each of *C. chacoense* ('PI 273419' from Argentina) and *C. chinense* ('USDA-Grif.9303' from Colombia).

Redinbough *et al* (2004) identified and mapped the genes and *quantitative trait loci* (QTL) associated with virus resistance in maize (*Zea mays*). Genes or major QTL for resistance to *maize dwarf mosaic virus, maize streak virus, sugarcane mosaic virus, wheat streak mosaic virus, maize mosaic virus, high plains virus* and *maize chlorotic dwarf virus* were mapped in the maize genome. Resistance to other viruses including *maize rayado fino virus, maize fine streak virus* and *necrotic streak virus* were identified but not characterized genetically. Clusters of maize virus resistance genes are located on chromosomes 3, 6 and 10. The clusters on chromosome 6 ('*bin 6.01*' near '*umc 85*') carry resistance to 3 members of the Potyviridae (MDMV, SCMV and WSMV). The clusters on chromosome 3 ('*bin 3.05*' near '*umc102*') and 10 ('*bin10.05*' near '*umc44*') carry resistance to phylo-genetically diverse viruses as well as bacterial and fungal pathogens.

Gragera *et al* (2003) developed some breeding programmes to obtain cultivars with resistance to *tomato spotted wilt virus* (TSWV) and found '*sw-5*' gene confers resistance to TSWV. Czech *et al* (2003) confirmed that '*sw-5*' gene confers resistance to one of the Polish isolates of TSWV. Cebolla-Cornejo *et al* (2003) identified the resistance conferred by the '*Tsw*' locus from *Capsicum chinensis*.

Syller (2003) identified 2 clones of potato (M62759, PS 1706) that showed resistance to *potato leaf roll virus* (PLRV) and showed high level of resistance to virus multiplication. Strong inhibition to virus transport to tubers was found with M 62759. Three valuable components of the resistance to PLRV are probably closely linked in the genotype, a combination that seems to occur rather rarely in potato clones. Kai and Amnon [International Service for the Acquisition of Agribiotech Application (ISAAA), 05 October 2007] screened over 200 bottle gourd accessions from different parts of the world and found 36 ZYMV resistant cultivars, 33 of which are from India. When popular ZYMV susceptible watermelon was grafted on the resistant bottle gourds, resistant watermelons were obtained. *Brassica* species, broccoli, cauliflower, rape, cabbage and mustard are susceptible to TuMV. The genes that include '*retro 1*' and '*Con Th 01*' determine the plant response to virus attack; plants can kill infected cells, thereby restricting viral infection in certain areas. They can also restrict viral spread from leaf to leaf. These genes are now bred to raise commercial varieties. A group of international researchers has identified the precise location and chemical nature of the first receptor for a non-circulative virus, CaMV. The receptors were located in the tip of the aphid's maxillary stylets as non-sugar coated proteins embedded in the stylet's chitin matrix. CaMV spread can now be controlled either by targetting the aphid receptors or P2 proteins, found to be responsible in mediating virus-host interaction (ISAAA, 16 November 2007).

There are now several commercial crops that have been made resistant through resistance-breeding programmes in different countries and a few examples of them are given in Table 8.2.

Table 8.2 Examples of resistance to viruses present in commercial varieties of some cultivated species (Astier *et al* 2007)

Crop	Virus
Bean	BCMV
Cucumber	CMV, PRSV, WMV, ZYMV
Lettuce	BWYV, LMV, TuMV
Melon	CMV, MNSV, resistance to aphid transmission
Pea	BYMV, PSbMV
Pepper	CMV, PMMoV, PVY, TMV
Potato	PLRV, PVX, PVY
Squash	CMV, ZYMV
Tomato	TMV, ToMV, TYLCV, TSWV

Genetic Engineering for Plant Virus Resistance

Resistant varieties developed through conventional breeding are normally desirable as the resistance is durable, environmentally safe and can be made available to the farmers easily. However, the process takes a long time and in some cases difficulties arise. As, for example, it is difficult to bring *Tm-22* gene into breeding programmes, as undesirable traits are tightly linked to the single partially dominant *Tm-22* gene; resistance factors have been identified to CMV in several wild tomato species (Watterson 1993) but a resistant variety could not be developed due to the polygenic nature of the resistance and plant infertility. Conventional breeding developed potatoes resistant to PVX, PVY, PVS, PVM, PVA, but the varieties were resistant only to the single viruses but multiple infections are a common occurrence in crops. Moreover these varieties were not preferred for commercial cultivation (Foxe 1992). The application of conventional breeding to potato is difficult because the cultivated potato is an autotetraploid and highly heterozygous and incorporation of resistance to PLRV is difficult because of the polygenic nature of the resistance.

Genetic engineering technologies can help to overcome some of the limitations of conventional breeding and can manipulate the resistance genes to develop durable transgenics for commercial cultivation.

Powell-Abel *et al* (1986) first discovered the usefulness of transgressing portions of the viral genome for imparting resistance. When capsid protein (CP) of TMV was transgressed to *N. tabacum*, the plant becomes resistant to TMV. Since then, numerous papers and reviews have been published on Pathogen Derived Resistance (PDR), and the concept has been proven valid for a wide variety of plants, for a number of genes or portion of genes, and for members of virtually every genus of plant viruses (Freitas-Astúa *et al* 2002).

Viral coding sequences have been used as sense, antisense, full length, truncated or untranslatable constructs. In addition, several viral non-coding sequences have been used, including satellite RNA, defective interfering RNA, terminal untranslatable sequences and ribozyme.

Protein-mediated resistance

Protein-mediated resistance depends on the synthesis of the transgene-encoded protein particularly capsid, replicase, movement protein, polyprotein, protease, etc. The concept of pathogen-derived resistance was first demonstrated in transgenic tobacco plants expressing the coat protein gene of TMV (Powell-Abel *et al* 1986). Since then, genetically engineered resistance has been reported for a number of plant viruses involving virus-derived genes or genome fragments that have been reviewed by Beachy (1993), Grumet (1994), Lomonosoff (1995), Fuchs *et al* (1997), Kaniewski and Lawson (1998), Hull (2002) and Astier *et al* (2007).

(i) *Coat Protein (CP)-mediated resistance*

The capsid gene is one of the most widely studied genes of many viruses, belonging to more than 15 taxonomic groups. The level of protection is linked to the level of expression of the transgene measured by the concentration of the capsid protein. The protection is stronger if the similarity between the capsid expressed by the plant and that of the infecting virus is very high. The protection is generally effective against the viral strains from which the transgenes come, and their homologues.

The expression of the capsid of RSV in two Japanese varieties protects the transformed plants and their progeny against this virus (Hayakama *et al* 1992). The same is true with the outer coat protein of RDV (Zheng *et al* 1997).

CP-mediated resistance to diverse virus strains or isolates have also been reported and engineered to produce transgenics. Some of these viruses are: TMV (Nelson *et al* 1988), ToMV (Sanders *et al* 1992), CMV (Namba *et al* 1991, Quemada *et al* 1991, Xue *et al* 1994), AlMV (Loesch-Fries *et al* 1987, Tumer *et al* 1987), PVX (Hemenway 1988), PVY (Perlak *et al* 1994; Makeshkumar *et al* 2002, Ghosh *et al* 2002), PLRV (Kawchuk *et al* 1991, Barker *et al* 1992, Kaniewaski *et al* 1993, Brown *et al* 1995, Presting *et al* 1995, Thomas *et al* 2000), TSWV (de Haan *et al* 1992), TBRV (Pacot *et al* 1999), WSMV (Sivamani *et al* 2002), RgMV (Xu *et al* 2001), CyMV-Ca (Lim *et al* 1999), TNV (Hackland *et al* 2000), ZYMV & WMV-2 (Fuchs and Gonsalves 1995, Arce-Achoa *et al* 1995, Clough and Hamm 1995, Tricoli *et al* 1995), PRSV (Fitch *et al* 1992, Liues *et al* 1996) and others.

Capsid protein normally confers resistance to selected viruses within a virus group, as for example, CMV-C capsid protein gives resistance to strains belonging to subgroup I (Quemada *et al* 1991) but there may be differences of expression of different strains of the same virus as for example PRSV. Coat protein of this virus confers resistance only to some strains (Gonsalves and Slightom 1993). Coat protein precursor gene can also confer resistance. Kodotani and Ekegami (2002) obtained *Patchouli mild mosaic virus* resistant *Patchouli* plants by genetic engineering of coat protein precursor gene.

It has been claimed that the coat proteins of some viruses also confer resistance to their vectors, as for example, potatoes that express coat protein of PVX and PVY are resistant to aphid transmitted PVY (Lawson *et al* 1990). Tobacco, tomato and cucumber expressing CMV coat protein are highly resistant to aphid transmission (Gonsalves and Slightom 1993, Quemada *et al* 1991, Tomassoli *et al* 1995). Coat protein of TSWV confers resistance to thrips transmission (Goldbach and de Haan 1993). There is a role of outer capsid in the transmission of phytoreoviruses by insect vectors (Omura and Yan 1999). It is however, not certain that this is true vector resistance or it just that the plants cannot be infected by the vector route.

Because of the risks and strict regulations for commercialization of transgenic crops, very few virus resistant transgenics are commercially available. Trangenic squash hybrid ZW 20 (renamed 'Freedom I') resistant to ZYMV and WMV-2 was approved as the first CP- mediated virus-resistant genetically engineered crop to be deregulated by the USDA-APHIS (Animal Plant Health Inspection Service) (Medley 1994). Transgenic squash line CZW-3 resistant to ZYMV, WMV-2 and CMV (multiple resistance conferred by the respective CP genes of the viruses) also received exemption status from the USDA-APHIS (Acord 1996). Transgenic cantaloupe has also been developed using multiple constructs of CP genes from ZYMV, WMV-2 and CMV (Fuchs *et al* 1997). Transgenic papaya line '55-1' expressing CP gene of PRSV-HA 5-1 named '*SunUp*' has been deregulated by the USDA-APHIS in 1996 (Starting 1996) but this variety is effective only in Hawaii. The development of multiple resistance is in the pipeline using PRSV isolates from the Caribbean, South America, and Asia.

(ii) *Replicase mediated resistance*

Replicase mediated resistance has been reviewed by Palukaitis and Zaitlin (1997). Nunome *et al* (2002) developed breeding materials of transgenic tomato plants with a truncated replicase gene of CMV for resistance to the virus. Plants expressing a non-functional AMV replication protein mutated in the polymerase motif GDD, present a resistance linked to a strong expression of the transgene (Brederode *et al* 1995). The expression of defective replicase ahead of the virus cycle could be the source of resistance. However, the characteristics of resistance in plants expressing replicase gene sequences depend on the homologous resistance linked to the RNA. The resistance is thus limited to strains in which the sequence is very similar to that of the transgene (Carr and Zaitlin 1993, Baulcombe 1996, Jones *et al* 1998).

In case of RTSV on the other hand, rice plants expressing the replicase gene of the virus in the anti-sense form, are protected up to 60% against infection. The plants expressing the complete or truncated replicase gene RTSV regularly manifest a total resistance to the virus even with high doses of inoculum, by a phenomenon of silencing, moreover, these plants are no longer capable of assisting RTBV for its transmission (Huet *et al* 1999).

Resistance against RYMV has been obtained by expressing a construct derived from the replicase gene, a gene in which the sequence is particularly conserved between different isolates. This is a resistance by 'homology' between transgene RNA and the viral RNA. The resistance seems exceptionally wide-ranging, and it is manifested again st isolates of different serological groups (Pinto *et al* 1999).

In some viruses where a protein (*Rep* protein) is recruited from a cellular system for their replication, the mutated or truncated form of it often acts as dominant negative competitor in viral replication and regulates its own transcription from the C1 gene. This type is often found in members of the Geminiviridae (Sangaré *et al* 1999, Brunetti *et al* 1997).

Movement protein-mediated resistance

Movement protein is expressed in a truncated form which confers resistance to a wide spectrum of viruses. For example, the expression of truncated movement protein of BMV confers resistance against TMV (Malyshenko 1993) and the expression of truncated movement protein of TMV confers resistance against CaMV, AMV, tobamoviruses similar to TMV and TRSV (Cooper *et al* 1995). The levels of protection of resistant plants are not the same in the above cases, which may be due to the competition between the viral and the mutated transgenic movement proteins, either at the level of plasmodesmata or at the level of attachment to viral RNA. The non-functional proteins behave like dominant negative mutants blocking the long-distance movement of the virus with high efficiency producing broad-spectrum resistance (Beachy 1997). Martin *et al* (2001) induced engineering resistance to raspberry ring spot against RBDV infected materials in the field and showed that 53 out of 141 transgenic lines were free from RBDV after 2 rounds of grafting.

Ribosome inactivating protein

Ribosome inactivating proteins (RIP) inactivate the ribosomal RNA and arrest protein synthesis. Their activity is exerted with a clear antiviral specificity that was shown by the clear antiviral specificity with RIP extracted from *Mirabilis jalapa*, *Phytolacca americana*, and *Dianthus caryophyllus*. Parveen *et al* (2001) characterized an inducer-protein (*Crip-31*) from *Clerodendron inerme* that induced systemic resistance against CMV, PVY and ToMV. This protein was considered as a ribosome inactivating protein possessing antiviral properties. The synthesis of such a protein put under the control of a viral promoter leads to cell death during the induction of the promoter by the viral infection (Astier *et al* 2007).

RNA-mediated resistance

According to Prins and Goldbach (1996), the RNA-mediated resistance (RMR) approach arose as an unexpected spin-off from the concept of PDR. In contrast to what was expected, the resistance observed in several transgenic lines, especially those transformed with the replicase gene had no direct correlation with the levels of protein produced (Anderson *et al* 1992, Audy *et al* 1994, Baulcombe 1994). In addition, in the early 1990s, several research groups reported that plants transformed with untranslatable sequences of viruses were resistant to their challenge homologous viruses (Lindbo and Dougherty 1992a, b, van der Vlugt *et al* 1992). Lindbo *et al* (1993) demonstrated through run-on analysis that tissues which exhibited a typical 'recovery phenotype' had high levels of transcription of the transgenes in the nucleus and very low levels in the cytoplasm. These observations led them to propose the presence of an RNA-degradation mechanism that would be activated by the presence of the high levels of a specific transcript. Canto and Palukaitis (2001) demonstrated that the CMV RNA1 transgene mediates suppression of the homologous viral RNA1 constitutively and prevents CMV entry into the phloem.

Vaslin *et al* (2001) transformed tobacco (*N. tabacum*) plants with the 3'-untranslated region of the *Andean potato mottle virus* (APmoV) genome RNA2 using three strategies:

(i) introduction of this region in sense, and anti-sense orientation and (ii) a fragment comprising the entire 3'-untranslated region from RNA2, and (iii) part of the CP22 coat protein sequence. The transgenic lines on inoculation showed a co-suppression-mediated mechanism towards the transgenic transcripts and the viral RNA. Izhar (1999) obtained transgenic tomato with the expression of the pathogen-derived nucleoprotein gene from the Bulgarian tobacco isolate L3 and the mitochondrial MnSOD gene from *Nicotiana plumbaginifolia*. Transgenic tomato plants carrying MnSOD were immune or tolerant to Bulgarian greenhouse tomato TSWV isolate ID upon mechanical inoculation. Transgenic expresser and non-expresser plants carrying L3 were completely resistant to heterologous greenhouse isolate ID-94 of TSWV. Yang *et al* (2004) field-evaluated TSWV resistance in transgenic peanuts (*Arachis hypogea*) under Tfton, Georgia, USA; they observed the expression of the nucleocapsid protein of TSWV and found significant tolerance in the transgenic plants. Melander *et al* (2001) identified a triple gene block (TGB) that consists of a set of three overlapping genes RNA of PMTV. These genes encode three proteins necessary for viral cell-to-cell movement. The gene encoding for the second triple gene block (TGB2) was mutated in a region that was highly conserved among TGB2 proteins from different viruses. The mutated TGB2 gene when transferred into potato can confer increased resistance to PMTV.

Smith *et al* (1994) reviewed the characteristics of RNA-mediated protection or resistance.

This type of protection is usually narrow and found against strains of the virus that has similar sequences to that of the transgene. Inoculating RNA cannot overcome this resistance that is not dose-dependent and operates at high levels of inoculum. The insert in the host genome comprises multiple copies of the transgene, particularly with direct repeats of coding regions (Sijen *et al* 1996). Copies of the transgene may be truncated and/or in an antisense orientation (Waterhouse *et al* 1998, Kohli *et al* 1999). Transgenic sequences and sometimes their promoter(s) may be methylated (Jones *et al* 1999, Kohli *et al* 1999, Sonoda *et al* 1999). RNA-mediated resistance is normally operated by three mechanisms:

(i) Anti-sense RNA-mediated

(ii) Satellite-RNA-mediated resistance

(iii) Ribozyme-mediated resistance

(i) Anti-sense RNA-mediated resistance

The anti-sense technology is based on the capability of complementary nucleic acids to form double-stranded helices. Any single stranded nucleic acid present in a living cell is therefore susceptible to bind via Watson-Crick base pairing to a nucleic acid of complementary (anti-sense) polarity. Specific binding of the 'anti-sense' nucleic acid results in interference with the biological function of the 'sense' nucleic acid, either by arresting the sense nucleic acid or by induced degradation of the double-stranded complex (Tabler *et al* 1998). Lindbo *et al* (1993) demonstrated that extensive methylation of the transgene sequence was likely associated with induction of the specific cytoplasmic RNA degradation mechanism. These observations were typical of 'gene silencing' (Napoli *et al* 1990, Van der Krol *et al* 1990). Since the degradation occurred after transcription took place, it was named 'post-transcription gene silencing' (PTGS).

All RNA-silencing mechanisms involves the cleavage of double-stranded RNA (ds RNA) following by an RNAase III-like protein known as '*Dicer*' into 21-26 nucleotides (nt) short RNAase (sRNAs) with 2-nt overhangs at the 3' ends. The two strands of these sRNAs are separated, presumably by a helicase and one of the two strands is recruited as the guide RNA of a silencing effect. There are variations on this basic silencing pathway: the dsRNA input, for example, can be derived from inverted repeat transcripts from complementary RNAs that anneal the base pairing or from single stranded RNA-dependent RNA polymerase (RDR) (Baulcombe 2005). Silencing pathways can also vary because there are different types of effect or complex that recruits the sRNAs (Baulcombe 2004).

RNA silencing is a genetic regulatory mechanism that targets host genes and RNA species. In many instances, this regulatory RNA silencing involves a subset of sRNAs referred to as micro RNAs (*MiRNAs*) produced by RNA polymerase II. The mature *MiRNAs* target mRNA either by cleavage or by poorly understood translational control mechanism. RNA silencing can reduce the expression of specific genes through post-transcriptional gene silencing the micro-RNA pathway, and also through transcriptional silencing. Post-transcriptional silencing also acts as an anti-virus mechanism (MacDiarmid 2005).

Gene silencing protects cells against invading nucleic acids of viruses or mobile genetic elements. dsRNA corresponding to the invader which is produced by an RDR or other appropriate methods and cleaved into sRNAs that target DNA or RNA of the invader (Baulcombe 2004). The first applications of silencing were with tomatoes in which transgene targetted RNA silencing prevented expression of a cell-wall-softening enzyme in ripe fruit. Virus-resistant transgenic plants were also an early product of RNA silencing. The transgenes in these plants included virus-derived-sequences, thus the plants were specifically resistant against viral strains containing an almost perfectly matched nucleotide sequence (Baulcombe 1999). The principal difficulty of this strategy pertains to the effective concentration of the transgenes in the cytoplasm in which it faces a high concentration of viral sense RNA molecules. There are, however, some examples where this strategy has been found to be successful such as TRSV (Yepes *et al* 1996), potyviruses (Hammond and Kamo 1995), RSTV (Huet *et al* 1999).

(ii)   Satellite-RNA-mediated resistance

Satellite RNA mediated resistance has been extensively reviewed by Jacquemond and Tepfer (1998) and Fuchs *et al* (1997). This type of resistance is normally found in cross protection that may naturally occurs between a severe and a mild strain of a virus. Several groups of researchers undertook extensive studies on the mechanism of action of satellite RNA found in CMV that, under certain circumstances, may act as a natural parasite of the virus, leading most often to symptom attenuation and reduction of the synthesis of genomic RNAs. Such cross protection is normally conferred by non-necrotic satellite RNAs or *R* satellite RNAs.

Kawbe *et al* (2002) obtained virus resistance in transgenic tomato expressing satellite RNA of CMV1. Tao and Zhao (2004) identified that a modified satellite DNA of TYLCCNV-Y10 suppresses gene expressions in *N. benthamiana* and the silencing persisted for more than a month. This property has been utilized in China in a commercial for covered cultivation of tomato.

(iii) Ribozyme-mediated resistance

Ribozymes are RNA enzymes derived from naturally occurring self-cleaving RNAs. They can specifically bind and cleave a particular RNA target and are normally considered as an extension and improvement of anti-sense technology. Like anti-sense RNA, ribozymes bind to their target, but unlike anti-sense RNAs they also induce specific cleavage of the complementary RNA (Tabler *et al* 1998). A number of examples have been described where the ribozyme approach has been used for gene suppression *in vivo* (Heinrich *et al* 1993, Koizumi *et al* 1992, Sarver *et al* 1990, Scanlon *et al* 1991, Sioud and Drlica 1991), but their successes are yet to resolve many difficulties.

Breakdown of Resistance

Durability of resistance generally depends on the host-virus combination, variability of the viral genome and vectors. In some cases it is durable, such as for dry and snap beans resistant to BCMV (Zaumeyer and Meiners 1995), sugarbeet resistant to BCTV (Duffus 1987), the original selection of which was made in the early 1920s, and TuMV resistance in lettuce (Duffus 1987, Robbins *et al* 1994). For other viruses and hosts the breakdown of resistance with time and changes in the ecosystem, is very common. Certain strains of CCMV overcome the resistance of cowpeas to this virus (Paguio *et al* 1988) and resistance-breaking strains of PVX infecting potatoes have been described from the UK (Jones 1985).

In plant viruses, genomic variation caused by mutation is enhanced by recombination, pseudo-recombination and acquisition of extra-genomic components. Genomic RNA typically has a high mutation rate, and individual virus isolates consist of swarms of mutants, with the consensus sequence changing in response to selection pressure. Recombination is found among field isolates of some RNA viruses especially Tobraviruses. Among the DNA viruses, field isolates of geminiviruses have much nucleotide diversity and frequency of recombination and some of it is interspecific. Recombination among whitefly-transmitted geminiviruses (begomoviruses) occurring in the same geographical region contributes to their evolutionary divergence. Several (perhaps potentially all) viral genes can encode a a-virulence factor that elicits resistance controlled by a cognate dominant host gene: such factors include viral coat protein, RNA polymerase, movement protein and proteinase. Some resistance-breaking virus variants have merely a single non-synonymous nucleotide replacement in their a-virulence gene, but with more durable resistances, virulence necessitates two or even multiple replacements. The probability of a resistance-breaking variant appearing and spreading also depends on its biological fitness in the absence of the host resistance gene, and on the type of resistance and number of resistance genes to be overcome. Viruses can have particular difficulty in overcoming strong resistance controlled either by multiple recessive genes or by coat protein transgenes. Resistance acting at the RNA level is exemplified by post-transcriptional gene silencing induced by transgenic viral sequences. This resistance may be broken by variants with > 10% nucleotide non-identity distributed along the sequence (Harrison 2002). With the advance of the production of transgenics, the probability of the resistance breaking is also on the rise.

Qiu and Moyer (1999) recorded the adaptation of TSWV to 'N' gene derived resistance by genome reassortment. Elements from two or more segments of the genome are involved in suppression of the resistant reaction.

Diaz *et al* (2002) recorded that Spanish *melon necrotic spot virus* isolate overcomes the resistance conferred by the recessive '*nsv*' gene of melon.

Co-infection by two viruses may also break the resistance to one virus by the other.

Resistance to CMV in cucumber cv. Delila was manifested as a very low level of accumulation of viral RNA and capsid protein, and an absence of symptoms. Resistance is observed even in single cell systems with a reduction of accumulation of CMV RNAs. However, this resistance breaks after co-infection with ZYMV. Resistance breakdown is manifested by an increase in the accumulation of (+) and (–) CMV RNA as well as CMV capsid protein with no increase in the level of the accumulation of ZYMV. This resistance breakage also occurs at the single cell level (Wang *et al* 2004).

European pathotype-2 isolate of BaYMV-2 infects barley genotype with '*rym 4*' resistance gene. The pathogenicity determinant of this virus is on RNA1. The complete ORFs of RNA1 of both the pathotypes (1 and 2) are similar to one another. The only consistent amino acid difference between the pathotypes 1 and 2 is the RNA1 encoded polyproteins (Kuhne *et al* 2003).

Fargette *et al* (2002) recorded resistance-breaking isolates of RYMV during serial inoculations; pronounced changes in pathogenicity occurred with serial passage of virus isolates inoculated to partially or highly resistant cultivars. The high resistance of the *O. sativa* indica cv. Gygante was overcome and the partial resistance of the *O. sativa* japonica cv. Azucena broke down. The ability of the isolates to break the resistance was not linked to a high initial pathogenicity of the isolates.

TSWV-A overcame the '*Sw-5*' resistance gene in tomato and TSWV-N resistance gene in tobacco, but TSWV-D cannot overcome the resistance conferred by these genes. The ability to overcome both forms of resistance is associated with the MRNA segment of TSWV-A (MA). Overcoming the '*Sw-5*' gene is linked solely to the presence of MA, and the ability of MA to overcome the TSWV-N gene is modified by the L RNA and the sRNA of TSWV-A. Sequence analysis of the MRNA segment of TSWV-A, -D and type isolate BR-01 revealed multiple differences in the coding and non-coding regions that prevent identification of the resistance breaking nucleotide sequences (Hoffmann *et al* 2001).

Resistance to vectors may also breakdown. Breeding red raspberry for resistance to the large raspberry aphid (*Amphorophora idaei*) using single major genes or polygenic minor genes was successful in controlling this virus vector aphid for more than 30 y. Later biotypes appeared with the ability to break the most resistant gene 'A1'. Breakdown of the formerly strongest resistant gene '110' was also reported (Williamson 2002).

The mechanism of break down may be diverse. The most common is the suppression of the silencing process. CMV-encoded '*2b*' protein ('*Cmv-2b*') is a nuclear protein that suppresses transgene RNA silencing in *N. benthamiana*. '*Cmv-2b*' is an important virulence determinant but is non-essential for systemic spread in *N. glutinosa*, in contrast to its indispensable role for systemic infection in cucumber. '*Cmv-2b*' becomes essential for systemic infections in older *N. glutinosa* plants or in young seedlings pretreated with salicylic acid (SA). Expression of '*Cmv2-b*' from the genome of either CMV or TMV significantly reduces the inhibitory effect of the SA on virus accumulation. According to Ji *et al* (2001) a close correlation occurs between '*Cmv-2b*' expression and a reduced SA-dependent induction of the alternative oxidase gene, a component of the SA-regulated anti-viral defense.

RNA silencing reduces the expression of specific genes through post-transcriptional gene silencing (PTGS), the microRNA pathway and also through transcriptional gene silencing. By suppressing the virus defence mechanism, viruses affect all these silencing pathways, in addition to the intercellular signalling mechanism that transmits RNA-based messages throughout the plant. Productive virus infection may therefore disrupt the normal gene expression patterns in plants, resulting, at least in part, in a symptomatic phenotype (McDiarmid 2005).

PTGS is an intrinsic plant defense mechanism efficiently triggered by dsRNA producing transgenes and can provide high level of virus resistance by specific targeting of cognate viral RNA. But such resistance is also unstable due to the virus-encoded suppressors of PTGS. CMV is able to suppress dsRNA-induced PTGS and the associated PVY immunity in tobacco (Mitter *et al* 2003). Several RNA plant viruses have been shown to encode suppressors of PTGS and overcome this host defense. Coat protein (CP) of TCV strongly suppresses PTGS. CP of TCV eliminates small interfering RNAs associated with PTGS, but it must be present at the time of silencing initiation to be an effective suppressor. The CP is able to suppress PTGS induced by sense, anti-sense and ds RNAs (Qu and Morris 2003).

P19 of tombusviruses is a potent silencing suppressor that prevents the spread of the mobile silencing signal by sequestering the PTGS-generated 21-25nt dsRNAs thus depleting the specificity determinants of PTGS (Sithav *et al* 2002). P19-mediated suppression of virus-induced gene silencing is controlled by genetic and dosage features that influences pathogenicity (Qiu *et al* 2002). Park *et al* (2004) provided evidence for a model in which virus spread through suppression of defense related gene silencing. P19 encoded by TBSV interacts with itself to predominantly form dimers, and with a novel host protein Hin19. The suppression of defense related gene silencing involves the formation of complex including P19 and Hin19.

P19 therefore provides an insight into how plant viruses can defeat their host's antiviral RNAi. Zamore (2004) furthermore provided evidences that P19 binds the virus inactivating RNAi and breaks the silencing mechanism.

Biosafety Issues

The commercialization of transgenic crops has generated intensive debates worldwide about biosafety and the impact of this powerful technology on agriculture, human health and the environment. Biosafety concerns about transgenics owe their origin to the fact that the tools of recombinant DNA technology have made it possible to have access to genes from completely unrelated or sexually incompatible organisms. Therefore, the transgenic technology, while providing potentially unlimited scope for crop improvement and pest and disease management, also imposes tremendous responsibility on the scientific community and regulatory authorities towards ensuring the safety of the plants produced. Biosafety needs to be considered from all angles before introducing transgenics for large-scale cultivation, especially if any toxins are present in the plants, their toxin retention and genotype × gene interactions. Thorough regulatory oversight and a scientifically designed transparent regimen are essential before any GM crop is introduced in the field. To allay such fears, 135 countries support the 'Cartagena Protocol' on biosafety, an international legal agreement under the UN Convention on Biological Diversity (CDB). It is the only international law that deals specifically with genetic engineering and genetically modified organisms but the debates still continue all over the world.

Transgenic CP-mediated virus resistance crops field tested for a long period of time include: (i) transgenic squash resistant to ZYMV, WMV-2 and CMV, and (ii) transgenic papaya resistant to PRSV. Fuchs and Gonsalves (2007) studied the benefits of these transgenic crops and also provided guidance for the safety assessment required for cultivation of transgenic crops in fields.

There are many potential benefits of virus resistant transgenic crops:

- virus resistance can be incorporated into a plant without changing its phenotype;
- the same resistant gene can be incorporated into different plant genera and species that are affected by a given virus and are amenable to transformation and regeneration;
- resistance can be incorporated into vegetatively propagated plants that overcomes gametic incompatability or linkage to undesired traits.

The commercial release of squash cultivars derived from lines ZW-20 and CZW-3 has demonstrated the stability and durability of their engineered resistance over more than a decade. Virus-resistant transgenic squash allowed growers to restore their initial yields in the absence of viruses with a net benefit of US \$22 million in 2005 (Shankula 2006). Transgenic squash resistant to ZYMV and WMV does not serve as virus source for secondary spread (Klas *et al* 2006). Papaya production in Hawaii from 1940 to 1998 provides a convincing success story of transgenics. The parent transgenic line was deregulated by the USDA-APHIS in 1996 and by the Environmental Protection Agency and consultation with the Food and Drug Administration was completed in 1997. Licences to commercialize transgenic papaya were obtained by the Papaya Administrative Committee (PAC) in April 1998 (Gonsalves 1998). From 1998, adoption of PRSV-resistant papaya by farmers in Hawaii was overwhelming and fast and did not show any breakdown of resistance.

Safety assessment

Fuchs and Gonsalves (2007) recently reviewed the safety of virus resistant transgenic plants giving emphasis to crops that have been commercialized or extensively tested in the field such as squash, papaya, plum, grape and sugarbeet. The topics that are commonly perceived to be of concern to the environment and to human health are: heteroencapsidation, recombination, synergism, gene flow, impact on non-target organisms, and food safety in terms of allergenicity. During the last two decades resistance has been achieved against nearly all families of plant viruses in numerous crops by applying PDR. Potential safety consideration relate directly to the fact that resistance to viruses in plants is achieved through the anti-viral pathways of RNA silencing that are triggered by expressing constitutively viral sequences (DeZoeten 1991, Martelli 2001, Rissler and Mellon 1996, Robinson 1996, Tepfer 2002).

Heteroencapsidation: Heteroencapsidation refers to the encapsidation of the genome of one virus by the coat protein of another virus. This may occur when more than one virus infects the same plant. In a transgenic plant, CP subunits can carry determinants for pathogenicity and vector specificity, among other key features that change the properties of the virus in the transgene (Callaway *et al* 2001). Vector non-transmissible virus may be vector- transmissible, virus could infect otherwise non-host plants and theoretically new virus epidemics may occur. Heteroencapsidation is common in transgenic herbaceous plants (Candelier-Harvey and Hull 1993, Farinelli *et al* 1992, Hammond and Dienelt 1997, Holt and Beachy 1991, Lecoq *et al* 1993, Osburn *et al* 1990). There may be encapsidation of the RNA genome of the challenging virus by the CP subunits of the transgene, but heteroencapsidation is generally not a permanent event as the viral genome in not normally affected.

Recombination: Recombination may occur between the transcript of a viral transgene and the genome of the challenge virus during replication in a transgenic plant cell. Recombinant viruses may have chimeric genomic molecule consisting of segments from both the challenge genome and the transgene transcripts. Chimeric viruses often show new biological properties that may be stable and that may perpetuate through progenies. Occurrence of recombination is common in transgenic herbaceous plants expressing viral genes (Adair and Kearney 2000, Allison *et al* 1996, Borja *et al* 1999, Frischmuth and Stanley 1998, Gal *et al* 1992, Green and Allison 1994, 1996, Jakab *et al* 1997, Lommel and Xiong 1991, Teycheney *et al* 2000, Varrelmann *et al* 2000, Wintermantel and Schoelz 1996). The presence of a complete viral 3′ noncoding region in some transgene constructs favour recombination (Allison *et al* 1996, Green and Allison 1996, Varrelmann *et al* 2000). Recombinants may have adverse ecological and environmental impacts.

Gene flow: Gene flow from the virus-resistant transgenic crops to wild relatives is an important ecological and environmental issue. If the gene transferred, provides a selective advantage, there may be evolution and development of weed species that may disrupt the natural ecosystems; it may also be a threat to genetic diversity of wild populations, even their extinction (Ellstrand 2003, Ellstrand *et al* 1999, Stuart *et al* 2003). An understanding of the transgene effect in wild population is necessary in addition to an understanding of the transgene movement before introducing it in the fields (Stuart *et al* 2003, Ellstrand *et al* 1999).

Synergism: In transgenic plants expression of viral genes can increase the susceptibility to a synergistic heterologous virus and affect the rate of the disease spread through the inhibition of the plant's PTGS defense responses to virus infection (Pruss *et al* 1997, Ryang *et al* 2004, Savenkov and Valkonen 2001). It, however, does not modify existing viruses with new characteristics.

Effect on non-target organisms: Viral genes that confer resistance in transgenic crops could provide soil microorganisms with a selective advantage upon horizontal gene transfer (Rissler and Mellon 1996).

Food safety and allergenicity: There may be potential allergenic properties of proteins encoded by viral sequences that are expressed in transgenic plants. Virus-derived transgene protein products can have stretches of amino acid sequences that are identical to potential immunoglobulin-E-binding linear epitopes of allergen proteins affecting food safety and nutrition.

Two processes are broadly involved in the production of transgenics: production of appropriate gene constructs for expression and regeneration of the transgenic plant. The gene construct is to allow a stable expression of the information in the plant. To be expressed in the plant cell, the sequence chosen is to be placed between particular DNA sequences: promoter, initiation and termination codons, 'leader' sequence for recognition by ribosomes, 'enhancer' consensus sequence. These act as signals recognized by the plant cellular mechanism, regulating its expression. The choice of sequence should also take into account the phenomenon of silencing (De Wilde *et al* 2000). The promoter most widely used at present is 35S RNA of CaMV. The gene construct is normally inserted in a bacterial cloning plasmid and multiplied in *E. coli*. To easily select the bacteria in which the plasmid and the target gene have been inserted, a gene for resistance to an antibiotic is used as a selection agent or marker. The transgenic cassette containing these constituents, on incorporation into crop plant cell, is expected to produce the desired proteins. There may be the production of allergen in food crops, and long-term effects of antibiotic resistant marker genes (horizontal gene transfer to produce antibiotic resistant bacteria) and of the viral promoters (integration with the host genome, etc.) may enter the food chain, crop and weed biodiversity, ecology and the environment.

Evidence suggests that heteroencapsidation and recombination between viral transgene products and challenge viruses are not real risks and can be minimized. Likewise, the allergenecity of CP can be minimized or overcome. People have been consuming virus-infected fruits and vegetables with no ill effects caused by the plant virus components. In contrast, gene flow and all its ramifications need to be considered case by case. An accumulation of objective case-by-case analyses will likely provide a solid framework to determine the amount of information necessary to make realistic regulatory decisions. Virus-resistant transgenic crops that have the scope to offer numerous benefits to growers and consumers need to be deployed safely after due realistic assessment of safety considerations (Fuchs and Gonsalves 2007).

Cultivation of Cross-protected Planting Materials

Lecoq (1998) and Fuchs *et al* (1997) reviewed the control of plant virus diseases by cross protection. The phenomenon of cross protection was discovered in tobacco plants that remained symptomless after inoculation with a virulent yellow mosaic strain, when pre-infected by a mild green strain of TMV (McKinney 1929). This observation led to new opportunities for making the planting materials resistant to severe strains of some viruses inoculating seedlings/planting materials with a mild strain that does not affect the quality of the product before planting them in the field.

Commercial cross protection has been introduced in a few crops and viruses, such as *Citrus tristeza virus, papaya ring spot virus, zucchini yellow mosaic virus, tobacco (tomato) mosaic virus, cucumber mosaic virus*-satellite RNA (Fuchs *et al* 1997, Lecoq 1998). The use of this technology has gradually declined because of inherent risk factors and development of resistant varieties.

Citrus Tristeza Virus (CTV)

CTV in nature, particularly at the points of origin of *Citrus*, occurs as mixtures of different strains. The effects of some of them are mild, some moderate while others are severe. Generally, after inoculating a mild strain, if a citrus plant is challenged by a severe strain, the symptoms remain unexpressed, but the completeness of the protection varies in different rootstock-scion combinations. Symptoms of mild strains normally appear in sour orange and Eureka lemon; intermediate symptoms (without stem pitting) with shortening of leaves and deficiency symptoms appear in grapefruit, 'Galego lime' (Key lime) and *Citrus webberii*, but no symptom appears in 'Pera sweet orange' or 'Rio mandarin' . The complexities of protective interference may occur in different CTV strains. The occurrence of such interference may depend both on the donor and the receptor plant species. When 'Galego lime' is pre-immunized with a mild isolate from 'Pera sweet orange' , protection does not occur, but pre-immunization of 'Pera orange' with the mild isolate from the same source exhibits protection. Pre-immunization of 'Ruby Red' grapefruit with mild isolate from 'Pera orange' on the other hand gives no protection. The breakdown of protection in pre-immunized plants also occurs with time (Thornton *et al* 1980). Cross protection has been commercially used in different countries to protect *Citrus* crops from CTV, particularly in Brazil, South Africa, USA, Reunion Island, etc. (Muller 1980, Urban *et al* 1990).

Papaya Ring Spot Virus (PRSV)

Yeh and Gonsalves (1984) isolated two mild variants by random mutagenesis with nitrous acid treatment of crude virus preparations of *papaya ring spot virus* (PRSV) from Hawaii and after selection from 663 single local lesion isolates. Pre-immunization of papaya plants with these mild variants protected the plantings in Hawaii. Similar protection also occurs in the plantations in Taiwan (Yeh *et al* 1988). Cross protection to protect papaya cultivation has been commercially applied in Hawaii, and Taiwan, but such protection may break under heavy inoculum pressure and the level of protection depends upon the geographical conditions where the isolates are being collected, and applications are made.

Hawaiian isolate HA 5-1 provides excellent protection in Hawaii and the aphids poorly transmit this isolate (Yeh and Gonsalves 1984, Ferreira *et al* 1993, Mau *et al* 1990).

Zuccini Yellow Mosaic Virus (ZYMV)

Lecoq *et al* (1991) obtained a mild variant of ZYMV-WK from an auxiliary branch of melon severely infected with the virus. The concerned branch showed very attenuated symptoms while most other parts of the plant showed severe symptoms. This variant also produced mild symptoms in cucumber, watermelon and squash and also in some non-cucurbitaceous plants like *Chenopodium amaranticolor*. In field condition, this variant provides effective protection in temperate, Mediterranean and tropical environments (Lecoq *et al* 1991, Walkey *et al* 1992, Urban *et al* 1990, Cho *et al* 1992, Wang *et al* 1991). The extent of protection, however, depends upon the geographical condition of the natural occurrence of the isolate. ZYMV-WK is effective against severe isolates from the Mediterranean basin, northern Europe, Africa, North and Central America, France, Hawaii, Taiwan, the UK but it fails to protect the plants against the severe strains from Reunion, Mauritius, Madagascar, etc. This variant is now commercially used to protect the crops from the infection by the severe strain in Europe, Israel, Taiwan, Hawaii and Tonga. Zammar *et al* (2001) observed the protection of squash cv. California in Lebanon. But in the protected crops, plants may start to show the breakage of protection producing moderate or severe symptoms with time. This isolate also protects cantaloupe in California, USA (Perring *et al* 1995).

Tobacco Mosaic/Tomato Mosaic Tobamovirus (TMV/ToMV)

Fulton (1986) obtained a mild strain (MII-16) from naturally occurring strains and also by culturing a severe strain in tissue held above 34°C. This mutant can also be obtained by nitrous acid mutagenesis followed by single local lesion selection (Rast 1972). MII-16 can protect tomato crops from the infection of ToMV and were in wide use for cross protection of tomatoes in Europe, Canada, New Zealand, France and Japan (Broadbent 1976, Fletcher 1978, Novatel *et al* 1983, Marrou and Migliori 1972, Novatel 1977, Oshima 1975). The major drawbacks of using the mild strain for cross protection against ToMV are: (i) break down of protection with time, (ii) the mild strain belongs to ToMV pathotype I; it induces necrotic reactions on plants possessing $Tm2_2$ gene in the heterozygous form; (iii) there remains the risk of synergism with other viruses particularly with cucumber mosaic virus.

Cucumber Mosaic Cucumovirus (CMV)

Rodriguez-Alvardo (2001) observed cross protection between CMV isolates collected from field grown pepper (*Capsicum annuum*). CMVSD2 strain (subgroup I) and CMV-35 do not elicit disease symptoms when inoculated to pepper plants previously inoculated with CMV-S (subgroup II). In the cross-protected plants there was hardly any detectable replication of the CMV challenge strain. The protective strain prevented replication of the challenge strain.

Tien Wu (1991) developed CMV-satellite RNA isolates and designated them as 'Biological Control Agents' (BCA). Extensive field tests with pepper, tomato and tobacco plants carried out over different regions in China showed the economic benefits of BCA. Quality control of the BCA was found to be the key factor for the success. The prevailing concern with the use of BCA has been related to the inherent danger of using satellite RNA since it is known that they can cause severe damage and new satellite RNA are easily obtained, especially when CMV is passed successively through hosts such as tobacco.

Gallitelli *et al* (1991) and Montasser *et al* (1991) conducted a limited trial with BCA in Italy and the USA respectively. These trials aimed at reducing damage caused by the CMV necrosis satellite RNA. Both sets of trial showed excellent protection against the satellite RNA-induced necrosis. But in most cases, the protection breaks down after 2-3 y.

Simon *et al* (2004) recently reviewed the potentials of the plant virus satellite and the defective interfering RNAs in the suppression of symptom expressions. Many sub-viral RNAs reduce disease symptoms caused by the helper virus. New models for DI RNA-mediated reduction in helper virus levels and symptom attenuation include DI RNA enhancement of PTGS, the antiviral defense mechanism in plants. But these studies elucidated more of the mechanisms of their activities rather than the scope of their use in cross protection.

Risks associated with using mild strains/sat RNAS for cross protection

1. Though there may be several mild strains of a pathogenic virus only a few of them are suitable for field application.
2. Purity of the mild strains/sat RNAs is crucial. Well-equipped laboratories are essential for their mass production: the cross protection of papaya in Hawaii used to be done by bringing the mild strains from Cornell University, USA.
3. Mild strains or a benign satellite RNA may mutate to a severe form.
4. The use of a 'live' virus deliberately to inoculate healthy plants is against the intuitive thinking of most plant pathologists and agriculturists.
5. Break down of the protection is very common in later parts of the crop cycle.

6. Specificity of mild strains/sat RNAs and varietal interactions are common in their application

Agronomic/Cultural Strategies and Tactics

Pre-sowing/planting considerations

Agronomic/cultural strategies need to start from the very beginning of the cultivation or plantation. The first and foremost requirement is to know the presence or absence of soil-borne viruses and the proneness of the area for the incidence of specific virus diseases. Crops and weeds around the selected fields also need to be monitored for the presence of viruses, including symptomless or latent forms. If there are viruses or viruliferous vectors in soil those must be eliminated. Depending on the type of the vector, fumigation may be necessary with a nematicide or a fungicide; which can be applied together if necessary. To fumigate the soil, chemicals need to be injected. Such injections require special equipment and should be done with great care since it involves the use of toxic chemicals. The soil should be rolled after fumigation. Cool months are preferred for fumigation. After 4 to 6 mon a 'watercress test' should be carried out to reveal the presence of any toxic residues. For solarization, the soil surface is punctured with a board to make holes (1.5 cm wide and 5 cm long). A transparent plastic sheet 70 -micron thick ethyl acetate vinyl) is placed over the surface. The punctures help diffuse solar heat into the soil. The soil becomes ready for planting after 2.5 to 3 mon of solar exposure. This technique is suitable in hot tropical countries. The alternative method is to remove the vectors and the viruses by crop rotation, cultivation of non-host crops or keeping the land fallow for some years or to break the cropping cycle or infection cycle.

Careful weeding around nurseries and within and around the fields is normally recommended but difficult to operate. It may be practicable with the viruses that have a narrow host range but not with those having a wide host range. Total destruction of weeds is also not desirable for environmental reasons. Intensive use of herbicides may affect the soil and the subsequent growth of the changed weed population.

Private flower and kitchen gardens can also be sources of viruses and may need to be monitored for potentially damaging viruses. Sample surveys may be made to determine the extent of sources of the anticipated viruses around the selected fields. While planting, phytosanitation is to be strictly followed and resistant varieties are to be grown as far as practicable.

Most of the viruses of diverse crops are insect transmitted. Some viruses are adapted to cultivated crops and survive in them having limited host range including wild or weed hosts. While other viruses are adapted to wild or weed hosts and occasionally invade susceptible crops. Suitable strategies are to be designed for these two types of viruses to reduce their risks. Breaking the crop cycle through successive crops stops the perpetuation of crop-adaptive viruses whereas viruses adapted to wild plants or weeds need separate treatments. Viruses again may be non-persistent or persistent in vectors. Different strategies are also necessary to deal with these two categories of vectors. Contact insecticides usually do not take effect quickly enough to prevent transmission of non-persistent viruses but can be effective for persistently transmitted viruses.

Breaking of the infection cycle: Breaking of the infection cycle can be an efficient method of reducing the spread of crop-adapted viruses having a limited host range. Introduction of a host-crop-free period in the cropping pattern may eliminate those viruses that survive through successive crops. *Celery mosaic virus* is adapted to crops belonging to the plant family Umbelliferae and occurs normally in celery and carrot. There is year-round cultivation of celery in many countries particularly the USA. Following the initiation of the programme to keep a 'celery-free' period in each year, the incidence of *celery mosaic virus* disease was dramatically reduced. Similarly sugarbeet growers in California broke the infection cycle of virus yellows by introducing a 'beet-free' period in the field. During the 'beet-free' period attention was also given to prevent sugarbeets being left in the ground and a cooperative effort was made to destroy beet weeds (Duffus 1983). Recently establishment of a crop-free period (mid-June to mid-July) in the Arava valley in Israel considerably reduced the attacks aphids and whitefly transmitted viruses in vegetable crops (Ucko *et al* 1998). Wisler and Duffus (2000) introduced 'lettuce-free' and 'beet-free' periods to eliminate *lettuce mosaic virus* and *beet yellows* and *beet western yellows virus* respectively in the Salinas Valley, California using resistant varieties to introduce certification programmes of these crops.

Eradication: Eradication of viruses (early detection and burning of the infected plants) needs to be pursued before new crops are planted. It is still in use for seed production of vegetatively propagated crops such as potato, ornamental bulbs, etc. Eradication is more or less an obligatory process for plantation crops and it proved its success in establishing new plantations of citrus in different countries.

On-field manipulation and tactics
Once the virus is excluded from the fields, the next approach would be to manage the crop so as to avoid the arrival of the virus from outside through vectors or to delay their arrival till the crop attains sufficient age to resist infection. Viruses may arrive in the field by mechanical contacts (particularly with mechanically transmitted viruses like *tobamoviruses, potexvirus* and *carmoviruses* where mechanical contacts with the operator, equipment and appliances, etc., and also contacts between the plant parts can spread infection) or via vectors particularly insects and mites. Most of the mechanically transmitted viruses are also spread by vectors. There may also be the arrival and spread of viruses transmitted by certain fungi and nematodes through irrigation or drainage water. In case of

any risk from these viruses, monitoring of the irrigation and drainage for the presence of vectors is to be conducted and appropriate treatment is to be made to remove them and also to avoid their spread in the field/greenhouse.

Arrival of persistently transmitted viruses is normally seasonal when they are introduced with the migrating viruliferous vectors. Their arrival may be curtailed by setting up traps around the field or growing 'wind breaks' around the plantation that acts as a physical barrier against the vector insects and mites. Lines of wind-breaks should be perpendicular to the prevailing winds and should be planted at least 8 m away from the first line of the crop. Types of wind-break plants are normally selected depending upon the climatic conditions of the planting area. For humid tropical areas, *Leucaena glauca* and *Erythrina fusca* are normally suitable for growing as wind break. In protected cultivation, viruliferous insects can be kept out by use of woven or unwoven insect-proof nets to cover the openings or the use of UV-blocking plastics.

Operational tactics

Measures against spread

(a) *Field spacing*

The principle of field spacing is to grow plants at a safe distance from infected fields. It is a simple but very effective means to minimize the chances of the crop getting infected by viruliferous vectors. It is most effective for the non-persistently transmitted viruses. The minimum isolation distance for growing a crop from the infected plant can be estimated from quantification of the disease gradient (Thresh 1976). In many cases, a distance of 100 m from the source can prevent virus infection. This technique is routinely employed in seedbeds and nurseries of fruits, vegetables and ornamentals. In determining the field spacing, one should consider virus risk diseases, vector flight duration and speed and direction of wind. For rice seedbeds, they may remain 'vector free' if they are raised at an isolation distance of 250 m from weeds and grasses. *Nephotettix virescens* vector of *rice tungro virus* usually moves 41 m/day. As these insects can keep the virus for 6 d (laboratory limit), seedbeds would remain beyond their reach if no weed and grasses remain within the limit of 246 m from the closest source of the vector (Ling *et al* 1983) unless assisted by the wind.

However, the estimated distance is a variable and the value is a function of the amount and potency of the virus source, the level of vector populations, prevailing wind, the aspect of inter-lying landscape, etc. These functions may also change from season to season due to weather and agricultural practices.

(b) *Adjustment of planting period*

Adjustment of planting period for the annual crops is an environmentally safe almost no-cost agronomic method to escape diseases that are spread by air-borne insect vectors. This is most true in tropical conditions and when temperature and water are not limiting. In temperate regions the optimum time for plant growth usually closely matches suitable conditions for virus vectors. So changing sowing/planting date to avoid vectors may incur a yield penalty. Most of the insect vectors have seasonal life-tables. If the crop is so planted that its early stage of growth escapes vector build up then the spread of the disease could be avoided. As for example, life-table of whiteflies is temperature dependent and the completion of generation and the emergence of this insect require 303.28-336.80 accumulated degree-days under West Bengal conditions that may again differ from year to year. So the planting time of the target crop is to be adjusted according to the medium range agrometeorological forecasts.

Agriculturists in different countries used this technique as an important component of the integrated system. Miao *et al* (2001) utilized the relationship between different sowing periods and the spread of *maize rough dwarf virus* and identified the safe planting month and date in Yongnian, Hebei Province, China. Fritts *et al* (1999) observed that in the high plains of Texas, USA, the incidence of diseases (HPD) was related to maize planting dates and winter wheat maturity dates. The producer could reduce the incidence of HPD by planting maize before or after the peak migration of wheat curl mite, the vector of the disease, from wheat. Huth (2003) found that a later sowing time (October instead of September) reduces BYDV infection in cereals in Austria by discouraging aphid population at that time. Harpaz (1982) controlled the incidence of *maize rough dwarf virus* by manipulating the planting date of maize in Israel. The plant hopper vectors cease to transmit the virus after exposure to high temperature. The sowing date was so adjusted that the mean daily temperature was higher than 24°C when maize seedlings emerged. Postponing the sowing date from its normal time in April to late May dropped the incidence of the virus from 45% to 3%.

Modifying Vector Flight and Settling Behaviour

(a) *Increase in crop density*

Aerial vectors visit closely spaced plants less frequently than when they are widely-spaced. Dense crop stands or high plant densities often escape infection or reduce virus incidence as well as spread of viruses such as *sugarbeet yellows, tomato curly top, groundnut rosette,* etc. Reddy *et al* (1983) observed that early planting and increased plant densities reduce the incidence of bud necrosis in groundnut.

(b) *Increase in area of field*

Incoming aerial vectors alight mostly on plants along edges of the fields. The border area of a small field has higher proportion of plants than that of a larger one. Accordingly, total virus incidence in a few large fields will be lower than that in a comparable area made up of small fields.

(c) *Cover crops and other barriers*

Cover crops protect the under-sown economically more important crops from insect vectors. The plant species to be used as a cover should be non-susceptible to the virus or the viruses of the crop to be protected but it should be attractive to the vectors. Taller susceptible plants can also be planted with regular monitoring and spraying of insecticide. The selection of a crop to be used as a cover depends on various factors such as crop to be protected, the virus involved, the season of cultivation, local agronomic practices, etc. Accordingly, barrier crops are planted alternately or at greater intervals with rows of the susceptible crop. Spraying of the barrier crops with insecticide makes them doubly effective. Barley is a well-known cover crop. Surrounding cauliflower seedbeds with narrow strips of barley (3 rows, 1.0 ft apart) could effectively reduce the virus incidence. Incoming aphids first land on barley that is resistant to cruciferous viruses and vectors, after feeding for a while on barley plants they fly off. Complete control of tulip breaking and lily symptomless viruses was obtained by using a barley barrier crop combined with weekly spray of albolineum oil (Harrison *et al* 1976).

Judicious intercropping also modifies vector flight and settling behaviour and reduces virus spread by the vector, as for example spread of viruses in cucurbits can be reduced by growing them intermixed with maize. Intercropping tomato and cucumber and planting cucumber 1 mon before tomato can reduce the incidence of tomato yellow leaf curl disease. Cucumber is a preferred host of whiteflies and it is immune to TYLCV. Insecticides are applied when adult whitefly populations are at high levels (usually 2 wk after planting of cucumber and the second one before tomato planting under Jordanian conditions) (Sharaf 1986).

Mixed cropping of tomato, French bean, and brinjal (eggplant) reduces the incidence of tomato leaf curl disease in tomato. Growing maize as a border crop and mulching the field with paddy straw after the crop emergence and intercropping with cowpea reduces the incidence of *mungbean yellow mosaic virus* in blackgram (Mukhopadhyay 1994).

(d) *Use of reflective surfaces*

Use of reflective surfaces or mulching is an important option in IPM. It is found to be a useful means for repelling vectors in commercial cultivation of several crops. Enhanced management in flower bulb production may be achieved by incorporating reflective mulches in the virus control strategies (Wilson 1999). Various authors have tried different types of materials for mulching. Commonly used materials and methods are: aluminum strips between rows; sticky yellow polyethylene (0.5 m wide) 0.7 m above the soil surrounding the plots; white polyethylene (1.60 m, 60 μ thick) on bare ground after weeding the edges keeping circular holes (15 cm in diameter) in the mulch for transplantation (Lecoq and Pitrat 1983), straw, etc.

For aphids, aluminum foil is usually effective and better than plastic. Unlike aphids, whiteflies tend to be attracted to aluminum foil and to the blue/UV part of the spectrum. They are also strongly attracted to yellow plastic or straw mulches. In hot dry climates, whiteflies attracted to such mulches probably stay there long enough to be killed by the reflected heat. Mulching of tomatoes and cucumber fields with saw-dust, straw or yellow polyethylene sheets markedly reduced the incidence of TYLCV and CVYV and the populations of the whitefly vector (Cohen and Melamed-Madjar 1978). Kemp (1978) observed that mulches can protect peppers from viruses. The virus incidence was reduced by 45% with sawdust, 35% with wood chips and only 12% with corncob mulches in the first year but in the next year the reduction increased to 72%, 63% and 53% respectively.

Aphid infestation on leaves, as well as PVY incidence significantly reduced by straw mulch (4-5 tons/ha) in organically produced potatoes (cvs Christa, Nicola) in Germany. Straw mulch was most effective when vector pressure was concentrated early in the year acting as a PVY protector for young plants. Combined mulching and pre-sprouting had a synergistic, complementary effect on reduction of PVY incidence (Saucke and Doring 2004).

The reflective surface treatments have limitations. Firstly, the control effect of mulches normally lasts for only 20-30 d. During this period however, mulches are usually more effective than insecticides (Cohen 1982). Besides, due to fading of the mulches because of weather, there is progressive shading with developing foliage. Secondly, the cost may be high and sometimes uneconomic. Moreover, the usefulness of any particular measure may vary widely with the combinations of crops, viruses and vectors. As for example, yellow sticky traps afforded excellent protection to pepper crop against aphid-borne PVY and CMV. Yellow polyethylene sheets, covered with transparent glue are erected vertically 70 cm above the ground at a distance of 6 m from the windward border of the pepper field, 4 d before germination. The in-flying aphids are attracted to these sheets and caught in large numbers resulting in more disease incidence (Cohen and Marco 1973).

For ornamental and vegetable crops, various types of plastic film are used to repel the activity of air-borne vectors. Woven or non-woven polyethylene or polypropylene covers form a physical barrier that prevents vectors landing on crops or entering a protected area (nurseries and greenhouses). The use of UV-absorbing polypropylene films for construction of plastic tunnels is highly effective at protecting plants from pests and vectors and the viruses they transmit (Antignus *et al* 1996). The elimination of UV rays from the light spectrum interferes with the capacity of insects to orient their movement under the tunnel.

Chemical control of vectors

As the spread of the vector-borne viruses is directly related to the vector activity in fields, vector control by chemicals often receives much attention. Pesticides are normally synthetic chemicals that may have either systemic or contact poisoning effects to the target vector organisms. The time of application is critical to their effectiveness, efficient use and the avoidance of unwanted side effects. Some propose the concept of threshold levels while others propose the action level concept. In any case it is usually a defensive technique and pesticides are applied only after the arrival of the vectors and aim to prevent the increase of the vectors in crops. So the main question which arises is that once the vectors are in the field under climatologically favourable conditions, how likely is it for economically important damage to occur. When the effect of pesticides on the control of a non-persistent (e.g. PVY) and a persistently transmitted virus (e.g. potato leaf roll) is compared, pesticides did not reduce the spread of PVY whereas timely application of systemic insecticides could reduce the spread of PLRV. If the application is not at the best time, however, it may have little or no benefit. The main reason why pesticides fail to control non-persistently-transmitted viruses is that a vector can infect its host within a few seconds, much more quickly than the pesticide can kill. However, they can decrease populations and spread of vectors but are ineffective if there is continual immigration of potentially virus transmitting vectors into the crop

When a crop is grown, vectors emigrate from local sources and settle in the crop, develop residential population and move to the nearby hosts as the crop is harvested. In another situation, vectors migrate from a long distance and settle in crops and take off for a long flight at the harvesting stage of the crops. Emigration from local sources is related to synoptic weather phenomena beyond the surface wind layer of the atmosphere. In temperate countries, where severe winter occurs, migration of vectors from hosts on which they survive the cold period, as adults or eggs, to crop fields is common in the growing season. In the tropical countries on the other hand where severe cold and high temperature is uncommon, residential populations survive throughout the year and emigrate on the onset of the favourable climate and cropping season.

Long distance migration may also occur in these countries depending upon the weather, vector species and crop growth stage. Using all these features of the arrival and population build up of vectors in a crop field, a need-based application of pesticides can be devised to reduce the incidence of viruses. Thus a pesticide can only be successfully applied to control vectors (particularly airborne ones) if their arrival and build up could be correctly forecasted along with the appropriate knowledge on vector and virus biology and ecology.

Pesticides and their Limitations

The application of pesticides has limitations and hazards. Among the pesticides that are commonly used are the organochlorines. These chemicals have high insecticidal activity, low cost and high resistance to degradation. They bioaccumulate in nature and contaminate animals, birds, fish, human health, food chain and the global ecosystem. They induce resistance to target organisms and cost pest resurgence. They also kill natural enemies of pests and beneficial insects and microorganisms. As a consequence of the properties, and the emergence of effective but environmentally safer materials, they are now widely banned.

Organophosphates were initially used to replace organochlorines as they were perceived as less hazardous. Some of these chemicals are fast acting and quickly degradable while others are quite stable. They are more acutely toxic and less persistent in the environment. However, their use leads to the resistance in target organisms, and they kill natural enemies of pests and beneficial organisms. Therefore continuous use of them causes rapid resurgence and secondary pests/ vectors outbreak. Many are now also banned Carbamates, formamidines and synthetic pyrethroids have largely replaced organophosphates. Carbamates share many properties of organophosphates but cause less permanent damage. Formamidines are effective against those pests that are resistant to organophosphates, but they are acutely toxic to fish and cause health hazards to users. A new approach to pest control is by the use of semio-chemicals, specific anti-metabolites, etc. to exploit or change insect behaviour or inhibit growth or reproductive processes of specific organisms and that have little adverse effect to natural ecosystem or biosphere. Development of bio- and botanical pesticides is also in progress to make completely safe vector control systems. Whatever chemicals are applied, they should be used only when necessary and be guided by knowledge of the pest/vector/host biology and interactions.

New Chemicals, Botanical Pesticides

Efforts are also being made in different countries to test the efficacy of new chemicals and plant product to control plant virus diseases. Some of them are:

1. Anfoka (2000) determined the potential of benzo-(1,2,3)-thiadiazole-7-carbothioic acid S-methyl ester (BTH) to trigger systemic acquired resistance (SAR) in tomato (*Lycopersicon esculentum* Mill cv. Vollendun) plants against a yellow strain of CMV-Y . Application of BTH as a drench 7 d before inoculation of the virus protects the plants from necrosis.

2. Smith *et al* (2003) observed resistance to CMV in cantaloupe induced by cibenzolar S methyl (ASM). ASM is a synthetic analogue of salicylic acid and developed for use in a novel strategy crop protection through systemic acquired resistance (SAR). It induces the systemic accumulation of chitinase, a marker protein for SAR in both greenhouse and field grown seedlings. ASM at 50 or 100 μg/ml provides almost complete protection against *Colletotrichum lagenarium* and effectively delayed the spread of CMV.

3. Pappu *et al* (2000) observed the use of plant defense activators such as 'Actiguard' (acibezolar-5'-methyl and imidacloprid) potentially offers an alternative and effective management tool for TSWV suppression in flue-cured tobacco. Csinos *et al* (2001) determined the effect of this compound (0.2 to 8 g a. i / 7000 plants) on pre-transplant application to suppress TSWV.

4. Parveen *et al* (2001) isolated and characterized an inducer protein (Crip-31) from *Clerodendrum inerme* leaves that induces systemic resistance when applied at the rate of ~25 μg/ml against CMV, PVY, ToMV in *Nicotiana tabacum* cv. White Burley. Resistance induction was detectable between 40 and 60 min after challenge inoculations.

5. Chen *et al* (1991) studied the effect of pokeweed antiviral protein (PAP) on the protection of plant viruses. PAP from *Phytolacca americana* inhibits *tobacco mosaic tobamovirus* infection of tobacco cultivar xanthi-ne. At 0.4 μg purified PAP/ml, the formation of local lesion is completely inhibited. Antiviral activity is dependent on the concentration of PAP and not of the virus. PAP also protects indicator plants from infection of six other viruses: *cucumber mosaic cucumovirus, alfalfa mosaic alfamovirus, potato virus X potexvirus, potato virus Y potyvirus, African cassava mosaic begomovirus* and *cauliflower mosaic caulimovirus.* PAP infiltration into the intercellular spaces through the lower surface partially prevents potato virus Y transmission by aphids.

6. Jones (2001) developed a new biological pesticide 'Messenger', that activates natural plant defense and growth systems. The active ingredient of 'Messenger' is 'hairpin', a protein produced in nature by certain bacterial plant pathogens. It is water-soluble and granular formulations degrade rapidly after application leaving no detectable residues on plants or in the soil. As 'hairpin' does not act by killing pests or pathogens,

the likelihood of resistance development is reduced. 'Messenger' has been tested on 40 crop groups and found effective on a number of crops including cotton, wheat, cucumber, citrus, tobacco, strawberry, tomato and peppers. It can repel insect pests and provides effective control of some viral diseases including *tomato mosaic* and *cucumber mosaic* in tomato and pepper and *beet curly top virus* in peppers.

B. SOME EXAMPLES OF CURRENTLY OPERATIONAL IPM

The thrust of epidemiological research is the application of IPM and the development of models for the application of safe pesticides. Some examples of this thrust are:

1. Krell *et al* (2004) developed bean leaf beetle management to control *bean pod mottle virus* in soybean (VE-VC). Early season mid-season insecticide sprays (lamda cyhalothrin) at over-wintering and ovipositioning beetles respectively, were targetted to lower the population of beetles in the main crop.

2. Riley and Pappu (2000) evaluated different management practices for TSWV in Georgia, USA, and observed that host plant resistance and application of reflective mulch significantly reduced thrips numbers and TSWV. Early planting on black plastic with an intensive insecticide treatment resulted in the highest yield.

3. Culbreath *et al* (2003) developed an integrated system that made use of moderately resistant cultivars and chemical and cultural practices each of which helps to suppress tomato spotted wilt epidemics in tomato.

4. Birch *et al* (2004) evaluated white sticky traps, specific flower volatile attractants, resistant varieties, synergism with natural enemies, and predictive models, to control raspberry beetle, the vector of several raspberry viruses; trap design and volatile release rate were optimized with spatial arrangement and fitted into the IPM system. The system was modelled on software and developed sprays for first generation midges and a computer-based prediction model was made available throughout the UK, and has been modified for use in Italy and Switzerland.

5. Thackray *et al* (2004) developed a simulation model to forecast aphid outbreaks and epidemics of CMV in lupin crops growing in the 'grain belt' of South West Australia which has a Mediterranean climate. The model uses rainfall during summer and early autumn to calculate an index of aphid build up on weeds, crop volunteers, and self-regenerated annual pastures in each 'grain belt' locally before the growing season commences in late autumn. The index is used to forecast the timing of

aphid immigration into crops. The model evaluates the effects of different sowing dates, percentage of CMV infection in seed sown and plant population densities on virus spread. The model was incorporated into a decision support system (DSS) for use of lupin farmers in planning CMV management and targetting sprays against aphids.

6. Lombard (2002) determined the effect of green manure crops and organic amendments on the incidence of nematode-borne tobacco rattle virus. Organic matter amendment of soil at 1% dry weight has been shown to reduce the host finding activity of *Paratrichodorus teres* under laboratory conditions. Household waste compost (GFT compost) was applied as a soil mix or planting furrow treatment at 12 t dry weight/ha for tulip and gladiolus. Spent mushroom compost (champost) was added as planting furrow treatment at 17 or 12 t dw/ha respectively for the same crops. Champost in the planting furrow and fodder radish as a preceding crop reduced the percentage of infection in tulip under favourable conditions TRV infection.

C. PHYTOSANITATION AND QUARANTINE

Phytosanitation is the key requirement for healthy crop production. All sanitary practices are to be followed during cultivation, harvest, sorting, grading, packaging, and transport while dealing with plant products. Otherwise, yield and cost are affected and also the scope for the spread of pests, diseases and noxious weeds increases. With the increasing movement of agricultural products throughout the world regulatory measures have been imposed in all countries.

Background

Trade is the primary vehicle of international economy. Two major constraints are normally recognized that limit international trade. These are 'Tariffs' and 'Barriers'. To regulate 'Tariffs', a 'General Agreement on Tariffs and Trade' (GATT) first became operative in 1947 involving only a few industrially advanced countries. But the 'Barriers' persist, and the agriculture and allied products were not in the purview of this Agreement at that stage. In non-member countries trade was mostly operative through bilateral agreements and the import decisions were absolute discretions of a particular country. With rapid changes in the world economy and the pressing necessity of a formulation of a dynamic global economy, a fresh GATT Agreement was made in 1994. This item became the primary cause of discussions in the Uruguay Round of Multinational Negotiations and led to the formation of 'Multi-Trade Organization'. This Organization in 1995 was transformed into the 'World Trade Organization' (WTO). The Articles of this Organization intended to remove the 'Barriers' and bring agriculture and allied products within the purview of the WTO.

International trade on agricultural and allied products involves risks of the spread of pests, pathogens, weeds, and also unwanted chemical residues having impacts on the health of plants, animals, human beings and the environment. Each country has its own protection system to avoid such importation, the Rules and Regulations of Quarantines. But there was no harmony in the rules and regulations and their implementation in different countries or transparency. The WTO reaffirmed that no Member State should be prevented from adopting or enforcing measures necessary to protect human, animal and plant life or health, subject to the requirement that these measures are not applied in a manner that would constitute a means of arbitrary or unjustifiable discrimination between Members where the same conditions prevail or a disguised restriction on international trade. WTO made an Agreement on the Application of Sanitary and Phytosanitary (SPS) measures uniformly applicable to all the Member countries.

Sanitary and Phytosanitary Measures

WTO defined the sanitary and phytosanitary measure as any measure applied: (a) to protect animal or plant life or health within the territory of the Member from risks arising from the entry, establishment or spread of pests, diseases, disease-carrying organisms or disease causing organisms. (b) to protect human and animal life or health within the territory of the Member from risks arising from additives, contaminants, toxins or disease causing organism in foodstuff, beverages or feedstuffs. (c) to protect human life or health within the territory of the Member from risks arising from diseases carried by the animals, plants or products thereof, or from the entry, establishment or spread of pests or (d) to prevent or limit of other damage within the territory of the Member from the entry, establishment or spread of pests.

Sanitary and phytosanitary measures include all relevant laws, decrees, regulations, requirements and procedures including *inter alia*, end product criteria, processes and production methods, testing, inspection, certification and approval procedures, quarantine treatments including relevant requirements associated with the transport of animals or plants or with the materials necessary for their survival during transport, provision of relevant statistical methods, sampling procedures and method of risk assessment, and packaging and labelling requirements directly related to food safety.

The Agreement consists of 14 Articles obligatory to all Members. These Articles include: General Provisions, Basic Rights and Obligations, Harmonization, Equivalence, Assessment of risks and determination of appropriate level of sanitary and phytosanitary protection, Adaptation to regional conditions including 'Pest-Disease-free' Areas of low pest and disease prevalence, Transparency, Control, Inspection and Approved Procedures,

Technical Assistance, Special and Differential Treatment, Consultation and Dispute Settlement, Administration, Implementation and Final Provisions. The WTO made the Agreement so that harmonious approaches are made by each Member country to avoid unwanted hazards along with the imports. The implementation of such approaches requires substantial backup of modern technologies that most of the developing countries lack.

Pre-agreement status of SPS in different countries

The term 'Quarantine' normally covers sanitation and phytosanitation. Plant Protection and Quarantine (PP & Q) is the umbrella term for the strategies undertaken for phytosanitation in different countries. It is basically a disease control strategy practised primarily by Government Agencies usually at the National but also at the State or Provincial or District level. These Agencies are empowered by Rules and Regulations promulgated by the Government for the purpose of reducing the risks of inadvertently introducing hazardous pests, pathogens, weeds and chemicals from foreign areas. The legal basis is either Legislation enacted by National and sometimes by State or Provincial Governments or enabling Legislation that directs the Minister of Agriculture to issue necessary Rules, Orders or Directives. Some countries are bound by Legislation enacted by Regional Parliaments, representing groups of countries often make Recommendations to their Member countries, but those Recommendations do not have any Legal or Binding status. The legal foundation of international plant quarantine matters is the International Plant Protection Convention (IPPC) of 1951 also known as the 'Rome Convention'. This Convention made a Treaty that is administered by the Food and Agricultural Organization (FAO) of the United Nations (UNO). Eighty nine countries are signatories in the IPPC, and as such, adhere to the standardized procedures and documentations which involve permits, phytosanitary certificates, and interception.

Characteristics of these Regulations are: 1. Specify prohibitions, exclusions, or other Regulatory actions. 2. Require import permits for the entry of named non-prohibited articles, and allow entry, without permits of other low risk named articles. 3. Allow entry, under a special permit, of prohibited items whose importation under specified safeguards is approved for scientific purposes. 4. Require phytosanitary certificates, and depending on permit specifications, need added declarations on certificates. 5. Stipulate inspection upon arrival. 6. Prescribe treatment at origin, during transit, or upon arrival. 7. Prescribe post-entry greenhouse growing, field isolation, or other safeguards after entry.

With the advance of agricultural systems, globalization and rapid increase in the inter-country movement of plant materials relevant advances in SPS have also been made at the international level. These are:

1974-79: Inclusion of SPS concerns for food safety and health of human beings, animals, and plants in the Agreements on the Technical Barriers of the Trade (TBT) during the Tokyo Round.

1986: Call for increased discipline in Agricultural Trade including SPS measures as per the *Punta del Este Declaration*, launched the Uruguay Round in September 1986.

1988: Mid-Term Review of the Uruguay Round in which the following priorities areas related to the SPS measures were identified: (a) harmonization of the SPS measures on the basis of the International Standards, (b) transparency through effective notification of National Regulations, and (c) improvement in the dispute settlement process including bilateral resolution of dispute and scientific support.

1990. Foundation of the Working Group on SPS Regulations. Inclusion of SPS measures in a separate draft Agreement.

1991: The Director General of GATT issued the 'Dunkel Draft'. This Draft included SPS issues. The final text of the Agreement on the Application of SPS measures was approved at the end of the Uruguay Round in 1994.

1993: A Secretariat of the IPPC was established that started the Standard setting process related to the SPS measures.

1995: Establishment of a Dispute Settlement Mechanism under the WTO that provides mutually acceptable solutions or adoption of a 'Panel/Appellate Body' ruling by the Dispute Settlement Body (DSB).

2000: (a) SPS Committee completes draft on the risk consistency. The Draft provides guidelines (not legally binding) on the levels of health protection. (b) Finalization of the United Nations' Agreement concerning trade in genetically modified organisms (GMOs).

Current scenario

The IPPC was developed as an International Treaty to secure action to prevent the spread and introduction of pests and plant products, and to promote appropriate measures for their control. It is governed by the Commission on Phytosanitary Measures (CPM) which adopts International Standards for Phytosanitary Measures (ISPMs). The CPM has confirmed the IPP as the preferred forum for National IPPC reporting and the exchange of more general information among the phytosanitary community. The IPPC Secretariat coordinates the activities of the Convention and is provided by the FAO.

As per Agreement each country is to have its own National Board/Council to deal with Plant Quarantine, Inspection, Risk analysis, Certification, etc. The USA has its own National Plant Board, but the quarantine in the European and Mediterranean countries is operated through European and Mediterranean Plant Protection Organization (EPPO). There is also a Pacific Plant Protection

Organization (PPPO) and Inter-African Phytosanitary Council to regulate the SPS in Pacific island countries and territories of African countries respectively.

The National Plant Board in the USA provides Plant Quarantine, Nursery Inspection and Certification Guidelines. The Guidelines mostly cover: A. Elements of Pest Prevention (pest exclusion, pest detection, eradication, pest diagnosis and record keeping and public information and education), B. Enforcement Mechanism, C. Guidelines for Pest Surveillance, Pest Survey, and Pest Rating List, D. Risk Characterization, E. Principles of Plant Quarantine, F. Principles of Pest Control, G. Initiating and Discontinuing State/Federal Plant Protection Programs, H. Interstate Origin Inspection, I. National Plant Board Standard for Phytosanitary Certification, J. Nursery Stock Certification, K. Good Nursery Practices, L. Nursery License/Registration/Certification Information.

EPPO is an intergovernmental organization responsible for European plant health. Founded in 1951 by 15 European countries, EPPO now has 49 members covering almost all the countries of Europe and the Mediterranean. Its objectives are to protect plants, to develop international strategies against the introduction and spread of dangerous pests and to promote safe and effective control methods. EPPO has produced a large number of standards and publications on plant pests, phytosanitary regulations and plant protection products. EPPO's priority objectives are: (i) Encourage harmonization of phytosanitary regulations and all other areas of official plant protection, (ii) Develop international strategy against the introduction and spread of pests that damage cultivated and wild plants, in natural and agricultural ecosystems, (iii) Promote the use of modern, safe, and effective pest control methods, (iv) Provide a documentation service on plant protection. The activities of EPPO concern any aspect of plant protection in agriculture, forestry and horticulture with an international dimension, in which National Plant Protection Organizations are involved. An important part of these activities is plant quarantine, as one of the aims of EPPO is to prevent entry or spread of dangerous pests including: (i) Information on quarantine pests, (ii) Pest Risk Analysis, (iii) EPPO Alert List, (iv) Phytosanitary regulations, (v) EPPO Project on quarantine pests for forestry, (vi) List of biological control agents widely used in the EPPO region, etc.

In the international phytosanitary and quarantine system, the most difficult task is to inspect, identify and eliminate viruses and to issue certificates because of the difficulties in their detection and identification. Even today the inter-country and inter-state movement and spread of viruses are common particularly in developing countries where detection facilities are very limited. Stringent regulatory measures along with technological development are necessary in these countries to pursue international quarantine regulations. Strict quarantine measures that are necessary with respect to viruses can be visualized through the practices undertaken in advanced countries particularly the USA and the EPPO countries. This aspect could be better illustrated by citing the legally bound quarantine practices in an appropriate country/state, for example, Washington State Department of Agriculture.

In the state of Washington the risk viruses are blueberry scorch, all grape viruses, peach mosaic, peach rosette, peach yellows and potato virus Yn. So the plant materials entering the state of Washington are subject to quarantine with respect to these viruses. The details of the quarantine practices followed in the state of Washington are given in Appendix VIII.1.

Each country/state is to have such legally bound regulatory features for all exporting and importing plants/plant parts, etc., to avoid the entry of viruses in that country. It is however necessary to develop a network of appropriate infrastructures operated by properly trained staff to grow the plants following the recommended practises to make the products under a complete SPS umbrella to overcome the barriers of exporting those items.

References

Abad, J., Anastasio, G., Fraile, A. and Garcia Arenal, F. 2000. A search for resistance to cucumber mosaic virus in the genus *Lycopersicon*. J. Plant Path. 82: 39-48

Acord, B.R. 1996. Availability of determination of nonregulated status for squash line genetically engineered for virus resistance. Federal Register 61: 33485-33486

Adair, T.L. and Kearney, C.M. 2000. Recombination between a 3-kilobase tobacco mosaic virus transgene and a homologous viral construct in the restoration of viral and nonviral genes. Arch. Virol. 145: 867-1883

Allison, R.F., Schneider, W.L. and Green, A.E. 1996. Recombination in plants expressing viral transgenes. Semin. Virol. 7: 417-422

Anderson, J.M., Palukaitis, P. and Zaitlin, M. 1992. A defective replicase gene induces resistance to cucumber mosaic virus in transgenic tobacco plants. Proc. Natl. Acad. Sci. USA 89: 8759-8763

Anfoka, G.H. 2000. Benzo-(1,2,3)-thiodiazole-7-carbothioic acid 5-methyl ester induces systemic resistance in tomato (*Lycopersicon esculentum* Mill. cv. Vollendung) to cucumber mosaic virus. Crop Protection 19: 401-405

Antignus, Y., Mor, N., Ben-Joseph, R., Lapidot, M. and Cohen, S. 1996. UV-absorbing plastic sheets protect crops from insect pests and from virus diseases vectored by insects. Env. Entomol. 25: 919-924

Arce-Achoa, J.P., Dainello, F., Pike, L.M. and Drews, D. 1995. Field performance comparison of two transgenic summer squash hybrids to their parental hybrid line. HortScience 30: 492-493

Astier, S., Albouy, J., Maury, Y., Robaglia, C. and Lecoq, H. 2007. Principles of Plant Virology: Genome, Pathogenenicity, Virus Ecology. Science Publishers Inc. Enfield, USA

Audy, P., Palukaitis, P., Slack, S.A. and Zaitlin, M. 1994. Replicase-mediated resistance to potato virus Y in transgenic tobacco plants. Molecular Plant Microbe Interactions 7: 15-22

Bacon, R.K., Kelly, J.T., Milus, E.A. and Parsons, C.E. 2002. Registration of 'Sabbe' wheat. Crop Science 42: 1375-1376

Balamuralikrishnan, M., Doraisamy, S., Ganapathy, T. and Viswanathan, R. 2002. Combined effect of chemotherapy and meristem culture on sugarcane mosaic virus elimination in sugarcane. Sugar Tech. 4: 19-25

Barker, H., Reavy, B., Kumar, A., Webster, K.D. and Mayo, M.A. 1992. Restricted virus multiplication in potatoes transformed with the coat protein gene of potato leaf roll luteovirus similarities with a type of host-gene-mediated resistance. Annals Appl. Biol. 120: 55-64

Barlass, M., Skene, K.G.M., Wooham, R.C. and Krabe, L.R. 1982. Regeneration of virus grapevines using *in vitro* apical culture. Ann. Appl. Biol. 101: 291-295

Baulcombe, D. 1994. Replicase-mediated resistance: A novel type of virus resistance in transgenic plants. Trends in Microbiology 2: 60-63

Baulcombe, D.C. 1996. Mechanisms of pathogen-derived resistance to viruses in transgenic plants. Plant Cell 8: 1833-1844

Baulcombe, D.C. 1999. Viruses and gene silencing in plants. Arch. Virol. 15 (Suppl): 189-201

Baulcombe, D.C. 2004. RNA silencing in plants. Nature 431: 356-363

Baulcombe, D.C. 2005. RNA silencing. Trends in Biochemical Sciences 30: 290-293

Baulcombe, D.C. and Molnar, A. 2004. Crystal structure of P19— a universal suppressor of RNA silencing. Trends in Biochemical Sciences 29: 279-281

Beachy, R.N. 1993. Transgenic resistance to plant viruses. Semin. Virol. 4: 327-416

Beachy, R.N. 1997. Mechanisms and applications of pathogen-derived resistance in transgenic plants. Curr. Opin. Biotechnol. 8: 215-220

Birch, A.N.E., Gordon, S.C., Fenton, B., Malloch, G., Jones, A.T., Griffiths, D., Graham, J., Brennan, R. and Woodford, J.A.T. 2004. Developing a sustainable IPM system for high value *Rubus* crop (raspberry, blackberry) for Europe. Acta Horticulturae No. 649: 289-292

Borja, M., Rubio, T., Scholthof, H.B. and Jackson, A.O. 1999. Restoration of wild-type virus by double recombination of tombusvirus mutants with a host transgene. Molecular Plant Microbe Interactions 12: 153-162

Bosland, P.W. 2000. Sources of curly top virus resistance in capsicum. Hort Science 35: 1321-1322

Boxus, P., Quoirin, M. and Laine, J.M. 1977. Large scale propagation of strawberry plants from tissue culture. In: Plant Cell, Tissue and Organ Culture. J. Reinert and Y.P.S. Bajas (eds). Springer-Verlag, Berlin, Germany. pp. 130-143

Brederode, F.T., Taschner, P.E.M., Posthumus, E. and Bol, J.F. 1995. Replicase-mediated resistance to alfalfa mosaic virus. Virology 207: 467-474

Bressan, P.H., Kim, Y.J., Hyndman, S.E., Hassegawa, P.M. and Bressan, R.A. 1982. Factors affecting the *in vitro* propagation of rose. J. Am. Hortic. Sci. 107: 979-990

Broadbent, L. 1976. Epidemiology and control of tomato mosaic virus. Annu. Rev. Phytopathol. 14: 75-96

Brown, C.R., Smith, O.P., Damsteegt, V.D., Yang, C.P., Fox, L. and Thomas, P.E. 1995. Suppression of PLRV titer in transgenic Russet Burbank and Ranger Russet. Am. Potato J. 72: 589-597

Brunetti, A., Tavazza, M., Noris, E., Tavazza, R., Ancora, G., Crepsi, S. and Accotto, G.P. 1997. High expression of truncated viral Rep protein confers resistance to tomato yellow leaf curl virus in transgenic tomato plants. Molecular Plant Microbe Interactions 10: 571-579

Buzzell, R.I. and Tu, J.C. 1984. Inheritance of soybean resistance to soybean mosaic virus. J. Hered. 75: 82

Callaway, A., Giesman-Cookmeyer, D., Gillock, E.T., Sit, T.L. and Lommel, S.A. 2001. The multifunctional capsid proteins of plant RNA viruses. Annu. Rev. Phytopathol. 39: 419-460

Candelier-Harvey, P. and Hull, R. 1993. Cucumber mosaic virus genome is encapsidated in transgenic plants. Trans. Res. 2: 277-285

Canto, T. and Palukaitis, P. 2001. A cucumber mosaic virus (CMV) RNA1 transgene mediates suppression of the homologous viral RNA1 constitutively and prevents CMV entry into the phloem. J. Virol. 75: 9114-9120

Carr, J.P. and Zaitlin, M. 1993. Replicase-mediated resistance. Seminars in Virology 4: 339-447

Carrol, T.W. 1983. Certification schemes against barley stripe mosaic. Seed Sci. Technol. 11: 1033-1042

Carver, B.F., Krenzer, E.G., Hunger, R.M., Martin, T.J., Klatt, A.R., Porter, D.R., Verchot, J., Rayas Duarte, P., Guenze, A.C., Martin, B.C. and Bai, G. 2003. Registration of 'Intrada' wheat. Crop Science 43: 1135-1136

Cebolla-Cornejo, J., Soler, S., Gomar, B., Soria, M.D. and Nuez, F. 2003. Screening *Capsicum* germplasm for resistance to tomato spotted wilt virus (TSWV). Annals Appl. Biol. 143: 143-152

Celebi, T.F., Slack, S.A. and Russo, P. 2003. Potato resistance to cucumber mosaic virus is temperature sensitive and virus strain specific. Breeding Science 53: 69-75

Chelkowski, J., Tyrke, M. and Sobkewicz, A. 2003. Resistance genes in barley (*Hordeum vulgare* L.) and their identification with moleculer markers. J. Appl. Genetics 44: 291-309

Chen, Q., Conner, R.L., Li, H.J., Sun, S.C., Ahmad, F., Laroche, A., Graf, R.J. and Chen, Q. 2003. Molecular cytogenetic discrimination and reaction to wheat streak mosaic virus and the wheat curl mite in Zhong series of wheat-*Thinopyrum intermedium* partial amphiploids. Genome 46: 135-145

Chen, Z.C., White, R.F., Antoniw, J.F. and Lin, Q. 1991. Effect of pokeweed antiviral protein (PAP) on the protection of plant viruses. Plant Pathology 40: 612-620

Cho, J.J., Ullman, D.E., Wheatley, E., Holly, J. and Gonsalves, D. 1992. Commercialization of ZYMV cross protection for zucchini production in Hawaii. Phytopathology 82: 1073

Cirullii, and Alexander, L. 1969. Influence of temperature and strain of tobacco mosaic virus on resistance of a tomato breeding line derived from *Lycopersicon peruvianum*. Phytopathology 59: 1287-1297

Clough, G.H. and Hamm, P.B. 1995. Coat protein transgenic resistance to watermelon mosaic and zucchini yellows mosaic virus in squash and cantaloupe. Plant Dis. 79: 1107-1109

Cohen, S. 1982. Control of whitefly vectors of viruses by color mulches. In: Pathogens, Vectors and Plant Diseases. K.F. Harris and K. Maramorosch (eds). Academic Press, London, UK. pp. 45-56

Cohen, S. and Marco, S. 1973. Reducing the spread of aphid-transmitted viruses I peppers by trapping the aphids on sticky yellow polyethylene sheets. Phytopathology 63: 1207-1209

Cohen, S. and Melamed-Madjar, V. 1978. Prevention by soil mulching of the spread of the tomato yellow leaf curl virus transmitted by *Bemisia tabaci* (Gennadius) in Israel. Bull. Entomol. Res. 68: 465-470

Cooper, B., Lapidot, M., Heick, J.A., Dodds, J.A. and Beachy, R.N. 1995. A defective movement protein of TMV in transgenic plants confers resistance to multiple virus whereas the functional analog increases susceptibility. Virology 206: 307-317

Criley, R. 2003. Production of virus tested gladioli through in vitro and in vivo techniques. In: Elegant Science in Floriculture. M. Mangal, S.V. Bhardwaj, A. Handa, K.K. Jindel and T. Blom (eds). Acta Horticulturae No. 624: 511-514

Csinos, A.S., Pappu, H.R., McPherson, R.M. and Stephenson, M.G. 2001. Acibenzolar S methyl (0.2 to 8 g a. i. /7000 plants) pretransplant application for suppression of TSWV. Plant Dis. 85: 292-296

Culbreath, A.K., Todd, J.W. and Brown, S.L. 2003. Epidemiology and management of tomato spotted wilt in peanut. Annu. Rev. Phytopath. 41: 53-75

Czech, A.S., Szklarczyk, M., Gajewski, Z., Zukowska, E., Michelik, B., Kobyko, T. and Strzalk, K. 2003. Selection of tomato plants resistant to a local Polish isolate of tomato spotted wilt virus (TSWV). J. Appl. Genetics 44: 473-480

Dardick, C.D. and Culvar, J.N. 1997. Tobamovirus coat proteins: elicitors of the hypersensitive response in *Solanam melongana* (eggplana). Molecular and Plant Microbe interactions 10: 776p778

Dardick, C.D., Taraporewala, Z., Lu, B. and Culver, J.N. 1999. Comparison of tobamovirus coat protein structural features that affect elicitor activity in pepper, eggplant and tobacco. Molecular Plant Microbe Interactions 121: 247-251

Daryono, B.S., Somowiyarjo, S. and Natsuaki, K.T. 2003. New source of resistance to cucumber mosaic virus in melon. SABRAO J. Breeding and Genetics 35: 19-26

de Haan, P., Gielen, J.J.L., Prins, M., Wijkamp, I.G., Van Schepen, A., Peters, D., Van Grinsven, M.Q.J.M. and Goldbach, R. 1992. Characterization of RNA-mediated resistance to tomato spotted wilt virus in transgenic tobacco plants. Bio/Technology 10: 1133-1137

De Wilde, C., Van Houdt, H., De Buck, S., Angenon, G., De Jaegger, G. and De Picker, A. 2000. Plants as bioreactors for protein production avoiding the problem of transgene silencing. Plant Mol. Biol. 43: 347-359

DeZoeten, G.A. 1991. Risk assessment: Do we let history repeat itself? Phytopathology 81: 585-586

Diaz, J.A., Nieto, C., Moriones, E. and Aranda, M.A. 2002. Spanish Melon necrotic spot virus isolate overcomes the resistance conferred by the recessive niv gene of melon. Plant Disease 86: 694

Dogimont, C., Bussemakers, A., Martin, J., Siama, S., Lecoq, H. and Pitrat, M. 1997. Two complementary recessive genes conferring resistance to Cucurbit Aphid borne Yellows Luteovirus in an Indian melon line (*Cucumis melo* L.). Euphytica: 391-395

Duffus, J.E. 1987. Durability of resistance. Ciba Foundation Symposium 133: 196-199

Ellastrand, N.C. 2003. Current knowledge of gene flow in plants: implications for transgene flow. Philos. Trans. R. Soc. London Ser B 358: 1163-1170

Ellstrand, N., Prentice, H.C. and Hancock, J.F. 1999. Gene flow and introgression from domesticated plants into their wild relatives. Annu. Rev. Ecol. Syst. 30: 539-563

Faccioli, G. and Marani, F. 1998. Virus Elimination by Meristem Tip Culture and Tip Micrografting. In: Plant Virus Disease Control. A. Hadidi, R.K. Khetarpal and H. Koganezawa (eds). APC Press, American Phytopathological Society. St. Paul, Minnesota, USA. pp. 346-388

Fauquet, C.M. Mayo, M.A., Maniloff, J., Desselberger, U. and Ball, L.A. 2005. Virus Taxonomy. VIIIth Report of the International Committee on Taxonomy of Viruses. Elsevier. Academic Press, San Diego, USA

Fargette, D., Pinel, A., Traore, O., Ghesquiere, A. and Konate, G. 2002. Emergence of resistance-breaking isolates of Rice yellow mottle virus during serial inoculations. Euro. J. Plant Path. 108: 585-591

Fedak, G., Chen, Q., Conner, R.L., Laroche, A., Comeau, A. and St Pierre, C.A. 2001. Characterization of wheat-Thinopyrum partial amphiploids for resistance to barley yellow dwarf virus. Euphytica 120: 373-378

Ferreira, S.A., Mau, R.F.L., Manshardt, R.M. and Gonsalves, D. 1993. Field evaluation of papaya ring spot virus cross protection. Proceedings of the 28th Annual Hawaii Papaya Industry Association Conference: 14-19

Farinelli, L., Malnöe, P. and Collect, G.F. 1992. Heterologous encapsidation of potato virus Y strain P (PVY0) with the transgenic coat protein of PVY strain N (PVYN) in *Solanum tuberosum* cv. Bintje. Bio/Technology 10: 1020-1025

Finlay, K.W. 1953. Inheritance of spotted wilt resistance in tomato. II. Five genes controlling spotted wilt resistance in four tomato types. Aust. J. Biol. Sci. 6: 153-163

Fitch, M.M.M., Manshardt, R.M., Gonsalves, D., Slightom, J.L. and Sanford, J.C. 1992. Virus resistant papaya derived from tissues bombarded with the coat protein gene of papaya ring spot virus. Bio/Technology 10: 1466-1472

Fletcher, J.T. 1978. The use of avirulent strains to protect plants against the effects of virulent strains. Ann. Appl. Biol. 89: 110-114

Foxe, M.J. 1992. Breeding for viral resistance: Conventional methods. Neth. J. Plant Path. 98: 13-20

Fraser, R.S.S., Loughlin, S.A.R. and Connor, J.C.1980. Resistance to tobacco mosaic virus in tomato: effects of the Tm-1 gene on symptom formation and multiplication of virus strain1. J. Gen. Virol. 50: 221-224

Freitas-Astúa, J., Purcifull, D.E., Polson, J.E. and Hiebert, E. 2002. Traditional and transgenic strategies for controlling tomato-infecting begomoviruses. *Fi topathol. Bras.* Vol. 27, No. 5, Brasilia Sept./Oct. 2002.

Frischmuth, T. and Stanley, J. 1998. Recombination between viral DNA and the transgenic coat protein gene of African cassava mosaic geminivirus. J. Gen. Virol. 79: 1265-1271

Fritts, D.A., Michel, G.L. Jr. and Rush, C.M. 1999. The effects of planting date and insecticide treatments on the incidence of high plains disease in corn. Plant Dis. 83: 1125-1128

Fuchs, M. and Gonsalves, D. 1995. Resistance of transgenic hybrids squash ZW-20 expressing the coat protein genes of zucchini yellow mosaic virus and watermelon virus 2 to mixed infections by both potyviruses. Bio/Technology 13: 1466-1473

Fuchs, M. and Gonsalves, D. 2007. Safety of Virus-Resistant Transgenic Plants Two Decades After Their Introduction: Lessons from Realistic Field Risk Assessment Studies. Annu. Rev. Phytopathol. 45: 173-202

Fuchs, M., Ferreira, S. and Gonsalves, D. 1997. Management of Virus Diseases by Classical and Engineered Protection. Mol. Plant Path. On-line, [http://www.bspp.org.uk/mppol/]1997/0116fuchs

Fulton, R.W. 1986. Practices and the precautions in the use of cross protection for plant virus disease control. Annu. Rev. Phytopathol. 24: 67-81

Gal, S., Pisan, B., Hohn, T., Grimsley, N. and Hohn, B. 1992. Agroinfection of transgenic plants leads to variable Cauliflower mosaic virus by intermolecular recombination. Virology 187: 525-533

Gallitelli, D., Vovlas, C., Martelli, G., Montasser, M.S., Tousignant, M.E. and Kaper, Z.M. 1991. Satellite-mediated protection of tomato against cucumber mosaic virus: II. Field test under natural epidemic conditions in Southern Italy. Plant Dis. 75: 93-95

Garcheva, P. 2002. Elimination of plum pox virus (PPV) in plum (*Prunus domestica* L.) cvs. In: Kyustendilska Sinya and Valjevka through in vitro techniques. L. Nacheva, S. Milusheva, K. Ivanova, V. Djouvinov and D. Dotchev (eds). Acta Horticulturae No. 577: 289-291

Ghosh, S.B., Nagi, L.H.S., Ganapathi, T.R., Khurana, S.M.P. and Bapat, V.A. 2002. Cloning and sequencing of potato virus Y coat protein gene from Indian isolate and development of transgenic tobacco for PVY resistance. Current Science 82: 855-859

Goethals, M. and van Hoof, P. 1971. Régénération des oellets par la culture de méristèms combinée à la thermotherapie. Parasitica 27: 36-41

Goldbach, R. and de Haan, P. 1993. Prospects of engineered forms of resistance against tomato spotted wilt virus. Sem. Virol. 4: 381-367

Gonsalves, D. 1998. Control of papaya ring spot virus in papaya: a case study. Annu. Rev. Phytopathol. 36: 415-437

Gonsalves, D. and Slightom, J.L. 1993. Coat protein-mediated protection: analysis of transgenic plants for resistance in a variety of crops. Sem. Virol. 4: 397-406

Gragera, J., Soler, S., Diez, M.J., Catala, M.S., Rodriguez, M.C., Esparrago, G., Rosello, S., Costa, J., Rodriguez, A. and Nuez, F. 2003. Evaluation of some processing tomato lines with resistance to tomato spotted wilt virus for agricultural and processing characters. Spanish J. Agric. Res. 1: 31-39

Greene, A.E. and Allison, R.F. 1994. Recombination between viral RNA and transgenic plant transcripts. Science 263: 1423-1425

Greene, A.E. and Allison, R.F. 1996. Deletions in the 3' untranslated region of cowpea chlorotic mottle virus transgene reduce recovery of recombinant viruses in transgenic plants. Virology 225: 231-234

Grube, R.C., Zhang, Y.P., Murphy, J.F., Loaiza Figueroa, F., Lackney, V.K., Provvidenti, R., Jahn, M.K. and Zhang, Y.P. 2000. New source of resistance to cucumber mosaic virus in *Capsicum frutescens*. Plant Disease 84: 885-891

Grumet, R. 1994. Development of virus resistant plants via genetic engineering. In: Plant Breeding Reviews. Vol. 12 J. Janick (eds). John Wiley & Sons, New York, USA. pp. 47-79

Hackland, H.F., Coetzer, C.T., Rybicki, E.P. and Thompson, J.A. 2000. Genetically engineered resistance in tobacco against South African strains of tobacco necrosis and cucumber mosaic virus. South African Journal of Science 96: 33-38

Hammond, J. and Kamo, K.K. 1995. Effective resistance to potyvirus infection conferred by expression of antisense RNA in transgenic plants. Plant Microbe Interactions 8: 674-682

Hammond, J. and Dienelt, M.M. 1997. Encapsidation of potyviral RNA in various forms of transgene coat protein is not correlated with resistance in transgenic plants. Molecular Plant Microbe Interaction 10: 1023-1027

Hammond, J., Lecoq, H. and Raccah, B. 1999. Epidemiological risks from mixed infections and transgenic crops expressing viral genes. Adv. Virus Res. 54: 180-314

Harrison B.D. 2002. Virus variation in relation to resistance-breaking in plants. Euphytica 124: 181-192

Harrison, B.D., Mayo, M.A., Roberts, I.M., Barker, H., Mowat, W.P., Hutchinson, A.M., Woodford, J.A.T., Murant, A.F., Jones, A.T., Salazar, L., Randles, J.W. and Hanad, P.K. 1976. Virology. In: Scottish Horticultural Research Institute 22nd Annual Report for the Year 1973, pp. 65-80

Harpaz, I. 1982. Non-pesticidal control of vector-borne diseases. In: Pathogens, Vectors and Plant Diseases: Approaches to Control. K.F. Harris and K. Maramorosch (eds). Academic Press, London, UK. pp. 1-21

Harper, G., Hull, R., Lockhart, B. and Olszewski, N. 2002. Viral sequences integrated into plants genomes. Annu. Rev. Phytopathol. 40: 119-136

Hayakawa, T., Zhu, Y., Itoh, K., Kimura, Y., Izawa, T., Shimamoto, K. and Toryama, S. 1992. Genetically engineered rice resistant to rice stripe virus, an insect transmitted virus. Proc. Natl. Acad. Sci. USA 89: 9865-9869

Heinrich, J.-C., Tabler, M. and Louis, C. 1993. Attenuation of white gene expression in transgenic *Drosophila melonogaster*: Possible role of a catalytic antisense RNA. Dev. Genet. 14: 258-265

Helliot, B., Panis, B., Poumay, Y., Swennen, R., Lepovre, P. and Frison, E. 2002. Cryopreservation for the elimination of cucumber mosaic and banana streak viruses from banana (*Musa* spp.). Plant Cell Reports 20: 1117-1122

Hemenway, C., Fang, R.-X., Keniewski, W., Chua, N.-H. and Tumer, N. 1988. Analysis of the mechanism of protection in transgenic plants expressing the potato virus X coat protein or its antisense RNA. EMBO J. 7: 1273-1280

Hirai, G., Nakazumi, H., Yagi, R., Nakano, M., Shiga, Y. and Dohi, H. 2003. A new MNSV resistant melon rootstock variety 'Sorachi Dai Ko 3'. Bull. Hokkaido Prefectural Agricultural Experiment Stations. No. 84: 47-54

Hiroyuki, L. and Yamada, S. 1980. Inactivation of *Citrus tristeza virus* (CTV) with heat treatment: Heat tolerance and inactivation of CTV on root-scion combinations. Proc. 8th Conf. IOCV. E.C. Calavan, S.M. Garnsey and L.W. Timmer (eds). Univ. Calif. Riverside, CA, USA. pp. 20-24

Hoffman, K., Qiu, W.P. and Moyer, J.W. 2001. Overcoming host- and pathogen-mediated resistance in tomato and tobacco maps to the MRNA of tomato spotted wilt virus. Molecular Plant Microbes Interactions 14: 242-249

Hollings, M. 1965. Disease control through virus-free stock. Annu. Rev. Phytopathol. 3: 365-396

Hollings, M. and Stone, O.M. 1968. Heat treatment, meristem-tip culture and the production of virus-tested carnation and chrysanthemum. Rep. Glass. Crops Res. Inst. 1967: 106-107

Holmes, F.O. 1948. Elimination of spotted wilt from dahlia. Phytopathology 38: 314

Holmes, F.O. 1929. Loca lesions in tobacco mosaic. Bot. Gaz. (Chicago). 87: 39-55.

Holt, C.A. and Beachy, R.N. 1991. In vivo complementation of infectious transcripts from mutant tobacco mosaic virus cDNAs in transgenic plants. Virology 181: 109-117

Huet, H., Mahendra, S., Wang, J., Sivamani, E., Ong, C.A., Chen, L., Kochko, A., de Beachy, R.N., Fauquet, C. and de Kochko, A. 1999. Near immunity to rice tungro spherical virus achieved in rice by a replicase-mediated resistance strategy. Phytopathology 89: 1022-1027

Hull, R. 2000. Virus-detection-nucleic acid hybridization. In: Encyclopedia of Plant Pathology. O.C. Maloy and T.D. Murray (eds). John Wiley, New York, USA. pp. 1092-1095

Hull, R. 2002. Matthew's Plant Virology. Academic Press, London, UK

Huth, W. 2003. The spread of dwarf diseases in cereal-supporting factors with possibilities for containing damage. In: Bericht uberdie Arbeitsagung 2002 der veremingung der Pflanzenzuchter und Saatgutkaufleeute Osterreichs 26 bis 28 November 2002 in Gumpenstein. pp. 95-97

Izhar, S. 1999. Resistance to tomato spotted wilt virus in transgenic tomato varieties. In: Plant Biotechnology and in vitro Biology in the 21st Century. P. Stoeva, D. Hristova, P. Donkova, M. Petrova, N. Gorinova, M. Yankulova, D. Inze, W. Van Kamp, A. Atanossov, A. Altman and M. Ziv (eds). Proc. IXth Intl. Cong. of the International Society of Plant Tissue Culture and Biotechnology, Jerusalem, Israel, 14-19 June 1998. Current Plant Science and Biotechnology in Agriculture Vol. 36: 599-552

Jacquemond, M. and Tepfer, M. 1998. Satellite RNA-mediated resistance to plant viruses: are the ecological risks well assessed? In: Plant Virus Disease Control. A. Hadidi, R.K. Khetarpal and H. Koganezawa (eds). APC Press, American Phytopathological Society, St. Paul, Minnesota, USA. pp. 94-120

Jakab, G., Vaistij, F.E., Droz, E. and Manëe, P. 1997. Transgenic plant expressing viral sequences create a favourable environment for recombination between viral sequences. In: Virus Resistant Transgenic Plants: Potential Ecological Impact. M. Tepfer and E. Balazs (eds). INRA Ed.-Springer. Heidelberg, Germany. pp. 45-51

James, D. and Clark, M.F. 2001. Long term assessment of the effects of in vitro chemotherapy as a tool for apple stem grooving virus elimination. Acta Horticulturae 550(2): 459-462

Ji, L.H., Ding, S.W. and Ji, L.H. 2001. The suppressor of transgene RNA silencing encoded by cucumber mosaic virus interferes with salicylic acid-mediated virus resistance. Molecular Plant Microbe Interactions 14: 715-724

Jones, A.T. 1998. Control of virus infection in crops through breeding plants for vector resistance. In: Plant Virus Disease Control. A. Hadidi, R.K. Khetarpal and H. Kogazawa (eds). APS Press, American Phytopathological Society. St. Paul, MN, USA. pp. 41-55

Jones, J. 2001. Harpin. Pesticide Outlook 12: 134-135

Jones, R.A.C. 1982. Break-down of potato virus X resistance gene Nx: selection of a group four strain from strain group three. Plant Pathol. 31: 325-331

Jones, R.A.C. 1985. Further studies on resistance-breaking strains of potato virus X. Plant Pathol. 34: 182-189

Jones, O.P. and Vine, S.J. 1969: Tip culture of gooseberry shoot tips for eliminating virus. J. Hortic. Sci. 43: 289-292

Jones, L., Hamilton, A.J., Voinnet, O., Thomas, C.L., Maule, A.J. and Baulcombe, D.C. 1999. RNA-DNA interactions and DNA methylation in post-transcriptional gene silencing. Plant Cell 11: 2291-2301

Kadotani, M., Ikegami, M. 2002. Production of Patchauli mild mosaic virus resistant Patchauli plants by genetic engineering of coat protein precursor gene. Pest Management Science 58: 1137-1142

Kanievsky, W. and Lawson, C. 1998. Coat protein and replicase-mediated resistance to plant viruses. In: Plant Virus Disease Control. A. Hadidi, R.K. Khetarpal and H. Koganezawa (eds). APS Press, St. Paul, Minnesota, USA. pp. 65-78

Kaniewski, W., Lawson, C. and Thomas, P. 1993. Agronomically useful resistance in transgenic Russet Burbank potato containing a PLRV CP gene. Abstract IX International Congress of Virology. Glasgow, Scotland

Kanyuka, K., Lovell, D.J., Mitrofanova, O.P., Hammond-Kosack, K. and Adams, M.J. 2004. A controlled environment test for resistance to soil borne cereal mosaic virus (SBCMV) and its use to determine the mode of inheritance in wheat cv. Cadenza and for screening *Triticum monococcum* genotypes for sources of SBCMV resistance. Plant Pathology 53: 154-160

Kartha, K.K., Gamborg, O.L. and Coastabel, R. 1974. Regeneration of pea (*Pisum sativum* L.) plants from apical meristem. Z. Pflanzenphysiol. 72: 172-176

Kawbe, K., Iwasaki, M., Hayano Saito, Y., Honda, Y., Yoshida, K. and Maoke, T. 2002. Virus resistance in transgenic tomato expressing satellite RNA of cucumber mosaic virus1: Genetic aspects and virus resistance. Res. Bull. Hokkaido Natl. Agric. Expt. Station No. 177: 37053

Kawchuk, L.M., Martin, R.R. and McPherson, J. 1991. Sense and antisense-mediated resistance to potato leaf roll virus in Russet Burbank potato plants. Molecular Plant Microbe Interactions 4: 247-253

Kemp, W.G. 1978. Mulches protect peppers from viruses. Canada Agriculture 23: 22-24

Kimble, K.A., Grogen, R.G., Greathed, A.S., Paulus, A.O. and House, J.K. 1975. Development, application and comparison of methods for indexing lettuce seed mosaic for mosaic virus in California. Plant Dis. Rep. 59: 461-464

Klas, F.E., Fuchs, M. and Gonsalves, D. 2006. Comparative spatial spread overtime of *Zucchini yellow mosaic virus* (ZYMV) and *Watermelon mosaic virus* (WMV) in fields of transgenic squash expressing the coat protein genes of ZYMV and WMV, and in fields of nontransgenic squash. Trans. Res. 15: 527-541

Kohli, A., Gahawa, D., Vain, P., Laurie, D.A. and Christou, P. 1999. Transgene expression in rice engineered through particle bombardment molecular factors controlling stable expression and transgene silencing. Plants 208: 88-97

Koizumi, M., Kamiya, H. and Ohtsuka, E. 1992. Ribosomes to inhibit transformation of NIH 3T3 cells by the activated c-Ha-ras gene. Gene 117: 179-184

Kowalska, A. 1974. Freeing carnation plants from viruses by meristem tip culture. Phytopathol. Z. 79: 301-309

Krell, R.K., Pedigo, L.P., Hill, J.H. and Rice, N.E. 2004. Bean leaf beetle (Coleoptera: Chrysomelidae) management of reduction of bean pod mottle virus. J. Econ. Ento. 97: 192-202

Kuhne, T., Shi, N.N., Proeseler, G., Adams, M.J. and Kanyuka, K. 2003. The ability of a bymovirus to overcome the rym-4-mediated resistance in barley correlates with a codon change in the VPg coding region on RNA1. J. Gen. Virol. 84: 2853-2859

Lawson, C., Kaniewski, W., Haley, L., Rozman, R., Newell, C., Sanders, P. and Tumer, N. 1990. Engineering resistance to mixed virus infection in a commercial potato cultivar: resistance to potato virus X and potato virus Y in transgenic Russet Burbank. Bio/Technology 8: 127-134

Lecoq, H. 1998. Control of plant virus diseases by cross protection In: Plant Virus Disease Control. A. Hadidi, R.K. Khetarpal and H. Koganezawa (eds). APS Press, St. Paul, Minnesota, USA. pp. 33-40

Lecoq, H. and Pitrat, M. 1983. Field experiments on the integrated control of aphid-borne viruses in muskmelon. In: Plant Virus Epidemiology. J.M. Thresh and R.T. Plumb (eds). Blackwell, Oxford, UK. pp. 169-176

Lecoq, H., Lemaire, J.M. and Wipf-Scheibel, C. 1991. Control of zucchini yellow mosaic virus in squash by cross protection. Plant Dis. 75: 208-211

Lecoq, H., Ravelonandro, M., Wipf-Scheibel, C., Mansion, M., Raccah, B. and Dunez, J. 1993. Aphid transmission of a non-aphid transmissible strain of zucchini yellow mosaic virus from transgenic plants expressing the capsid protein of plum pox potyvirus. Molecular Plant Microbe Interactions 6: 403-406

Lewellen, R.T. 2002a. Registration of sugar beet CRO9-1 with dual resistance to cercospora and rhizomania. Crop Science 42: 672-673

Lewellen, R.T. 2002b. Registration of monogerm rhizomania resistant sugar beet parental lines C833-5 and C1333-5CMS. Crop Science 42: 321-322

Lewellen, R.T. 2004a. Registration of sugar beet germplasm lines C 67/2, C69/2, C78/3 and C80/2 with resistance to virus yellows and rhizomania. Crop Science 44: 358-359

Lewellen, R.T. 2004b. Registration of rhizomania resistant monogerm populations C869 and C869 CMS sugar beet. Crop Science 44: 358-359

Lim, S.H., Ko, M.K., Lee, S.J.K, La, Y.J. and Kim, B.D. 1999. Cymbidium mosaic virus coat protein gene in antisense confers resistance to transgenic *Nicotiana occidentalis*. Molecules and Cells 9: 603-608

Limasset, P. and Cornuet, P. 1949. Recherche de virus de la mosaïque du tabac dans les méristèms des plants infectées. CR Acad Sci. (Paris) 228: 1871-1872

Lindbo, J.A. and Dougherty, W.G. 1992a. Untranslatable transcripts of the tobacco etch virus coat protein gene sequence can interfere with tobacco etch replication in transgenic plants and protoplasts. Virology 189: 725-733

Lindbo, J.A. and Dougherty, W.G. 1992b. Pathogen-derived resistance to a potyvirus: immune and resistant phenotypes in transgenic tobacco expressing altered forms of a potyvirus coat protein. Plant Microbe Interact. 5: 144-153

Lindbo, J.A., Silva-Rosales, L., Proebsting, W.M. and Dougherty, W.G. 1993. Induction of a highly specific antiviral state in transgenic plants: implications for regulation of gene expression and virus resistance. Plant Cell 5: 1749-1759

Ling, K.C., Tiongco, R. and Flores, Z.M. 1983. Epidemiological studies of rice tungro. In: Plant Virus Epidemiology. R.T. Plumb and J.M. Thresh (eds). Blackwell Scientific Publications, Oxford, UK. pp. 249-258

Lius, S., Manshardt, R.M., Fitch, M.M.M., Slightom, J.L., Sanford, J.C. and Gonsalves, D. 1996. Pathogen-derived resistance provides papaya with effective protection against papaya ring spot virus. Molecular Breeding 3: 161-168

Lizarraga, R.E., Salazar, L.F., Roca, W.M.T. and Schildeó Rutschler, L. 1980. Elimination of potato spindle tuber viroid by low temperature and meristem culture. Phytopathology 70: 754-755

Loesch-Fries, L.S., Merlo, D., Sinnenin, T., Burhop, L., Hill, K., Krahn, K., Jarvis, N., Nelson, S. and Halk, E. 1987. Expression of alfalfa mosaic virus RNA 4 in transgenic plants confers virus resistance. EMBO J. 6: 1845-1851

Lombard, C. 2002. Effect of green manure crops and organic amendments on incidence of nematode-borne tobacco rattle virus. Acta Horticulturae No. 570: 287-292

Lommel, S.A. and Xiong, Z. 1991. Reconstitution of a functional red clover necrotic mosaic virus by recombinational rescue of the cell-to-cell movement gene expressed in a transgenic plant. J. Cell Biochem. 15A: 151

Lomonossoff, G.P. 1995. Pathogen-derived resistance to plant viruses. Annu. Rev. Phytopathology 33: 323-343

Lozoya-Saldana, H. and Dawson, W.O. 1982. The use of constant and alternating temperature regimes and tissue culture to obtain PVS-free potato plants. Am. Potato J. 59: 221-230

Lukicheva, L.A. and Mitrofanov, V.I. 2002. Cleaning up of cherry and plum by thermotherapy in vitro. Byulleten Gosudarst Vennogo Nikitskogo Botanicheskogo Sada No. 86: 59-61

McGram, G.R.D. and Adams, M.J. 2004. Investigating resistance to barley mild mosaic virus. Plant Pathology 53: 161-169

MacDonald, D.M. 1973. Heat treatment and meristem culture as a means of freeing potato varieties from viruses X and S. Potato Res. 16: 263-269

MacDiarmid, R. 2005. RNA silencing in productive virus infections. Annu. Rev. Phytopathol. 43: 523-544

Maene, L.J. and Debergh, P.C. 1985. Liquid medium additions to established tissue culture to improve elongation and rooting *in vivo*. Plant Cell Tissue Org. Cult. 5: 23-33

Maia, E., Beck, D. and Gaggelli, D. 1969. Obtension de clones d'oeillets méditerraneans indemnes de virus. Ann. Phytopathol. 1 (H.S.): 311-319

Makeshkumar, T., Varma, A., Singh, K.K., Malathi, V.G., Gupta, M.D. and Bhat, A.I. 2002. Coat protein gene related resistance of potato virus Y in transgenic tobacco. Indian Pl. Pathology 55: 187-194

Mallor Gimenez, C., Alvarez, J.M. and Arteaga, M.L. 2003. Inheritance of resistance to systemic symptom expression of melon necrotic spot virus (MNSV) in *Cucumis melo* L. 'Doublon'. Euphytica 134: 319-324

Malyshenko, S.I., Kondakova, O.A., Nazarova, J.V., Kaplan, I.B., Tallansky, M.E. and Atabekov, J.G. 1993. Reduction of tobacco mosaic virus accumulation in transgenic plants producing non-functional viral transport proteins. J. Gen. Virol. 76: 1149-1156

Marrou, J. and Migliori, A. 1972. La prémunition, une nouvelle méthode protection des cultures contre le virus de la mosaïque du tabac. P. H. M. Revue Horticole 124: 27-31

Martelli, G.P. 2001. Transgenic resistance to plant pathogens: benefits and risks. J. Plant Pathol. 83: 37-46

Martin, B., Kahbe, Y. and Fereres, A. 2003. Blockage of stylet tips as the mechanism of resistance to virus transmission by *Aphis gossypii* in melon lines bearing Vat gene. Annals Appl. Biol. 142: 245-250

Martin, R.R., Matthews, H. and Martin, R.R. 2001. Engineering resistance to Raspberry bushy dwarf virus. Proceedings of the Ninth International Symposium on Small Fruit Virus Diseases. Canterbury and HRI-East Malling, UK, 9-15 July 2000. Acta Horticulturae No. 551: 33-37

Mau, R.F.L., Gonsalves, D. and Bautista, R. 1990. Use of cross protection to control papaya ring spot virus at Waianae. Proceedings of the 25th Annual Papaya Industry Association Conference, 1989, pp. 79-84

McKinney, H.H. 1929. Mosaic diseases of Canary Islands, West Africa, and Gibraltar. J. Agric. Res. 39: 557-578

McGrann, G.R.D. and Adams, M.J. 2004. Investigating resistance to Barley mild mosaic virus. Plant Pathology 53: 161-169

McPhee, K.E. and Muchlbaur, F.J. 2002. Registration of 'Lifter' green dry pea. Crop Science 42: 1377-1378

Medley, T.L. 1994. Availability of determination of nonregulated status for virus resistant squash. Federal Register 59: 64187-64188

Melander, M., Lee, M. and Sandgren, M. 2001. Reduction of potato mop top virus accumulation and incidence in tubers of potato transformed with a modified triple gene block gene of PMTV. Molecular Breeding 8: 197-206

Melchers, G., Jockusch, H. and von Sengbusch, P. 1966. A tobacco mutant with a dominant allele for hypersensitivity against some TMV strains. Phytopathol. Z. 55: 86-88

Mellor, F.C. and Stace-Smith, R. 1977. Virus-free potatoes by tissue culture. In: Applied and Fundamental Aspects of Plant Cell, Tissue and Organ Culture. J. Reinert and Y.P.S. Bajaj (eds). Springer-Verlag, Berlin, Germany. pp. 616-635

Miao, H.Q., Chen, X.Z., Cao, K.Q., Yang, Y.J. and Li, S.Y. 2001. The relationship between different sowing periods and diseased plant rate of maize rough dwarf virus. Plant Protection 27: 9-12

Mink, G.I., Wample, R. and Howell, W.E. 1998. Heat Treatment of Perennial Plants to Eliminate. Phytoplasmas, Viruses, and Viroids While Maintaining Plant Survival. In: Plant Virus Disease Control. A. Hadidi, R.K. Khetarpal and H. Koganezawa (eds). APC Press, American Phytopathological Society, St. Paul, Minnesota, USA. pp. 332-345

Mitter, N., Sulistyowati, E. and Dietzgen, R.G. 2003. Cucumber mosaic virus infection transiently breaks dsRNA-induced transgenic immunity to potato virus Y in tobacco. Molecular Plant Microbe Interactions 16: 936-944

Monette, P.L. 1986. Elimination *in vitro* of two grapevine nepoviruses by an alternating temperature regime. J. Phytopathol. 116: 88-91

Móntasser, M.S., Tousignant, M.E. and Kaper, J.M. 1991. Satellite-mediated protection of tomato against cucumber mosaic virus: I. Greenhouse experiments and simulated epidemic conditions in the fields. Plant Dis. 75: 86-92

Morel, G. 1948. Recherches sur la cultu re associée de parasities obligatoires et de tissues végétaux. Ann. Epiphyt. I: 123-234

Morel, G. and Martin, C. 1952. Guerison de dahlias atteints d'une maladie a virus. C.R. Acad. Sci. (Paris) 235: 1324-1325

Morel, G. and Martin, C. 1955. Guerison de pommes de terre atteints de maladies a virus. C. R. Acad. Agr. Fr. 41: 472-475

Moskovets, S.N., Gorbarenko, N.I. and Zhuk, I.P. 1973. The use of the method of the culture of apical meristems in the combination with low temperature for the sanitation of potato against mosaic virus. Sel-skokhozyaiastvennaya Biol. 8: 271-275

Mukhopadhyay, S., Nath, P.S., Das, B.K., Basu, D. and Verma, A.K. 1994. Comparative epidemiology and cultural control of some whitefly transmitted viruses in West Bengal. In: Virology in the Tropics. N. Rishi, K.L. Ahuja and B.P. Singh (eds). Malhotra Publishing House, New Delhi, India. pp. 285-304

Mukhopadhyay, S., Rai, J., Sharma, B.C., Gurung, A., Sengupta, R.K. and Nath, P.S. 1997. Micro-propagation of Darjeeling orange (*Citrus reticulata* Blanco) by shoot-tip grafting. J. Hort. Sci. 72: 493-499

Muller, G.W. 1980. Use of mild strain of citrus tristeza virus (CTV) to reestablish commercial production of 'Pera' sweet orange in Sao Paulo, Brazil. Proc. Flo. State Hort. Soc. 93: 62-64

Mullin, R.H., Smith, S.H., Frazier, N.W., Schlegel, D.E. and McCall S.R. 1974. Meristem culture freed strawberries of mild yellow edge, pallidosis and mottle diseases. Phytopathology 64: 1425-1429

Munger, H.M. 1993. Breeding for virus disease resistance in cucurbits. In: Resistance to Viral Diseases of Vegetables: Genetics and Breeding. M.M. Kyle (ed). Timber Press, Oregon, USA. pp. 44-60

Mur, G. 1979. Thermothérapie de varieties de *Vitis vinifera* par la méthode de culture *in vitro*. Prog. Agric. Vitic. 96: 148-151

Murashige, T., Bitters, W.B., Rangan, T.S., Naver, E.M., Roistacher, C.N. and Holliday, P.B. 1972. A technique of shoot apex grafting and its utilization towards recovering virus-free citrus clone. HortScience: 118-119

Namba, S., Ling, K., Gonsalves, C., Gonsalves, D. and Slightom, J.L. 1991. Expression of the gene encoding the coat protein of cucumber mosaic virus (CMV) strain WL appears to provide protection to tobacco plants against infection by several different CMV strains. Gene 107: 181-188

Napoli, C., Lemieux, C. and Jorgensen, R.A. 1990. Introduction of a chimeric chalcone synthase gene into *Petunia* results in reversible co-suppression of homologous genes in trans. Plant Cell 2: 279-289

Navarro, L., Roistacher, C. and Murashige, T. 1975. Improvement of shoot tip grafting *in vitro* for virus-free Citrus. J. Am. Soc. Hortic. Sci. 100: 471-479

Ndjiondjop, M.N., Albar, L., Fargette, D., Fauquet, C. and Ghesquiere, A. 1999. The genetic basis of high resistance to rice yellow mottle virus (RYMV) in cultivars of two cultivated rice species. Plant Disease 83: 931-935

Ndowora, T., Dahal, G., Lafleur, D., Harper, G., Hull, R., Olszewski, N.E. and Lockhart, B. 1999. Evidence that badnavirus infection in *Musa* can originate from integrated pararetroviral sequences. Virology 255: 214-220

Nelson, R.S., McCormick, S.M., Delanney, X., Dube, P., Layton, J., Anderson, E.J., Kaniewaska, M., Proksch, R.K., Horsch, R.B., Rogers, S.G., Fraley, R.T. and Beachy, R.N. 1988. Virus tolerance, plant growth, and field performance of transgenic tomato plants expressing coat protein from tobacco mosaic virus. Bio/Technology 6: 109-119

Nogueira, M.S.R., Brioso, P.S.T. and Freire Filho, F.R. 2001. Resistance and heredity study in cowpea 'Costelao' to cowpea severe mosaic virus serotype I infection. Advancos technologicos no feijao coupi. V. Reuniao Naclonal de Pesquisa de Caupi, Teresina, Brazil, 4-7 dezembre 2001. Documentos Embrapa Meio Norte No 56: 63-66

Novatel, J.C. 1977. Lutte contre la mosaïque du tabac sur tomate: prémunition et avariétés résistantes. CTFL Documents N° 53: 109-119

Novatel, J.C., Trapaleau, M. and Marchoux, G. 1983. La prémunition: méthode de protection contre les virus. Bilan de la lutte contre in mosïque du tabac, quelques perspectives nouvelles. CR Col. ACTA 'Faune et flore auxilliare en agriculture, Paris, France. pp. 36-42

Nunome, T., Fukumoto, F., Hanada, K. and Hirai, M. 2002. Development of breeding materials of transgenic tomato plants with a truncated replicase gene of cucumber mosaic virus for resistance to the virus. Breeding Science 52: 219-223

Ogundiwin, B.A., Thottappilly, G., Aken' Ova, M.P., Ekp, E.J.A. and Fatokun, C.A. 2002. Resistance to cowpea mottle carmovirus in *Vigna vexillata*. Plant Breeding 121: 517-520

Okada, Y., Kashiwazaki, S., Kantani, R., Arai, S. and Ito, K. 2003. Effects of barley yellow mosaic disease resistant rym1 on the infection by strains of Barley yellow mosaic virus and Barley mild mosaic virus. Theoretical and Applied Genetics 106: 181-189

Omura, T. and Yan, J. 1999. Role of outer capsid protein in transmission of phytoreovirus by insect vectors. Adv. Virus Res. 54: 15-39

Ordon, F., Friedt, W., Scheurer, K., Pellio, B., Werner, K., Neuhaus, G., Huth, W., Habekuss, A. and Graner, A. 2004. Molecular markers in breeding for virus resistance in barley. J. Appl. Gen. 45: 145-159

Osburn, J.K., Sarkar, S. and Wilson, T.M.A. 1990 Complementation of coat protein-defective TMV mutants in transgenic tobacco plants expressing TMV coat protein. Virology 179: 921-925

Oshima, N. 1975. The control of tomato mosaic virus disease with attenuated virus of tomato strain of TMV. Rev. Plant Prot. Res. 8: 126-135

Pacot, H.C., Gall, O le, Candresse, T., Delpos, R.P., and Dunez, J. 1999. Transgenic tobaccos transformed with a gene encoding a truncated form of coat protein of tomato black ring nepovirus are resistant to viral infection. Plant Cell Reports 19: 203-209

Paguio, O.R., Kuhn, C.W. and Boerma, H.R. 1988. Resistance-breaking variants of cowpea chlorotic mottle virus in soybean. Plant Dis. 72: 768-770

Palukaitis, P. and Zaitlin, M. 1997. Replicase-mediated resistance to plant virus disease. Adv. Virus Res. 48: 349-377

Pappu, H.R., Csinos, A.S., McPherson, R.M., Jones, D.C. and Stephenson, M.G. 2000. Effect of acibezolar-5'-methyl and imidacloprid on suppression of tomato spotted wilt tospovirus in flue-cured tobacco. Crop Protection 19: 349-354

Paris, H.S., Burger, Y. and Yoseph, R. 1988. Single gene resistance to zucchini mosaic virus in *Cucurbita moschtata*. Euphytica 37: 27-29

Park, J.W., Faure, R.S., Robinson, M.A., Desvoyes, B. and Scholthop, H.B. 2004. The multifunctional plant virus suppressor of gene silencing P19 interacts with itself and an RNA binding host protein. Virology 323: 49-58

Park- Insook, A., Choi, J.D., Goo, D.H. and Kim, K.W. 2002. Elimination of viruses from virus-infected gladiolus plants through cormal tip and callus culture. J. Korean Hort. Sci. 43: 531-535

Parrella, G., Moretti, A., Gognalons, P., Lesage, M.L., Marchoux, G., Gebre-Selassie, K. and Caranta, C. 2004. Am gene controlling resistance in alfalfa mosaic virus in tomato is located in the cluster of dominant resistance genes on chromosome 6. Phytopathology 94: 345-350

Parveen, S., Tripathi, S. and Varma, A. 2001. Isolation and characterization of an inducer protein (Crip-31) from *Clerodendrum inerme* leaves responsible for induction of systemic resistance against viruses. Plant Science 161: 453-459

Perlak, F., Kaniewski, W., Lawson, C., Vincent, M. and Feldman, J. 1994. Genetically improved potatoes: Their potential role in integrated pest management. In: Proc. 3rd EFPP Conf. M. Manka (ed). pp. 451-454

Perring, T.M., Farrar, C.A., Blua, M.J., Wang, H.L. and Gonzalves, D. 1995. Cross protection of cantaloupe with a mild strain of zucchini yellow mosaic virus: effectiveness and application. Crop Protection 14: 601-606

Pico, B., Villar, C. and Nuez, F. 2003. Screening *Cucumis sativus* land races for resistance to cucumber vein yellowing virus. Plant Breeding 122: 426-430

Pilowski, M., Frankel, R. and Cohen, S. 1961. Studies of the variable reaction at high temperature of F1 hybrid tomato plants resistant to tobacco mosaic virus. Phytopathology 71: 319-323

Pinto, Y.M., Kok, R.A. and Baulcombe, D.C. 1999. Resistance to rice yellow mottle virus (RYMV) in cultivated African rice varieties containing RYMV transgenes. Nature Biotechnol. 17: 702-707

Powell-Abel, P., Nelson, R.S., De, B., Hoffmann, N., Rogers, S.G., Fraley, R.T. and Beachy, R.N. 1986. Delay of disease development in transgenic plants that express the tobacco mosaic virus coat protein gene. Science 232: 738-743

Preil, W. 1982. Elimination of poinsettia virus (PoiMV) 1 and poinsettia cryptic virus (PoiCV) from *Euphorbia pulcherrima* Willd by cell suspension culture. Phytopathol. Z. 105: 193-197

Preitus-Astúa, J., Purcifull, D.E., Polston, J.E. and Hiebert, E. 2002. Traditional and transgenic strategies for controlling tomato-infecting begomoviruses. Fitopatol. Bras. Vol. 27, No. 5 Brasilia

Presting, G.G., Smith, O.P. and Brown, C.R. 1995. Resistance to potato leaf roll virus in potato plants transformed with the coat protein gene or with vector control constructs. Phytopathology 85: 436-442

Prins, M. and Goldbach, R. 1996. RNA-mediated virus resistance in transgenic plants. Arch. Virol. 141: 2259-2276

Provvidenti, R. and Schroeder, W.T. 1973. Resistance in *Phaseolus vulgaris* to the severe strain of bean yellow mosaic virus. Phytopathology 63: 196-197

Pruss, G., Ge, X., Shi, X.M., Carrington, J.C. and Vance, V.B. 1997. Plant viral synergism: the poty viral genomes encode a broad-range pathogenicity enhancer that transactivates replication of heterologous viruses. Plant Cell 9: 859-868

Qiu, W.F. and Moyer, J.W. 1999. Tomato spotted wilt tospovirus adapts to the TSWV N-gene derived resistance by genome re-assortment. Phytopathology 89: 575-582

Qiu, W.F., Park, J.W. and Scholthof, H.B. 2002. Tombus P9-mediated suppression of virus-induced gene silencing is controlled by genetic and dosage features that influence pathogenicity. Molecular and Plant Microbe Interactions 15: 269-280

Qu, F. and Morris, T.J. 2003. The coat protein of Turnip crinkle virus suppresses post-transcriptional gene silencing at early initiation step. J. Virol. 77: 511-522

Qu, F.N., Zhu, M., Jiang, Z.W. and Kong, Q.X. 2003. Study on the improved method for eliminating the latent viruses of apple trees. China Fruits No. 5: 30-32

Quak, F. 1977. Meristem culture and virus-free plants. In: Applied and Fundamental Aspects of Plant Cell, Tissue and Organ Culture. J. Reinert and Y.P.S. Bajaj (eds). Springer-Verlag, Berlin, Germany. pp. 598-615

Quemada, H.D., Gonsalves, D. and Slightom, J.L. 1991. Expression of coat protein gene from cucumber mosaic virus strain C in tobacco: Protection against infections by CMV strains transmitted mechanically or by aphids. Phytopathology 81: 794-802

Rast, A.T.B. 1972. MII-16, an artificial symptomless mutant of tobacco mosaic virus for seedling inoculation of tomato crops. Neth. J. Plant Pathol. 78: 110-112

Reddy, D.V.R., Amin, P.W., McDonald, D. and Ghanekar, A.M. 1983. Epidemiology and control of groundnut bud necrosis and other diseases of legume crops in India caused by tomato spotted wilt virus. In: Plant Virus Epidemiology. R.T. Plumb and J.M. Thresh (eds). Blackwell Scientific Publications, Oxford, UK. pp. 93-102

Redinbough, M.G., Jones, M.V. and Gingery, R.E. 2004. The genetics of virus resistance in maize (*Zea mays* L.). Maydica 49: 183-190

Riley, D.G. and Pappu, H.R. 2000. Evaluation of tactics for management of thrips-vectored tomato spotted wilt virus in tomato. Plant Dis. 84: 847-852

Rissler, J. and Mellon, M. 1996. The Ecological Risks of Engineered Crops. MIT Press, Cambridge, MA. USA

Robbins, M.A., Witsenboer, H., Michelmore, R.W., Laliberte, J.F. and Fortin, M.G. 1994. Genetic mapping of turnip mosaic virus resistance in *Lactua sativa*. Theor. Appl. Genet. 89: 583-589

Robinson, D.J. 1996. Environment risk assessment of releases of transgenic plants containing virus-derived inserts. Trans. Res. 5: 359-362

Rocha, M. de M., Lima, J.A. de A, Freire Finho, F.R., Rosal, C.J. de S., Lopes, A.C. de A 2003. Resistance of white seed coat cowpea (*Vigna uniquiculata* L. Walp.) genotype to virus strains Bromoviridae. Revista Cientificia Rural 8: 85-92

Rodriguez-Alvardo, G., Kurath, G. and Dodds, J.A. 2001. Cross protection between and within subgroups I and II of cucumber mosaic virus isolates from pepper. Agrociencia Montecillo 35: 563-573

Russo, F. and Starrantino, A. 1973. Ricerche sulla, tecnologia dei microinnesti nel quadro del miglioramento genetico sanitario degli agrumi. Ann. Ist. Sperim. Agrum. 6: 209-222

Ryang, B.S., Kobori, T., Matsumoto, T., Kosaka, Y. and Ohki, S.T. 2004. *Cucumber mosaic virus* 2b protein compensates for restricted systemic squash of potato virus Y in doubly infected tobacco. J. Gen. Virol. 85: 3405-3414

Sanders, P.R., Sammons, B., Kaniewski, W., Haley, L., Layton, J., Lavallee, B.J., Delannay, X. and Tumer, N.E. 1992. Field resistance of transgenic tomatoes expressing the tobacco mosaic or tomato mosaic virus coat protein genes. Phytopathology 82: 683-690

Sangaré, A., Deng, D., Fauquet, C.M. and Beachy, R.N. 1999. Resistance to *African cassava mosaic virus* conferred by a mutant of the putative NTP-binding domain of the Rep gene in *Nicotiana benthamiana*. Mol. Biol. Rep. 5: 95-102

Sarvar, N., Cantin, E.M., Chang, P.S., Zaia, J.A., Ladne, P.A., Stephens, D.A. and Rossi, J.J. 1990. Ribozymes as potential anti-HIV-I therapeutic agent. Science 247: 1222-1225

Savenkov, E.I. and Valkonen, J.P. 2001. Potyviral helper-component proteinase expressed in transgenic plants enhances titers of Potato leafroll virus but does not alleviate its phloem limitation. Virology 10: 285-293

Saucke, H. and Doring, T.F. 2004. Potato virus Y reduction by straw mulch in organic potatoes. Annals Appl. Biol. 144: 347-355

Scanlon, K.J., Jiao, L., Funato, T., Wang, W., Tone, T., Rossi, J.J. and Kashani-Sabet, M. 1991. Ribozyme-mediated cleavage of c-Pos mRNA reduces gene expression of DNA synthesis enzymes and metallothionein. Proc. Natl. Acad. Sci. USA 88: 10591-10595

Shankula, S. 2006. Quantification of the impacts on US agriculture of biotechnology-derived crops planted in 2005. http://www.ncfap.org

Schoelz, J.E. and Wintermantel, W.M. 1993. Expansion of viral host range through complementation and recombination in transgenic plants. Plant Cell 5: 1669-1679

Schroeder, W.T. and Provvidenti, R. 1968. Resistance to bean (*Phaseolus vulgaris*) to the PV2 strain of bean yellow mosaic virus conditioned by the single dominant gene *By*. Phytopathology 58: 1719

Sears, R.G., Martin, T.J., McCluskey, P.J., Paulsen, G.M., Heer, W.F., Long, J.H., Witt, M.D. and Brown Guedira, G. 2001a. Registration 'Hyne' wheat. Crop Science 41: 1367

Sears, R.G., Martin, T.J., Mccluskey, P.J., Paulson, G.M., Heer, W.F., Long, J.H., Witt, M.D. and Guedira, G. 2001b. Registration of 'Betty' wheat. Crop Science 41: 1366-1367

Sharaf, N. 1986. Chemical control of *Bemisia tabaci*. Agric. Ecosystem Environ. 17: 111-127

Sijen, T., Welleck, J., Hendriks, J., Verner, J. and van Kammen, A. 1995. Replication of cowpea mosaic virus RNA1 or RNA2 is specifically blocked in transgenic *Nicotiana benthamiana* plants expressing the full-length replicase or movement protein genes. Mol. Plant-Microbe Interact. 8: 340-347

Simon, A.E., Roossinck, M.J. and Havelda, Z. 2004. Plant Virus Satellite and Defective Interfering RNAs: New Paradigms for a New Century. Annu. Rev. Phytopathol. 42: 415-437

Sioud, M. and Drlica, K. 1991. Prevention of human immunodeficiency virus type I integrase expression in *Escherichia coli* by a ribozyme. Proc. Natl. Acad. Sci. USA 88: 7303-7307

Sisler, W.W. and Timian, R.G. 1956. Inheritance of barley stripe mosaic virus resistance of Modjo (CI 3212) and CI 3212-1. Plant Dis. Rep.40: 1106-1108

Sithav, Y.D., Molner, A., Lucioli, A., Szitty, G., Hornyik, C., Tavazza, M. and Burgyan, J. 2002. A viral protein suppresses RNA silencing and binds silencing-generated 21-25-nucleotide double-stranded RNAs. EMBO J. 21: 3070-3080

Sivamani, E., Brey, C.W., Talbert, L.E., Young, M.A., Dyer, W.E.E., Kaniewski, W.K. and Qu, R. 2002. Resistance to wheat streak mosaic virus in transgenic wheat engineered with the viral coat protein gene. Transgenic Research 11: 31-41

Smith, B.J., Keen, N.T. and Becker, J.O. 2003. Acibenzolar S methyl induces resistance to *Colletotrichum lagenarium* and cucumber mosaic virus in cantaloupe. Crop Protection 22: 769-774

Smith, H.A., Swancy, S.I., Parks, T.D., Wernsman, E.A. and Dougherty, W.G. 1994. Transgenic plant resistance mediated by untranslatable RNAs: expression, regulation and fate of nonessential RNAs. Plant Cell 6: 1441-1453

Snir, I. and Stein, A. 1985. *In vitro* detection and elimination of necrotic ringspot virus in sweet cherry. Riv. Ortoflorofruttic. Ital. 69: 191-194

Sobcykiewicz, D. 1979. Heat treatment and meristem culture for production of virus-free strawberry plants. Acta Hortic. 95: 79-82

Sonoda, S., Mori, M. and Nishiguchi, M. 1999. Homology dependent virus resistance in transgenic plants with coat protein gene of sweet potato feathery mottle potyvirus target specificity and transgene methylation. Phytopathology 89: 385-391

Stace-Smith, R. and Mellor, F. 1968. Eradication of potato virus X and S by thermotherapy and axillary bud culture. Phytopathology 58: 199-203

Starting, A. 1996. Availability of determination of nonregulated status of papaya lines genetically engineered for virus resistance. Federal Register 61: 48663-48664

Stone, O.M. 1963. Factors affecting the growth of carnation plants from shoot apices. Annals Appl. Biol. 52: 199-209

Stuart, C.N., Halfhill, M.D. and Warwick, S.I. 2003. Transgene introgression from genetically modified crops to their wild relatives. Nat. Rev. Genet. 4: 806-817

Superak, T.H., Scully, B.T., Kyle, M.M. and Munger, H.M. 1993. Interspecific transfer of plant viral resistance in *Cucurbita*. In: Resistance to Viral Diseases in Vegetables: Genetics and Breeding. M.M. Kyle (ed). Timber Press, Oregon, USA. pp. 217-236

Syller, J. 2003. Inhibited long-distance movement of potato leaf roll virus to tubers in potato genotypes expressing combined resistance to infection, virus multiplication and accumulation. J. Phytopathol. 151: 492-499

Tabler, M., Tsagirs, M. and Hammond, J. 1998. Antisense RNA and ribozyme-mediated resistance to plant viruses. In: Plant Virus Disease Control. A. Hadidi, R.K. Khetarpal and H. Koganezawa (eds). APC Press, American Phytopathological Society. St. Paul, Minnesota, USA. pp. 79-93

Tao, X.R. and Zhou, X.P. 2004. A modified viral satellite DNA that suppresses gene expression in plants. Plant Journal 38: 850-860

Tepfer, M. 2002. Risk assessment of virus-resistant transgenic plants. Annu. Rev. Phytopathol. 40: 467-491

Teycheney, P.Y., Aaziz, R., Dinant, S., Salanki, K., Tourneur, C, Balåzs, E., Jacquemond, M. and Tepfer, M. 2000. Synthesis of (–) strand RNA from the 3′ untranslated region of plant viral genome expressed in transgenic plants upon infection with related viruses. J. Gen. Virol. 81: 1121-1126

Thackray, D.J., Diggle, A.J., Berlandier, F.A., Jones, R.A.C. 2004. Forecasting aphid outbreaks and epidemics of cucumber mosaic virus in lupin crops in Mediterranean type environment. In: Plant Virus Epidemiology. First steps in the new millennium. Virus Research 100: 67-82

Thomas, P.E., Lawson, E.C., Zalewski, J.C., Reed, G.L. and Kaniewski, W.K. 2000. Extreme resistance to potato leaf roll virus cv. Russet Burbank mediated by viral replicase gene. Special Issue: Plant Virus Epidemiology: Challenges for the twenty-first Century. Almeria, Spain, 11-16 April 1999. Virus Research 71: 49-62

Thornton, I.R., Emmett, R.W. and Stubbs, L.L. 1980. A further report on the grapefruit tristeza pre-immunization trial Mildara. Proc. 8th Conf. IOCV. E.C. Calvan, S.M. Granser and L.W. Timmer (eds). Univ. Calif. Riverside, CA, USA. pp. 51-53

Thresh, J.M. 1976. Gradients of plant virus diseases. Ann. Appl. Biol. 82: 381-406

Tien, P. and Wu, G.S. 1991. Satellite RNA for the biocontrol of plant disease. Adv. Virus. Res. 39: 321-339

Tomassoli, L., Kaniewaski, W., Ilardi, V. and Barba, M. 1995. Produzione di piante di pomodoro geneticamente modificate per la resistenza al virus del mosaico del cetriolo. L Informatore Agrario 4: 1-8

Toyoda, H., Oishi, Y., Matsuda, Y., Chatani, K. and Hirai, T. 1985. Resistance mechanism of cultured plant cells to tobacco mosaic virus. IV. Changes in tobacco mosaic virus concentrations in somaclonal tobacco callus tissues and production of virus-free plantlets. Phytopathol. Z. 114: 126-133

Tricoli, D.M., Carney, K.J., Russell, P.F., McMaster, J.R., Groff, D.W., Hadden, K.C., Himmel, P.T., Hubbard, J.P., Boeshore, M.L., Reynolds, J.F. and Quemada, H.D. 1995. Field evaluation of transgenic squash containing single or multiple virus coat protein gene constructs for resistance to cucumber mosaic virus, watermelon mosaic virus 2, and/or zucchini yellow mosaic virus. Bio/Technology 13: 1458-1465

Tumer, N., Hemenway, C., O'Connell, K., Cuozzzo, M., Fang, R.-X., Kaniewski, W. and Chua, N.-H. 1987. Expression of coat protein genes in transgenic plants confers protection against alfalfa mosaic virus, cucumber mosaic virus and potato virus X. In: Plant Molecular Biology. D. von Wettstein and N.-H. Hua (eds). Plenum Publishing Corporation, New York, USA. pp. 351-356

Ucko, O., Cohen, S. and Ben Joseph, R. 1998. Prevention of virus epidemics by a crop-free period in the Arava region of Israel. Phytoparasitica 26: 313-321

Urban, L.A., Sherwood, J.L., Rezende, J.A.M. and Melcher, U. 1990. Examination of mechanisms of cross protection with non-transgenc plants. In: Recognition and Response in Plant-Virus Interactions. R.S.S. Fraser (ed). Springer-Verlag, Berlin, Germany. pp. 415-426

van der Krol, A.R., Mur, L.A., Beld, M., Mol, J.N.M. and Stultje, A.R. 1990. Flavenoid genes in Petunia: addition of a limited number of gene copies may lead to a suppression of gene expression. Plant Cell 2: 291-299

van der Vlugt, R.A.A., Ruiter, R.K. and Goldbach, R. 1992. Evidence for sense RNA-mediated resistance to PVTV in tobacco plants transformed with the viral coat protein cistron. Plant Mol. Biol. 20: 631-639

van Os, H. 1964. Production of virus-free carnations by means of meristem culture. Neth. J. Plant Pathol. 70: 18-26

Varrelmann, M., Palkovics, L. and Maiss, E. 2000. Transgenic or plant expressing vector-mediated recombination of plum pox virus. J. Virol. 74: 7462-7469

Vaslin, M.F.S., Vidal, M.S., Alves, E.D., Farinelli, L. and de Oliveira, D.E. 2001. Co-suppression-mediated virus resistance in transgenic tobacco plants harboring the 3'-untranslated region of Andean potato mottle virus. Transgenic Research 10: 489-499

Verma, N., Ram, R., Hallan, V., Kumar, K. and Zaidi, A.A. 2004. Production of cucumber mosaic virus-free chrysanthemum by meristem tip culture. Crop Protection 23: 469-473

Walkey, D.G.A. 1976. High temperature inactivation of cucumber and alfalfa mosaic viruses in *Nicotiana rustica* cultures. Ann. Appl. Biol. 84: 183-192

Walkey, D.G.A. and Cooper, V.C. 1972. Comparative studies on the growth of healthy and virus infected rhubarbs. J. Hortic. Sci. 47: 37-41

Walkey, D.G.A., Lecoq, H., Collier, R. and Dobson, S. 1992. Studies on the control of zucchini yellow mosaic virus in courgettes by mild strain protection. Plant Pathol. 4: 762-771

Wang, H.L., Gonsalves, D., Provvidenti, R. and Lecoq, H. 1991. Effectiveness of cross-protection by a mild strain of zucchini yellow mosaic virus in cucumber, melon and squash. Plant Dis. 75: 203-207

Wang, Q., Mawassi, M., Li, P., Gafny, R., Sela, I., Tanne, E., Wang, Q.C. and Li, P. 2003. Elimination of grapevine virus A (GVA) by cryopreservation of in vitro-grown shoot tips of *Vitis vinifera* L. Plant Science 165: 321-327

Wang, Y., Lee, K.C., Gaba, V., Wong, S.M., Palukaitis, P. and Gal, O.A. 2004. Breakage of resistance to cucumber mosaic virus by infection with Zucchini yellow mosaic virus: enhancement of CMV accumulation independent of symptom expression. Archiv. Virol. 149: 379-396

Waterhouse, P.M., Graham, M.W. and Wang, M.-B. 1998. Virus resistance and gene silencing in plants can be induced by simultaneous expression of sense and antisense RNA. Proc. Natl. Acad. Sci. USA 95: 13959-13964

Watterson, J.C. 1993. Development and breeding of resistance to pepper and tomato viruses. In: Resistance to Viral Diseases of Vegetables: Genetics and Breeding. M.M. Kyle (ed). Timber Press, Oregon, USA. pp. 80-101

Webster, A.D. 2000. Elimination of apple chlorotic leaf spot virus (ACLSV) from pear cv. Pierre by in vitro thermotherapy and chemotherapy. Acta Horticulturae No. 596: 481-484

Werner, K., Ronicke, S., Gouis, J. le, Friedt, W. and Ordon, F. 2003. Mapping of a new BaMMV-resistance gene derived from the variety 'TaihokuA'. Zeischrift fur Planzenkrankheiten und Pfllanzenschutz 110: 304-311

White, P.R. 1934. Multiplication of the viruses of tobacco roots. Phytopathology 24: 1003-1011

Williamson, B. 2002. Resistance breaking raspberry aphid biotypes: constraints to sustainable control through plant breeding. Proceedings of the 8th Intl. Rubus and Ribes Symp. Invergourie, Dundee, UK, 4-12 July 2001. Vol. 1, A.N.E. Birch, A.T. Jones, B. Fenton, G. Malloch, I. Geoghegan, S.C. Gordon, J. Hillier, G. Begg, R.M. Brennan and S.L. Gordon (eds). Acta Horticulturae No. 585: 315-317

Wilson, C.R. 1999. The potential of reflecting mulching in combination with insecticides sprays for control of aphid-borne viruses of iris and tulip in Tasmania. Annals Appl. Biol. 134: 293-297

Wintermantel, W.M. and Schoelz, J.E. 1996. Isolation of recombinant viruses between Cauliflower mosaic virus and a viral gene in transgenic plants under conditions of moderate selection pressure. Virology 223: 156-164

Wisler, G.C. and Duffus, J.E. 2000. A century of plant virus management in the Salinas Valley of California, 'East of Eden'. Special Issue: Plant Virus Epidemiology: Challenges for the twenty-first century, Almeria, Spain 11-16 April, 1999. Virus Res. 71: 161-169

Wren, J.D., Roosinck, M.I., Nelson, R.S., Scheets, K., Palmer, M.W. and Melcher, U. 2006. Plant virus biodiversity and ecology. PloS Biol. 4(3): e80

Xu, J., Schubert, J., Altpeter, F. and Xu, J.P. 2001. Dissection of RNA-mediated ryegrass mosaic virus resistance in fertile transgenic perennial ryegrass (*Lolium perenne* L.). Plant Journal 26: 265-274

Xue, B., Gonsalves, C., Provvidenti, R., Slinghtom, J.L., Fuchs, M. and Gonsalves, D. 1994. Development of transgenic tomato expressing a high level of resistance to cucumber mosaic virus strains of subgroups I and II. Plant Dis. 78: 1038-1041

Yang, H.J. and Clore, W.J. 1973. Rapid vegetative propagation of asparagus through lateral bud culture. HortScience 8: 141-143

Yang, H., Ozaias-Akins, P., Culbreath, A.K., Gorbet, D.W., Weeks, J.R., Mandal, B. and Pappu, H.R. 2004. Field evaluation of Tomato spotted wilt virus resistance in transgenic peanut (*Arachis hypogaea*). Plant Disease 88: 259-264

Yeh, S.D. and Gonsalves, D. 1984. Evaluation of induced mutants of papaya ring spot virus for control by cross protection. Phytopathology 74: 1086-1091

Yeh, S.D., Gonsalves, D., Wang, H.L., Namba, R. and Chiu, R.J. 1988. Control of papaya ring spot virus by cross protection. Plant Dis. 72: 375-380

Yepes, L.M., Fuchs, M., Slightom, J.L. and Gonsalves, D. 1996. Sense and antisense coat protein gene constructs confer high level of resistance to tomato ring spot nepovirus in transgenic *Nicotiana* species. Phytopathology 86: 417-424

Yi, T.S., Hu, H. and Luo, G.F. 2001. Studies on preventing state of carnation viruses in Kunming seedlings. Acta Botanica-Yunnanica 23: 345-349

Zammar, E.I., Abou, S., Jawdah, Y. and Sobh, H. 2001. Management of virus diseases of squash in Lebanon. J. Plant Path. 83: 21-25

Zamore, P.D. 2004. Plant RNAi: how a viral silencing suppressor inactivates siRNA. Current Biology: R198-R200

Zaumeyer, W.J. and Meiners, J.P. 1995. Disease resistance in beans. Annu. Rev. Phytopathol. 13: 313-334

Zheng, H., Li, Y., Yu, Z., Li, W., Chen, M., Ming, X., Casper, R. and Chen, Z. 1997. Recovery of transgenic rice plants expressing the rice dwarf virus outer coat protein gene. Theor. Appl. Gen. 94: 522-527

Appendix I

APPENDIX I.1

Cryptogamic notations of paired characters for plant virus nomenclature (Gibbs 1968)

First Pair:	Type of nucleic acid and number of strands R = RNA; D = DNA; 1 = Single-stranded; 2 = Double-stranded
Second Pair:	Molecular weight of the nucleic acid (in millions) and percentage of Nucleic acid in infective particle
Third Pair:	Outline of the particle and outline of the Nucleocapsid. (**S** = essentially spherical; **E** = elongated with parallel sides, ends not rounded; **U** = elongated parallel sides and end(s) rounded; **X** = complex or none above).
Fourth Pair:	Kind of host infected and kind of vector
	Kind of host: A = alga; B = bacterium; F = fungus; I = invertebrate; P = pteridophyte; S = seed plant; M = mycoplasma; V = vertebrate. *Kind of vector*: Ac = mite and tick; Al = whitefly; Ap = aphid; Au = leaf-, plant- or tree hopper; Cc = mealy bug; Cl = beetle; Di = fly and mosquito; Fu = fungus; Gy = mirid, piesmid or tingid bug; Ne = nematode; Ps = psylla; Ve = vector known but none above; 0 = spreads without a vector; Si = flea; Th = thrips

*Properties of the virus not known; ( )enclosed information doubtful.

APPENDIX I.2

Characters of plant viruses (Harrison *et al* 1971) considered for cluster analysis and to separate them into distinct groups.

Behaviour in Hosts

Host range, predominant symptoms, symptom types, tissues to which restricted, approximate concentration in sap, type of inclusion bodies, intracellular location of inclusions, effectiveness of heat therapy, seed transmissibility, geographical distribution and cross-protecting viruses.

Vector Relations

Taxa of vectors, number of vector species, threshold feeding/access period, inoculation feeding/access period, latent period, persistence in the vectors, multiplication in vectors, transmission through eggs/resting spores, vector stage able to acquire and inoculate the virus.

Properties of the Virus Particles

Shape, symmetry, size, number and arrangements of morphological subunits, arrangements of nucleic acid, sedimentation co-efficient, molecular weight, electrophoretic mobility, isoelectric point, thermal inactivation point, retention of infectivity of sap at 20°C, dilution end point, serologically related viruses, major properties of accessory particles.

Particle Composition

Nucleic acid: RNA or DNA; Nucleotide ratio (mole%); Molecular weight; Percentage of particle weight; Number of strands; Ring form or not; Sedimentation co-efficient. *Proteins*: Number of protein species in a particle; Molecular weight of subunits; Number of chemical subunits per particle; Number of amino acid residues per subunit; Noteworthy amino-acids included or excluded; Enzymatic activity; Percentage of particle weight. *Lipid*: Percentage of particle weight.

Other compounds: Type and amount.

APPENDIX I.3

Groups of plant viruses derived by Harrison *et al* (1971) through computer based cluster analysis of data on 113 plant virus diseases

Group	Type member
1. Tobravirus	*Tobacco rattle virus*
2. Tobamovirus	*Tobacco mosaic virus*
3. Potexvirus	*Potato virus X*
4. Carlavirus	*Potato virus S & M*
5. Potyvirus	*Potato virus Y*
6. Cucumovirus	*Cucumber mosaic virus*
7. Tymovirus	*Turnip yellow mosaic virus*

8.	Comovirus	*Cowpea mosaic virus*
9.	Nepovirus	*Tobacco ring spot virus*
10.	Bromovirus	*Bromegrass mosaic virus*
11.	Tombusvirus	*Tomato bushy stunt virus*
12.	Caulimivirus	*Cauliflower mosaic virus*
13.	Alfalfa mosaic virus	*Alfalfa mosaic virus*
14.	Pea enation mosaic virus	*Pea enation mosaic virus*
15.	Tobacco necrosis virus	*Tobacco necrosis mosaic virus*
16.	Tomato spotted wilt virus	*Tomato spotted wilt virus*

APPENDIX I.4

Monothetic hierarchial system of classification adopting the Group concept and proposing Family/Group status to them, approved by the ICTV (Francki *et al* 1991) and the type members of different groups

Key characters	Family/Group	Type member	Number of members	Number of probable members
ds DNA, non-enveloped	Caulimovirus	*Cauliflower mosaic virus*	11	6
	Commelina yellow mottle virus group	*Commelina yellow mottle virus*	3	11
ssDNA, non-enveloped	Geminivirus	*Bean golden mosaic virus*	35	13
dsRNA, non-enveloped	Cryptovirus	*Carnation cryptic virus*	20	10
	Reoviridae		6	2
	Phytoreovirus	*Wound tumor virus*		
	Fiji virus	*Fiji virus*		
	Oryzavirus	*Rice ragged stunt virus*		
ssRNA, enveloped	**Rhabdoviridae**		14	76
	Subgroup A	*Lettuce necrotic yellow virus*		
	Subgroup B	*Potato yellow dwarf virus*		
	Bunyaviridae			
	Tospovirus	*Tomato spotted wilt virus*	1	0
ssRNA, non-enveloped; **Monopartite**	**Isometric particle**			
	Carmovirus	*Carnation mottle virus*	8	9
	Luteovirus	*Barley yellow dwarf virus*	12	12
	Marafivirus	*Maize raydo fino virus*	3	0
	MCDV group	*Maize chlorotic dwarf virus*	1	2

	Ncrovirus	*Tobacco necrosis virus*	2	2
	PYFV group	*Parsnip yellow fleck virus*	2	1
	Sobemovirus	*Southern bean mosaic virus*	10	6
	Tombusvirus	*Tomato busy stunt virus*	12	0
	Tyovirus	*Turnip yellow mosaic virus*	18	1
	Rod-shaped particle			
	Capillovirus	*Apple stem grooving virus*	2	2
	Closterovirus	*Beet yellows virus*	10	12
	Carlavirus	*Carnation latent virus*	27	29
	Potexvirus	*Potato virus X*	18	21
	Potyvirus	*Potato virus Y*	74	84
	Tobamovirus	*Tobacco mosaic virus*	12	2
ssRNA, non-enveloped, bipartite genome	**Isometric particle**			
	Comovirus	*Cowpea mosaic virus*	14	0
	Dianthovirus	*Carnation ring spot virus*	3	0
	Fabavirus	*Broadbean wilt virus*	3	0
	Nepovirus	*Tobacco ring spot virus*	28	8
	PEMV group	*Pea enation mosaic virus*	0	1
	Rod-shaped particle			
	Furovirus	*Soil-borne wheat mosaic virus*	5	6
	Tobravirus	*Tobacco rattle virus*	3	0
ssRNA, non-enveloped, Tripartite genome				
	Bromovirus	*Brome mosaic virus*	6	0
	Cucumovirus	*Cucumber mosaic virus*	3	1
	Ilarvirus	*Tobacco streak virus*	20	0
ssRNA, non-enveloped, Tripartite genome				
	Isometric & Bacilliform particle			
	AMV group	*Alfalfa mosaic virus*	1	0
	Hordeivirus	*Barley stipe mosaic virus*	4	0
Total Families	**37**		**334**	**321**
Total Members & Probable ones				**655**

APPENDIX I.5

Classification of viruses (van Regenmortel *et al* 2000)

Type of the genome/family	Genus	Type member	Number of members	Number of probable members
dsDNA (RT)/				
1.Caulimoviridae	1. Caulimovirus	*Cauliflower mosaic virus*	9	4
	2. Soybean chlorotic mottle virus	*Soybean chlorotic mottle virus*	2	0
	3. Cassava vein mosaic virus	*Cassava vein mosaic virus*	1	0
	4. Petunia vein clearing virus	*Petunia vein clearing virus*	1	0
	5. Badnavirus	*Commelina yellow mosaic virus*	12	4
	6. Rice tungro bacilliform virus	*Rice tungro bacilliform virus*	1	0
SsDNA				
2. Geminiviridae	7. Mastrevirus	*Maize streak virus*	12	2
	8. Curtovirus	*Beet curly top virus*	3	1
	9. Begomovirus	*Bean golden mosaic virus*	76	8
3. Circoviridae	10. Topocuvirus			
	11. Nanovirus	*Subterranea clover stunt virus*	4	1
DsRNA				
4. Reoviridae	12. Fijivirus	*Fiji virus*	8	0
	13. Oryzavirus	*Rice ragged stunt virus*	2	0
	14. Phytoreovirus	*Wound tumor virus*	3	1
5. Partitiviridae	15. Alfacryptovirus	*White clover cryptic virus 1*	16	10
	16. Betacryptovirus	*White clover cryptic virus 2*	4	1
No Family	17. Varicosavirus	*Lettuce big vein virus*	1	3
ssRNA (−)				
6. Rhabdoviridae	18. Cytorhabdovirus	*Lettuce necrotic yellow virus*	8	0
	19. Nucleorhabdo-virus	*Potato yellow dwarf virus*	7	0
7. Bunyaviridae	20. Tospovirus	*Tomato spotted wilt virus*	8	5
No Family	21. Tenuivirus	*Rice stripe virus*	6	5

	22. Ophiovirus	*Citrus psorosis virus*	3	0
ssRNA (+)				
8. Bromoviridae	23. Bromovirus	*Brome mosaic virus*	6	0
	24. Cucumovirus	*Cucumber mosaic virus*	3	0
	25. Alfamovirus	*Alfalfa mosaic virus*	1	0
	26. Ilarvirus	*Tobacco streak virus*	17	0
	27. Oleavirus	*Olive latent virus 2*	1	0
9.Comoviridae	28. Comovirus	*Cowpea mosaic virus*	15	0
	29. Fabavirus	*Broadbean wilt virus*	4	0
	30. Nepovirus	*Tobacco ring spot virus*	31	9
10. Potyviridae	31. Potyvirus	*Potato virus Y*	91	88
	32. Ipomovirus	*Sweet potato mild mottle virus*	1	1
	33. Macluravirus	*Maclura mosaic virus*	2	0
	34. Rymovirus	*Ryegrass mosaic virus*	4	1
	35. Tritimovirus	*Wheat streak virus*	2	0
	36. Bymovirus	*Barley yellow mosaic virus*	6	0
11. Tombusviridae	37. Tombusvirus	*Tomato bushy stunt virus*	13	0
	38. Aureusvirus	*Pothos latent virus*	1	0
	39. Avenavirus	*Oat chlorotic stunt virus*	1	0
	40. Carmovirus	*Carnaton mottle virus*	12	0
	41. Machlomovirus	*Maize chlorotic mottle virus*	1	0
	42. Necrovirus	*Tobacco necrosis virus A*	5	2
	43. Dianthovirus	*Carnation ring spot virus*	4	1
	44. Panicovirus	*Panicum mosaic virus*	1	1
12. Sequiviridae	45. Sequivirus	*Parsnip yellow fleck virus*	2	0
	46. Waikavirus	*Rice tungro spherical virus*	3	0
13. Closteroviridae	47. Closterovirus	*Beet yellows virus*	11	16
	48. Crinivirus	*Lettuce infectious yellows virus*	1	0
14. Luteoviridae	49. Luteovirus	*Barley yellow dwarf virus*	2	0
	50. Polerovirus	*Potao leaf roll virus*	5	0
	51. Enamovirus	*Pea enation mosaic virus 1*	1	0
No Family	52. Tobamovirus	*Tobacco mosaic virus*	16	1
	53. Tobravirus	*Tobacco rattle virus*	3	0
	54. Potexvirus	*Potato virus X*	26	19
	55. Carlavirus	*Carnation latent virus*	31	29
	56. Allexivirus	*Shallot virus X*	7	3
	57. Capillovirus	*Apple stem grooving virus*	3	1
	58. Foveavirus	*Apple stem pitting virus*	2	1
	59. Trichovirus	*Apple chlorotic leaf spot virus*	3	0
	60. Vitivirus	*Grapevine virus A*	4	1

61.	Furovirus	*Soil-borne wheat mosaic virus*	1	4
62.	Pecluvirus	*Peanut clump virus*	2	0
63.	Pomovirus	*Potato mop top virus*	4	0
64.	Benyvirus	*Beet necrotic yellow vein virus*	2	0
65.	Hordeivirus	*Barley stripe mosaic virus*	4	0
66.	Marafivirus	*Maize rayado fino virus*	3	0
67.	Sobemovirus	*Southern bean mosaic virus*	11	3
68.	Tymovirus	*Turnip yellow mosaic virus*	21	2
69.	Idaeovirus	*Raspberry bushy stunt virus*	1	0
70.	Ourmiavirus	*Ourmia melon virus*	3	0
71.	Umbravirus	*Carrot mottle virus*	7	4

(−) ssRNA: unassigned   58

(+) ssRNA: unassigned   11

Total: Families 14   **Genera: 71**   **Members: 657**   **Probable Members: 85**

Total Members including Probable & Unassigned   **742**

APPENDIX I.6

Families and genera of plant viruses described by Fauquet *et al* (2005)

Family	*Genera*
1. Caulimoviridae	1. *Caulimovirus* 2. *Soymovirus* 3. *Caemovirus* 4. *Petuvirus* 5. *Badnavirus* 6. *Rice tungro virus*
2. Geminiviridae	7. *Mastrevirus* 8. *Curtovirus* 9. *Begomovirus* 10. *Topocuvirus*
3. Nanoviridae	11. *Nanovirus* 12. *Babuvirus*
4. Reoviridae	13. *Fijivirus* 14. *Oryzavirus* 15. *Phytoreovirus*
5. Partitiviridae	16. *Alfacryptovirus* 17. *Betacryptovirus* 18. *Endomavirus*
6. Flexiviridae	19. *Allexivirus* 20. *Capillovirus* 21. *Carlavirus* 22. *Foveavirus* 23. *Mandarivirus* 24. *Potexvirus* 25. *Trichovirus* 26. *Vitivirus*
7. Closteroviridae	27. *Closterovirus* 28. *Crinivirus* 29. *Ampelovirus*
8. Potyviridae	30. *Potyvirus* 31. *Ipomovirus* 32. *Tritimovirus* 33. *Bymovirus* 34. *Rymovirus*
9. Rhabdoviridae	35. *Cytorhabdovirus* 36. *Nucleorhabdovirus*
10. Tombusviridae	37. *Aureusvirus* 38. *Avenavirus* 39. *Carmovirus* 40. *Dianthovirus* 41. *Machlomovirus* 42. *Necrovirus* 43. *Panicovirus* 44. *Tombusvirus*
11. Bromoviridae	45. *Bromovirus* 46. *Cucumovirus* 47. *Ilarvirus* 48. *Oleavirus* 49. *Alfamovirus*

12. Comoviridae	50. *Comovirus* 51. *Fabavirus* 52. *Nepovirus*
13. Luteoviridae	53. *Luteovirus* 54. *Enamovirus* 55. *Poleroirus*
14. Tymoviridae	56. *Tymovirus* 57. *Marafivirus*
15. Sequiviridae	58. *Sequivirus* 59. *Waikavirus*
16. Bunyaviridae	60. *Tospovirus*
17. Pseudoviridae	61. *Sirevirus* 62. *Pseudovirus*
18. Metaviridae	63. *Metavirus*
No Family	64. *Benyvirus* 65. *Furovirus* 66. *Hordeivirus* 67. *Macluravirus*
	68. *Ophiovirus* 69. *Pecluvirus* 70. *Pomovirus* 71. *Tenuivirus*
	72. *Tobamovirus* 73. *Tobravirus* 74. *Varicosavirus*
	75. *Ourmiavirus* 76. *Cheravirus* 77. *Idaeovirus* 78. *Sadwavirus*
	79. *Sobemovirus* 80. *Umbravirus*

APPENDIX I.7

Alphabetical list of plant viruses, acronyms and their families (Fauquet *et al* 2005)

Name of the virus	Acronym	Genus	Family
1 *Abutilon mosaic virus*	AbMV	*Begomovirus*	Geminiviridae
2 *Abutilon yellows virus*	AbYV	*Crinivirus*	Closteroviridae
3 *Aconitum latent virus*	AcLV	*Carlavirus*	Flexiviridae
4 *African cassava mosaic virus*	ACMV	*Begomovirus*	Geminiviridae
5 *Ageratum enation virus*	AEV	*Begomovirus*	Geminiviridae
6 *Ageratum yellow vein China virus*	AYVCNV	*Begomovirus*	Geminiviridae
7 *Ageratum yellow vein Sri Lanka virus*	AYVSLV	*Begomovirus*	Geminiviridae
8 *Ageratun yellow vein Taiwan virus*	AYVTV	*Begomovirus*	Geminiviridae
9 *Ageratum yellow vein virus*	AYVV	*Begomovirus*	Geminiviridae
10 *Aglaonema bacilliform virus*	ABV	*Badnavirus*	Caulimoviridae
11 *Agropyron mosaic virus*	AgMV	*Rymovirus*	Potyviridae
12 *Ahlum waterborne virus*	AWBV	*Carmovirus*	Tombusviridae
13 *Alfalfa cryptic virus 1*	ACV-1	*Alphacryptovirus*	Partitiviridae
14 *Alfalfa mosaic virus*	AMV	*Alfamovirus*	Bromoviridae
15 *Alligatorweed stunting virus*	AWSV	Unassigned	Closteroviridae
16 *Alstroemeria mosaic virus*	AlMV	*Potyvirus*	Potyviridae
17 *Alternanthera mosaic virus*	AltMV	*Potexvirus*	Flexiviridae
18 *Amaranthus leaf mottle virus*	AmLMV	*Potyvirus*	Potyviridae
19 *American hop latent virus*	AHLV	*Carlavirus*	Flexiviridae
20 *American plum line pattern virus*	APLPV	*Ilarvirus*	Bromoviridae
21 *Andean potato latent virus*	APLV	*Tymovirus*	Tymoviridae
22 *Andean potato mottle virus*	APMoV	*Comovirus*	Comoviridae
23 *Anthoxanthum latent blanching virus*	ALBV	*Hordeivirus*	Unallocated
24 *Anthriscus yellows virus*	AYV	*Waikavirus*	Sequiviridae

25	*Apium virus Y*	ApVY	*Potyvirus*	Potyviridae
26	*Apple chlorotic leafspot virus*	ACLSV	*Trichovirus*	Flexiviridae
27	*Apple latent spherical virus*	ALSV	*Cheravirus*	Unallocated
28	*Apple mosaic virus*	ApMV	*Ilarvirus*	Bromoviridae
29	*Apple stem grooving virus*	ASGV	*Cappilovirus*	Flexiviridae
30	*Apple stem pitting virus*	ASPV	*Foveavirus*	Flexiviridae
31	*Apricot latent ringspot virus*	ALRSV	*Nepovirus*	Comoviridae
32	*Apricot latent virus*	ApLV	*Foveavirus*	Flexiviridae
33	*Arabidopsis thaliana Art1 virus*	AthArt1V	*Pseudovirus*	Pseudoviridae
34	*Arabidopsis thaliana Athila virus*	AthAthV	*Metavirus*	Metaviridae
35	*Arabidopsis thaliana AtRE1 virus*	AthAtRE1V	*Pseudovirus*	Pseudoviridae
36	*Arabidopsis thaliana Endovir virus*	AthEndV	*Sirevirus*	Pseudoviridae
37	*Arabidopsis thaliana Evelknievel virus*	AthEveV	*Pseudovirus*	Pseudoviridae
38	*Arabidopsis thaliana Ta1 virus*	AthTa1V	*Pseudovirus*	Pseudoviridae
39	*Arabidopsis thaliana Tat4 virus*	AthTat4V	*Metavirus*	Metaviridae
40	*Arabis mosaic virus*	ArMV	*Nepovirus*	Comoviridae
41	*Araujia mosaic virus*	ArjMV	*Potyvirus*	Potyviridae
42	*Arracacha virus A*	AVA	*Nepovirus*	Comoviridae
43	*Artichoke Aegean ringspot virus*	AARSV	*Nepovirus*	Comoviridae
44	*Artichoke Italian latent virus*	AILV	*Nepovirus*	Comoviridae
45	*Artichoke latent virus*	ArLV	*Potyvirus*	Potyviridae
46	*Artichoke mottled crinkle virus*	AMCV	*Tombusvirus*	Tombusviridae
47	*Artichoke yellow ringspot virus*	AYRSV	*Nepovirus*	Comoviridae
48	*Asparagus virus-1*	AV-1	*Potyvirus*	Potyviridae
49	*Asparagus virus-2*	AV-2	*Ilarvirus*	Bromoviridae
50	*Asparagus virus-3*	AV-3	*Potexvirus*	Flexiviridae
51	*Atropa belladonna virus*	AtBV	Unassigned	Rhabdoviridae
52	*Bamboo mosaic virus*	BaMV	*Potexvirus*	Flexiviridae
53	*Bamboo bract mosaic virus*	BBrMV	*Potyvirus*	Potyviridae
54	*Banana bunchy top virus*	BBTV	*Babuvirus*	Nanoviridae
55	*Banana mild mosaic virus*	BanMMV	Unassigned	Flexiviridae
56	*Banana streak GF virus*	BSV-GF	*Badnavirus*	Caulimoviridae
57	*Banana streak Mysore virus*	BSV-Mys	*Badnavirus*	Caulimoviridae
58	*Banana streak OL virus*	BSV-OL	*Badnavirus*	Caulimoviridae
59	*Barley mild mosaic virus*	BaMMV	*Bymovirus*	Potyviridae
60	*Barley stripe mosaic virus*	BSMV	*Hordeivirus*	Unallocated
61	*Barley yellow dwarf virus-GPV*	BYDV-GPV	Unassigned	Luteoviridae
62	*Barley yellow dwarf virus-MAV*	BYDV-MAV	*Luteovirus*	Luteoviridae
63	*Barley yellow dwarf virus -PAS*	BYDV-PAS	*Luteovirus*	Luteoviridae
64	*Barley yellow dwarf virus-PAV*	BYDV-PAV	*Luteovirus*	Luteoviridae
65	*Barley yellow dwarf virus-RMV*	BYDV-RMV	Unassigned	Luteoviridae
66	*Barley yellow dwarf virus-SGV*	BYDV-SGV	Unassigned	Luteoviridae
67	*Barley yellow mosaic virus*	BaYMV	*Bymovirus*	Potyviridae

68	*Barley yellow striate mosaic virus*	BYSMV	*Cytorhabdovirus*	Rhabdoviridae
69	*Bean calico mosaic virus*	BcaMV	*Begomovirus*	Geminiviridae
70	*Bean common mosaic necrosis virus*	BCMNV	*Potyvirus*	Potyviridae
71	*Bean common mosaic virus*	BCMV	*Potyvirus*	Potyviridae
72	*Bean dwarf mosaic virus*	BDMV	*Begomovirus*	Geminiviridae
73	*Bean golden mosaic virus*	BGMV	*Begomovirus*	Geminiviridae
74	*Bean golden yellow mosaic virus*	BGYMV	*Begomovirus*	Geminiviridae
75	*Bean leaf roll virus*	BLRV	*Luteovirus*	Luteoviridae
76	*Bean mild mosaic virus*	BMMV	*Carmovirus*	Tombusviridae
77	*Bean pod mottle virus*	BPMV	*Comovirus*	Comoviridae
78	*Bean rugose mosaic virus*	BRMV	*Comovirus*	Comoviridae
79	*Bean yellow dwarf virus*	BeYDV	*Mastrevirus*	Geminiviridae
80	*Bean yellow mosaic virus*	BYMV	*Potyvirus*	Potyviridae
81	*Beet black scorch virus*	BBSV	*Necrovirus*	Tombusviridae
82	*Beet chlorosis virus*	BChV	*Polerovirus*	Luteoviridae
83	*Beet cryptic virus-1*	BCV-1	*Alphacryptovirus*	Partitiviridae
84	*Beet cryptic virus-2*	BCV-2	*Alphacryptovirus*	Partitiviridae
85	*Beet cryptic virus-3*	BCV-3	*Alphacryptovirus*	Partitiviridae
86	*Beet curly top virus*	BCTV	*Curtovirus*	Geminiviridae
87	*Beet leaf curl virus*	BLCV	Unassigned	Rhabdoviridae
88	*Beet mild curly top virus*	BMCTV	*Curtovirus*	Geminiviridae
89	*Beet mild yellowing virus*	BMYV	*Polerovirus*	Luteoviridae
90	*Beet mosaic virus*	BMV	*Potyvirus*	Potyviridae
91	*Beet necrotic yellow vein virus*	BNYVV	*Benyvirus*	Unallocated
92	*Beet pseudoyellows virus*	BPYV	*Crinivirus*	Closteroviridae
93	*Beet ringspot virus*	BRSV	*Nepovirus*	Comoviridae
94	*Beet severe curly top virus*	BSCTV	*Curtovirus*	Geminiviridae
95	*Beet soil-borne mosaic virus*	BSBMV	*Benyvirus*	Unallocated
96	*Beet soil-borne virus*	BSBV	*Pomovirus*	Unallocated
97	*Beet virus Q*	BVQ	*Pomovirus*	Unallocated
98	*Beet western yellows virus*	BWYV	*Plerovirus*	Luteoviridae
99	*Beet yellow stunt virus*	BYSV	*Closterovirus*	Closteroviridae
100	*Beet yellows virus*	BYV	*Closterovirus*	Closteroviridae
101	*Belladonna mottle virus*	BeMV	*Tymovirus*	Tymoviridae
102	*Bermuda grass etched-line virus*	BELV	*Marafivirus*	Tymoviridae
103	*Bhendi yellow vein mosaic virus*	BYVMV	*Begomovirus*	Geminiviridae
104	*Bidens mottle virus*	BiMoV	*Potyvirus*	Potyviridae
105	*Black raspberry necrosis virus*	BRNV	Unassigned	Unallocated
106	*Black currant reversion virus*	BRRV	*Nepovirus*	Comoviridae
107	*Blueberry leaf mottle virus*	BLMoV	*Nepovirus*	Comovridae
108	*Blueberry red ringspot virus*	BRRV	*Soymovirus*	Caulimoviridae
109	*Blueberry scorch virus*	BlScV	*Carlavirus*	Flexiviridae
110	*Blueberry shock virus*	BlShV	*Ilarvirus*	Bromoviridae

111	*Blueberry shoestring virus*	BSSV	*Sobemovirus*	Unallocated
112	*Brachypodium yellow streak virus*	BraYSV	Unassigned	Unallocated
113	*Brassica oleracea Melmoth virus*	BolMelV	*Pseudovirus*	Pseudoviridae
114	*Broad bean mottle virus*	BBMV	*Bromovirus*	Bromoviridae
115	*Broad bean necrosis virus*	BBNV	*Pomovirus*	Unallocated
116	*Broad bean stain virus*	BBSV	*Comovirus*	Comoviridae
117	*Broad bean true mosaic virus*	BBTMV	*Comovirus*	Comoviridae
118	*Broad bean wilt virus-1*	BBWV-1	*Fabavirus*	Comoviridae
119	*Broad bean wilt virus-2*	BBWV-2	*Fabavirus*	Comoviridae
120	*Broccoli necrotic yellows virus*	BNYV	*Cytorhabdovirus*	Rhabdoviridae
121	*Brome mosaic virus*	BMV	*Bromovirus*	Bromoviridae
122	*Brome streak mosaic virus*	BStV	*Tritimovirus*	Potyviridae
123	*Burdock yellows virus*	BuYV	Unassigned	Closteroviridae
124	*Cabbage leaf curl virus*	CaLCuV	*Begomovirus*	Geminiviridae
125	*Cacao necrosis virus*	CaNV	*Nepovirus*	Comoviridae
126	*Cacao swollen shoot virus*	CSSV	*Badnavirus*	Caulimoviridae
127	*Cacao yellow mosaic virus*	CYMV	*Tymovirus*	Tymoviridae
128	*Cactus virus 2*	CV-2	*Carlavirus*	Flexiviridae
129	*Cactus virus X*	CVX	*Potexvirus*	Flexiviridae
130	*Cajanus cajan Panzee virus*	CcaPanV	*Pseudovirus*	Pseudoviridae
131	*Calanthe mild mosaic virus*	CalMMV	*Potyvirus*	Potyviridae
132	*Callistephus chinensis chlorosis virus*	CCCV	Unassigned	Rhabdoviridae
133	*Calopogonium yellow vein virus*	CalYVV	*Tymovirus*	Tymoviridae
134	*Canna yellow mottle virus*	CaYMV	*Badnavirus*	Caulimoviridae
135	*Caper latent virus*	CapLV	*Carlavirus*	Flexiviridae
136	*Cardamine chlorotic fleck virus*	CCFV	*Carmovirus*	Tombusviridae
137	*Cardamom mosaic virus*	CdMV	*Macluravirus*	Potyviridae
138	*Carnation bacilliform virus*	CBV	Unassigned	Rhabdoviridae
139	*Carnation cryptic virus-1*	CCV-1	*Alphacryptovirus*	Partitiviridae
140	*Carnation etched ring virus*	CERV	*Caulimovirus*	Caulimoviridae
141	*Carnation Italian ringspot virus*	CIRV	*Tombusviruss*	Tombusviridae
142	*Carnation latent virus*	CLV	*Carlavirus*	Flexiviridae
143	*Carnation mottle virus*	CarMV	*Carmovirus*	Tombusviridae
144	*Carnation necrotic fleck virus*	CNFV	*Closterovirus*	Closteroviridae
145	*Carnation ringspot virus*	CRSV	*Dianthovirus*	Tombusviridae
146	*Carnation vein mottle virus*	CVMoV	*Potyvirus*	Potyviridae
147	*Carrot latent virus*	CtLV	Unassigned	Rhabdoviridae
148	*Carrot mottle mimic virus*	CmoMV	*Umbravirus*	Unallocated
149	*Carrot mottle virus*	CmoV	*Umbravirus*	Unallocated
150	*Carrot temperate virus 1*	CTeV-1	*Alphacryptovirus*	Partitiviridae
151	*Carrot temperate virus 2*	CTeV-2	*Betacryptovirus*	Partitiviridae
152	*Carrot temperate virus 3*	CTeV-3	*Alphacryptovirus*	Partitiviridae
153	*Carrot temperate virus 4*	CTeV-4	*Alphacryptovirus*	Partitiviridae

154	*Carrot thin leaf virus*	CTLV	*Potyvirus*	Potyviridae
155	*Carrot virus Y*	CarVY	*Potyvirus*	Potyviridae
156	*Carrot yellow leaf virus*	CYLV	*Closterovirus*	Closteroviridae
157	*Cassava American latent virus*	CsALV	*Nepovirus*	Comoviridae
158	*Cassava brown streak virus*	CBSV	*Ipomovirus*	Potyviridae
159	*Cassava common mosaic virus*	CsCMV	*Potexvirus*	Flexiviridae
160	*Cassava green mottle virus*	CsGMV	*Nepovirus*	Comoviridae
161	*Cassava Ivorian bacilliform virus*	CsIBV	Unassigned	Unallocated
162	*Cassava symptomless virus*	CsSLV	Unassigned	Rhabdoviridae
163	*Cassava vein mosaic virus*	CsVMV	*Cavemovirus*	Caulimoviridae
164	*Cassava virus C*	CsVC	*Ourmiavirus*	Unallocated
165	*Cassava virus X*	CsVX	*Potexvirus*	Flexiviridae
166	*Cassia yellow blotch virus*	CYBV	*Bromovirus*	Bromoviridae
167	*Cauliflower mosaic virus*	CaMV	*Caulimovirus*	Caulimoviridae
168	*Celery mosaic potyvirus*	CeMV	*Potyvirus*	Potyviridae
169	*Ceratobium mosaic virus*	CerMV	*Potyvirus*	Potyviridae
170	*Cereal chlorotic mottle virus*	CCMoV	Unassigned	Rhabdoviridae
171	*Cereal yellow dwarf virus-RPS*	CYDV-RPS	*Polerovirus*	Luteoviridae
172	*Cereal yellow dwarf virus-RPV*	CYDV-RPV	*Polerovirus*	Luteoviridae
173	*Chara australis virus*	CAV	Unassigned	Unallocated
174	*Chayote mosaic virus*	ChMV	*Tymovirus*	Tymoviridae
175	*Chayote yellow mosaic virus*	ChaYMV	*Begomovirus*	Geminiviridae
176	*Chenopodium necrosis virus*	ChNV	*Necrovirus*	Tombusviridae
177	*Cherry green ring mottle virus*	CGRMV	Unassigned	Flexiviridae
178	*Cherry leaf roll virus*	CLRV	*Nepovirus*	Comoviridae
179	*Cherry mottle leaf virus*	ChMLV	*Trichovirus*	Flexiviridae
180	*Cherry necrotic rusty mottle virus*	CNRMV	Unassigned	Flexiviridae
181	*Cherry rasp leaf virus*	CRLV	*Cheravirus*	Unallocated
182	*Cherry virus A*	CVA	*Cappilovirus*	Flexiviridae
183	*Chickpea stunt disease associated virus*	CpSDaV	Unassigned	Luteoviridae
184	*Chicory yellow mottle virus*	ChYMV	*Nepovirus*	Comoviridae
185	*Chilli leaf curl virus*	ChiLCuV	*Begomovirus*	Geminiviridae
186	*Chilli veinal mottle virus*	ChiVMV	*Potyvirus*	Potyviridae
187	*Chinese wheat mosaic virus*	CWMV	*Furovirus*	Unallocated
188	*Chino del tomate virus*	CdTV	*Begomovirus*	Geminiviridae
189	*Chloris striate mosaic virus*	CSMV	*Mastrevirus*	Geminiviridae
190	*Chrysanthemum frutescens virus*	CFV	Unassigned	Rhabdoviridae
191	*Chrysanthemum vein chlorosis virus*	CVCV	Unassigned	Rhabdoviridae
192	*Chrysanthemum virus B*	CVB	*Carlavirus*	Flexiviridae
193	*Citrus leaf blotch virus*	CLBV	Unassigned	Flexiviridae
194	*Citrus leaf rugose virus*	CiLRV	*Ilarvirus*	Bromoviridae
195	*Citrus leprosis virus*	CiLV	Unassigned	Rhabdoviridae
196	*Citrus mosaic virus*	CiMV	*Badnavirus*	Caulimoviridae

197	*Citrus psorosis virus*	CPsV	*Ophiovirus*	Unallocated
198	*Citrus tristeza virus*	CTV	*Closterovirus*	Closteroviridae
199	*Citrus variegation virus*	CVV	*Ilarvirus*	Bromoviridae
200	*Clitoria virus Y*	ClVY	*Potyvirus*	Potyviridae
201	*Clitoria yellow vein virus*	CYVV	*Tymovirus*	Tymoviridae
202	*Clover enation virus*	ClEV	Unassigned	Rhabdoviridae
203	*Clover yellow mosaic virus*	ClYMV	*Potexvirus*	Flexiviridae
204	*Clover yellow vein virus*	ClYVV	*Potyvirus*	Potyviridae
205	*Cocksfoot mottle virus*	CfMV	*Sobemovirus*	Unallocated
206	*Cocksfoot streak virus*	CSV	*Potyvirus*	Potyviridae
207	*Coconut foliar decay virus*	CFDV	Unassigned	Nanoviridae
208	*Coffee ringspot virus*	CoRSV	Unassigned	Rhabdoviridae
209	*Cole latent virus*	CoLV	*Carlavirus*	Flexiviridae
210	*Colocasia bobone disease virus*	CBDV	Unassigned	Rhabdoviridae
211	*Colombian datura virus*	CDV	*Potyvirus*	Potyviridae
212	*Commelina mosaic virus*	ComMV	*Potyvirus*	Potyviridae
213	*Commelina virus X*	ComVX	*Potexvirus*	Flexiviridae
214	*Commelina yellow mottle virus*	ComYMV	*Badnavirus*	Caulimoviridae
215	*Coriander feathery red vein virus*	CFRVV	Unassigned	Rhabdoviridae
216	*Cotton leaf crumple virus*	CLCrV	*Begomovirus*	Geminiviridae
217	*Cotton leaf curl Alabad virus*	CLCuAV	*Begomovirus*	Geminiviridae
218	*Cotton leaf curl Gezira virus*	CLCuGV	*Begomovirus*	Geminiviridae
219	*Cotton leaf curl Kokhran virus*	CLCuKV	*Begomovirus*	Geminiviridae
220	*Cotton leaf curl Multan virus*	CLCuMV	*Begomovirus*	Geminiviridae
221	*Cotton leaf curl Rajasthan virus*	CLCuRV	*Begomovirus*	Geminiviridae
222	*Cow parsnip mosaic virus*	CpaMV	Unassigned	Rhabdoviridae
223	*Cowpea aphid-borne mosaic virus*	CABMV	*Potyvirus*	Potyviridae
224	*Cowpea chlorotic mottle virus*	CCMV	*Bromovirus*	Bromoviridae
225	*Cowpea golden mosaic virus*	CPGMV	*Begomovirus*	Geminiviridae
226	*Cowpea green vein banding virus*	CGVBV	*Potyvirus*	Potyviridae
227	*Cowpea mild mottle virus*	CPMMV	*Carlavirus*	Flexiviridae
228	*Cowpea mosaic virus*	CPMV	*Comovirus*	Comoviridae
229	*Cowpea mottle virus*	CPMoV	*Carmovirus*	Tombusviridae
230	*Cowpea severe mosaic virus*	CPSMV	*Comovirus*	Comoviridae
231	*Crimson clover latent virus*	CCLV	*Nepovirus*	Comoviridae
232	*Croton yellow vein mosaic virus*	CYVMV	*Begomoviru*	Geminiviridae
233	*Cucumber Bulgarian latent virus*	CBLV	*Tombusvirus*	Tombusviridae
234	*Cucumber fruit mottle mosaic virus*	CFMMV	*Tobamovirus*	Unallocated
235	*Cucumber green mottle mosaic virus*	CGMMV	*Tobamoviru*	Unallocated
236	*Cucumber leafspot virus*	CLSV	*Aureusvirus*	Tombusviridae
237	*Cucumber mosaic virus*	CMV	*Cucumovirus*	Bromoviridae
238	*Cucumber necrosis virus*	CuNV	*Tombuvirus*	Tombusviridae
239	*Cucumber soil-borne virus*	CuSBV	*Carmovirus*	Tombusviridae

240	Cucumber vein yellowing virus	CVYV	Ipomovirus	Potyviridae
241	Cucurbit aphid-borne yellows virus	CABYV	Polerovirus	Luteoviridae
242	Cucurbit leaf curl virus	CuLCuV	Begomovirus	Geminiviridae
243	Cucurbit yellow stunting disorder virus	CYSDV	Crinivirus	Closteroviridae
244	Cycas necrotic stunt virus	CNSV	Nepovirus	Comoviridae
245	Cymbidium mosaic virus	CynMV	Potexvirus	Flexiviridae
246	Cymbidium ringspot virus	CymRSV	Tombusvirus	Tombusviridae
247	Cynara virus	CraV	Unassigned	Rhabdoviridae
248	Cypripedium virus Y	CypVY	Potyvirus	Potyviridae
249	Dahlia mosaic virus	DMV	Caulimovirus	Caulimoviridae
250	Dandelion latent virus	DaLV	Carlavirus	Flexiviridae
251	Dandelion yellow mosaic virus	DaYMV	Sequivirus	Sequiviridae
252	Daphne virus X	DVX	Potexvirus	Flexiviridae
253	Dasheen mosaic virus	DsMV	Potyvirus	Potyviridae
254	Datura shoestring virus	DSSV	Potyvirus	Potyviridae
255	Datura yellow vein virus	DYVV	Nucleorhabdovirus	Rhabdoviridae
256	Dendrobium leaf streak virus	DLSV	Unassigned	Rhabdoviridae
257	Desmodium yellow mottle virus	DYMoV	Tymovirus	Tymoviridae
258	Dicliptera yellow mottle virus	DiYMoV	Begomovirus	Geminiviridae
259	Digitaria streak virus	DSV	Mastrevirus	Geminiviridae
260	Digitaria striate virus	DiSV	Unassigned	Rhabdoviridae
261	Dioscorea bacilliform virus	DBV	Badnavirus	Caulimoviridae
262	Diurus virus Y	DiVY	Potyvirus	Potyviridae
263	Dolichos yellow mosaic virus	DoYMV	Begomovirus	Geminiviridae
264	Dalcamara mottle virus	DuMV	Tymovirus	Tymoviridae
265	East African cassava mosaic Cameroon virus	EACMCV	Begomovirus	Geminiviridae
266	East African cassava mosaic Malawi virus	EACMMV	Begomovirus	Geminiviridae
267	East African cassava mosaic virus	EACMV	Begomovirus	Geminiviridae
268	East African cassava mosaic Zanzibar virus	EACMZV	Begomovirus	Geminiviridae
269	Echinochloa hoja blanca virus	EHBV	Tenuivirus	Unallocated
270	Echinochloa ragged stunt virus	ERSV	Oryzavirus	Reoviridae
271	Eggplant mosaic virus	EMV	Tymovirus	Tymoviridae
272	Eggplant mottled crinkle virus	EMCV	Tombusvirus	Tombusviridae
273	Eggplant mottled dwarf virus	EMDV	Nucleorhabdovirus	Rhabdoviridae
274	Elderberry symptomless virus	EISLV	Cariavirus	Flexiviridae
275	Elm mottle virus	EMoV	Ilarvirus	Bromoviridae
276	Endive necrotic mosaic virus	ENMV	Potyvirus	Potyviridae
277	Epirus cherry virus	EpCV	Ourmiavirus	Unallocated
278	Erysimum latent virus	ErLV	Tymovirus	Tymoviridae

279	*Euonymus fasciation virus*	EFV	Unassigned	Rhabdoviridae
280	*Eupatorium yellow vein virus*	EpYVV	*Begomovirus*	Geminiviridae
281	*Euphorbia leaf curl virus*	EuLCV	*Begomovirus*	Geminiviridae
282	*Faba beans necrotic yellows virus*	FBNYV	*Nanovirus*	Nanoviridae
283	*Festuca leaf streak virus*	FLSV	*Cytorhabdovirus*	Rhabdoviridae
284	*Figwort mosaic virus*	FMV	*Caulimovirus*	Caulimoviridae
285	*Fiji disease virus*	FDV	*Fijivirus*	Reoviridae
286	*Finger millet mosaic virus*	FMMV	Unassigned	Rhabdoviridae
287	*Flame chlorosis virus*	FlCV	Unassigned	Unallocated
288	*Foxtail mosaic virus*	FoMV	*Potexvirus*	Flexiviridae
289	*Fragaria chiloensis latent virus*	FCILV	*Ilarvirus*	Bromoviridae
290	*Frangipani mosaic virus*	FrMV	*Tobamovirus*	Unallocated
291	*Freesia mosaic virus*	FreMV	*Potyvirus*	Poytviridae
292	*Galinsoga mosaic virus*	GaMV	*Carmovirus*	Tombusviridae
293	*Garlic common latent virus*	GarCLV	*Carlavirus*	Flexiviridae
294	*Garlic dwarf virus*	GDV	*Fijivirus*	Reoviridae
295	*Garlic mite-borne filamentous virus*	GarMbFV	*Allexivirus*	Flexiviridae
296	*Garlic virus A*	GarV-A	*Allexivirus*	Flexiviridae
297	*Garlic virus B*	GarV-B	*Allexivirus*	Flexiviridae
298	*Garlic virus C*	GarV-C	*Allexivirus*	Flexiviridae
299	*Garlic virus D*	GarV-D	*Allexivirus*	Flexiviridae
300	*Garlic virus E*	GarV-E	*Allexivirus*	Flexiviridae
301	*Garlic virus X*	GarV-X	*Allexivirus*	Flexiviridae
302	*Gerbera symptomless virus*	GeSLV	Unassigned	Rhabdoviridae
303	*Gloriosa stripe mosaic virus*	GSMV	*Potyvirus*	Potyviridae
304	*Glycine max SIRE1 virus*	GmaSIRV	*Sirevirus*	Pseudoviridae
305	*Glycine max Tgmr virus*	GmaTgmrV	*Pseudovirus*	Pseudoviridae
306	*Glycine mosaic virus*	GMV	*Comovirus*	Comoviridae
307	*Gomphrena virus*	GoV	Unassigned	Rhabdoviridae
308	*Gooseberry vein banding associated virus*	GVBAV	*Badnavirus*	Caulimoviridae
309	*Grapevine Algerian latent virus*	GALV	*Tombusvirus*	Tombusviridae
310	*Grapevine berry inner necrosis virus*	GINV	*Trichovirus*	Flexiviridae
311	*Grapevine Bulgarian latent virus*	GBLV	*Nepovirus*	Comoviridae
312	*Grapevine chrome mosaic virus*	GCMV	*Nepovirus*	Comoviridae
313	*Grapevine fanleaf virus*	GFLV	*Nepovirus*	Comoviridae
314	*Grapevine fleck virus*	GFkV	*Maculavirus*	Tymoviridae
315	*Grapevine leafroll-associated virus1*	GLRaV-1	*Ampelovirus*	Closteroviridae
316	*Grapevine leafroll-associated virus2*	GLRaV-2	*Closterovirus*	Closteroviridae
317	*Grapevine leafroll-associated virus3*	GLRaV-3	*Ampelovirus*	Closteroviridae
318	*Grapevine leafroll-associated virus5*	GLRaV-5	*Ampelovirus*	Closteroviridae
319	*Grapevine leafroll-associated virus7*	GLRaV-7	Unassigned	Closteroviridae
320	*Grapevine Tunisian ring spot virus*	GTRSV	*Nepovirus*	Comoviridae
321	*Grapevine virus A*	GVA	*Vitivirus*	Flexiviridae

322	*Grapevine virus B*	GVB	*Vitivirus*	Flexiviridae
323	*Grapevine virus D*	GVD	*Vitivirus*	Flexiviridae
324	*Groundnut bud necrosis virus*	GBNV	*Tospovirus*	Bunyaviridae
325	*Groundnut eye spot virus*	GEV	*Potyvirus*	Potyviridae
326	*Groundnut ring spot virus*	GRSV	*Tospovirus*	Bunyaviridae
327	*Groundnut rosette assistor virus*	GRAV	Unassigned	Luteoviridae
328	*Groundnut rosette virus*	GRV	*Umbravirus*	Unallocated
329	*Groundnut yellow spot virus*	GYSV	*Tospovirus*	Bunyaviridae
330	*Guinea grass mosaic virus*	GGMV	*Potyvirus*	Potyviridae
331	*Hart's tongue fern mottle virus*	HTFMoV	Unassigned	Unallocated
332	*Hawaiian rubus leaf curl virus*	HRLCV	Unassigned	Unallocated
333	*Helenium virus S virus*	HVS	*Carlavirus*	Flexiviridae
334	*Helenium virus Y virus*	HVY	*Potyvirus*	Potyviridae
335	*Henbane mosaic virus*	HMV	*Potyvirus*	Potyviridae
336	*Heracleum latent virus*	HLV	*Vitivirus*	Flexiviridae
337	*Hibbertia virus Y*	HiVY	*Potyvirus*	Potyviridae
338	*Hibiscus chlorotic ring spot virus*	HCRSV	*Carmovirus*	Tombusviridae
339	*Hibiscus latent Fort Pierce virus*	HLFPV	*Tobamovirus*	Unallocated
340	*Hibiscus latent ring spot virus*	HLRSV	*Nepovirus*	Comoviridae
341	*Hibiscus latent Singapore virus*	HLSV	*Tobamovirus*	Unallocated
342	*Hippeastrum mosaic virus*	HiMV	*Potyvirus*	Potyviridae
343	*Holcus lanatus yellowing virus*	HLYV	Unassigned	Rhabdoviridae
344	*Hollyhock leaf crumple virus*	HLCrV	*Begomovirus*	Geminiviridae
345	*Honeysuckle latent virus*	HnLV	*Carlavirus*	Flexiviridae
346	*Honeysuckle yellow vein mosaic virus*	HYVMV	*Begomovirus*	Geminiviridae
347	*Honeysuckle yellow vein virus*	HYVV	*Begomovirus*	Geminiviridae
348	*Hop latent virus*	HpLV	*Carlavirus*	Flexiviridae
349	*Hop mosaic virus*	HpMV	*Carlavirus*	Flexiviridae
350	*Hop trefoil cryptic virus 1*	HTCV-1	*Alphacryptovirus*	Partitiviridae
351	*Hop trefoil cryptic virus 2*	HTCV-2	*Betacryptovirus*	Partitiviridae
352	*Hop trefoil cryptic virus 3*	HTCV-3	*Alphacryptovirus*	Partitiviridae
353	*Hordeum mosaic*	HoMV	*Rymovirus*	Potyviridae
354	*Hordeum vulgare BARE-1 virus*	HVuBar1V	*Pseudovirus*	Pseudoviridae
355	*Horseradish curly top virus*	HrCTV	*Curtovirus*	Geminiviridae
356	*Horseradish latent virus*	HrLV	*Caulimovirus*	Caulimoviridae
357	*Hosta virus X*	HVX	*Potexvirus*	Flexiviridae
358	*Humulus japonicus latent virus*	HJLV	*Ilarvirus*	Bromoviridae
359	*Hyacinth mosaic virus*	HyaMV	*Potyvirus*	Potyviridae
360	*Hydrangea latent virus*	HdLV	*Carlavirus*	Flexiviridae
361	*Hydrangea ring spot virus*	HdRSV	*Potexvirus*	Flexiviridae
362	*Impatiens necrotic spot virus*	INSV	*Tospovirus*	Bunyaviridae
363	*Indian cassava mosaic virus*	ICMV	*Begomovirus*	Geminiviridae
364	*Indian citrus ringspot virus*	ICRSV	*Mandarivirus*	Flexiviridae

365	*Indian peanut clump virus*	IPCV	*Pecluvirus*	Unallocated
366	*Indonesian soybean dwarf virus*	ISDV	Unassigned	Luteoviridae
367	*Ipomoea yellow vein virus*	IYVV	*Begomovirus*	Geminiviridae
368	*Iris fulva mosaic virus*	IFMV	*Potyvirus*	Potyviridae
369	*Iris germanica leaf stripe virus*	IGLSV	Unassigned	Rhabdoviridae
370	*Iris mild mosaic virus*	IMMV	*Potyvirus*	Potyviridae
371	*Iris severe mosaic virus*	ISMV	*Potyvirus*	Potyviridae
372	*Ivy vein clearing virus*	IVCV	Unassigned	Rhabdoviridae
373	*Japanese iris necrotic ring virus*	JINRV	*Carmovirus*	Tombusviridae
374	*Japanese yam mosaic virus*	JYMV	*Potyvirus*	Potyviridae
375	*Johnsonggrass mosaic virus*	JGMV	*Potyvirus*	Potyviridae
376	*Kalanchoë latent virus*	KLV	*Carlavirus*	Flexiviridae
377	*Kalanchoë mosaic virus*	KMV	*Potyvirus*	Potyviridae
378	*Kalanchoë top-spotting virus*	KTSV	*Badnavirus*	Caulimoviridae
379	*Kennedya yellow mosaic virus*	KYMV	*Tymovirus*	Tymoviridae
380	*Konjac mosaic virus*	KoMV	*Potyvirus*	Potyviridae
381	*Kyuri green mottle mosaic virus*	KGMMV	*Tobamovirus*	Unallocated
382	*Laelia red leafspot virus*	LRLV	Unassigned	Rhabdoviridae
383	*Lamium mild mosaic virus*	LMMV	*Fabavirus*	Comoviridae
384	*Lato river virus*	LRV	*Tombusvirus*	Tombusviridae
385	*Launea arborescens stunt virus*	LArSV	Unassigned	Rhabdoviridae
386	*Leek white stripe virus*	LWSV	*Necrovirus*	Tombusviridae
387	*Leek yellow stripe virus*	LYSV	*Potyvirus*	Potyviridae
388	*Lemon scented thyme leaf chlorosis virus*	LSTCV	Unassigned	Rhabdoviridae
389	*Lettuce big-vein associated virus*	LBVaV	*Varicosavirus*	Unallocated
390	*Lettuce chlorosis virus*	LCV	*Crinivirus*	Closteroviridae
391	*Lettuce infectious yellows virus*	LIYV	*Crinivirus*	Closteroviridae
392	*Lettuce mosaic virus*	LMV	*Potyvirus*	Potyviridae
393	*Lettuce necrotic yellows virus*	LNYV	*Cytorhabdovirus*	Rhabdoviridae
394	*Lettuce ring necrosis virus*	LRNV	*Ophiovirus*	Unallocated
395	*Lettuce speckles mottle virus*	LSMV	*Umbravirus*	Unallocated
396	*Lilac chlorotic leafspot virus*	LiCLV	*Capillovirus*	Flexiviridae
397	*Lilac mottle virus*	LiMoV	*Carlavirus*	Flexiviridae
398	*Lilac ring mottle virus*	LiRMoV	*Ilarvirus*	Bromoviridae
399	*Lilium henryi De1 virus*	LheDe1V	*Metavirus*	Metaviridae
400	*Lily mottle virus*	LMoV	*Potyvirus*	Potyviridae
401	*Lily symptomless virus*	LSV	*Carlavirus*	Flexiviridae
402	*Lily virus X*	LVX	*Potexvirus*	Flexiviridae
403	*Little cherry virus 1*	LChV-1	Unassigned	Closteroviridae
404	*Little cherry virus 2*	LChV-2	*Ampelovirus*	Closteroviridae
405	*Lolium ryegrass virus*	LoRV	Unassigned	Rhabdoviridae
406	*Lotus stem necrosis virus*	LoSNV	Unassigned	Rhabdoviridae
407	*Lucerne Australian latent virus*	LALV	*Nepovirus*	Comoviridae
408	*Lucerne enation virus*	LEV	Unassigned	Rhabdoviridae

409 Lucerne transient streak virus	LTSV	Sobemovirus	Unallocated
410 Luffa yellow mosaic virus	LYMV	Begomovirus	Geminiviridae
411 Lupin yellow vein virus	LYVV	Unassigned	Rhabdoviridae
412 Lychnis ringspot virus	LRSV	Hordeivirus	Unallocated
413 Lycopersicon esculentum ToRTL1 virus	LesToRV	Sirevirus	Pseudoviridae
414 Lycoris mild mottle virus	LyMMoV	Potyvirus	Potyviridae
415 Maclura mosaic virus	MacMV	Macluravirus	Potyviridae
416 Macroptilium mosaic Puerto Rico virus	MaMPRV	Begomovirus	Geminiviridae
417 Macroptilium yellow mosaic Floridavirus	MaYMFV	Begomovirus	Geminiviridae
418 Macroptilium yellow mosaic virus	MaYMV	Begomovirus	Geminiviridae
419 Maize chlorotic dwarf virus	MCDV	Waikavirus	Sequiviridae
420 Maize chlorotic mottle virus	MCMV	Machlomovirus	Tombusviridae
421 Maize dwarf mosaic virus	MDMV	Potyvirus	Potyviridae
422 Maize fine streak virus	MFSV	Unassigned	Rhabdoviridae
423 Maize mosaic virus	MMV	Unassigned	Rhabdoviridae
424 Maize rayado fino virus	MRFV	Marafivirus	Tymoviridae
425 Maize rough dwarf virus	MRDV	Fijivirus	Reoviridae
426 Maize sterile stunt virus	MSSV	Unassigned	Rhabdoviridae
427 Maize streak virus	MSV	Mastrevirus	Geminiviridae
428 Maize stripe virus	MSpV	Tenuivirus	Unallocated
429 Maize white line mosaic virus	MWLMV	Unassigned	Unallocated
430 Mal de Rio Cuarto virus	MRCV	Fijivirus	Reoviridae
431 Malva silvestris virus	MaSV	Unassigned	Rhabdoviridae
432 Malvastrum yellow vein virus	MYVV	Begomovirus	Geminiviridae
433 Megakepasma mosaic virus	MegMV	Unassigned	Closteroviridae
434 Melandrium yellow fleck virus	MYFV	Bromovirus	Bromoviridae
435 Melilotus latent virus	MeLV	Unassigned	Rhabdoviridae
436 Melon chlorotic leaf curl virus	MCLCuV	Begomovirus	Geminiviridae
437 Melon necrotic spot virus	MNSV	Carmovirus	Tombusviridae
438 Melon rugose mosaic virus	MRMV	Tymovirus	Tymoviridae
439 Melon variegation virus	MVV	Unassigned	Rhabdoviridae
440 Milk vetch dwarf virus	MDV	Nanovirus	Nanoviridae
441 Mirabilis mosaic virus	MiMV	Caulimovirus	Caulimoviridae
442 Mirafiori lettuce virus	MiLV	Ophiovirus	Unallocated
443 Miscanthus streak virus	MiSV	Mastrevirus	Geminiviridae
444 Moroccan pepper virus	MPV	Tombusvirus	Tombusviridae
445 Moroccan watermelon mosaic virus	MWMV	Potyvirus	Potyviridae
446 Mulberry latent virus	MLV	Carlavirus	Flexiviridae
447 Mulberry ringspot virus	MRSV	Nepovirus	Comoviridae
448 Mungbean yellow mosaic India virus	MYMIV	Begomovirus	Geminiviridae
449 Mungbean yellow mosaic virus	MYMV	Begomovirus	Geminiviridae

450	*Muskmelon vein necrosis virus*	MuVNV	*Carlavirus*	Flexiviridae
451	*Myrolaban latent ringspot virus*	MLRSV	*Nepovirus*	Comoviridae
452	*Narcissus common latent virus*	NCLV	*Carlavirus*	Flexiviridae
453	*Narcissus degeneration virus*	NDV	*Potyvirus*	Potyviridae
454	*Narcissus late season yellows virus*	NLSYV	*Potyvirus*	Potyviridae
455	*Narcissus latent virus*	NLV	*Macluravirus*	Potyviridae
456	*Narcissus mosaic virus*	NMV	*Potexvirus*	Flexiviridae
457	*Narcissus yellow stripe virus*	NYSV	*Potyvirus*	Potyviridae
458	*Neckar River virus*	NRV	*Tombusvirus*	Tombusviridae
459	*Nerine latent virus*	NeLV	*Carlavirus*	Flexiviridae
460	*Nerine virus X virus*	NVX	*Potexvirus*	Flexiviridae
461	*Nerine yellow stripe virus*	NeYSV	*Potyvirus*	Potyviridae
462	*Nicotiana tabacum Tnt1 virus*	NtaTnt1V	*Pseudovirus*	Pseudoviridae
463	*Nicotiana tabacum Tto1 virus*	NtaTto1V	*Pseudovirus*	Pseudoviridae
464	*Nicotiana velutina mosaic virus*	NVMV	Unassigned	Unallocated
465	*Nilaparvata lugens reovirus*	NLRV	*Fijivirus*	Reoviridae
466	*Northern cereal mosaic virus*	NCMV	*Cytorhabdovirus*	Rhabdoviridae
467	*Nothoscordum mosaic virus*	NoMV	*Potyvirus*	Potyviridae
468	*Oat blue dwarf virus*	OBDV	*Marafivirus*	Tymoviridae
469	*Oat chlorotic stunt virus*	OCSV	*Avenavirus*	Tombusviridae
470	*Oat golden stripe virus*	OGSV	*Furovirus*	Unallocated
471	*Oat mosaic virus*	OMV	*Bymovirus*	Potyviridae
472	*Oat necrotic mottle virus*	ONMV	*Tritimovirus*	Potyviridae
473	*Oat sterile dwarf virus*	OSDV	*Fijivirus*	Reoviridae
474	*Oat striate mosaic virus*	OSMV	Unassigned	Rhabdoviridae
475	*Obuda pepper virus*	ObPV	*Tobamovirus*	Unallocated
476	*Odontoglossum ringspot virus*	ORSV	*Tobamovirus*	Unallocated
477	*Okra mosaic virus*	OkMV	*Tymovirus*	Tymoviridae
478	*Okra yellow vein mosaic virus*	OYVMV	*Begomovirus*	Geminiviridae
479	*Olive latent ringspot virus*	OLRSV	*Nepovirus*	Comoviridae
480	*Olive latent virus 1*	OLV-1	*Necrovirus*	Tombusviridae
481	*Olive latent virus 2*	OLV-2	*Oleavirus*	Bromoviridae
482	*Olive leaf yellowing-associated virus*	OLYaV	Unassigned	Closteroviridae
483	*Onion yellow dwarf virus*	OYDV	*Potyvirus*	Potyviridae
484	*Ononis yellow mosaic virus*	OYMV	*Tymovirus*	Tymoviridae
485	*Orchid fleck virus*	OFV	Unassigned	Rhabdoviridae
486	*Ornithogalum mosaic virus*	OrMV	*Potyvirus*	Potyviridae
487	*Ornithogalum virus 2*	OrV2	*Potyvirus*	Potyviridae
488	*Ornithogalum virus 3*	OrV3	*Potyvirus*	Potyviridae
489	*Oryza australiensis RIRE 1 virus*	OauRIRE1V	*Pseudovirus*	Pseudoviridae
490	*Oryza longstaminata Retrofit virus*	OloRetrofit1V	*Pseudovirus*	Pseudoviridae
491	*Oryza rufipogon endornavirus*	ORV	*Endornavirus*	Unallocated
492	*Oryza sativa endornavirus*	OSV	*Endornavirus*	Unallocated

493	Ourmia melon virus	OuMV	Ourmiavirus	Unallocated
494	Pangola stunt virus	PaSV	Fijivirus	Reoviridae
495	Panicum mosaic virus	PMV	Panicovirus	Tombusviridae
496	Panicum streak virus	PanSV	Mastrevirus	Geminiviridae
497	Papaya leaf curl China virus	PaLCuCNV	Begomovirus	Geminiviridae
498	Papaya leaf curl Guandong virus	PalCuGDV	Begomovirus	Geminiviridae
499	Papaya leaf curl virus	PaLCV	Begomovirus	Geminiviridae
500	Papaya leaf distortion mosaic virus	PLDMV	Potyvirus	Potyviridae
501	Papaya mosaic virus	PapMV	Potexvirus	Flexiviridae
502	Papaya ringspot virus	PRSV	Potyvirus	Potyviridae
504	Paprika mild mottle virus	PaMMV	Tobamovirus	Unallocated
505	Parietaria mottle virus	PmoV	Ilarvirus	Bromoviridae
506	Parsley latent virus	PaLV	Unassigned	Unknown genome
507	Parsley virus	PaV	Unassigned	Rhabdoviridae
508	Parsnip mosaic virus	ParMV	Potyvirus	Potyviridae
509	Parsnip yellow fleck virus	PYFV	Sequivirus	Sequiviridae
510	Passiflora latent virus	PLV	Carlavirus	Flexiviridae
511	Passion fruit woodiness virus	PWV	Potyvirus	Potyviridae
512	Passion fruit yellow mosaic virus	PFYMV	Tymovirus	Tymoviridae
513	Pea early-browning tobravirus	PEBV	Tobravirus	Unallocated
514	Pea enation mosaic virus-1	PEMV-1	Enamovirus	Luteoviridae
515	Pea enation mosaic virus-2	PEMV-2	Umbravirus	Unallocated
516	Pea green mottle virus	PGMV	Comovirus	Comoviridae
517	Pea mild mosaic virus	PMiMV	Comovirus	Comoviridae
518	Pea seed-borne mosaic virus	PSbMV	Potyvirus	Potyviridae
519	Pea stem mosaic virus	PSMV	Carmovirus	Tombusviridae
520	Pea streak virus	PeSV	Carlavirus	Flexiviridae
521	Peach mosaic virus	PMoV	Trichovirus	Flexiviridae
522	Peach rosette mosaic virus	PRMV	Nepovirus	Comoviridae
523	Peanut chlorotic streak virus	PCSV	Soymovirus	Caulimoviridae
524	Peanut clump virus	PCV	Pecluvirus	Unallocated
525	Peanut mottle virus	PeMoV	Potyvirus	Potyviridae
526	Peanut stunt virus	PSV	Cucumovirus	Bromoviridae
527	Peanut yellow mosaic virus	PeYMV	Tymovirus	Tymoviridae
528	Pear latent virus	PeLV	Tombusvirus	Tombusviridae
529	Pelargonium flower break virus	PFBV	Camovirus	Tombusviridae
530	Pelargonium leaf curl virus	PLCV	Tombusvirus	Tombusviridae
531	Pelargonium zonate spot virus	PZSV	Unassigned	Bromoviridae
532	Pepino mosaic virus	PepMV	Potexvirus	Flexiviridae
533	Pepper golden mosaic virus	PepGMV	Begomovirus	Geminiviridae
534	Pepper huasteco yellow vein virus	PHYVV	Begomovirus	Geminiviridae
535	Pepper leaf curl Bangladesh virus	PepLCBV	Begomovirus	Geminiviridae
536	Pepper leaf curl virus	PepLCV	Begomovirus	Geminiviridae

537	*Pepper mild mottle virus*	PMMoV	*Tobamovirus*	Unallocated
538	*Pepper mottle virus*	PepMoV	*Potyvirus*	Potyviridae
539	*Pepper ringspot virus*	PepRSV	*Tobravirus*	Unallocated
540	*Pepper severe mosaic virus*	PepSMV	*Potyvirus*	Potyviridae
541	*Pepper veinal mottle virus*	PVMV	*Potyvirus*	Potyviridae
542	*Pepper yellow mosaic virus*	PepYMV	*Potyvirus*	Potyviridae
543	*Peru tomato mosaic virus*	PTV	*Potyvirus*	Potyviridae
544	*Petunia asteroid mosaic virus*	PetAMV	*Tombusvirus*	Tombusviridae
545	*Petunia vein banding virus*	PetVBV	*Tymovirus*	Tymoviridae
546	*Petunia vein clearing virus*	PVCV	*Petuvirus*	Caulimoviridae
547	*Phalaenopsis chlorotic spot virus*	PhCSV	Unassigned	Rhabdoviridae
548	*Phaseolus vulgaris endornavirus*	PVuV	*Endomavirus*	Unallocated
549	*Phaseolus vulgaris Tpv2-6 virus*	PvuTpvV	Unassigned	Pseudoviridae
550	*Physalis mottle virus*	PhyMV	*Tymovirus*	Tymoviridae
551	*Pigeon pea proliferation virus*	PPPV	Unassigned	Rhabdoviridae
552	*Pigeon pea sterility mosaic virus*	PPSMV	Unassigned	Unknown genome
553	*Pineapple chlorotic leaf streak virus*	PCLSV	Unassigned	Rhabdoviridae
554	*Pineapple mealy bug wilt-associated virus 1*	PMWaV-1	*Ampelovirus*	Closteroviridae
555	*Pineapple mealy bug wilt-associated virus 2*	PMWaV-2	*Ampelovirus*	Closteroviridae
556	*Piper yellow mottle virus*	PYMoV	*Badnavirus*	Caulimoviridae
557	*Pisum virus*	PisV	Unassigned	Rhabdoviridae
558	*Plantago asiatica mosaic virus*	PlAMV	*Potexvirus*	Flexiviridae
559	*Plantago mottle virus*	PlMoV	*Tymovirus*	Tymoviridae
560	*Plantago severe mottle virus*	PlSMoV	*Potexvirus*	Flexiviridae
561	*Plantain mottle virus*	PlMV	Unassigned	Rhabdoviridae
562	*Plantain virus X*	PlVX	*Potexvirus*	Flexiviridae
563	*Plantain virus Y*	PlVY	*Potyvirus*	Potyviridae
564	*Plum pox virus*	PPV	*Potyvirus*	Potyviridae
565	*Poa semilatent virus*	PSLV	*Hordeivirus*	Unallocated
566	*Poinsettia latent virus*	PnLV	Unassigned	Unallocated
567	*Poinsettia mosaic virus*	PnMV	Unassigned	Tymoviridae
568	*Pokeweed mosaic virus*	PkMV	*Potyvirus*	Potyviridae
569	*Poplar mosaic virus*	PopMV	*Carlavirus*	Flexiviridae
570	*Potato aucuba mosaic virus*	PAMV	*Potexvirus*	Flexiviridae
571	*Potato black ringspot virus*	PBRSV	*Nepovirus*	Comoviridae
572	*Potato latent virus*	PotLV	*Carlavirus*	Flexiviridae
573	*Potato leafroll virus*	PLRV	*Polerovirus*	Luteoviridae
574	*Potato mop-top virus*	PMTV	*Pomovirus*	Unallocated
575	*Potato virus A*	PVA	*Potyvirus*	Potyviridae
576	*Potato virus M*	PVM	*Carlavirus*	Flexiviridae
577	*Potato virus S*	PVS	*Carlavirus*	Flexiviridae
578	*Potato virus T*	PVT	Unassigned	Flexiviridae

579	*Potato virus U*	PVU	*Nepovirus*	Comoviridae
580	*Potato virus V*	PVV	*Potyvirus*	Potyviridae
581	*Potato virus X*	PVX	*Potexvirus*	Flexiviridae
582	*Potato virus Y*	PVY	*Potyvirus*	Potyviridae
583	*Potato yellow dwarf virus*	PYDV	*Nucleorhabdovirus*	Rhabdoviridae
584	*Potato yellow mosaic Panama virus*	PYMPV	*Begomovirus*	Geminiviridae
585	*Potato yellow mosaic Trinidad virus*	PYMTV	*Begomovirus*	Geminiviridae
586	*Poto yellow mosaic virus*	PYMV	*Begomovirus*	Geminiviridae
587	*Pothos latent virus*	PoLV	*Aureusvirus*	Tombusviridae
588	*Prune dwarf virus*	PDV	*Ilarvirus*	Bromoviridae
589	*Prunus necrotic ringspot virus*	PNRSV	*Ilarvirus*	Bromoviridae
590	*Quail pea mosaic virus*	QPMV	*Comovirus*	Comoviridae
591	*Radish mosaic virus*	RaMV	*Comovirus*	Comoviridae
592	*Radish yellow edge virus*	RYEV	*Alphacryptovirus*	Partitiviridae
593	*Ranunculus repens symptomless virus*	RaRSV	Unassigned	Rhabdoviridae
594	*Ranunculus white mottle virus*	RWMV	*Ophiovirus*	Unallocated
595	*Raphanus virus*	RaV	Unassigned	Rhabdoviridae
596	*Raspberry bushy dwarf virus*	RBDV	*Idaeovirus*	Unallocated
597	*Raspberry ringspot virus*	RpRSV	*Nepovirus*	Comoviridae
598	*Raspberry vein chlorosis virus*	RVCV	Unassigned	Rhabdoviridae
599	*Red clover cryptic virus 2*	RCCV-2	*Betacryptovirus*	Partitiviridae
600	*Red clover mosaic virus*	RClMV	Unassigned	Rhabdoviridae
601	*Red clover mottle virus*	RCMV	*Comovirus*	Comoviridae
602	*Red clover necrotic mosaic virus*	RCNMV	*Dianthovirus*	Tombusviridae
603	*Red clover vein mosaic virus*	RCVMV	*Carlavirus*	Flexiviridae
604	*Rhopalanthe virus Y*	RhVY	*Potyvirus*	Potyviridae
605	*Rhynchosia golden mosaic virus*	RhGMV	*Begomovirus*	Geminiviridae
606	*Ribgrass mosaic virus*	RMV	*Tobamovirus*	Unallocated
607	*Rice black streaked dwarf virus*	RBSDV	*Fijivirus*	Reoviridae
608	*Rice dwarf virus*	RDV	*Phytoreovirus*	Reoviridae
609	*Rice gall dwarf virus*	RGDV	*Phytoreovirus*	Reoviridae
610	*Rice grassy stunt virus*	RGSV	*Tenuivirus*	Unallocated
611	*Rice hoja blanca virus*	RHBV	*Tenuivirus*	Unallocated
612	*Rice necrosis mosaic virus*	RNMV	*Bymovirus*	Potyviridae
613	*Rice ragged stunt virus*	RRSV	*Oryzavirus*	Reoviridae
614	*Rice stripe virus*	RSV	*Tenuivirus*	Unallocated
615	*Rice tungro bacilliform virus*	RTBV	*Tungrovirus*	Caulimoviridae
616	*Rice tungro spherical virus*	RTSV	*Waikavirus*	Sequiviridae
617	*Rice yellow mottle virus*	RYMV	*Sobemovirus*	Unallocated
618	*Rice yellow stunt virus*	RYSV	*Nucleorhabdovirus*	Rhabdoviridae
619	*Rubus yellow net virus*	RYNV	*Badnavirus*	Caulimoviridae

620	*Rupestris stem pitting-associated virus*	RSPaV	*Foveavirus*	Flexiviridae
621	*Ryegrass cryptic virus*	RGCV	*Alphacryptovirus*	Partitiviridae
622	*Ryegrass mosaic virus*	RGMV	*Rymovirus*	Potyviridae
623	*Ryegrass mottle virus*	RGMoV	*Sobemovirus*	Unallocated
624	*Saguaro cactus virus*	SgCV	*Carmovirus*	Tombusviridae
625	*Saintpaulia leaf necrosis virus*	SLNV	Unassigned	Rhabdoviridae
626	*Sambucus vein clearing virus*	SVCV	Unassigned	Rhabdoviridae
627	*Sammons's Opuntia virus*	SOV	*Tobamovirus*	Unallocated
628	*Sarcochilus virus Y*	SaVY	*Potyvirus*	Potyviridae
629	*Sarracenia purpurea virus*	SPV	Unassigned	Rhabdoviridae
630	*Satsuma dwarf virus*	SDV	*Sadwavirus*	Comoviridae
631	*Scallion mosaic virus*	ScMV	*Potyvirus*	Potyviridae
632	*Scallion virus X*	ScaVX	*Potexvirus*	Flexiviridae
633	*Schefflera ringspot virus*	SRV	*Badnavirus*	Caulimoviridae
634	*Schrophularia mottle virus*	ScrMV	*Tymovirus*	Tymoviridae
635	*Sesbania mosaic virus*	SeMV	*Sobemovirus*	Unallocated
636	*Shallot latent virus*	SLV	*Carlavirus*	Flexiviridae
637	*Shallot virus X*	ShVX	*Allexivirus*	Flexiviridae
638	*Shallot yellow stripe virus*	SYSV	*Potyvirus*	Potyviridae
639	*Sida golden mosaic Costa Rica virus*	SiGMCRV	*Begomovirus*	Geminiviridae
640	*Sida golden mosaic Florida virus*	SiGMFV	*Begomovirus*	Geminiviridae
641	*Sida golden mosaic Honduras virus*	SiGMHV	*Begomovirus*	Geminiviridae
642	*Sida golden mosaic virus*	SiGMV	*Begomovirus*	Geminiviridae
643	*Sida golden yellow vein virus*	SiGYVV	*Begomovirus*	Geminiviridae
644	*Sida mottle virus*	SiMoV	*Begomovirus*	Geminiviridae
645	*Sida yellow mosaic virus*	SiYMV	*Begomovirus*	Geminiviridae
646	*Sida yellow vein virus*	SiYVV	*Begomovirus*	Geminiviridae
647	*Sint-Jan's onion latent virus*	SJOLV	*Carlavirus*	Flexiviridae
648	*Sitke water-borne virus*	SWBV	*Tombusvirus*	Tombusviridae
649	*Soil-borne cereal mosaic virus*	SBCMV	*Furovirus*	Unallocated
650	*Soil-borne wheat mosaic virus*	SBWMV	*Furovirus*	Unallocated
651	*Solanum nodiflorum mottle virus*	SNMoV	*Sobemovirus*	Unallocated
652	*Solanum tuberosum Tst1 virus*	StuTst1V	*Pseudovirus*	Pseudoviridae
653	*Sonchus virus*	SonV	*Cytorhabdovirus*	Rhabdoviridae
654	*Sonchus yellow net virus*	SYNV	*Nucleorhabdo-virus*	Rhabdoviridae
655	*Sorghum chlorotic spot virus*	SrCSV	*Furovirus*	Unallocated
656	*Sorghum mosaic virus*	SrMV	*Potyvirus*	Potyviridae
657	*Sorghum virus*	SrV	Unassigned	Rhabdoviridae
658	*Soursop yellow blotch virus*	SYBV	Unassigned	Rhabdoviridae
659	*South African cassava mosaic virus*	SACMV	*Begomovirus*	Geminiviridae
660	*Southern bean mosaic virus*	SBMV	*Sobemovirus*	Unallocated

661	*Southern cowpea mosaic virus*	SCPMV	*Sobemovirus*	Unallocated
662	*Sowbane mosaic virus*	SoMV	*Sobemovirus*	Unallocated
663	*Sowthistle yellow vein virus*	SYVV	*Nucleorhabdo-virus*	Rhabdoviridae
664	*Soybean chlorotic mottle virus*	SbCMV	*Soymovirus*	Caulimoviridae
665	*Soybean crinkle leaf virus*	SCLV	*Begomovirus*	Geminiviridae
666	*Soybean dwarf virus*	SbDV	*Luteovirus*	Luteoviridae
667	*Soybean mosaic virus*	SMV	*Potyvirus*	Potyviridae
668	*Spartina mottle virus*	SpMV	Unassigned	Potyviridae
669	*Spinach latent virus*	SpLV	*Ilarvirus*	Bromoviridae
670	*Spinach temperate virus*	SpTV	*Alphacryptovirus*	Partitiviridae
671	*Spiraea yellow leafspot virus*	SYLSV	*Badnavirus*	Caulimoviridae
672	*Spring beauty latent virus*	SBLV	*Bromovirus*	Bromoviridae
673	*Squash leaf curl China virus*	SLCCNV	*Begomovirus*	Geminiviridae
674	*Squash leaf curl Philippines virus*	SLCPHV	*Begomovirus*	Geminiviridae
675	*Squash leaf curl virus*	SLCV	*Begomovirus*	Geminiviridae
676	*Squash leaf curl Yunnan virus*	SLCYNV	*Begomovirus*	Geminiviridae
677	*Squash mild leaf curl virus*	SMLCV	*Begomovirus*	Geminiviridae
678	*Squash mosaic virus*	SqMV	*Comovirus*	Comoviridae
679	*Squash yellow mild mottle virus*	SYMMoV	*Begomovirus*	Geminiviridae
680	*Sri Lankan cassava mosaic virus*	SLCMV	*Begomovirus*	Geminiviridae
681	*Stachytarpheta leaf curl virus*	StaLCV	*Begomovirus*	Geminiviridae
682	*Strawberry crinkle virus*	SCV	*Cytorhabdovirus*	Rhabdoviridae
683	*Strawberry latent ringspot virus*	SLRSV	*Sadwavirus*	Unallocated
684	*Strawberry mild yellow edge virus*	SMYEV	*Potexvirus*	Flexiviridae
685	*Strawberry mottle virus*	SmoV	*Sadwavirus*	Unallocated
686	*Strawberry pseudo mild yellow edge virus*	SPMYEV	*Carlavirus*	Flexiviridae
687	*Strawberry vein banding virus*	SVBV	*Caulimovirus*	Caulimoviridae
688	*Subterranean clover mottle virus*	SCMoV	*Sobemovirus*	Unallocated
689	*Subterranean clover stunt virus*	SCSV	*Nanovirus*	Nanoviridae
690	*Sugarcane bacilliform IM virus*	SCBV-IM	*Badnavirus*	Caulimoviridae
691	*Sugarcane bacilliform Mor virus*	SCBV-Mor	*Badnavirus*	Caulimoviridae
692	*Sugarcane mosaic virus*	SCMV	*Potyvirus*	Potyviridae
693	*Sugarcane streak Egypt virus*	SSEV	*Mastrevirus*	Geminiviridae
694	*Sugarcane streak mosaic virus*	SCSMV	Unassigned	Potyviridae
695	*Sugarcane streak Reunion virus*	SSRV	*Mastrevirus*	Geminiviridae
696	*Sugarcane streak virus*	SSV	*Mastrevirus*	Geminiviridae
697	*Sugarcane striate mosaic-associated virus*	SCSMaV	Unassigned	Flexiviridae
698	*Sugarcane yellow leaf virus*	SCYLV	*Polerovirus*	Luteoviridae
699	*Sunflower mosaic virus*	SuMV	*Potyvirus*	Potyviridae
700	*Sunn-hemp mosaic virus*	SHMV	*Tobamovirus*	Unallocated
701	*Sweet clover necrotic mosaic virus*	SCNMV	*Dianthovirus*	Tombusviridae

702	*Sweet potato chlorotic stunt virus*	SPCSV	*Crinivirus*	Closteroviridae
703	*Sweet potato feathery mottle virus*	SPFMV	*Potyvirus*	Potyviridae
704	*Sweet potato latent virus*	SPLV	*Potyvirus*	Potyviridae
705	*Sweet potato leaf curl Georgia virus*	SPLCGV	*Begomovirus*	Geminiviridae
706	*Sweet potato leaf curl virus*	SPLCV	*Begomovirus*	Geminiviridae
707	*Sweet potato leaf speckling virus*	SPLSV	Unassigned	Luteoviridae
708	*Sweet potato mild mottle virus*	SPMMV	*Ipomovirus*	Potyviridae
709	*Sweet potato mild speckling virus*	SPMSV	*Potyvirus*	Potyviridae
710	*Sweet potato virus*	G SPVG	*Potyvirus*	Potyviridae
711	*Tamus red mosaic virus*	TRMV	*Potexvirus*	Flexiviridae
712	*Taro bacilliform virus*	TaBV	*Badnavirus*	Caulimoviridae
713	*Telfairia mosaic virus*	TeMV	*Potyvirus*	Potyviridae
714	*Thistle mottle virus*	ThMoV	*Caulimovirus*	Caulimoviridae
715	*Tobacco bushy top virus*	TBTV	*Umbravirus*	Unallocated
716	*Tobacco curly shoot virus*	TbCSV	*Begomovirus*	Geminiviridae
717	*Tobacco etch virus*	TEV	*Potyvirus*	Potyviridae
718	*Tobacco latent virus*	TLV	*Tobamovirus*	Unallocated
719	*Tobacco leaf curl Japan virus*	TbLCJV	*Begomovirus*	Geminiviridae
720	*Tobacco leaf curl Kochi virus*	TbLCKoV	*Begomovirus*	Geminiviridae
721	*Tobacco leaf curl Yunnan virus*	TbLCYV	*Begomovirus*	Geminiviridae
722	*Tobacco leaf curl Zimbabwe virus*	TbLCZV	*Begomovirus*	Geminiviridae
723	*Tobacco mild green mosaic virus*	TMGMV	*Tobamovirus*	Unallocated
724	*Tobacco mosaic virus*	TMV	*Tobamovirus*	Unallocated
725	*Tobacco mottle virus*	TmoV	*Umbravirus*	Unallocated
726	*Tobacco necrosis virus A*	TNV-A	*Necrovirus*	Tombusviridae
727	*Tobacco necrosis virus D*	TNV-D	*Necrovirus*	Tombusviridae
728	*Tobacco necrotic dwarf virus*	TNDV	Unassigned	Luteoviridae
729	*Tobacco rattle virus*	TRV	*Tobravirus*	Unallocated
730	*Tobacco ringspot virus*	TRSV	*Nepovirus*	Comoviridae
731	*Tobacco streak virus*	TSV	*Ilarvirus*	Bromoviridae
732	*Tobacco vein banding mosaic virus*	TVBMV	*Potyvirus*	Potyviridae
733	*Tobacco vein clearing virus*	TVCV	*Cavemovirus*	Caulimoviridae
734	*Tobacco vein mottling virus*	TVMV	*Potyvirus*	Potyviridae
735	*Tobacco yellow dwarf virus*	TYDV	*Mastrevirus*	Geminiviridae
736	*Tomato aspermy virus*	TAV	*Cucumovirus*	Bromoviridae
737	*Tomato black ring virus*	TBRV	*Nepovirus*	Comoviridae
738	*Tomato bushy stunt virus*	TBSV	*Tombusvirus*	Tombusviridae
739	*Tomato chino La Paz virus*	ToChLPV	*Begomovirus*	Geminiviridae
740	*Tomato chlorosis virus*	ToCV	*Crinivirus*	Closteroviridae
741	*Tomato chlorotic mottle virus*	ToCMoV	*Begomovirus*	Geminiviridae
742	*Tomato chlorotic spot virus*	TCSV	*Tospovirus*	Bunyaviridae
743	*Tomato curly stunt virus*	ToCSV	*Begomovirus*	Geminiviridae
744	*Tomato golden mosaic virus*	TGMV	*Begomovirus*	Geminiviridae
745	*Tomato golden mottle virus*	ToGMoV	*Begomovirus*	Geminiviridae

746	Tomato infectious chlorosis virus	TICV	*Crinivirus*	Closteroviridae
747	Tomato leaf curl Bangalore virus	ToLCBV	*Begomovirus*	Geminiviridae
748	Tomato leaf curl Bangladesh virus	ToLCBDV	*Begomovirus*	Geminiviridae
749	Tomato leaf curl China virus	ToLCCNV	*Begomovirus*	Geminiviridae
750	Tomato leaf curl Gujarat virus	ToLCGV	*Begomovirus*	Geminiviridae
751	Tomato leaf curl Indonesia virus	ToLCIDV	*Begomovirus*	Geminiviridae
752	Tomato leaf curl Iran virus	ToLCIRV	*Begomovirus*	Geminiviridae
753	Tomato leaf curl Karnataka virus	ToLCKV	*Begomovirus*	Geminiviridae
754	Tomato leaf curl Laos virus	ToLCLV	*Begomovirus*	Geminiviridae
755	Tomato leaf curl Malaysia virus	ToLCMV	*Begomovirus*	Geminiviridae
756	Tomato leaf curl virus New Delhi virus	ToLCNDV	*Begomovirus*	Geminiviridae
757	Tomato leaf curl Philippines virus	ToLCPV	*Begomovirus*	Geminiviridae
758	Tomato leaf curl Sri Lanka virus	ToLCSLV	*Begomovirus*	Geminiviridae
759	Tomato leaf curl Sudan virus	ToLCSDV	*Begomovirus*	Geminiviridae
760	Tomato leaf curl virus Taiwan virus	ToLCTWV	*Begomovirus*	Geminiviridae
761	Tomato leaf curl virus Vietnam virus	ToLCVV	*Begomovirus*	Geminiviridae
762	Tomato mild mottle virus	TomMMoV	Unassigned	Potyviridae
763	Tomato mosaic Havana virus	ToMHV	*Begomovirus*	Geminiviridae
764	Tomato mosaic virus	ToMV	*Tobamovirus*	Unallocated
765	Tomato mottle Taino virus	ToMoTV	*Begomovirus*	Geminiviridae
766	Tomato mottle virus	ToMoV	*Begomovirus*	Geminiviridae
767	Tomato pseudo-curly top virus	TPCTV	*Topocovirus*	Geminiviridae
768	Tomato ringspot virus	ToRSV	*Nepovirus*	Comoviridae
769	Tomato rugose mosaic virus	ToRMV	*Begomovirus*	Geminiviridae
770	Tomato severe leaf curl virus	ToSLCV	*Begomovirus*	Geminiviridae
771	Tomato severe rugose mosaic	ToSRV	*Begomovirus*	Geminiviridae
772	Tomato spotted wilt virus	TSWV	*Tospovirus*	Bunyaviridae
773	Tomato yellow leaf curl China virus	TYLCCNV	*Begomovirus*	Geminiviridae
774	Tomato yellow leaf curl Kanchanaburi virus	TYLCKaV	*Begomovirus*	Geminiviridae
775	Tomato yellow leaf curl Malaga virus	TYLCMAIV	*Begomovirus*	Geminiviridae
776	Tomato yellow leaf curl Sardinia virus	TYLCSV	*Begomovirus*	Geminiviridae
777	Tomato yellow leaf curl Thailand virus	TYLCTHV	*Begomovirus*	Geminiviridae
778	Tomato yellow leaf curl virus	TYLCV	*Begomovirus*	Geminiviridae
779	Triticum aestivum chlorotic spot virus	TACSV	Unassigned	Rhabdoviridae
780	Triticum aestivum WIS-2 virus	TaeWis2V	*Pseudovirus*	Pseudoviridae
781	Tropaeolum mosaic virus	TrMV	*Potyvirus*	Potyviridae
782	Tuberose mild mosaic virus	TuMMV	*Potyvirus*	Potyviridae
783	Tulare apple mosaic virus	TAMV	*Ilarvirus*	Bromoviridae
784	Tulip breaking virus	TBV	*Potyvirus*	Potyviridae

785	*Tulip mild mottle mosaic virus*	TMMMV	*Ophiovirus*	Unallocated
786	*Tulip streak virus*	TStV	Unassigned	Unallocated
787	*Tulip virus X virus*	TVX	*Potexvirus*	Flexiviridae
788	*Turnip crinkle virus*	TCV	*Carmovirus*	Tombusviridae
789	*Turnip mosaic virus*	TuMV	*Potyvirus*	Potyviridae
790	*Turnip rosette virus*	TroV	*Sobemovirus*	Unallocated
791	*Turnip vein clearing virus*	TVCV	*Tobamovirus*	Unallocated
792	*Turnip yellow mosaic virus*	TYMV	*Tymovirus*	Tymoviridae
793	*Turnip yellows virus*	TuYV	*Polerovirus*	Luteoviridae
794	*Ullucus mild mottle virus*	UMMV	*Tobamovirus*	Unallocated
795	*Ullucus virus C*	UVC	*Comovirus*	Comoviridae
796	*Urochloa hoja blanca virus*	UHBV	*Tenuivirus*	Unallocated
797	*Velvet tobacco mottle virus*	VTMoV	*Sobemovirus*	Unallocated
798	*Verbena latent virus*	VeLV	*Carlavirus*	Flexiviridae
799	*Vicia cryptic virus*	VCV	*Alphacryptovirus*	Partitiviridae
800	*Vicia faba endomavirus*	VFV	*Endomavirus*	Unallocated
801	*Vigna sinensis mosaic virus*	VSMV	Unassigned	Rhabdoviridae
802	*Voandzeia necrotic mosaic virus*	VNMV	*Tymovirus*	Tymoviridae
803	*Wasabi mottle virus*	WMoV	*Tobamovirus*	Unallocated
804	*Watercress yellow spot virus*	WYSV	Unassigned	Unallocated
805	*Watermelon chlorotic stunt virus*	WmCSV	*Begomovirus*	Geminiviridae
806	*Watermelon leaf mottle virus*	WLMV	*Potyvirus*	Potyviridae
807	*Watermelon mosaic virus*	WMV	*Potyvirus*	Potyviridae
808	*Watermelon silver mottle virus*	WSMoV	*Tospovirus*	Bunyaviridae
809	*Weddel water-borne virus*	WWBV	*Carmovirus*	Tombusviridae
810	*Wheat American striate mosaic virus*	WASMV	*Cytorhabdovirus*	Rhabdoviridae
811	*Wheat chlorotic streak virus*	WCSV	Unassigned	Rhabdoviridae
812	*Wheat dwarf virus*	WDV	*Mastrevirus*	Geminiviridae
813	*Wheat rosette stunt virus*	WRSV	Unassigned	Rhabdoviridae
814	*Wheat spindle streak mosaic virus*	WSSMV	*Bymovirus*	Potyviridae
815	*Wheat streak mosaic virus*	WSMV	*Tritimovirus*	Potyviridae
816	*Wheat yellow leaf virus*	WYLV	*Closterovirus*	Closteroviridae
817	*Wheat yellow mosaic virus*	WYMV	*Bymovirus*	Potyviridae
818	*White clover cryptic virus 1*	WCCV-1	*Alphacryptovirus*	Partitiviridae
819	*White clover cryptic virus 2*	WCCV-2	*Betacryptovirus*	Partitiviridae
820	*White clover cryptic virus 3*	WCCV-3	*Alphacryptovirus*	Partitiviridae
821	*White clover mosaic virus*	WClMV	*Potexvirus*	Flexiviridae
822	*White clover virus L*	WClVL	Unassigned	Unallocated
823	*Wild cucumber mosaic virus*	WCMV	*Tymovirus*	Tymoviridae
824	*Wild potato mosaic virus*	WPMV	*Potyvirus*	Potyviridae
825	*Winter wheat Russian mosaic virus*	WWRMV	Unassigned	Rhabdoviridae
826	*Wisteria vein mosaic virus*	WVMV	*Potyvirus*	Potyviridae
827	*Wound tumor virus*	WTV	*Phytoreovirus*	Reoviridae
828	*Yam mild mosaic virus*	YMMV	*Potyvirus*	Potyviridae

829	*Yam mosaic virus*	YMV	*Potyvirus*	Potyviridae
830	*Youcai mosaic virus*	YoMV	*Tobamovirus*	Unallocated
831	*Zantedeschia mosaic virus*	ZaMV	*Potyvirus*	Potyviridae
832	*Zea mays Hopscotch virus*	ZmaHopV	*Pseudovirus*	Pseudoviridae
833	*Zea mays Opie-2 virus*	ZmaOp2V	*Sirevirus*	Pseudoviridae
834	*Zea mays Prem-2 virus*	ZmaPr2V	*Sirevirus*	Pseudoviridae
835	*Zea mays Sto-4 virus*	ZmaStonorV	*Pseudovirus*	Pseudoviridae
836	*Zea mays virus*	ZMV	Unassigned	Rhabdoviridae
837	*Zea mosaic virus*	ZeMV	*Potyvirus*	Potyviridae
838	*Zucchini green mottle mosaic virus*	ZGMMV	*Tobamovirus*	Unallocated
839	*Zucchini lethal chlorosis virus*	ZLCV	*Tospovirus*	Bunyaviridae
840	*Zucchini yellow fleck virus*	ZYFV	*Potyvirus*	Potyviridae
841	*Zucchini yellow mosaic virus*	ZYMV	*Potyvirus*	Potyviridae

Appendix II

APPENDIX II.1

Salient features of the structure of different genera of plant viruses (Brunt 1996, Hull 2002, Fauquat *et al* 2005, ICTVdB Management 2006, Astier *et al* 2007)

Virus	Shape & other particulars	Number of the particle	Size	Capsomeric arrangement
1. Caulimovirus (ds DNA-RT)	Capsid round, icosa-hedral, isometric, appear round or hexagonal; shell composed of multiple layers; non-enveloped	1	35-50 nm in diameter	Clearly visible or not obvious; T=7 (12 pentamers and 60 hexamers)
2. Badnavirus (ds DNA-RT)	Capsid elongated, icosahedral, bacil-liform; shell composed of multiple layers; non-enveloped	1	95-130 nm or 60-900 nm long, 24-35 nm in diameter	—
3. Ricetungro-virus (ds DNA-RT)	Capsid elongated, icosahedral, bacil-liform; shells compo-sed of multiple layers; not enveloped	1	95-130 nm or 60-900 nm long, width 24-35 nm	T=1/T=3 pentamers and hexamers
4. Soymovirus (ds DNA-RT)	Capsid elongated, icosahedral, bacilli-form; shells multilay-ered; not enveloped	1	Length 130 nm or 60-95 (–900) nm, width 24-35 nm	Conspicuous
5. Cavemovirus (ds DNA-RT)	Capsid round to elongated, icosahe-dral, isometric; not enveloped	1	Length 95-130 nm or 60-900 nm, width 24-35 nm	Not conspicuous
6. Mastrevirus	Capsid elongated,	Twinned	Length 30 nm,	Conspicuous, 22

(ss DNA)	icosahedral, hexag-onal in outline, gem-inate; not enveloped		diameter 18 nm	capsomers/nucleocapsid
7. Curtovirus (ss DNA)	Capsid elongated, icosahedral, gemin-ate; not enveloped	Twinned	Length 30 nm, diameter 16 nm	Conspicuous, 22 capsomers/nucleocapsid
8. Begomovirus (ss DNA)	Capsid elongated, icosahedral, gemin-ate or prolate; round or hexagonal in out-line; not enveloped	Twinned	Length 25-30 nm, diameter 15-20 nm	Conspicuous or not, 22 capsomers/nucleocapsid
9. Topocuvirus (ss DNA)	Capsid elongated, icosahedral, gemin-ate; not enveloped	Twinned	Length 30 nm, diameter 18 nm	Conspicuous, 22 capsomers/nucleocapsid
10. Nanovirus (ss DNA)	Capsid isometric, icosahedral; nucleo-capsid round; not enveloped	1	Diameter 17-26 nm	Inconspicuous
11. Babuvirus (ss DNA)	Capsid isometric, icosahedral, round in outline; not enveloped	1	Diameter 18-20 nm	Inconspicuous
12. Fijivirus (ds RNA)	Virion consists of capsid, a core and a nucleoprotein complex; capsid/nucleocapsid round, icosahedral, isome-tric; not enveloped; capsid shells two-layered, appears round; surface projections distinct "A" types protruding from 12 vertices (11 nm in length and bre-adth); core spherical	1	Capsid diameter 65-80 nm; inner capsid diameter 55 nm; core diameter 35 nm	Clearly visible or inconspicuous
13. Oryzavirus (ds RNA)	Virion consists of capsid, a core and a nucleoprotein complex; capsid/nucleocapsid round, icosahedral, isometric; capsid shells composed of a single inner capsid layer; capsid appears round; not enveloped; surface projections distinct "B" type protruding from 12 vertices; spikes 8-10 nm long, 23-26 nm wide at	1	Diameter of the inner capsid 65-80 nm; diameter of the core 50 nm	Clearly visible or inconspicuous

	the base, and 14-17 nm at the top that overlie the spherical core with no spikes			
14. Phytoreo-virus (ds RNA)	Virion consists of a capsid, a core and a nucleoprotein complex; isosahedral, isometric; capsid shells 3-layered; not enveloped; surface projections distinct "B" type spikes protruding from 12 vertices (8 nm long, 12 nm dia-meter), core spherical, smooth	1	Diameter of the inner capsid layer 55 nm; diameter of the core 35-40 nm	Distinctly visible or not obvious; outer shell T=13 (260 trimeric clusters); inner shell T=1 (60 dimeric clusters)
15. Alfacrypto-virus (ds RNA)	Capsid icosahedral, isometric, round or hexagonal in outline; not enveloped	1	28-30 nm in diameter	Clearly visible or not obvious; T=1 (30 dimers)
16. Betacrypto-virus (ds RNA)	Capsid round, icosa-hedral, isometric, round or hexagonal in outline, not enveloped	2 types	30-38 nm in diameter	Clearly visible
17. Endomavirus (ds RNA)	–	–	–	–
18. Varicosavirus (ss RNA–)	Anisometric, rod; axial canal 3-10 nm in diameter	–	120-360 nm long; width 18-22-30 nm	–
19. Cytorhabdo-virus (ss RNA–)	Virion consists of an envelope and a nucleocapsid; bullet shaped/bacilliform/pleomorphic; surface projections densely dispersed, peplomers that are drumstick-shaped, cover evenly the surface; elongated, helical; nucleocapsid uncurled filamentous or coiled, straight or cylindrical with cross-bands and herring bone pattern; axial canal 14-35 nm dia-meter; basic helix obvious or obscure, pitch 4.2-5.5 nm	1	100-360 nm in length, 42-130 nm in diameter	Spike surface subunits in hexamers associated with G protein with trimer subunits

20. Nucleorhab-dovirus (ss RNA–)	Virion consists of an envelope and a nucleocapsid; bullet/bacilliform/pleomorphic; surface projections are distinctive, densely dispersed pepilomas that are drumstick-shaped cover evenly the surface of the capsid; nucleocapsid elongated, helical, uncoiled, filamentous or cylindrical (coiled, cross-banded); axial canal distinct or indistinct; basic helix obvious or obscure, pitch 4-5 nm	1	95-500 nm in length, 42-100 nm in diameter	Spike surface subunits in hexamers associated with G protein with trimer subunits
21. Tospovirus (ss RNA–)	Virion consists of an envelope and nucleocapsid; spherical to plpleomorphic; surface projections surface spikes (5-10 nm long) surrounded by prominent fringe; projections embedded in a lipid bilayer (5 nm thick); Capsid/nucleocapsid elongated, helical; ribonucleocapsid filamentous or circular	1	Enveloped capsid 70-90 nm in diameter; ribonucleocapsid 200-3000 nm long, 2-2.5 nm in diameter	–
22. Tenuivirus (ss RNA negative sense or ambisense)	Capsid elongated, helical, filamentous, flexuous, spiral, branched, circular, highly coiled; axial canal distinct (length proportional to the size of the RNA), width 8 nm; basic helix obvious	More than one type of particle	950-1350 long, 3-10 nm wide	Inconspicuous
23. Ophiovirus (ss RNA– & ds RNA)	Virion consists of a nucleocapsid, elongated, helical, filamentous, flexuous;	More than one type of particle	Modal length 1500-2500, 760 nm, width 3 nm or 9 nm	–

	segmented, kinked circles; not enveloped		(duplex structure)	
24. Bromovirus (ss RNA+)	Capsid round, icosahedral, isometric; not enveloped	3	25-28 nm in diameter	Conspicuous or not obvious; T=3 (32 capsomers/ nucleocapsid)
25. Cucumovirus (ss RNA+)	Capsid round, icosahedral, isometric; round or hexagonal in outline; not enveloped	Three	29-30 nm in diameter	Conspicuous or not obvious; T=3 (32 capsomers/ nucleocapsid), 180 subunits
26. Alfamovirus (ss RNA+)	Capsid round to elongated, icosahedral, quasi-isometric to bacilliform; not enveloped	More than one type of particle	Capsids of different lengths: 56 nm, 30 nm (bacilliform or ellipsoidal), width 18 nm	Conspicuous; icosahedrons: 30 dimers, 60 subunits; 240, 186, 150, 132 subunits in quasi-isometric to bacilliform
27. Ilarvirus (ss RNA+)	Capsid round to elongated, icosahedral, isometric to bacilliform; appears round or hexagonal in outline or slightly angular; not enveloped	3	20-55 nm long, diameter 19-36 nm	Conspicuous or not obvious
28. Oleavirus (ss RNA+)	Capsid round to elongated, polyhedral symmetry, isometric to bacilliform; not enveloped	More than one type of particle (four)	36 nm or a length of 37, 43, 48 and 55 nm, width 18 nm	–
29. Comovirus (ss RNA+)	Capsid round, icosahedral, isometric; appears round or hexagonal in outline; not enveloped	2	20-30 nm in diameter	Conspicuous or not obvious; 32 capsomers/ nucleocapsid (12 pentamers and 20 trimers)
30. Fabavirus (ss RNA+)	Capsid round, icosahedral, isometric; appears hexagonal in outline; not enveloped	2	(22–) 30-(–32) nm in diameter	Conspicuous or inconspicuous
31. Nepovirus (ss RNA+)	Capsid round, icosahedral, isometric; appears round or hexagonal in outline;	More than one type of particle; associated	(22–) 28 (–33) nm in diameter	Conspicuous or not obvious

	not enveloped		with a satellite virus (TNV)	
32. Potyvirus (ss RNA +)	Nucleocapsid filamentous, flexuous, helical; axial canal obvious or obscure, 2-3 nm in diameter; basic helix obvious or obscure; pitch of the helix 3.3-4 nm; not enveloped	1	Modal length 650-900 nm, diameter– 11-20 nm	–
33. Ipomovirus (ss RNA +)	Capsid / nucleo-capsid elongated, helical, filamentous; pitch of the helix 3.4 nm; not enveloped	1	Modal length 800-950 nm, diameter 12-15 nm	
34. Macluravirus (ss RNA+)	Capsid elongated with helical symmetry, filamentous, flexuous; axial canal indistinct; basic helix obscure; pitch of the helix 3.4 nm; not enveloped	1	Modal length 672 nm, width 13-16 nm	–
35. Rymovirus (ss RNA+)	Capsid elongated with helical symmetry, filamentous; axial canal indistinct; basic helix obscure; pitch of the helix 3.4 nm; not enveloped	1	Modal length 693-760 nm, width 11-15.5 nm	–
36. Tritimovirus (ss RNA+)	Capsid elongated with helical symmetry, filamentous; pitch of the helix 3.4 nm; not enveloped	1	Modal length 700 nm, 12-15 nm in diameter	–
37. Bymovirus (ss RNA+)	Capsid/nucleocapsid elongated wth helical symmetry, filamentous, flexuous; axial canal distinct, helix obscure; pitch of the helix 3.4 nm; not enveloped	More than one type of particle	Predominating modal length 500-600 nm, 200-300 nm, width 12-15 nm	–
38. Tombusvirus (ss RNA+)	Capsid round, icosahedral, isometric; appears round or hexagonal in outline or slightly angular; not enveloped	1	32-35 nm in diameter	Conspicuous or not obvious; T = 3; 32 capsomers/ nucleocapsid (180 protein subunits)

39. Aureusvirus (ss RNA+)	Capsid round, icosahedral, isometric; hexagonal in outline; not enveloped	1	30 nm in diameter	Conspicuous or not obvious; T = 3 (180 protein subunits)
40. Avenavirus (ss RNA+)	Capsid round, icosahedral, isometric; hexagonal in outline; not enveloped	1	30 nm in diameter	Not conspicuous
41. Carmovirus (ss RNA+)	Capsid round, icosahedral, isometric; hexagonal in outline or slightly angular; not enveloped	1	27-34 nm in diameter	Conspicuous or not obvious; T = 3; 32 capsomers/ nucleocapsid (180 protein subunits)
42. Machlomovirus (ss RNA+)	Capsid round with icosahedral symmetry, isometric; hexagonal in outline; not enveloped	1	30 nm in diameter	Not conspicuous
43. Necrovirus (ss RNA+)	Capsid round with icosahedral symmetry, isometric; appears round or hexagonal in outline; not enveloped	More than one type of particle (associated with satellite TNV)	26-30 nm in diameter	Conspicuous or not obvious; 32 capsomers/ nucleocapsid
44. Panicovirus (ss RNA+)	Capsid round with icosahedral symmetry, isometric; appears hexagonal in outline; not enveloped	1	30 nm in diameter	Conspicuous or not obvious: T=3 (180 protein subunits)
45. Sequivirus (ss RNA+)	Capsid/nucleocapsid isometric; appears round in outline; not enveloped	1	31 nm in diameter	Not conspicuous
46. Waikavirus (ss RNA+)	Capsid round, polyhedral; appears round or hexagonal in outline; not enveloped	1	29-30 nm in diameter	Conspicuous or inconspicuous; T = 7
47. Closterovirus (ss RNA+)	Capsid elongated with helical symmetry; filamentous, flexuous; cross-banded, longitudinally striated/obliquely striated, crisscrossed appearance, rope-like features; axial canal indistinct; basic helix obvious or obscure. Pitch of the helix 3.4-3.9 not enveloped	1	Length 600-1536 nm, width 12-19 nm;	5 turns/repeat in the helix, 9-10 subunits/turn

48. Crinivirus (ss RNA+)	Capsid elongated with helical symmetry, filamentous, very flexuous; pitch of the helix 3.4-3.7; not enveloped	1	Length 1200-2000 nm, width 10-13 nm	–
49. Luteovirus (ss RNA+)	Isometric, round or angular in outline; not enveloped	1	23-30 nm in diameter	Conspicuous, 32 capsomers/nucleotide (180 protein subunits
50. Polerovirus (ss RNA+)	Capsid round, icosahedral, isometric; hexagonal in outline; not enveloped	1	24 nm in diameter	Not conspicuous
51. Tobamovirus (ss RNA+)	Capsid elongated with helical symmetry; rod; straight, herringbone pattern; axial canal distinct, basic helix obvious or obscure, pitch of the helix 2.3-4 nm; not enveloped	More than one type of particles	Clear modal length 303.3-320 nm; short particles 70-100 nm, width 15.18.2 nm	16/one-third subunits/turn in the helix
52. Tobravirus (ss RNA+)	Capsid elongated with helical symmetry; rod usually straight; distinct central canal, (4-5 nm) , axial canal obvious or obscure, pitch of the helix 2.5 nm; not enveloped	More than one type of particles	Modal length 180-215 nm (long particle), 46-115 nm (short particles), 20-22 nm in diameter	Conspicuous, 76 subunits in 3 turns in the helix
53. Potexvirus (ss RNA+)	Capsid elongated with helical symmetry, filamentous, flexuous; axial canal 3.3-6 nm in diameter; pitch of the helix 2.8-3.5 nm; not enveloped	1	Modal length 470-580 nm, width 13 nm	4-11 turns/repeat in the helix; 8-9 subunits/turn
54. Carlavirus (ss RNA+)	Capsid elongated with helical symmetry, filamentous, flexous or straight; pitch of the helix 2.8-3.5 nm; not enveloped	1	Modal length 500-720 nm long, 11-18 nm wide	Conspicuous
55. Allexivirus (ss RNA+)	Capsid/nucleocapsid elongate with helical symmetry, filamentous or flexuous; not enveloped	1	Modal length 700 nm, width 13 nm	—
56. Capillovirus (ss RNA+)	Capsid/nucleocapsid elongated with helical	1	Modal length 600-1536 nm,	9-10 turns/repeat in the helix

	symmetry, filamentous, flexuous, cross-banded, longitudinally/obliquely striated, crisscrossed appearance, rope-like feature; axial canal indistinct; basic helix obvious or obscure; pitch of the helix 3.4-3.9 nm; not enveloped		width 12-19 nm	
57. Fobeavirus (ss RNA+)	Capsid elongated with helical symmetry, filamentous, very flexuous; pitch of the helix 3.4-3.7 nm; not enveloped	1	Modal length 1200-2000 nm, width 10-13 nm	
58. Trichovirus (ss RNA+)	Capsid elongated with helical symmetry, filamentous, flexuous, cross-banded, obliquely striated with a crisscrossed appearance and rope-like features; axial canal indistinct; basic helix obvious; pitch of the helix 3.1-3.9 nm; not enveloped	1	Modal length 640-800 nm, width 10.2-12.5 nm	10 turns/repeat in the helix, 9.3-9.8 subunits/turn
59. Vitivirus (ss RNA+)	Capsid elongated with helical symmetry, filamentous, flexuous; basic helix obvious; pitch of the helix 8.30 nm; not enveloped	1	Modal length 800 nm, width 12 nm	5 turns/repeat in the helix; 8-9 subunits/turn
60. Furovirus (ss RNA+)	Capsid/nucleocapsid elongated with helical symmetry, rod, straight (often fragile); distinct central canal; axial canal obvious or obscure, 4.5 nm in diameter; basic helix obvious or obscure; pitch of the helix 2.6-2.9 nm; not enveloped	More than one type of particles	Modal length 65-532 nm, 250-300 nm, 18-25 nm in diameter	Conspicuous
61. Pecluvirus (ss RNA+)	Capsid elongated with helical symmetry; rod,	More than one type	Modal length 245 nm,	–

	straight; not enveloped	of particles	190 nm, 160 nm (in some preparations)	
62. Pomovirus (ss RNA+)	Capsid elongated with helical symmetry; rod, straight; axial canal distinct; basic helix obvious; pitch of the helix 2.5 nm; not enveloped	More than one type of particles	Modal length 100-150 nm or 250-300 nm, width 18-20 nm	–
63. Benyvirus (ss RNA+)	Capsid/nucleocapsid elongated with helical symmetry; rod, straight, herringbone pattern; distinct central canal; pitch of the helix 2.6 nm; not enveloped	More than one type of particles	Predominating modal length 390 nm, 265 nm, 100 nm, 70 nm, width 20 nm	Conspicuous
64. Hordeivirus (ss RNA+)	Capsid elongated with helical symmetry; nucleocapsid rigid rod, longitudinally striated; axial canal obvious, 3-4 nm in diameter; basic helix obvious; pitch of the helix 2.5-2.6 nm; not enveloped	More than one type of particles	Modal length 144-175 nm, 109-126 nm, width 18-25 nm	–
65. Sobemovirus (ss RNA+)	Capsid isometric, icosahedral; appears to be round or angular; not enveloped	1	25-33 nm in diameter	Conspicuous or not obvious; T = 3; 32 capsomers (180 protein subunits)
66. Marafivirus (ss RNA+)	Capsid round, icosahedral, isometric; appears round; not enveloped	More than one type of particles	28-33 nm in diameter	Conspicuous
67. Tymovirus (ss RNA+)	Capsid isometric, icosahedral, appears round or angular; not enveloped	1 (empty capsids often found)	25-32 nm in diameter	Conspicuous,; T=3; 32 capsomers/ (12 pentamers and 20 hexamers)
68. Idaeovirus (ss RNA+)	Capsid round, icosahedral, isometric; appears hexagonal, not enveloped	More than one type of particles	(22-) 30 (-32) nm in diameter	Conspicuous or not obvious
69. Ourmiavirus (ss RNA+)	Capsid elongated, polyhedral, bacilliform; enveloped	Multipartite nucleocap-sids surrounded	26-76 nm in length, width 18-26 nm	Conspicuous; T = 1/ T = 3; subunits in dimers/ trimers

		by one envelope		
70. Umbravirus (ss RNA+)	Capsid vesicle-like; nucleocapsid round	1	50-90 nm in diameter	–
71. Dianthovirus (ss RNA+)	Capsid round, icosahedral, isometric; round or hexagonal in outline; not enveloped	2	31-35 nm in diameter	Distinct or not obvious
72. Enamovirus (ss RNA+)	Capsid round, polyhedral, isometric; round or hexagonal in outline; not enveloped	2	25-28 nm in diameter	Not conspicuous
73. Petuvirus (ds RNA-RT)	Capsid round to elongated icosahedral symmetry; bacilliform; capsid shells composed of multiple layers	1	95-130 in length, width 24-25 nm	–
74. Metavirus (ss RNA-RT)	LRT-retrotransposon-like particles with complex nucleoprotein; capsid not enveloped	–	–	–
75. Pseudovirus (ss RNA-RT)	LRT-retrotransposon-like particles; round, icosahedral, isometric to quasi-isometric, roughly spherical; not enveloped	1	30-40 nm in diameter	Conspicuous; T = 3/4
76. Cheravirus (ss RNA+)	Isometric, icosahedral; not enveloped	1	30 nm in diameter	–
77. Sadwavirus (ss RNA+)	Capsid round, icosahedral, isometric; not enveloped	2	26-30 nm in diameter	–
78. Varicosavirus (ss RNA–)	Nucleocapsid rod with helical symmetry, usually straight; axial canal obvious; pitch of the helix 5-10 nm	1	Modal length 120-360, 18-30 nm in diameter	–
79. Cilevirus Citrus leprosis virus (ss RNA+)	Capsid bullet-shaped	2	–	–
80. Sirevirus (ss RNA-RT)	LRT-retrotransposon-like particles; capsid round, icosahedral, isometric to quasi-isometric; appears roughly spherical to slightly elongated; not enveloped	1	50 nm in diameter	Conspicuous; T = 3 and 4

81. Mandarivirus (ss RNA+)	Capsid/ nucleocapsid elongated with helical symmetry; filamentous; not enveloped	1	–	–
82. Maculavirus (ss RNA+)	Capsid with icosahedral symmetry, isometric; appears round; not enveloped	More than one type of particles	30 nm in diameter	Conspicuous; hexamers and pentamers

Appendix III

APPENDIX III.1

Physico-chemical structures of viruses containing double-stranded nucleic acid (Brunt *et al* 1996, Hull 2002, Fauquet *et al* 2005, ICTVdB Management 2006)

	Virus	Family	Shape	Genome size (kb)	Presence of non-genomic nucleic acid	Nature of 5'/3' end of the genome	Number of structural/ Non-structural protein
1.	*Alphacrypto-virus* (ds RNA)	Bromoviridae	Bacilliform/ quasi-isometric parts	1.7-2 in 2-3 parts	–	–	1-2
2.	*Badnavirus* (ds DNA-RT)	Caulimoviridae	Bacilliform	7.5-8; 2, circular with gaps	–	Poly A at 3'end	2
3.	*Betacrypto-virus* (ds RNA)	Partitiviridae	Isometric	2'1-2.35 in 2-3 parts	–	–	–
4.	*Caulimo-virus* (ds DNA-RT)	Caulimoviridae	Isometric	8; 2 circular with gap	Absent	Poly A at 3'end	2-4
5.	*Cavemo-virus* (ds DNA-RT)	Caulimoviridae	Isometric	–	–	–	2
6.	*Fijivirus* (ds RNA)	Reoviridae	Isometric	23.89-28.91	Absent	Conserved region at 5'and 3' ends	3 or 6
7.	*Oryzavirus* (ds RNA)	Reoviridae	Isometric	26.64-26.66	Absent	Conserved region at 5' and 3' ends	8
8.	*Phytoreo-virus* (ds RNA)	Reoviridae	Isometric	25.13-25.27	Absent	Conserved region at 3' and 5' ends	6-7

9. *Tungrovirus* (ds DNA-RT)	Caulimoviridae	Bacilliform	–	Absent	Terminally redundant sequences	2
10. *Soymovirus* (ds DNA-RT)	Caulimoviridae	Isometric	–	–	–	8
11. *Petuvirus* (ds DNA-RT)	Caulimovirus	Isometric	–	Absent	Terminally redundant sequences	1
12. *Endomavirus* (ds RNA)	No family	–	–	–	–	–

APPENDIX III.2

Physico-chemical properties of the viruses having single-stranded DNA (Brunt *et al* 1996, Hull 2002, Fauquet *et al* 2005, ICTVdB Management 2006)

Virus	*Family*	*Shape*	*Genome size (kb)*	*Presence of non-genomic nucleic acid*	*Nature of 3′, 5′ ends*	*Number of proteins (structural and non-structural)*
1. *Babuvirus*	Nanoviridae	Isometric	–	Absent	–	1
2. *Begomovirus*	Geminiviridae	Geminate	5.096-5.57, in 2 parts	Present or absent	No Poly A at 3′ end	6-7
3. *Curtovirus*	Geminiviridae	Geminate	2.8	Absent	–	–
4. *Mastrevirus*	Geminiviridae	Geminate	2.672-2.9	Absent	No intergenic Poly A	2
5. *Nanovirus*	Nanoviridae	Isometric	1 in 6 parts	Absent	Intergenic Poly A present	1
6. *Topocuvirus*	Geminiviridae	Isometric	2.8	–	–	2

In addition to the above ss DNA viruses, association of ss DNA beta is found with the following virus species, the physico-chemical properties of which are yet to be known:

1. *Ageratum yellow vein virus*	10. *Hollyhock leaf crumple virus*
2. *Bhendi yellow vein mosaic virus*	11. *Honey suckle yellow vein virus*
3. *Chilli leaf curl virus*	12. *Malvastrum yellow vein virus*
4. *Coconut foliar decay*	13. *Papaya leaf curl virus*
5. *Cotton leaf curl Gezira virus*	14. *Tobacco curly shoot virus*
6. *Cotton leaf curl Kokhran virus*	15. *Tomato leaf curl virus*
7. *Cotton leaf curl Multan virus*	16. *Tomato yellow leaf curl China virus*
8. *Cotton leaf curl Rajasthan virus*	17. *Cotton leaf curl Alabad virus*
9. *Eupatorium yellow vein virus*	

APPENDIX III.3

Physico-chemical properties of single-stranded RNA containing rods/ filamentous viruses (Brunt *et al* 1996, Hull 2002, Fauquet *et al* 2005, ICTVdB Management 2006)

Virus	Family	Shape size (kb)	Genome of non-	Presence ends genomic nucleic acid	Nature of 3', 5' structural and	Number of non-structural proteins
1. Allexivirus	Flexiviridae	Filamentous/flexuous	9.0	Present	Poly A at the 3' end	3
2. Ampelovirus	Flexiviridae	Filamentous/flexuous	–	–	–	–
3. Benyvirus	No family	Filamentous/flexuous	1.4-6.7	–	Poly A, very long conserved region at 3' end, short conserved region at 5' end	–
4. Bymovirus	Potyviridae	Filamentous/flexuous	11.21-12	Absent	Poly A at 3' end, VPg at 5' end	2-4
5. Capillovirus	Flexiviridae	Filamentous/flexuous	6.5	–	–	3
6. Carlavirus	Flexiviridae	Flexuous	6.48-8.5	Absent	Methylated cap at 5' end, Poly A at 3' end	1
7. Closterovirus	Closteroviridae	Filamentous/very flexuous	7.5-20	Absent	Methylated cap at 5' end, potential hairpin structure at 3' end	1
8. Crinivirus	Closteroviridae	Filamentous/very flexuous	–	–	Methylated cap at 5' end, potential hairpin structure at 3' end	–
9. Foveavirus	Flexiviridae	Filamentous/flexuous	8.4-9.3	–	Poly A at 3' end	–
10. Furovirus	No family	Rod	9.7-13.5 in 2 parts	Present or absent	Methylated cap at 5' end, Poly at 3' end or not	8

Table Contd...

Table Contd..

	Genus	Family	Shape	Size		Genome	No.
11.	*Hordeivirus*	No family	Rod	9-15.6 in 3-4 parts	Present	Methylated cap at 5' end, Poly A region present. T-RNA-like structure at 3' end	4
12.	*Ipomovirus*	Potyviridae	Filamentous	10.8	–	VPg at 5' end, Poly A at 3' end	1
13.	*Macluravirus*	No family	Filamentous/ flexuous	8.0	—	VPg at 5' end, Poly A at 3' end	1
14.	*Ophiovirus*	No family	Filamentous/flexuous	11.3-12.0	—	3' end complementary to 5' end, forms a panhandle	
15.	*Pecluvirus*	No family	Rod	—	—	t-RNA like structure at 3' end	1
16.	*Pomovirus*	No family	Rod	2.5-6	—	—	1
17.	*Potexvirus*	No family	Rod	5.845-8.1	Absent	Methylated cap at 5' end, Poly A at 3' end	
18.	*Potyvirus*	Potyviridae	Filamentous/flexuous	9-12	Absent	VPg at 5' end, Poly A at 3' end	2-12
19.	*Rymovirus*	Potyviridae	Filamentous	—	—	VPg at 5' end, Poly A at 3' end	2
20.	*Tenuivirus*	No family	Filamentous/flexuous/ spiral/coiled/circular	15.46-18.6	—	—	2
21.	*Tobamovirus*	No family	Rod	6.355-6.5	Present or absent	Methylated cap at 5' end, t-RNA like structure at 3' end	1-2
22.	*Tobravirus*	No family	Rod	9.09-12.0	Present	Methylated cap at 5' end, Poly A at 3' end	3-4
23.	*Trichovirus*	Flexiviridae	Filamentous/flexuous	7.5-8.2	—	Methylated cap at 5' end, Poly A at 3' end	3
24.	*Tritimovirus*	Potyviridae	Filamentous	—	—	VPg at 5' end	1
25.	*Vitivirus*	Allexiviridae	Filamentous/flexuous	7.6	—	Methylated cap at 5' end, Poly A at 3' end	—
26.	*Varicosavirus*	No family	—	—	—	—	

APPENDIX III.4

Physico-chemical properties of RNA-containing isometric viruses (Brunt *et al* 1996, Hull 2002, Anonymous 2002, Fauquet *et al* 2005, ICTVdB Management 2006)

Virus	Family	Genome size (kb)	Presence or absence of non-genomic nucleic acid	Nature of 3'/5' ends	Number of proteins, structural and non-structural
1. *Aureus-virus*	Tombusviridae	4.4	Present, may contain sub-genomic RNA	Methylated cap at 5' end	1
2. *Avena-virus*	Tombusviridae	4.4	Present, may also contain sub-genomic RNA	Methylated cap at 5' end	1
3. *Bromo-virus*	Bromoviridae	8.216-9.5, in 3 parts	Present, also sub-genomic RNA	Methylated cap at 5' end, tRNA-like structure at 3' end	2
4. *Carmo-virus*	Tombusviridae	3.94-5.3	Present or not, may contain sub-genomic RNA and satellite RNA	Methylated cap at 5' end	1
5. *Chera-virus*	No family	–	–	VPg at 5' end, Poly A at 3' end	32 plypro-teins
6. *Como-virus*	Comoviridae	9.57-13.5	Absent	VPg at 5' end, poly A at 3' end, conserved sequences at 3' end	8-11
7. *Cucumo-virus*	Bromoviridae	8.5-8.695, in 3 parts	Present, also contain sub-genomic RNA and satellite RNA	VPg or methylated cap at 5' end, tRNA-like structure at 3' end, conserved sequences at 3' end	2-4
8. *Diantho-virus*	Tombusviridae	–	Present	Methylated cap at 5' end	1
9. *Enamo-virus*	Luteoviridae	–	–	VPg at 5' end, Poly A at 3' end	3
10. Faba-virus	Comoviridae	10.8, in 2 parts	–	VPg at 5' end, Poly A at 3' end	1 ploy-protein
11. *Idaeo-virus*	No family	8.6, in 3 parts	Present, may also contain sub-genomic RNA	–	1
12. *Ilarvirus*	Bromoviridae	7.717-12.22,	Present, may also contain sub-genomic	Methylated cap at 5' end,	2

Table Contd..

Table Contd..

		in 3-5 parts	RNA	intergenic Poly A present	
13. *Luteo-virus*	Luteoviridae	5.641	Present, may also contain sub-genomic RNA, satellite RNA	VPg at 5′ end, Poly A tail at 3′ end	3-6
14. *Machlo-movirus*	Tymoviridae	4.4	Present	Methylated cap at 5′ end	1
15. *Marafi-virus*	Tymoviridae	6.7-7.5, in more than 1 part	–	Sub-genomic promoter "Tymobox" at 5′ end, conserved region	3-4
16. *Necro-virus*	Tombusviridae	3.759-4.185	Absent	Methylated cap at 5′ end	1
17. *Nepo-virus*	Comoviridae	10.1-16.9	Present or absent, may also contain satellite RNA	VPg at 5′ end, Poly A at 3′ end, and conser-ved sequences	1-4
18. *Olea-virus*	Bromoviridae	–, in 3 parts	Present, may also contain sub-genomic RNA	–	–
19. *Panico-virus*	Tombusviridae	–	Present, may also contain sub-genomic RNA	Methylated cap at 5′ end	1
20. *Polero-virus*	Luteoviridae	–	Present, may also contain sub-genomic RNA	–	1
21. *Sadwa-virus*	No family	–	–	–	–
22. *Sequi-virus*	Sequiviridae	9.871	–	VPg at 5′ end, Poly A at 3′ end	3
23. *Sobemo-virus*	No family	4.1-5.7	Present or absent	VPg at 5′ end	4-6
24. *Tombus-virus*	Tombusviridae	4.114-4.8	Present or absent, may also contain sub-genomic and satellite RNA	Methylated cap at 5′ end	4
25. *Tospo-virus*	Bunyaviridae	17.2-17.6, in 3 parts	Absent	3′ and 5′ termini complementary forming a 'panhan-dle", both the ends have conserved regions	4
26. *Tymo-virus*	Tymoviridae	2-6.5	Present	Methylated cap at 5′ end, tRNA-like structure at 3′ end or not, Poly A region present or absent	3
27. *Umbra-virus*	No family	–		TRNA-like structure at 3′ end	
28. *Waika-virus*	Sequiviridae	11	–	VPg at 5′ end, Poly A at 3′ end	1/2/4

APPENDIX III.5

Physico-chemical properties of RNA containing bacilliform/bullet-shaped viruses (Brunt *et al* 1996, Hull 2002, Fauquet *et al* 2005, ICTVdB Management 2006)

Virus	Family	Genome size (kb)	Presence or absence of non-genomic nucleic acid	Nature of 5' and 3'ends	Proteins (structural and non-structural)
1. *Alfamovirus* (round/ elongated/bacilliform/ ellipsoidal)	Bromoviridae	8.27, in 3 parts	Present, may also contain sub-genomic RNA	Methylated cap at 5' end, conserved sequences at 3' end	2
2. *Cytorhabdovirus*	Rhabdoviridae	10.5-14	Present, may also contain nucleic acid of host origin	–	3/5/6 structural proteins
3. *Nucleorhabdovirus*	Rhabdoviridae	10.5-14	Absent	–	4/5/6 structural proteins
4. *Umbravirus* (vesicle-like)	No family	4	Present	–	Non-structural proteins
5. *Ourmiavirus*	No family	1.5-10.65	Present, may also contain satellite RNA	–	1
6. *Cileyvirus* (?) Citrus leprosis virus*	No family	–	–	rhabdo-like *cap at 5' end, Poly A at 3' end	–

*Chung and Bransky 2006

**Locali-Fabris, E.C., Freitas-Astira, I., Souza, A.A., Takita, M.A., Astira-Monge, G., Antonioli-Luizon, R., Rodrigues, V., Targon, M.L.P.N. and Machado, M.A. 2006. J. Gen. Vorol. 87: 2721-2729

APPENDIX III.6

GENOME ORGANIZATION AND EXPRESSION IN DIFFERENT GENERA OF VIRUSES

The genome and its expression is the most complex and organized system of biological information technology, much of which is yet to be explored, but is gradually being revealed through extensive genomic research.

Virus genomes consist of closely packed nucleotide sequences, or codons, that carry the information. Some genomes may also contain non-coding sequences. Some codons contain "start" and "stop" signals for the initiation and termination of the translation process and in some there may be "leaky" termination signals. There is also a "suppression" mechanism in the termination process. Codes are normally non-overlapping but some virus genomic codes do overlap causing the overlapping/linkage of different translational methods. In their initiation, termination, transcription and translation, there is a "coding dictionary" for viruses that helps in understanding of the translation and functional processes. In the functional operation there may also be interaction between the host-coded protein and virus-coded polypeptide.

A. General Methods for Genome Studies

The structure of genomes can be studied by physical, chemical and enzymatic methods. Among the physical methods, x-ray crystallography is very useful for studying the genomes of rod-shaped viruses where the nucleic acid is arranged in a regular helix within a cylinder of protein. However, this method is ineffective for crystallizable small isometric viruses, as the nucleic acid that remains inside the protein shell is not arranged in a regular manner.

Various methods such as nucleic acid hybridization, gel-electrophoresis, construction of restriction maps using restriction endonucleases, molecular techniques using DNA polymerase-I and labelled nucleic acid probes can be used to study genome structure and organization. Identification of the 5′ or 3′ ends of a linear genomic nucleic acid can be done by chemical and enzymatic methods. Molecular weights and the number of different size classes of split genomes can be determined by electrophoresis of purified nucleic acids.

The most important aspect of studying genomic organization is sequencing. There are two major methods for sequencing nucleotides in DNA (Matthews 1992). The "Sanger dideoxy sequencing" procedure is commonly used which has been modified and improved by adopting new techniques as they are developed. The original procedure involves the incorporation of a dideoxy nucleotide to terminate a growing chain in the 3′ direction, as this nucleotide cannot form a diphosphodiester bond. A short radioactively labelled primer at the 3′ end is used to anneal the DNA fragment to be sequenced. DNA polymerase-I from Escherichiaa coli is added for the incorporation of the triphosphates. The analysis of the copies synthesized reveals the pattern of the sequences. This basic method has been improved to make the separation quick and prominent by applying more efficient polymerases, different labelled colours for different bases and introducing automation.

For RNA viruses, a complementary DNA chain is first synthesized as an RNA template by using reverse transcriptase. RNAse then destroys the RNA template. DNA-dependent polymerase then converts the single stranded copy DNA into ds DNA.

These procedures have limitations for the complete sequencing of genomes, as only 300-500 nucleotides can be sequenced on a single gel. These limitations can be overcome by using several strategies, some of which have been illustrated by Matthews (1992).

The genomic or cloned cDNA can be cut into fragments with restriction endo-nucleases. These enzymes cleave recognition sequences (4-8 nucleotides long) at specific sites of the DNA. It is possible to cut the DNA at various sites using different enzymes. More than 100 enzymes can be used for this purpose. These sequences are then inserted into a vector for random sub-cloning and to obtain an overlapping set of sequences from which a full-length sequence can be derived. Alternatively specific fragments at known positions in the restriction map of genomic DNA, or cDNA from a RNA virus, can be cloned or purified and sequenced. Recently it has become possible to synthesize oligo-nucleotides primed at predetermined points to produce overlapping sequences. Rapid synthesis of oligo-nucleotides is possible by automated oligo-nucleotide synthesizing machines to obtain sequence information at the end of the sub-clone.

(a) *Viral Genomes*

The genome of a virus is normally described as sequence maps. Such maps are only available for a few viruses. Matthews (1992) reported that although the number is increasing all the time, so that in excess of 250 virus species, representing most of the virus genera had been fully sequenced by 2002 (Hull 2002, Astier *et al* 2007). Some conventions are normally followed in drawing up genome maps of some RNA viruses (Matthews 1992).

(b) *Preconditions for Replication*

Replication of a genome depends upon its structure, type of nucleic acid and physical integrity. Replication of the genomic nucleic acids should have no breakage of the poly-di-esterase bonds and the presence of virus coded enzymes or virus specific host enzymes depending upon the nature of the genome. Most plant viruses contain positive sense ssRNA (ssRNA+) and sub-genomic mRNA to initiate specific protein synthesis. A few of the (+) ssRNA containing viruses have one small protein co-valently linked to the 5′ end of the nucleotide that has a role in the replication process. Alfamoviruses contain (+) ssRNA, but require the presence of coat protein or coat protein mRNA to become functional. Satellite nucleic acids associated with several viruses may also have roles in the replication process.

In viruses containing negative sense (–) ssRNA, the nucleic acid has to be copied to (+) mRNA by a virus-coded enzyme that remains in the particle to become functional. Double stranded (ds) RNA viruses also contain viral coded enzymes to copy the genomic RNA into mRNAs. However single (ss) or double stranded (ds) DNA viruses use host enzymes to produce mRNA.

The replication process in ssRNA is normally taken place by a viral enzyme polymerase with the help of other viral proteinic factors (notably helicase, methyl transferase, and others) as well as cellular factors, all of which constitute the replicase. Replicase is closely associated to the viral RNAs it copies, and together with them forms the replicative complex. This complex may be associated with cellular structures such as endoplasmic reticulum (Tobamovirus, Potyvirus, etc.). The sites of synthesis are in spherical invaginations of the endoplasmic reticulum, in an invagination of the membrane of vacuoles or tonoplast (Cucumovirus), in an invagination of the mitochondria membrane (Tobravirus), in cluster of vesicles in a rosette derived from the endoplasmic reticulum or Golgi apparatus (Nepovirus) (see Astier *et al* 2007).

Scope of replication

A single nucleotide sequence of a viral genome can code for not more than 12 polypeptides. To code the polypeptides, the nucleotide sequence transcripts to Reading Frames that translate the information for the synthesis of polypeptides. Normally there are three types of Reading Frame, one of which may be an "Open Reading Frame" (ORF) responsible for translation. The number of ORFs in the genome depends on the genome size, as also do the number of translated products or proteins. More polypeptides can be formed through additional frame-shift mechanisms when a Frame possesses a leaky termination signal codon; but the possibility of generating additional proteins through such frame-shifts is limited.

(c) Translational Mechanism

In a eukaryotic protein synthesizing system mRNA is mono-cistronic. 80S ribosomes are adopted to translate Open Reading Frames of the mRNA from the 5′ region. The Reading Frames of the viral nucleotide sequences have to be converted to make them accessible to this route. Viruses have developed strategies to ensure the accessibility of all of their genes to the eukaryotic protein synthesizing system. The strategies to be adopted depend upon the nature of the virus and the number of proteins to be produced by them. Production of multiple proteins is possible from a single ORF by translating one "Polyprotein" that is subsequently cleaved to form different proteins normally by viral coded protease. Several viruses possess sub-genomic RNAs. These may act as mRNAs. The 5' ORF of these mRNAs may be translated to form additional proteins. In the case of multi-partite genomes the 5' ORF of each segment of the genome may be translated to proteins. Additional Read-Through proteins may also be formed by the Frame-shift mechanism due to the activity of the leaky Termination Codon or Signals of the 5' ORF. A portion of the ribosome carries on translation to another Stop Codon downstream from the first to a second functional protein. These proteins may be switched on near the termination codon of the 5' ORF allowing the translation of a second longer transformed protein.

All the viral genomes produce more coat proteins than any other gene products. There may be two pathways for coat protein production. Coat protein mRNA may originate from the 3' end of the genome. These small monocistronic RNAs are highly efficient messengers for coat protein. There may also be a mechanism to produce more coat protein by transcription of mRNA from a minus strand template.

B.   Genomes and their Translated Products

(a)   Double stranded (ds) Positive Sense (+) DNA Containing Viruses

(i)   Family: Caulimoviridae

1.   Caulimovirus [Cauliflower mosaic virus (CaMV) (ds DNA-RT)

The genome of the type species, Cauliflower mosaic virus (CaMV) is 8000 nucleotides long out of which most of the nucleotides have been sequenced. It is a single circular molecule of RTds DNA. The genome has single stranded discontinuity at specific sites and the transcribed strand has one or more discontinuities.

The genome has 8 genes encoded through 8 open reading frames (ORF 1- 37 kDa, ORF 2-18 kDa, ORF 3-15 kDa, ORF 4-57 kDa, ORF 5-79 kDa, ORF 6-8 kDa, ORF 7-11 kDa and ORF 8-12 kDa, lies within ORF-4Da). The genes are close-spaced without any overlap except in the 8th gene. Expression of the genes follows the mRNA strategy to produce 8 proteins responsible for cell-to-cell movement (37 kDa), aphid transmission (18 kDa), symptom expression (57 kDa), coat protein (58 kDa), reverse transcriptase (79 kDa), and other structural and functional proteins (Pfeiffer *et al* 1987, Hohn 1999). The virion-associated protein (VAP) is conserved and contains a coiled-coil structure that assembles as a tetramer. Tetramerization is a common property of all these proteins (Stavolone *et al* 2001). Haas *et al* (2002) recognized 6 proteins, a cell-to-cell movement protein (P1), 2 aphid transmission factors (P2, P3), the precursor of the capsid protein (P4), a poly-protein precursor of proteinase, reverse transcriptase, ribonuclease H (P5) and inclusion body protein/translation transactivator (P6).

Karsies *et al* (2002) observed that the mature CaMV capsid protein (CP), if expressed in the absence of other viral proteins, is translated into the plant cell nucleus by the action of a nuclear localization signal (NLS) close to the N-terminus. In contrast, virus particles do not enter the nucleus but dock at the nuclear membrane, a process inhibited by anti-NLS antibodies or by GTP gamma S, and apparently mediated by the interaction of CP with the host-importing alpha. They found the localization of the precursors in the cytoplasm and the existence of a control mechanism that ensures the use of CP precursor for virus assembly in the cytoplasm and transportation of the mature virus particle to the nuclear pore. Champagne *et al* (2004) further extended these observations and proposed a model for virus assembly. They found that CaMV ORF 4 encodes a coat protein precursor (pre-CP) harbouring an N-terminal extension cell that is cleaved off by the CaMV-encoded protease. In transfected cells, pre-CP is present in the cytoplasm, while the processed form (P44) of CP is targetted to the nucleus suggesting that N-terminal extension might be involved in keeping the pre-CP in the cytoplasm for virus assembly. Immunogold-labelling electron-microscopy using poly-clonal antibodies directed to the N-terminal extension of the pre-CP revealed that this region is closely associated with virus particles present in small aggregates, called "small bodies" adjacent to the main inclusion bodies typical of CaMV infection.

Symptom expression and replication
CaMV protein P6, a product of Gene VI or ORF6 plays a major role in eliciting symptom expression and viral propagation. This response leads to a regulation of viral multiplication cycle that is associated with leaf mosaics. The host regulation of CaMV appears to operate at the transcriptional level through an effect on the 35S promoter or at the post-transcriptional level by the process that is akin to gene silencing and can lead to host recovery depending upon the genetic background of the host. This protein is a translation and interacts with ribosomal protein L13 (Bureau *et al* 2004).

Li and Leisner (2002) observed that P6 is a multifunctional protein and at least four independent domains bind to the full length of it. Viruses lacking central domain D3 were incapable of producing systemic infection. D3 is capable of binding to at least two of the other domains but is unable to self-associate. D3 facilitates P6 self-association by binding to the other domains but not itself. The presence of multiple domains involved in P6 self-association explains the ability of the protein to form the intracellular inclusions characteristic of caulimoviruses.

Hii *et al* (2002) recorded synergic interaction between CaMV and Turnip vein clearing virus (TVCV). TVCV causes an almost imperceptible symptom in infected turnip (cv. Just Right) but simultaneous infection with cabbage S isolate of CaMV produces a severe symptom. Such synergy does not occur with all isolates such as the CM4-184 isolate of CaMV. Isolate specific synergy is not due to accumulation of viruses but it reflects subtle genetic interactions mapping to the ORF4-ORF5 region of the CaMV DNA.

Geri *et al* (2004) made transgenic Arabidopsis with the gene VI and found that it induces a symptom-like phenotype with chlorosis and stunting in the absence of infection. Seeds of the transgenics when mutagenized by gamma irradiation gave mutants showing suppressed phenotypes that were recessive and allelic. When the transgenes were eliminated by back crossing wild-type Arabidiopsis, the progeny lines homozygous for the putative suppressor mutant was less susceptible to CaMV and the phenotype was stably inherited with partial ethylene insensitivity. These observations suggest that the symptoms in P6 transgenics and CaMV infected plants are dependent on interactions between P6 and components involved in ethylene signalling and the suppressor gene product may function to augment these interactions.

The primary function associated with the gene VI product of CaMV is the production of Translational Trans-activator (TAV). TAV controls translation reinitiation of major ORFs on polycistronic RNA. TAV function depends upon its association with polysomes and eukaryotic initiation factor el F3 in vitro and vivo. TAV physically interacts with eiF3 and 60S ribosome L24 protein. TAV mediates efficient recruitment of eiF3 to polysomes allowing translation of the other CaMV genes on the polycistronic mRNAs (35S RNA) by reinitiation overcoming the normal cell barriers to this process (Park *et al* 2001). In addition gene VI protein can elicit a specific type of plant defense response called hypersensitive response (HR) in *Nicotiana edwardsonii* but the TAV function is separate from its role as an elicitor of HR (Palanichelvam and Schoelz 2002).

Kobayashi and Hohn (2004) observed that CaMV TAV is a multifunctional protein essential for basic replication of CaMV. It also plays a role in the pathogenesis in Cruciferous and Solanaceous plants. N- and C-terminal parts are not essential for replication in transfected protoplasts. A further TAV domain required for avirulence function in Solanaceous plants is not essential for CaMV infectivity but has a role in viral virulence in susceptible hosts. Park *et al* (2004) observed that interaction between TAV and eiF3g (eukaryotic initiation factor 3) is critical for the reinitiation process. eiF4B can preclude formation of the TAV/eiF3 complex via competition with TAV for eiF3g binding. There may be overlapping of eiF3B and TAV-binding site. eiF4B interferes with TAV/eiF3/ 40 S ribosome complex formation during the first initiation event. Transient over-expression of eiF4B in protoplasts specifically inhibits TAV-mediated reinitiation of a second ORF. TAV enters the host translation machinery at the eiF4B removal step to stabilize eiF3 on the translating ribosome, thereby allowing translation of polycistronic viral RNA. Ryabova *et al* (2004) proposed a model in which TAV enters the host translational machinery at the eiFB4-removal step to stabilize eiF3 within polysomes.

Ryabova *et al* (2000) observed that CaMV polysomal mRNA or 35S RNA leader is long (600 nucleotides) and contains several short ORFs and folds into an extended hairpin structure with three main stable stem sections. Translation initiation downstream of the leader is cap-dependent and occurs via the ribosomal shunt under the control of two cis elements, a short ORF A (SORFA) followed by stem section 1; a second similar configuration comprising SORFB followed by stem section 2 also allows shunting. The efficiency of the secondary shunt was greatly increased when stem section 1 is destabilized. A significant fraction of reinitiation-competent ribosomes that escape both shunt events migrate linearly via the structured central region but are intercepted by internal AUG start codons. Expression downstream of 35S RNA leader is largely controlled by its multiple SORFs. Similar observation was also made by Pooggin *et al* (2000) with the pre-genomic 35S RNA of CaMV belonging to the growing number of mRNAs known to have a complex leader sequence. According to Guerra *et al* (2000) the interaction of the CaMV pregenomic 35S RNA (Pg RNA) leader with the viral coat protein (CP), its precursor and a series of derivatives where purine rich domain in the centre of Pg RNA leader specifically interact with the coat protein. The zinc finger motif of coat protein and the preceding basic domain were essential for this interaction. Removal of the N-terminal protein of the basic domain leads to the loss of specificity but did not affect the strength of the interaction. Mutation of the zinc finger motif abolishes not only the interaction with the RNA but also viral infectivity.

Froissart *et al* (2004) observed that 70% of CaMV 35S RNA serves as a substrate both for reverse transcription and polycistronic mRNA. When it is spliced into four additional RNA species, one donor site is 5′ untranslated region and three within ORF 1. A unique acceptor site is found at position 1508 in ORF2. This site is vital for CaMV infectivity and expression of ORF3 and 4. It has been suggested that splicing may facilitate the expression of downstream CaMV ORFs. The vital role of splicing in CaMV is regulation of P2 expression and that P2 exhibits biological properties that, whilst indispensable for virus-vector interactions, can block in plant virus infection if this regulation is abolished.

Hull *et al* (2000) recorded that the 35S RNA promoter of CaMV is a component of transgenic constructs in more than 80% of genetically modified (GM) plants. Alarming reports have suggested that the 35S promoter might cause accidental activation of plant genes or endogenous viruses, promote horizontal gene transfer or might even recombine with mammalian virus such as HIV with unexpected consequences. According to Vlasak *et al* (2003) the activity of CaMV 35S in human cells may have an impact on the risk assessment for the environmental release of GM plants. Transient expression of several constructs containing beta-glucuronidase (GUS) gene driven by CaMV 35S promoter or by immediate early promoter of human cytomegalovirus (PCMV) when tested in both potato leaf protoplasts and cultured in human cells was low but measurable. Activity of the 35S promoter in human 293T-cells (0.01% of that revealed using PCMV) and in 293-cells that do not produce SV 40 T antigen, this activity was even lower. On the other hand, in potato protoplasts, PCMV displays nearly 1% activity with promoter 35S.

Movement protein
Gene I of CaMV encodes a protein that is required for virus movement. CaMV movement protein (MP) forms tubules and mediates the translocation of virus particles through plasmodesmata. Tubule formation is dependent on the MP domain. Mutant domains interfere with the ability of the wild-type MP to form tubules; they act as competitive rather than dominant negative or inhibitors (Thomas and Maule 1999). Huang *et al* (2000) observed that Gene I or ORF1 of CaMV encodes the MP and the tubule formation is located on overlapping domains of the MP. According to them, a large region of the MP extending to its N-terminus is required for tubule formation and a smaller region in or around the centre of the MP that overlaps the RNA-binding domain prompts targeting the MP.

Aphid transmission
Transmission of CaMV by aphids follows "helper strategy". It requires two viral non-structural proteins, the ORF2 and 3 products (P2 and P3). ORF3 encodes P3 (a 15 kDa protein). A tetramer may be the functional form of P3. This protein is also required for other biological functions in the natural infection under field condition (Tsuge *et al* 2001). P2 is the helper component encoded by ORF2. This protein is essential for virus replication and also required for aphid transmission. Secondary structure of P2 is 35% alpha-helical. Most alpha helices are located in the C-terminal domain. The active form of P2 assembles as a huge soluble oligomer containing 200-300 sub-units. P2 can also polymerize as long para-crystalline filaments. Self-interaction of P2 forms a parallel trimer (Hebrard *et al* 2001). P3 interacts through its C-terminus with the viral coat protein. The last three amino-acids of P3 are essential for the interaction and virus infectivity and the deletion of them decreases the strength of the interaction by 90% (Leclerc *et al* 2001).

An interaction between C-terminal domain of P2 and N-terminal domain of P3 is essential for transmission. P3 plays a crucial role in the formation of CaMV-transmissible complex by serving as a bridge between P2 and the virus particle (Leh *et al* 2001). Drucker *et al* (2002) using baculovirus expression of P2 and P3, and following electron-microscopy, surface plasmon resonance, affinity chromatography, and transmission assays developed a model describing the regulatory role of P3 in the formation of the transmissible CaMV complex in plants and during acquisition by aphids. They demonstrated that P3 must be previously virions in order that attachment to P2 will allow the aphid transmission of CaMV. P2:P3 complex exists in the absence of virions but is not functional in transmission. Unlike P2, P3 and virions the vector cannot sequentially acquire P3 and the virion. Immunogolds labelling revealed the predominance of spatially separated P2:P3 and P3 and virion complexes in infected cells. This specific distribution indicates that the transmissible complex P2:P3:virion does not form in infected cells but in aphids.

2. Soymovirus [Soybean Chlorotic mottle virus (SbCMV) (dsDNA-RT)]

The genome is a single molecule of circular RT dsDNA that forms an open circle. Complete genome of SbCMV is 7350-74900 nucleotides along with a terminally redundant sequence.

It has 8 major ORFs. Among these ORFs, 1a, 2, 3, 4 and 5 are essential for systemic infection. The product of ORF1b is non-essential but the putative tRNA Met primer binding site in ORF1b is essential. ORF1a is required for virus movement but not for replication. ORF4 encodes a capsid protein by peptide sequencing of the capsid. The genome gives rise to a pre-genomic RNA and an ORF6 mRNA (Takemoto and Hibi 2001). Reddy and Richins (1999) reviewed the genome organization of this virus. Analysis of the viral transcripts showed that the SbCMV DNA genome gives rise to 'pregenomic RNA' and an ORF6 mRNA is found on CaMV ((Takemoto and Hibi 2001). It was further discovered that self-interaction of ORFII protein through a leucine zipper is essential for SbCMV infectivity (Takemoto and Hibi 2005).

Like CaMV, SbCMV is also associated with a promoter. Such a fragment of 378 bp in ORF3 shows a strong expression activity in tobacco mesophyll protoplasts. This fragment contains CAAT and TATA boxes but no transcriptional enhancer signal as reported for the CaMV 35S promoter. Instead, it has sequences homologous to a part of the translational enhancer signal reported for the 5′-leader sequence of TMV RNA (Hasegawa *et al* 1989). The large non-coding region of the genome may also have a promoter fragment, the expression activity of which is five times stronger than that found with CaMV 35S promoter (Conci *et al* 1993).

3. Cavemovirus [Cassava vein mosaic virus (CsVMV) (dsDNA-RT)]

Calvert *et al* (1995) first characterized the genome of the Cassava vein mosaic virus. The genome was 8158 bp in size. A cytosolic initiator methionine tRNA (tRNAmet I)-binding site was found to be present that acts as a primer for minus strand synthesis. The genome contained 5 ORFs that potentially encode proteins with predicted molecular masses of 186 kDa, 9 kDa, 77 kDa, 24 kDa and 26 kDa. The putative 186 kDa protein had regions with similarity to the zinc finger-like RNA-binding domain (common element in the capsid proteins) and intercellular transport domain of pararetroviruses. The predicted 77 kDa protein had regions with similarity to aspartic proteases, reverse transcriptase and RNase H of pararetroviruses.

de Kochko *et al* (1998) established this new genus with the type Cassava vein mosaic virus (CaVMV). The genome of the virus is 8160 nucleotides long and completely sequenced. Genome organization, number and size of the ORFs and sequence comparisons of this virus differ from that found in caulimovirus. The genome has terminally redundant sequences that have direct terminal repeats, reiterated internally in inverted form. RNA dependent DNA polymerase codes 5 ORFs (ICTVdB Management 2006). Verdaguer *et al* (1997) found two promoter fragments in the two DNA fragments CPV1 and CPV2) of the virus. 388 nucleotides from position –368 to +20 of CPV1 and 511 nucleotides from position –443 to +72 when fused to the uidA reporter gene caused high levels of gene expression in protoplasts isolated from cassava and tobacco cell suspensions. These fragments acted in a near constitutive pattern of expression showing that the promotional activity is in different cell types and plant organs. Lockhart *et al* (2000) while working on the genomic analysis of tobacco vein clearing virus (TVCV), a member of the cavemovirus, found that the episomal form of this virus may arise from integrated pararetroviral elements present in the host genome.

4. **Petuvirus** [Petunia vein clearing virus (PVCV) (ds DNA-RT)]

The genome of this virus is a single molecule of circular-RT-ds DNA where the circle is open. The type species contains 7200 nucleotides and has been completely sequenced having terminally redundant sequences (ICTVdB Management 2006). Each genome is translated through 2 ORFs. Harper *et al* (2003) detected this virus and established a model for detection of DNA viruses in plants with homologous endogenous pararetrovirus sequences. The genome or a part of it often integrates with the host genome. There are two forms of integrant: one forms episomal virus infections and the other cannot (Harper *et al* 2002). Richert-Poggeler *et al* (2003) observed the induction of infectious PVCV from endogenous provirus hybrida genome.

5. **Badnavirus** [Banana streak virus) (BSV) (ds DNA-RT)]

The genome is a single molecule circular-RT. ds DNA that forms an open circle The total genomic length of this virus is 7200-76000 nucleotides with a 3200 Da structural protein. The genomic nucleotides have been completely sequenced and the genome is translated through 3 ORFs (ICTVdB Management 2006), of which I encodes the transmission protein and the II and III encode polyprotein carrying the protease, ribonuclease H, and reverse transcriptase. This transcript is also the pregenomic RNA that serves as a template for reverse transcriptase for the synthesis of viral DNA.

The virus is associated with a helper virus, but is independent of its functions during replication. The helper virus is required for transmission. Lockhart and Olszewski (1999) reviewed the genome organization of this virus.

Braithwaite *et al* (2004) found that a variable region of the SCBV genome can be used to generate promoters for transgene expression in sugarcane. The promoter regions of the plant pararetroviruses direct transcription of the full-length viral genome into a pregenomic RNA that is an intermediate in the replication of the virus. It serves as a template for reverse transcription and as polycistronic mRNA for translation to viral proteins. There are four regions that are important for activity of the BSV-Cmv promoter. These are the region containing an as—1-like element, the region around −141 and down to −77, containing several putative transcription factor binding site, the region including the CAAT box, and the leader region. The activity of these promoters provides an insight into the plant cell-mediated replication of the viral genome in banana streak disease (Remans *et al* 2005).

Ndowora *et al* (1999) showed that BSV DNA is integrated into the genome of banana and plantain (*Musa* spp). Although BSV does not employ integration during its infection cycle, the arrangement of the integrated DNA suggests that it can yield an infectious episomal genome via homologous recombination.

6. Rice Tungro Virus/Rice Tungro Bacilliform Virus [(RTBV) (ds DNA)

Hull (2002) proposed RTBV as a new genus that encapsidates a circular double-stranded DNA within a bacilliform particle. Hay *et al* (1991) and Dasgupta *et al* (1991) completely sequenced the genome from infectious clones collected from the Philippines. The size of the genome is 8002 bp. The genome has terminally redundant sequences and encodes proteins from four ORFs that are close to each other with one small (between ORFs 3 and 4 containing 23 and 20 nucleotides) and one large, intergenic region (between ORFs 4 and 1 containing 890 nucleotides). Several workers (Herzog *et al* 2000, Jacquot *et al* 1997, Laco *et al* 1995, Marmey *et al* 1999) determined the functions of some of the proteins encoded by the ORFs. Fan *et al* (1996) collected an Indian isolate, the genome of which showed a characteristic deletion of 64 residues in the non-coding region. Seven complete sequences of RTBV (Cabauatan *et al* 1999) showed that the length of the genomes vary from 8000 to 8016 bp and there are a number of nucleotide substitutions between them, the maximum being 394. Nath *et al* (2002) found numerous insertions, deletions and substitutions mostly in the intergenic region of the genome of the isolates collected from India. The lengths of the genomes of their two collections were 7907 and 7934. The novel feature of these genomes is the presence of an unconventional start codon for

ORF4. The genetic diversity of these collections from those of the classical ones establishes them as distinct strains of RTBV.

Herzog *et al* (2000) found that the four genes of the RTBV genome produce three proteins and a large poly-protein. Gene II product P2 interacts with the coat protein (CP) domain of the gene III poly-protein (P3). They identified the domains involved in the P2-CP association and mapped on both proteins and found a correlation between the virus viability with the ability of P2 to interact with CP domain of P3. P2 may participate in RTBV capsid assembly (Thole and Hull 2002).

Fütterer *et al* (1997) observed that ORFs 2 and 3 are translated from polycistronic pregenomic RNA. Rothnie *et al* (2001) characterized the polyadenylation signal of RTBV. The cis-acting signals required for direct polyadenylation conformed to the plant Poly (A) signals. In infected plants, ss-RNA is hardly detectable but is found in transgenic rice plants. Absence of ss-RNA in infected plants is attributed to Poly (A) site bypass in the viral context to ensure production of full length pregnomic viral RNA. RTBV Poly (A) site suppression thus depends both on context and the expression system.

He *et al* (2002) observed that downstream sequences influence the activity of RTBV promoter in protoplasts derived from cultured rice cells. A DNA element is located between positions +50 and +90 relative to the transcription start site to which nuclear proteins bind. Two rice nuclear proteins (from roots and not from shoots) bound specifically to this DNA element. The DNA element enhanced RTBV promoter activity in a copy number-dependent manner when transferred to a position upstream of the promoter. Petruccelli *et al* (2002) observed that the promoter of RTBV is expressed only in phloem tissues in transgenic rice plants. RF2a, a b-Zip protein from rice binds the box II cis element near TATA box of the promoter. According to them, full-length RTBV promoter and a truncated fragment E of the promoter comprising nucleotides –164 to +45 result in phloem-specific expression of beta-glucuronidase (GUS) reporter gene in transgenic tobacco plants. When a fusion gene comprising the CaMV 35S promoter and RF2a c DNA is co-expressed with the GUS reporter gene, GUS activity is increased by 2- to 20-fold. The increase is positively correlated with the amount of RF2a, and the expression pattern of RTBV promoter is altered from phloem-specific to constitutive type. Dai *et al* (2004) observed the regulation of phloem-specific promoter of RTBV, in part, by sequence-specific DNA-binding proteins that bind to box II, an essential cis element. B-Zip protein, RF2a is involved in transcriptional regulation of the RTBV promoter. They identified and functionally characterized a second b-Zip protein, RF2b. Like RF2a, RF2b activates RTBV promoter in transient assays and in transgenic tobacco plants. Both are predominantly expressed in vascular tissues but have different DNA-binding affinities to box II. Box II cis element and cognate host factors regulate the promoter of the virus (Dai *et al* 2006).

(b) Single Stranded (ss) Positive Sense (+) DNA Containing Viruses

(i) Family: Geminiviridae

7. Mastrevirus [Maize streak virus (MSV) (ss DNA)

Palmer and Rybicki (1998) reviewed the genome structure of Mastrevirus (Type species: Maize streak virus—MSV). The genome of this virus is composed of a single molecule of circular, ambisense ssDNA that forms a closed circle. The complete genome is 2672-2900 nucleotides long.and has been completely sequenced.

The genome nucleic acid is not infectious. The virion is associated with a helper virus but independent from its functions during replication. There are four coding regions and two non-coding intergenic regions. Two ORFs are on the viral strand and the other two are on the complementary strand. Liu *et al* (1999) observed that transport of MSV DNA into the nucleus of host cells is essential for virus replication and the presence of the virus particles in the nuclei of the infected cells. This implies that coat protein (CP) must enter the nuclei. In addition to entering the nucleus where it is required for encapsidation of the viral ssDNA, the MSV CP facilitates the rapid transport of viral (ss or ds) DNA into the nucleus.

Hefferon and Dugdale (2003) observed that Bean yellow dwarf virus (BeYDV) Mastrevirus contains four ORFs encoding a movement protein, a coat protein and two Rep gene products, Rep and Rep A (replication- associated protein) that are encoded by two overlapping ORFs. Constitutive expression of both Rep and Rep A supports replication and enhance gene expression.

Viral proteins regulate Viral DNA replication and transcription and also affect cellular gene expression. The virus replication and also other steps during infection rely heavily on the activity of the cellular factors. One of the primary events seems to be the inactivation of RBR (Retinoblast-related) protein that negatively regulates GL/S transition in cells. Mastreviruses have their own mechanism to interact with RBR. Here it depends on LXCXE amino acid motif present in protein, such as Rep A encoded by the virus. Additional interactions with other cell growth regulatory factors are also likely to be important (Gutierrez 2002).

Luque *et al* (2002) identified a novel interaction of Wheat dwarf virus (WDV) Rep, the replication initiation protein with the large subunit of the wheat replication factor C complex (TmRFC-1). WDV Rep stimulates binding of recombinant TmRFC-1 to a model substrate containing 3'-OH terminus and a WDV Rep binding site. WDF could participate in the recruitment of RFC to the newly formed primer. This pathway may represent an initial event to facilitate the assembly of other replication factors, a model that could also apply to other eukaryotic replicons, such as nanoviruses, circoviruses, parvoviruses with a similar DNA replication strategy.

Nikovics *et al* (2001) observed that the viral DNA replication is coupled to the cell cycle regulatory complex. Of the plant cell and the virus-early (complementary or C sense) gene products, REP and REPA may be able to manipulate the regulation of the cycle. They examined the expression from the promoters of MSV in transgenic maize plants and cells to determine whether they show cell cycle specificity. Histo-chemical staining with plant roots containing "long and short" C-sense promoter sequences upstream of the GUS (beta-glucuronidase) reporter gene shows that promoter activity is restricted to the meristematic region of the roots and is enhanced by 2,4-D application and all MSV promoters show cell cycle specificity. The coat protein gene promoters show highest activity in early interphase (G2) whereas C-sense promoter sequences produce two peaks of activity in the interphases (S and G2) of the cell cycle. Shepherd *et al* (2005) also confirmed the dependence of the infectivity on interactions between the virus replication-associated proteins Rep and Rep A and host retinoblastoma-related proteins (RBR) that control cell-cycle progression.

Munoz Martin *et al* (2003) observed that WDV Rep A activates both WDV and MSV virion (V)-sense expression in plant tissues. Rep alone does not have any effect on the silent WDV promoter and it represses the basal MSV promoter activity. MSV promoter activation depends on an intact Rep A retinoblastoma protein (RP)-binding domain. WDV promoter contains two sites that fit the consensus E2F-binding site. One site WDV1 binds human E2F-1 in one-hybrid assays in yeast. It also binds specifically to maize and wheat promoters in vivo and when fused to minimal CaMV 35S promoter, it confers responsiveness to Rep A only when Rep A, RB-binding domain and WDV1 site are intact. In the whole WDV V-sense promoter context, mutation of this sequence has no effect suggesting that additional sequences are important for Rep A-mediated promoter activation. McGivern *et al* (2005) did not find an intact RBR-binding motif necessary for the invasion of mesophyll cells. The replication- associated protein (Rep A) of MSV interacts in yeast with retinoblastoma-related protein (RBR), the regulator of cell-cycle progression. This may allow the virus to subvert cell-cycle control to provide an environment suitable for virus replication. MSV Rep A-RBR interaction is essential only in tissues with high levels of active RBR

Dinant *et al* (2004) observed that a large intergenic region (LIR) of WDV contains cis-acting elements essential for the replication of the genome as well as for the bi-directional transcription of viral genes. WDV V-sense promoter consistently induces an expression pattern restricted to the vascular tissues, predominantly in the phloem of all organs.

Liu *et al* (2001) identified a proline-cysteine-lysine motif at amino acids 180-182 of MSV CP that is conserved in most of the cereal-infecting mastreviruses. A mutant produced by substituting the lysine with a valine, cannot produce any systemic symptom in maize plants though the mutant is replicated around the inoculation site. Both the wild type and mutant CP localized to the nucleus, when expressed in insect and tobacco cells. *Escherichia coli*-expressed protein bound both ss DNA and ds DNA and interacted with movement protein in vitro.

Casado *et al* (2004) found that single T=1 isometric particles of Maize streak virus (MSV) contain sub-genomic (Sg) MSV DNA encapsidated by the MSV coat protein. The largest sg DNA is 1.65 kb and the smallest is 0.2 kb. These are not infectious and do not play any role in pathogenicity. Implication of these sub-genomic DNA opens a new vista for understanding the nature of these viruses.

8. Curtovirus [Beet curly top virus (BCTV) (ss DNA)]

Stanley *et al* (1986) established the sequence of the type species (Beet curly top virus—BCTV) of Curtovirus. Its genome is a single molecule, circular, ambisense ssDNA that forms a closed circle. The complete genome is 2672-2900 nucleotides long. There are four ORFs producing six to seven proteins. Three ORFs are in the positive sense and four are in the negative sense. Complementary sense ORFs produce four proteins involved in virus replication.

The products of the positive and negative sense ORFs are: V1+ produces coat protein (29-30 kDa; this coat protein is essential for infectivity (Briddon *et al* 1989) V2+ produces ss/ds RNA regulator (12-15 kDa); V3+ produces movement protein (10 kDa). C1– produces replication protein (39-44 kDa); C2– produces protein of unknown function (16-20 kDa); C3– produces a protein (16 kDa) similar to the replication enhancer protein of begomoviruses and C4– produces a protein (9-13 kDa) that initiates cell division and symptom development. Some members may lack C3– and some may have an additional positive sense ORF (V4+).

Stenger (1998) cloned the genomes of three different strains of beet curly top virus: CFH, Worland and Cal/Logan. These strains serve as helper viruses to trans-replicate defective (D) DNAs that are incapable of self-replication due to deletions within C1 ORF coding, the replication initiator (Rep) protein. This trans-replication function is specific to the respective strains. Logan Rep protein trans-replicates a Logan-derived D-DNA in a transient replication assay but fail to trans-replicate D-DNAs derived from CFH and Worland. Rep protein of the CFH and Worland strains trans-replicates the D-DNAs from these strains but not D-DNA of Logan. Comparative studies of the amino acid sequences of the Rep proteins from these strains suggest that transacting replication specificity element may reside in one or more of 12 amino acid residues that are identical. Studies on replication specificity, nucleotide sequence identity and symptom expression suggest these three strains are separate species, BCTV (Cal/Logan), Beet severe curly top virus (CFH and closely related Iranian isolate) and Beet mild curly top virus (Worland).

Briddon and Markham (2001) found a novel interrelationship between curtovrus, topocuvirus (mono-partite genomes) and begomovirus (bi-partite genomes). The DNA-A components of begomoviruses encode all viral functions required for replication and encapsidation but the gene products encoded on DNA-B is necessary for cell-to-cell movements in plants. Strangely, curto- and topocuviruses are able to trans-complement efficient movement of DNA-A of begomovirus in the absence of the corresponding DNA-B. The spread of DNA-A independent of DNA-B, following *Agrobacterium*-mediated inoculation is shown to require coat protein, whereas trans-complemented spread of DNA occurs independent of coat protein encoded on DNA-A.

Preiss and Jesoke (2003) found multitasking replication of beet curly top virus. The replication may follow a rolling circle replication mechanism or recombinant-dependent replication (RDR) route in analogy to T4 phages.

Baliji *et al* (2004) observed that natural genomic recombination is common among geminiviruses and identified a new curtovirus, spinach curly top virus, the genome of which contains 2925 nucleotides with 7 ORFs encoding proteins homologous to those produced by other curtoviruses. Three virion-sense genes were different from that of the four complementary-sense genes.

9. Begomovirus [African cassava mosaic virus (ACMV) (ss DNA)]

Stanley and Gay (1983) established that the genome of the type species of Begomovirus (ACMV) is bi-partite (A, B) and occurs as circles of similar size but the nucleotide sequences in them differ widely, excluding the 200 nucleotide-non-coding region. Lengths of DNA-A (component A) and DNA-B (component B) are 2647 nucleotides and 2585 nucleotides respectively. The expression of genome in DNA-A takes place through six ORFs, two in the virion sense (one encodes coat protein and movement protein) and four in the complementary sense (replication initiation protein, replication enhancer protein and symptom expression protein). The expression of genome in DNA-B takes place through two ORFs of which one is in the virion sense (producing nuclear shuttle protein) and another in the complementary sense (producing cell-to-cell movement protein). These expressions take place following mRNA strategy (Davies *et al* 1987, Buck 1999).

The non-coding region of the components is highly conserved amongst geminiviruses and has regions capable of forming hairpin loops. It may be involved in RNA transcription and DNA replication.

One of the virion sense ORFs of component A codes for coat protein and most of the ORFs of the complementary sense are involved in replication. Two ORFs of the component B encode products involved in virus spread within plants.

Diversity of ACMV and their mixed infection are common in different cassava growing regions indicating their epidemiological potentiality (Ariye *et al* 2005, Tadu *et al* 2006).

Up to 2005, 105 begomoviruses have been reported (Fauquet *et al* 2005) and newly evolved members are being reported. Variability arises through mutations, recombinations and pseudo-recombination. Heterologous recombinants containing parts of the host genome and/or sequences from satellite-like molecules associated with mono-partite begomoviruses provide unlimited evolutionary opportunities (Varma and Malathi 2003). The DNA satellites share no significant homology with begomovirus sequences and are of various types. The first to be discovered was 682 nucleotides in size and associated with Tomato leaf curl virus in Australia. The satellite depended on the helper virus for replication, movement and encapsidation but had no apparent effect on disease symptoms (Dry *et al* 1997).

Subsequently, similar but larger satellites (~1350 nucleotides) were found associated with other mono-partite begomoviruses (Ageratum yellow vein virus, Cotton leaf curl Multan virus) and these DNA-β molecules were found to be essential for sufficient accumulation of DNA-A and symptom development (Briddon *et al* 2001, Saunders *et al* 2000). DNAβ molecules are widespread in the old world (Briddon *et al* 2003, Bull *et al* 2004, Zhou *et al* 2003) and plants infected with these molecules usually contain another group of ssDNA satellites of similar size termed DNA1s (Briddon *et al* 2004, Saunders and Stanley 1999, Stanley 2004). DNA1s are, however, not required for disease symptom induction and reduce virus accumulation (Saunders *et al* 2001, Wu and Zhou 2005).

DNA-β satellite molecules are also probably associated in extending the host range of begomoviruses. For example, at least five diverse begomovirus species, including papaya leaf curl virus, can cause cotton leaf curl disease in Pakistan but only when associated with a particular DNA-β molecule (Mansoor *et al* 2003).

A similar function appears to be provided by an independently evolved mechanism in bi-partite begomoviruses, where the DNA-B component is usually essential for systemic symptom development. Defective DNA-B components occur commonly in some bi-partite begomoviruses and are considered to compete during replication with full-size functional components (Patil and Dasgupta 2006, Stanley *et al* 1990) (see Seal *et al* 2006).

10. Topocuvirus [(Tomato pseudo-curly top virus (TPCTV) (ss DNA)]
The genome of the type species of Topocuvirus (TPCTV) is a single component or molecule, circular, ambisense ss DNA. Each genome is 2860 nucleotides long.

The genome of this virus has the properties of both mastreviruses and begomoviruses (Briddon *et al* 1996). It contains six ORFs (two in the positive sense and four in the negative sense) producing different proteins. The virion-sense strand is close to the mastreviruses whereas the complementary strand is close to the begomoviruses.

The proteins produced by the virion-sense ORFs are: (i) 29-30 kDa coat protein, and (ii) 11-15 kDa-movement protein. Proteins produced by ORFs on the negative sense strand are: (i) 39-44 kDa replicase A, (ii) 16 kDa replicase B, (iii) 16-20 kDa similar to the replication enhancer protein of Begomoviruses, and (iv) 9-13 kDa protein that initiates cell division and symptom expression (Anonymous 2002).

(ii) Family Nanoviridae

11. Nanovirus [Fababean necrotic yellows virus (FBNYV) (ssDNA)]

Gronenborn (2004) illustrated the genome organization and protein function in nanoviruses. There are six fragments of circular ss DNA and the complete genome is 1290-3360 nucleotides long.

Multi-partite ssDNA genomes share similarities with members of the Circoviridae family that infect mammals or birds as well as with the Geminiviridae. Though the virions of the latter are unique and different from that of the Circoviruses, the mode of replication of viruses with mono-partite or multi-partite circular ssDNA genomes is strikingly similar. They multiply by rolling circle replication using virus encoded multi-functional replication initiator proteins (Rep proteins) that catalyze the initiation of ssDNA replication and resolution of replicative ss DNA into circular single stranded virion DNA. All these ss DNA viruses exploit host polymerases for virus replication.

Genomes of nanoviruses consist of multiple circular ssDNA components each of which encodes a single protein. The phylogeny of the replication proteins (Rep) of the closely related genera indicates that these small multi-gene groups have evolved by a process of duplication and subsequent loss of Rep-encoding genome components analogous to the "birth-and-death" process of evolution. By contrast repeated recombinational events between components are found to have homogenized the non-coding portions of several components encoding unrelated compounds. There is a process of concerted evolution that raises the possibility that certain non-coding regions are subject to functional constraint (Hughes 2004a, b).

Timchenko *et al* (2001) developed the "master rep" concept in replication of FBNYV and attempted to identify the missing genome components that are potential for natural genetic reassortment. Kakul *et al* (2000) studied the genome organization and variability of FBNYV. The genome consists of eight DNA components. The most protein is encoded by the 'master Rep DNA' and seven non-rep components form an integral part of the genome as found with other nanoviruses.

Aronson *et al* (2000) observed the interaction between "Clink", a 20 kDa protein of FBNYV with cell cycle regulator proteins. An LXCXE motif and an F-box of "Clink" mediate the interactions with respective regulator proteins (pRB and skP 1 homologue from *Medicago sativa*). The capacity of "Clink" to bind pRB correlates with its ability to stimulate viral replication. These regulators are constituents of the "ubiquitin"-protein pathway and appear to be a new class of viral cell cycle modulators.

Aronson *et al* (2002) assessed the protein-protein interactions in plants using the above FBNYV replication and expression system. A series of virus-based episomal vectors were constructed that were designed for transient replication and expression in plant cells and were found to be suitable for high level expression of proteins that allows isolation of recombinant proteins and protein complexes from plant tissues.

The presence of a promoter element was also found with another member of the genus subterranean clover stunt virus (SCSV). Each circular DNA component of it contains a promoter element, a single ORF and a terminator. Promoters derived from segments 4 and 7 direct high levels of expression in various plant tissues and organs. A suite of plant expression vectors derived from SCSV genome were constructed and used to produce herbicide and insect resistant cotton demonstrating their utility in the expression of foreign genes in dicotyledonous plants. Schurumann *et al* (2003a, b) observed that insertion of an "intron" between the promoter and the trans-gene ORF (analogous to rice "actin" and maize "ubiquitin" promoter system) increased trans-genic expression 50-fold. The expression patterns directed by the "intron"-modified SCSV (segments 4, 7) promoters were very similar to those directed by "actin" or "ubiquitin" types of promoters. The PPLEX vectors as described are an important counterpart to the dicot PPLEX series and have the potential to be useful in monocotyledonous plants.

Association with Other Geminiviruses

When *Nicotiana benthamiana* is coinoculated with the begomoviruses or curtoviruses and the DNA1 component, systemic movement of DNA1 occurs. This nanovirus-like component is transmissible between plants by *Circulifer tenellus*, a leafhopper. A second nanovirus-like DNA-2 component is also found with 47% nucleotide sequence identity with DNA1 but transmissible by aphids (Saunders *et al* 2002, Stanley 2004).

12. **Babuvirus** [Banana bunchy top (BBTV) (ssDNA)]

The genome of the virus consists of 6 segments of circular ss DNA. The complete genome is 13000 nucleotides. Wanit *et al* (2000) did functional analysis of proteins encoded by banana bunchy top virus DNA 4 to 6 by green fluorescence protein tagging (GFP). Protein encoded by BBTV DNA4 that possesses hydrophobic N-terminus is exclusively found in the cell periphery, while the proteins encoded by BBTV DNA3 and 6 are found in both nucleus and cytoplasm. It has been suggested that BBTV may utilize a system analogous to that of begomoviruses with the BBTV DNA6 protein acting as a nuclear shuttle protein (NSP) while DNA4 protein transports the NSP-DNA complexes to the cell periphery for intercellular transport. The protein encoded by BBTV DNA5 is found to contain an intact LXCXE motif and has retinoblastoma (Rb)-binding activity. Horser *et al* (2001) observed that BBTV DNA1 encodes the "master" replication initiation protein. This virus DNA genome comprises at least six integral components DNA1 to DNA6 that are consistently associated. One Taiwanese isolate exhibited three other components S1, S2 and Y but they are not integral components.

There may be plant protein mediated enhancement of the BBTV DNA6 promoter in embryonic cells of banana and other plants. Dudgale *et al* (2001) observed "Intron" containing fragments derived from 5′ untranslated regions of maize 'ub 1', maize 'adh 1', "rice act 1" and sugarcane 'rbcS' genes show their enhancement effects on the BBTV DNA6 promoter (BT 6.1) in banana embryonic cells. The rice "act 1" and maize "ub 1" introns provide the highest levels of promoter activity by 300-fold and 100-fold respectively and do not affect the tissue specificity of the promoter.

(c) Double-stranded (ds) Positive sense (+) RNA Containing Viruses

(i) Family: Reoviridae

Members of the Reoviridae infect vertebrates, invertebrates and plants. They all have a similar general structure and the particles are isometric. Three genera have so far been identified that infect plants. Two genera have been identified that infect insects and a fungus.

13. Fijivirus [Fiji disease virus (FDV) (ds RNA)]
Genomes of the Fiji viruses are monomeric, segmented (10 segments) ds RNAs. The genomes are 23890-28910 nucleotides long. Sizes and groupings of the genome species are characteristic and distinctive of the serogroups.

There are four members of this genus: Rice black-streaked dwarf virus (RBSDV), Maize rough dwarf virus (MRDV), Oat sterile dwarf virus (OSDV), Nilaparvata lugens virus (NLRV) and Mal de Rio Cuarto virus (MRCV). Van Regenmortel *et al* (2000) studied the sizes of the genome segments, their expression and the function of the proteins synthesized by them. According to them, the sizes of the segments, the proteins and their functions are not known with 1-5 segments of RBSDV, MRVD and OSDV. The 6th and 8th segments of RBSVD produce 2 proteins each and the 10th segment 1 protein. The protein produced by the 9th segment is not known. For MRDV the 6th, 7th, 8th and 10th segments produce 2, 1, 2 and 1 protein respectively. Protein produced by the 9th segment is unknown. The 8th and 10th segments of OSDV produce

2 proteins while 7th and 9th segments produce 1 protein each. The sizes of all the 10 segments of NLRV and the proteins produced by them are known. All the segments produce 1 protein each except the 9th that produces 2 proteins. Functions of most of these proteins have also been identified for all the viruses. The products of the segments 8, 9 and 10 are non-structural proteins. The "B" type spike on the inner core of the particle is encoded by segment 2.

McQualter *et al* (2003) analyzed the fijivirus genome segments 1 and 3. S1 comprised 4532 nucleotides and was predicted to encode a 170.6 kDa protein. S3 comprised 3623 nucleotides and was predicted to produce a 135.5 kDa protein. The predicted translation product of S3 was similar to that of S1 of RBSDV and predicted to produce viral RNA dependent RNA polymerase (RdRp). The predicted translation product of S3 was similar to that of S4 of RBSDV that encodes "B-spike" protein. McQualter *et al* (2004) made further molecular analysis of the genome segments 5, 6, 8, 10 of FDV and determined the number of nucleotides present in those segments. The number of nucleotides in S1, S6, S8 and S10 were 3150, 2831, 1959 and 1919 respectively.

Zhang *et al* (2001) characterized RBSDV. The total length of the complete genome is 29142 bp, the largest reovirus genome. Genome segment S1 encodes RNA-dependant RNA polymerase with similarities to that encoded by S1 of Nilaparvata lugens virus (NLRV). S2 and S3 appear to be homologous to S3 and S4 of both fijiviruses and NLRV. Protein encoded by S4 shows some similarity to that encoded by NLRV S2. The proteins encoded by S5 and S6, were similar to those encoded by NLRV S5 and S6, but there is no homology between them or any other known protein. Zhang *et al* (2002) analyzed the S10 of RBSDV. This segment contains 1801 nucleotides with only one ORF that encodes a polypeptide with molecular weight 63.1 kDa. Bai *et al* (2002) isolated polypeptides encoded on S10 of RBSDV that could bind specifically to the outer coat protein. They synthesized peptide1 that had a high affinity to RBSDV outer coat protein and fusion protein with glutathione S-transferase (GST) in an *E. coli* expression system. RBSDV could be detected with a high sensitivity using purified GST fusion protein. This protein functions only when it has the correct confrontation. The peptide binding specificity to it possibly disturbs the function of the virus outer coat protein and might be used to block the transmission pathway of the virus. These peptides showed high specificity and sensitivity and diagnostic potential for RBSDV; this opens up a novel strategy for the development of resistance to RBSDV. Wang *et al* (2003) and Sun *et al* (2004) made sequence analysis of the complete genome of this virus isolated from maize and obtained more or less similar properties of the segments. Each segment possesses the genus specific termini with conserved nucleotide sequences and a perfect or imperfect inverted repeat of 7 to 11 nucleotides immediately adjacent to the terminal conserved sequence. While the coding strand of most of the segments contains one ORF, there are two non-overlapping ORFs in S7 and S9, and one small overlapping ORF downstream of the major ORF in S5. Sun *et al*

(2004) made sequence analysis and prokaryotic expression of genomic segments 8 and 9 of this virus isolated from maize. S8 and S9 consist of 1936 and 1900 bp respectively. Single ORF of S8 encodes a 68kDa polypeptide containing 591 amino acids. S9 contains 2 non-overlapping ORFs encoding 347 (ORF1) and 209 (ORF2) amino acids respectively.

Distefanco *et al* (2003) made sequence and phylogenetic analysis of Mal de Rio Cuarto virus and confirmed it as a member of the Fijivirus genus.

14. Oryzavirus [Rice ragged stunt virus (RRSV) (ds RNA)]

Genomes of all the members have 10 segments. The complete genome is 26640-26660 nucleotides long. The respective lengths of the different segments of the RNA are: 3850-3900 (RNA1), 3820-3900 (RNA2), 3700-3800 (RNA3), 2700-2750 (RNA5), 2300-2600 (RNA6), 1900 (RNA8), 1200 (RNA9) and 1160 (RNA10) nucleotides (ICTVdB Management 2006).

Van Regenmortel *et al* (2000) studied the genome, sizes of the segments, their expression, sizes of the proteins produced and their functions with the type member Rice ragged stunt virus (RRSV). RNA polymerase is the largest gene product and is produced by the 4th segment of the genome.

Shao *et al* (2003) studied ectopic expression of the spike protein of RRSV in transgenic rice plants inhibiting the transmission of the virus to the insects. RRSV has a complex multi-component particle bearing spikes on its outer surface. Transgenic rice lines expressing 39 kDa spike protein show good resistance to infection by RRSV. Nilaparvata lugens fed on these plants, prior to feeding on RRSV-infected plants, is significantly protected against RRSV infection. The viral titer in insects initially fed on transgenic plants is inversely proportional to the levels of the 39 kDa protein expressed in transgenic plants. It appears that 39 kDa protein interferes with the interaction between the intact virus particles and insect cell receptors and the spike protein may contribute to the vector specificity.

15. Phytoreovirus [Rice dwarf virus (RDV) (ds RNA)]

Hillman and Nuss (1999) reviewed the organization of the genome of this virus. The genome is monometric and divided into 12 segments. The length of the complete genome is 25130-25270 nucleotides. The lengths of the segments are: 4131-4423 (RNA1), 3385-3532 (RNA2), 3195-3205 (RNA3), 2462-2565 (RNA4), 2470-2613 (RNA5) and 1699-1726 (RNA6) nucleotides (ICTVdB Management 2006).

Van Regenmortel *et al* (2000) studied the genomic properties of the type member Rice dwarf virus (RDV). They determined the sizes of the segments, proteins they produce and their function. Each segment has one ORF expressing one protein. Segments 2, 3 and 8 encode structural proteins. The product of segment 2 is also required for vector transmission. Non-structural proteins may be the products of segments 4, 6, 9, 11 and 12.

Encoded proteins with known functions are: P2—essential for virus transmission; P3—inner shell and capsid protein; P5—guanyltransferase; P7— nucleic acid binding protein; P8—major outer capsid protein; P11—nucleic acid binding protein. Li *et al* (2004) found that S6 encoded non-structural protein that has a cell-to-cell movement function.

S10 of rice dwarf virus encodes a protein Pns10 that exhibit RNA silencing suppressor (RSS) activity. This protein targets an upstream step of dsRNA formation in the RNA silencing pathway (Cao *et al* 2005).

Insect and Fungus Infecting Reoviruses

Nilaparvata lugens virus (NLRV) infecting the brown plant hopper *Nilaparvata lugens* is a Fijivirus. Other insects infecting Reoviruses belong to the genera Coltivirus and Cypovirus. Some of the members are: *Dendrolimus punctatus* cytoplasmic polyhydrosis (Zhao *et al* 2003), Colorado tick fever and European Eyach viruses (Wei *et al* 2004), *Bombyx mori* cytoplasmic polyhedrosis virus (Ikeda *et al* 2001) and a fungal virus Rosellinia anti-rot (Wei *et al* 2004).

Comparative sequence analyses of nucleotides and amino acids of these viruses, and those of other reoviruses infecting plants, suggest the origin of plant infecting reoviruses from those infecting insects or that both groups had a common ancestor (Rao *et al* 2003, Wei *et al* 2004, Ikeda *et al* 2001).

(ii)　Family: Partitiviridae

Milne and Marzachi (1999) reviewed the organization of the genomes of the viruses belonging to this family. Two viruses, Alphacryptovirus and Betacryptovirus have so far been identified belonging to this family. Genomes of these viruses are multi-partite. It is believed that larger segments of both the viruses encode viral-associated RNA polymerase and the smaller segments encode coat protein.

16.　Alfacryptovirus [White clover cryptic virus 1, 3 (WCCV-1, WCCV-3) (ds RNA)]

Boccardo and Candresse (2005a, b) analyzed the complete sequences of the RNA2 and RNA1 of white clover alfacryptivirus, type species of the genus. Full-length sequences of the RNA2 and RNA1 were determined from overlapping cDNA clones and shown to be 1708 nucleotides and 1955 nucleotides long, excluding the poly (A) tail found at the 3' end of the coding strand. The non-coding region (NCR) is 104 nucleotides long. The molecule harbours a single ORF ending with a UGA stop codon at positions 1566-1568 in RNA2 and 1923-1925 in RNA1, while the 3' NCR in RNA2 is 140 nucleotides long and that in RNA1 is only 30 nucleotides long. The encoded protein by RNA2 was of 487 amino acids with a molecular weight of 54.2 kDa whereas that by RNA1 was 616 amino acids with a molecular weight of 72.9 kDa. The protein encoded by RNA2 was identified as coat protein.

17. Betacryptovirus [White clover cryptic virus-2 (WCCV-2) (ds RNA)]

The genome of this virus has 3 segments. The complete length of the genome is 4310-5150 nucleotides and the lengths of the RNAs are: 2050-2550 (RNA1), 1650-2040 (RNA2) and 450 nucleotides (ICTVdB Management 2006).

18. Endomavirus [Oryza sativa endoma virus (OSV) (ds RNA)]

The genome organization and their expression are yet to be adequately understood.

(d) Single-stranded Rod/Bullet/Filamentous RNA Viruses

(i) Family: Flexiviridae

19. Allexivirus [Shallot virus X (ShVX) (ss RNA+)]

The genome of the virus is mono-partite, linear and the approximate length of the genome is 8000-9000 nucleotides. Genes are translated from a total of 6 ORFs.

Kanyuka *et al* (1992) observed that the length of the genome of ShVX was 8890 nucleotides and the sequence contained 6 ORFs that encoded putative proteins (in the 5' to 3' direction) of Mr 194528 (ORF1), 26333 (ORF2), 11245 (ORF3), 42209 (ORF4), 28486 (ORF5) and 14741 (ORF6). The ORF1 protein was highly homologous to the putative potexvirus RNA replicases, ORF3, -5 and -6 proteins also had analogues among the potex- and/or carlavirus-encoded proteins. ORF3 was followed by an AUG-lacking frame coding for an amino acid sequence homologous to that of the 7 K to 8 K proteins of the triple gene block of the above viruses. The putative ORF4 protein did not have any reliable homology with proteins in the database. Except for the unique 42 protein gene, the ShXV genome was found to combine a number of elements typical of both carla- and potexviruses.

Adams *et al* (2004) assessed the molecular criteria for the species demarcation of this genus and Chen *et al* (2004) emphasized the significance of differences in sequences to distinguish various species of this genus.

20. Capillovirus [Apple stem grooving virus (ASGV) (ss RNA+)]

The complete genome is 6500-7570 nucleotides long (ICTVdB Management 2006). There may be sequence variants that differ considerably from each other in nucleotide sequence (Magome *et al* 1999). The genome is mono-partite, linear and the 3' terminus has a Poly (A) tail. Viral RNA is translated into a poly-protein (240-270 kDa) that is subsequently processed with functional products with coat protein at the C-terminus. A separate protease, or movement protein, may also be translated directly form a different reading frame.

Yoshikawa *et al* (1992) determined the complete nucleotide sequences of ASGV. The length of the genome was 6496 nucleotides excluding a 3′-terminal Poly (A) tail and contained two overlapping ORFs. ORF1 began at nucleotide position 37 and was terminated at position 6341 encoding a protein with a molecular weight 241 kDa. ORF2 that was in a different reading frame within ORF1 began at position 4788 and could encode a 36 kDa protein. The 241 kDa protein contained two consensus sequences associated with RNA-dependent RNA polymerase and the NTP-binding helicase. The coat protein was located in the C-terminal region of the 241 kDa-polyprotein. Shim HyeKyong *et al* (2004) made more or less similar observations with a Korean isolate of ASGV. As regards the symptom expression Hirata *et al* (2003) observed that a single silent substitution in the genome causes symptom attenuation.

21.   Carlavirus [Carnation latent virus (CLV) (ss RNA+)]

The genome consists of one molecule of linear ss RNA and the genomic length is 6480-8535 nucleotides. The 5′ terminus end sometimes has a cap or monophosphate. The 3′ end has a Poly (A) tail. Encapsidated nucleic acid is solely genomic or both genomic and sub-genomic (in shorter particles).

Lukhovitskaya *et al* (2005) observed that the genome codes for a small cysteine-rich protein (CRP) with unknown function. This protein neither acts as an avirulence factor nor suppresses post-transcriptional gene silencing. The same protein (CRP) encoded by the genome of the Chrysanthemum virus B (CVB) was identified as a virus pathogenecity determinant that controls the interaction of the virus with the host plant in a way that depends on plant defence mediated by resistance genes such as N gene.

22.   Foveavirus [Apple stem pitting virus (ASPV) (ss RNA+)]

Martelli and Jelkmann (1998) established this genus which was typified by ASPV. The size of the genome was 8.4 to 9.3 kb and the size of the single type of the coat protein was 28 to 44 kDa. The presence of 5 ORFs was confirmed. These are replication-related proteins (ORF1), the putative movement proteins (ORFs2-4 constituting the triple block gene) and the coat protein (ORF5). The replication strategy was based on direct expression of the 5′-proximal ORF expression of downstream ORFs through subgenomic RNAs. The ORF1 the coat protein cistron are significantly larger than that of Potex-, Carla-, Allexiviruses. Yoshikawa *et al* (2001) observed sequence variants in 3′ terminal regions of different isolates of the virus.

23.   Mandarivirus [Indian citrus ring spot virus (ICRSV) (ss RNA+)]

The sequence of the genome of the Indian citrus ringspot virus (ICRSV) consists of 7560 nucleotides. It contains 6 ORFs that encode putative proteins of 187.3, 25, 12, 6.4, 34 and 23 kDa respectively. ORF1 encodes a polypeptide that contains all the elements of a replicase; ORF2, 3 and 4 compose a triple gene-block; ORF5 encodes the capsid protein; the function of the ORF6 is unknown. Phylogenetic analysis of the complete genome and each ORF separately and database researches (particularly the presence of ORF6, the genome and CP sizes, and particle morphology) indicate it as a separate genus (Rustici *et al* 2002). Shelly *et al* (2003) showed that ICRSV and cirtrus psorosis are completely different from one another.

24. Potexvirus [Potato virus X (PVX) (ss RNA+)]

Genomic RNA of a typical Potexvirus (white clover mosaic virus) contains 5835 nucleotides with a methylated nucleotide cap at the 5′ end and Poly (A) at the 3′ end. Genome expression takes place through 5 open reading frames (ORF1-147 kDa, ORF2-26 kDa, ORF3-13 kDa, ORF4-7 kDa and ORF5-21 kDa) and following only sub-genomic mRNA strategy producing 4-5 proteins for different functions.

The mRNA has ORF5 (21 kDa) that translates to 1 protein. There is also another large sub-genomic mRNA. Five genes translate to operate five functions of which two have been characterized (polymerase-147 kDa, and coat protein-21 kDa) (Sit *et al* 1994).

Two stem-loop structures were identified within the 5′ 230 nucleotides. Mutations in these 2 structures (SL1, nucleotides 32-196; and SL2, nucleotides 143-183) generally reduced genomic plus-strand RNA and coat protein accumulation in tobacco protoplasts. There are a number of feures of SL1 that are critical for PVX plus-strand accumulation (Miller *et al* 1998). Kwon *et al* (2005) observed that RNA secondary structural elements within SL1 and/or sequences therein are crucial for the formation of VLPs and are required for specific recognition by the CP subunit.

The intercellular movement of the virus 12 kDa, 8 kDa, and coat proteins (CP) is essential for cell-to-cell movement. Mutations of these proteins within the genome inhibited viral intercellular movement. The plasmodesmata gating may not be an essential function of these proteins for viral cell-to-cell movement (Krishnamurthy *et al* 2002). Mitra *et al* (2003) observed that ER (endoplasmic reticulum) association of TGBp2 protein of PVX may be required but not be sufficient for virus movement TGBp2 is likely to provide an activity for PVX movement beyond ER association. Serazev *et al* (2003) observed that PVX virions interact with MTs (microtubules) in vivo and that, consequently, cytoskeleton elements participate in the intracellular compartmentalization of the PVX genome quences and by viral and/or host proteins. RNA sequence element as well as a stable RNA stem-loop structure in the 5′ end of the genome affects accumulation of genomic RNA and subgenomic RNA (sgRNA). The putative sgRNA promoter regions upstream of the PVX triple gene-block (TB) and coat protein (CP) gene were critical for both TB and CP sgRNA accumulation. 5′ end of the positive-strand RNA formed an RNA-protein complex with cellular proteins suggesting possible involvement of cellular proteins for PVX replication (Kim and Kim 2001). Kagowada *et al* (2005) observed that a single amino acid residue of RNA-dependent RNA-polymerase in the PVX genome determines the symptoms in Nicotiana plants (Kwon SunJung *et al* 2005).

25. Trichovirus [Apple chlorotic leaf spot virus (ACLSV) (ss RNA+)]

Martelli *et al* (1994) first identified Trichovirus as a separate genus and described the genome as 3' polyadelynated ssRNA of 7.5 to 8.7 kb in size. The genome is mono-partite and linear. The complete genome length is 7500-8200 nucleotides. The 5' end has a methylated cap and 3' end has a poly (A) tail. Virus codes 5-7 ORFs (ICTVdB Management 2006). The ORFs may overlap. The largest ORF (ORF1) is probably involved in replication while ORF2 (putative movement protein) and ORF3 are expressed from sub-genomic mRNAs. The type member produces a poly-protein (216.5 kDa-putative replicase); a 50.4 kDa movement protein and 21.4 kDa coat protein. Five virus-specified ds RNA species is found in infected cells.

The virion is associated with a helper virus, but independent from its function during replication.

Yoshikawa *et al* (1999) observed that 50 kDa movement protein was localized on plasmodesmata and near cytoplasm in ACLSV-infected *Chenopodium quinoa* leaves. Yoshikawa *et al* (2000) observed increased susceptibility to ACLSV homologues but strong resistance to Grapevine berry inner necrosis virus (GINV), a strong heterologous virus when 50 kDa movement protein was expressed in transgenic *Nicotiana occidentalis* plants.

26. Vitivirus [Grapevine virus A (GVA) (ss RNA+)]

Martelli *et al* (1997) identified Vitivirus as a separate genus. The genome is mono-partite and linear ssRNA. The length of the genome is 7400-7600 nucleotides; the 5' terminus has a methylated cap and the 3' has a Poly (A) tail.

The genome is translated by five overlapping ORFs. The largest ORF (ORF1) is probably involved in replication. ORF3 is a putative movement protein. ORF4 is the coat protein and the ORF5 is a putative RNA binding protein.

The following proteins are produced by the 5 ORFs in the type member of the genus: (i) 195 kDa replication protein, (ii) 19 kDa protein of unknown function, (iii) 31 kDa movement protein, (iv) 22 kDa coat protein and (v) 10 kDa RNA binding protein . Dell'Orco *et al* (2002) observed that GVA particles carry a highly structured epitope centred on a common peptide region of the coat protein sequences.

Gali *et al* (2003a) studied the 10 kDa RNA-binding protein encoded by ORF5. It carries two distinct domains: a basic arginine-rich domain and a zinc-finger domain. This protein interacts with nucleic acid and shows the nucleic acid binding capabilities, but the zinc-finger domain did not affect binding. ORF mutations do not affect replication of GVA-RNA, but affect symptom development and markedly reduced expression of the movement protein (the product of ORF3).

Gali *et al* (2003b) found that two classes of sub-genomic RNA of GVA are produced by internal controller elements. The production of viral RNAs in GVA-infected *Nicotiana benthamiana* was characterized as a nested set of 3 5′-terminal sg RNAs of 5.1, 5.5 and 6.0 kb and another of 3′-terminal sg RNAs of 2.2, 1.8 and 1.0 kb that could serve for expression of ORF 2-3 respectively. Neither 3′ nor 3′-terminal sgRNAs that would correspond to ORFs, was detected, suggesting that a mechanism other than specific cleavage was involved in production of these sgRNAs. Apparently the production of the 5′ and 3′-terminal sgRNAs was controlled by sequence upstream of the 5′ terminus of each of the ORFs. Detection of plus and minus strands of the 5′ and 3′ terminal sg RNA, though at different levels suggested that each of these cis-acting elements is involved in the production of four RNAs: a 3′ terminal plus strand sg RNA which could act as an mRNA and the corresponding 3′ terminal minus strand RNA, a 5′ terminal plus strand sgRNA and the corresponding 5′ terminal minus strand RNA.

Galiakparov *et al* (2003a, b, c) attempted to define the roles of the various genes of GVA by inserting mutations in every ORF and studying the effect on viral replication, gene expression, symptoms and viral movement. Mutation in ORF1 abolished RNA replication but mutations in ORF2 did not affect any of the aforementioned parameters. Mutations in ORFs3 and 4 restricted viral movement. Mutations in ORF5 rendered the virus asymptomatic and partially restricted its movement.

(ii) Family: Closteroviridae

27. Closterovirus [Beet yellows virus (BYV) (ss RNA+)]

Dolja *et al* (1994), Bar-Joseph *et al* (1997) and German-Retana *et al* (1999) reviewed the organization of the genome of the closteroviruses. The size of the genomes of the members differs: genome size of BYV is 15.5 kb, CTV is 19.3 kb. The length of the genome may be up to 20,000 nucleotides. There may be a methylated cap at the 5′ end. The genome of the type member (BYV) consists of 15500 nucleotides with 9 ORFs. The genome of Citrus tristeza virus (CTV) on the other hand has 12 ORFs. These ORFs express themselves in four types of functions: replication, cell-to-cell movement, virion assembly and a fourth function yet to be determined. Sub-genomic mRNA and 1-6 virus-specified dsRNAs are found in infected cells. The virus is associated with a helper virus but independent from its function during replication. It may also act as a helper to other viruses. Ghorbel *et al* (2001) found a 23 kDa protein (P23) coded by the 3′ terminal gene of CTV is an RNA binding protein. It contains motif-rich cysteine and hystidine residues in the core of a putative zinc-finger domain.

It performs a regulatory role for CTV replication or gene expression and also performs a key role in pathogenesis.

28. Crinivirus [Lettuce infectious yellows (LIYV) (ss RNA+)]

The genome of the virus is bi-partite and of linear ssRNA. The length of the genome is 15,000-20,000 nucleotides. The 5′ terminus may have a methylated cap. There are sequences with potential hairpin structure near the 3′ end. Klaassan *et al* (1995) sequenced the genome of the type member of this virus (LIYV). It consists of two different RNA species.

Livicratos and Coutts (2002, 2003) conducted nucleotide sequence and phylogenetic analysis of cucurbit yellow stunting disorder virus (a crinivirus) RNA2 and RNA1. Its length was 7281 nucleotides and contained the closterovirus hallmark gene array with a similar arrangement to the prototype member of the genus. RNA2 contained ORFs potentialy encoding in a 5′ to 3′ direction of proteins of 5 kDa (ORF1: hydrophobic protein), 62 kDa (ORF2: heat shock protein 70 homologue HSP 70), 59 kDa (ORF3, protein of unknown function), 9 kDa (ORF4: protein of unknown function), 28.5 kDa (ORF5: coat protein P), 53 kDa (ORF6: coat protein minor) and 26.5 kDa (ORF7: protein of unknown function). RNA1 was 9126 nucleotides long and contained two overlapping ORFs that encode the replication module, consisting of the putative papain-like cysteine proteinase methyl transferase, helicase and polymerase domains, a small 5 kDa hydrophobic protein and two further downstream ORFs potentially encoding proteins respectively 25 and 22 kDa. The genomic position and homology of the four domains comprising the replication module appear to be similar for all sequenced criniviruses but there is divergence in the downstream ORFs in terms of number, size, position, and sequence homology.

Aguilar *et al* (2003) studied the sequence differences in the genomic organization of CYSDV, SPCSV, CUYV and LIYV. The most striking and novel features of CYSDV were a unique gene arrangement in the 3′ terminal region of RNA1, the identification in this region of an ORF potentially encoding a protein which has no homologues in any databases and the prediction of an unusually long 5′ non-coding region in RNA2. Additionally, several sub-genomic RNAs (sgRNAs) are found in CYSDV-infected plants. Generation of sgRNAs may be a strategy used for the expression of internal ORFs. Kreuze *et al* (2002) detected several sg RNAs in SPCSV-infected plants and obtained indication that the sgRNAs formed from RNA1 accumulated earlier than those of RNA2. The 5′ ends of seven sgRNAs were cloned and sequenced, provided compelling evidence that the sgRNAs are capped in infected plants, a novel finding for members of the closteroviridae.

29. Ampelovirus [Grapevine leaf roll virus-3 (GLRaV-3) (ss RNA+)]

Martelli *et al* (2002) separated closteroviridae into three genera: (i) Closterovirus, (ii) Ampelovirus and (iii) Crinivirus but little work has been done on

Ampelovirus. The genome of this genus is mono-partite, 16.9-19.5 kb in size and contains a 35-37 Da major CP. A CPd cistron is located downstream of the CP gene. Mealy bugs transmit this virus.

(iii) Family: Potyviridae

30. Potyvirus [Potato virus Y (PVY) (ss RNA+)]

The genome contains one molecule of ssRNA. The total genome length is 9000 to 12000 nucleotides. The 5′ end has a genome-linked protein (VPg) and the 3′ end has a Poly (A) tract or tail (ICTVdB Management 2006). There are three or five structural proteins (28300-54100 Da) and one or three or four or five or seven non-structural proteins (39000-81000 Da). Pinwheel inclusion bodies are found in infected cells.

Genomic RNA of a typical Potyvirus, Tobacco etch virus (TEV) contains 9704 nucleotides with VPg (24kDa) at the 5′ end and Poly (A) tail at the 3′ end. Genome expression takes place only through one ORF (ORF1 of 9161 nucleotides—346 kDa) and following only "Poly-protein" strategy. This ORF translates to a poly-protein. This poly-protein is cleaved at 5 different sites by proteinase to produce 4 sub-poly-proteins and two proteins (6 kDa, 49 kDa). One sub-poly-protein is cleaved at 3 sites to produce 3 proteins of which 2 have been characterized (31 kDa, 56 kDa). The second and third sub-poly-proteins are cleaved at 2 sites respectively and produce 4 proteins (50 kDa, 70 kDa and 58 kDa and 30 kDa respectively). The sub-poly-proteins are cleaved following cis- or trans-processing. Functions of these proteins are associated with cell-to-cell movement (31 kDa), insect transmission (56 kDa), proteinase (50 k, 49 kDa), replication (70 kDa), polymerase (50 kDa), and coat protein (30 kDa). In addition, several other proteins also accumulate from the poly-protein, such as cytoplasmic pinwheel inclusion (CI) protein (70 kDa), cytoplasmic amorphous inclusion protein, nuclear amorphous inclusion protein (NI) and helper protein (HC). López-Moya and Gracia (1999) reviewed the genome organization of this virus in detail.

Anindya *et al* (2004) and Anindya and Savithri (2003) studied the genome organization and particle assembly of pepper vein banding virus. The viral genome contained 9711 nucleotides excluding the Poly (A) tail and the lengths of the 5′ and 3′ untranslated regions (UTR) were 163 and 281 nucleotides respectively. The genome has a single open reading frame (ORF) starting at nucleotide 164 and ending at nucleotide 9430 that encodes a poly-protein of 3088 amino acid residues. There are nine putative conserved cleavage sites within the poly-protein that can result in ten functionally distinct protein products. Disassembly and reassembly of PVBV proceed via ring-like intermediate and electrostatic interactions that may be pivotal in stabilizing the particle. Saenz *et al* (2002) observed that long-distance movement and pathogenicity enhancement are related activities of the potyviral HC proteins and some components of the long-distance machinery of the potyviruses are not strictly virus-specific. The HC protein might be a relevant factor for controlling the host range of the potyviruses.

31. Ipomovirus [Sweet potato mild mottle virus (SPMMV) (ss RNA+)]

The genome is mono-partite and linear ssRNA and its length is about 11,000 nucleotides. The 5' end has a genome- linked protein (VPg) and the 3' end has a Poly (A) tail. The genome is translated into a single poly-protein (390 kDa) subsequently processed into 8, probably 10 products with coat protein at the C-terminus and produced by proteolytic processing. The ten products of the poly-proteins are: (i) P1—putative protease, 83 kDa; (ii) HC-Pro—putative helper component protease, 51 kDa; (iii) P3—unknown function, 34 kDa; (iv) 6 K1—unknown function, 5 kDa; (v) C1—cylindrical inclusion putative helicase, 71 kDa; (vi) 6K2—unknown function, 6 kDa; (vii) VPg—genome-linked protein, 20 kDa; (viii) N1a Pro—nuclear inclusion putative protease, 27 kDa; (ix) N1b—nuclear inclusion putative polymerase 57 kDa and (x) CP—coat protein (Anonymous 2002).

The low CP amino acid sequence similarity of SPMMV isolates with other characterized viruses of the family Potyviridae and the unusual putative proteolytic cleavage site at the Nib/CP junction demonstrate SPMMV as a very distinct virus in the family (Mukusa *et al* 2003).

32. Tritimovirus [Wheat streak mosaic (WSMV) (ss RNA+)]

The genome is mono-partite, linear ss RNA and 9000-10000 nucleotides long. The 5' terminus has a genome-linked protein (VPg) and the 3' end has a Poly (A) tract (tail). It is very close to the Rymoviruses. The genome is translated into a single poly-protein (344 kDa) from which an estimated 10 functional products are produced with the coat protein at the C-terminus. The estimated products are: (i) P1—putative protease, 40 kDa; (ii) Hc-Pro—putative helper component proteases, 44 kDa; (iii) P3—of unknown function, 32 kDa; (iv) 6 K1—of unknown function, 6 kDa; (v) C1—Cylindrical inclusion putative helicase, 73 kDa; (vi) 6 K2—of unknown function, 6 kDa; (vii) VPg—genome-linked protein, 23 kDa; (viii) N1a—nuclear inclusion putative protease, 26 kDa; (ix) N1b—nuclear inclusion putative polymerase and (x) CP—coat protein, 37 kDa.

Rabenstein *et al* (2002) determined phylogenetic relationships, strain diversity and biogeography of Tritimoviruses. Choi *et al* (2002) mapped the P1 proteinase cleavage site in the poly-protein of WSMV. Three virus-encoded proteinases cleaved the viral poly-protein into mature proteins. The amino-terminal region of the viral poly-protein is autocatalytically cleaved by the P1 proteinase and the cleavage site was mapped to a position downstream of amino acid residue 348 and upstream of amino acid residue 353 with the peptide bond between amino acid residues Y352 and G353 the most probable site of hydrolysis. An alignment of potyvirus poly-protein sequences in the C-terminal region of the P1 domain revealed WSMV P1 contained conserved H257, D2267, S303 and FVXG 325-329 residues upstream of the cleavage site that are typical of serine proteinases and shown by others to be required for P1 proteolysis in tobacco etch virus. Insertion of the glucuronidase (GUS) reporter gene immediately downstream of the P1 cleavage site in a full- length clone of WSMV resulted in a systemic infection and GUS expression upon inoculation of plants with *in vitro* transcripts. When cleaved by P1 at the amino acid terminus and Nic proteinase at a site engineered in the carboxy-terminus, active GUS protein expressed by WSMV in infected wheat had electrophoretic mobility similar to wild-type GUS protein.

Stenger and French (2004a, b) made a complete nucleotide sequence of oat necrotic mottle virus (ONMV) and confirmed it as a Tritimovirus and not a Rymovirus. The genome encoded a single poly-protein with proteinase cleavage sites very similar to those of WSMV. Pair-wise comparison of ONMV and WSMV cistrons revealed that P3 was most conserved whereas HC-Pro was most divergent.

33. Bymovirus [Barley mild mosaic virus (BaMMV) (ss RNA+)]
The genome is bi-partite, linear ss RNA with a genome length of 8500-10000 nucleotides. The 5′ end has a methylated cap and the 3′ end has a Poly (A) tail. There may be one or three structural protein (30000-36000 Da). There is one non-structural protein (69000-72000 Da). Virions are associated with a helper virus but independent from its function during replication. Kashiwazaki and Hibino (1996) sequenced the genome of barley mild mosaic bymovirus (BaMMV), the type member of the genus. The genome is divided into two segments RNA1 consists of 7632 nucleotides and the RNA2 consists of 3585 nucleotides. The expression of the genome follows Poly-protein strategy and RNA1 is involved in pathogenesis.

Peerenboom *et al* (1997) found a large duplication in the 3′-untranslated region of a subpopulation of RNA2 of the UK-M isolate that cannot be transmitted by fungi and shows a 1092 nucleotide deletion in the coding region. This subpopulation contains a direct imperfect sequence repeat of 552 nucleotides in the 3'-untranslated region.

Subr *et al* (2000) found that the genome of BaMMV (Aschersleben isolate) consists of 7263 nucleotides excluding the 3′-poly (A) tail. The 5′ and 3′ non-translated regions (NTR) are 148 and 338 nucleotides in length respectively and flank of a single large ORF coding for a precursor polypeptide with a calculated molecular mass of 256 kDa. Sequence comparison revealed a 96% amino acid (aa) identity to RNA1 translation products of Japanese and French BaMMV isolates. Conserved nucleotide motifs in the 3′-sense and 5′-complementary sense NTR of the two genomic RNAs were identified that may represent the polymerase recognition sites.

Substantial work has been done on two other members of the Bymovirus, Barley yellow mosaic virus (BYMV), wheat yellow mosaic virus (WYMV) and wheat spindle streak mosaic virus (WSSMV). Cheng *et al* (1999) made molecular analysis of barley yellow mosaic virus isolates from China. The complete sequences of both the RNAs resembled those found in the isolates from Japan and Germany. The greatest difference between the Japanese and Chinese isolates were in the 5'-UTRs of RNA1 and 2 and there were differences in P1 (RNA2) and P3, CI, N1a and the 5'-end of the coat protein (CP) (RNA1). Collecting the isolates from different geographical regions it appears that P2 fragment (RNA2) was more variable than the CP and phylogenetic analysis of both regions showed that Asian and European isolates formed distinct clusters.

Kuhne *et al* (2003) found that European pathotype 2 isolate of barley yellow mosaic virus (BaYMV-2) infects barley genotypes with the rym4 resistance gene whereas BaYMV-1 cannot. When examined at the molecular level, consistent amino acid differences in the RNA1-encoded poly-proteins were found between the two pathotypes. BaYMV-1 had lysine at aa 1307, whereas BaYMV-2 had arginine (histidine in some isolates). The polymorphism occurred in the central region of VPg, which has been shown to be required for pathogenicity of genotypes carrying recessive resistance genes in several Potyvirus/dicotyledonous plant pathosystems.

Chen *et al* (2000a, b) also studied the sequence diversity in the coat protein-coding region of wheat yellow mosaic virus isolates from China. Coat protein coding region of 879 nucleotides consensus amino acid sequences for all bymoviruses was constructed. The 3'-UTR sequences had nucleotides from 255 to 259.

34. Rymovirus [Ryegrass mosaic virus (RGMV) (ss RNA+)]

The genome is mono-partite, linear ss RNA. The length of the genome is 8500-10000 nucleotides. 5' end of the genome has a genome-linked protein (VPg) and the 3' end has a Poly (A) tail. The genome encodes a single poly-protein with proteinase cleavage sites demarcating protein products (French and Stenger 2005). The poly-protein subsequently processed into 8 to 10 products with the coat protein at the C-terminus. The 10 functional products of the poly-protein (348 kDa) produced by proteolytic processing are: (i) P1—putative protease, 29 kDa; (ii) HC-Pro—putative helper component protease, 52 kDa; (iii) P3— unknown function, 40 kDa; (iv) 6 K1—unknown function, 6 kDa; (v) C1—cylindrical inclusion putative helicase, 71 kDa; (vi) 6 K2—unknown function, 6 kDa; (vii) VPg—genome-linked protein, 22 kDa; (viii) N1a—nuclear inclusion putative protease, 27 kDa; (ix) N1b—nuclear inclusion putative protease, 50 kDa and (x) Coat protein—44 kDa.

(iv) Family: Rhabdoviridae

35. Cytorhabdovirus [Lettuce necrotic yellows (LNYV) (ss RNA)]

The virion contains one molecule of linear, negative sense ss RNAwith a genome length of 10500-14000 nucleotides with reverse complementary extremities. It is normally transcribed into six messenger RNAs by a viral transcriptase associated with the virion. These messengers are generally monocistronic, and their abundance decreases from the 3′ end to 5′ end. The 5′ end is transcribed and not translated. Encapsidated nucleic acid is genomic but may be associated with host nucleic acid like tRNA Subgenomic RNA may be present in the infected cell.

The virion is associated with a helper virus but independent from its function during replication (ICTVdB Management 2006).

Schoen *et al* (2004) made complete genomic sequence of strawberry crinkle virus. The total length of the genome was 14547 nucleotides. There were 7 ORFs on the viral complementary sequence. Four putative SCV proteins were postulated including nucleocapsid (N), matrix protein (M), glycoprotein (G) and polymerase (L). A series of two ORF3 (genes P2 and P3) were present between the G and L sequences. Klerks *et al* (2004) made partial sequences of the putative polymerase (L) protein that are common to other members of the genus.

36. Nucleorhabdovirus (Potato yellow dwarf (PYDV) (ss RNA–)]

The genome contains a single molecule of linear, negative sense ss RNA of length 7000-14000 nucleotides. Genome nucleic acid is not infectious. Encapsidated nucleic acid is solely genomic. Sub-genomic RNA may be present or absent. The virion is associated with a helper virus, but independent from its function during replication. The viral proteins synthesized transit through the nucleus, where the nucleocapsid is assembled. The mature virions accumulate in the space between the two nuclear membranes by budding through the internal nuclear membrane.

(v) No Family

37. Benyvirus [Beet necrotic yellow vein virus (BNYVV) (ss RNA+)]

The genome is linear and segmented. There may be 4-5 segments. RNA is polyadenylated and there is a triple gene-block on RNA 2. The 3′ end of all RNAs has a Poly (A) tract (tail). There is a single ORF on RNA1 while the coat protein, followed by a read-through domain which is at the 5′ end of RNA 2 followed by four other ORFs including the triple gene-block. RNA 3 has 1-2 ORFs and there is one each on RNA4 and RNA5 (not found in all isolates). The smaller RNAs (3-5) of BNYVV can be lost during mechanical transmission. There is a short conserved region at the 5′ end and a very long conserved sequence at the 3′ end.

The function of the genomes on these RNAs is not known for certain but RNA3 affects symptom production and RNA4 is necessary for efficient fungus transmission. RNA3 and RNA4 are not required for infection. RNA3 is associated with rhizomania and systemic movement in the beet and spinach species tested. RNA4 is associated with efficiency of transmission by fungal vector (Anonymous 2002). There may be wide variability of RNA3 in nature. During the infection process of BNYVV virus particles localize to mitochondria (Erhardt *et al* 2001).

Geographical distribution of BNYVV is very wide and there are several genotypic strains of this virus. Strains with a fifth RNA appears to be more aggressive (Rush 2003).

38. Furovirus [Soil-borne wheat mosaic virus (SBWMV) (ss RNA+)]

The genome is a linear ssRNA with two, four or five segments. The length of the genome is 9400-13500 nucleotides. Respective lengths of the RNAs in a 5 segmented genome are: (i) RNA 1—5900-7342 nucleotides; (ii) RNA 2-3000—7342 nucleotides; (iii) RNA3-2100-2400—nucleotides; (iv) RNA 4—not known; (v) RNA 5—1.45 nucleotides. There is a methylated cap at the 5′ end and there may or may not be any Poly (A) tract at the 3′ end (ICTVdB Management 2006). RNA has three ORFs. Products (3) for replication are encoded on RNA1. The coat proteins are at 5′ end of the RNA 2. The products of the RNA 1 are: (i) 150 kDa replication protein containing methylesterase and helicase motifs; (ii) 209 kDa polymerase read through of P1, incorporating the RNA polymerase motif and 37 kDa movement (cell-to-cell transport) protein. The products of RNA2 are: (i) 19 kDa coat protein (ii) 84 kDa coat protein and (iii) 19 kDa cysteine-rich protein (Anonymous 2002).

Yamamiva *et al* (2000) constructed full-length cDNA clones of RNA1 and RNA2, pjS1 and pjS2 respectively, from which infectious RNA could be transcribed in vitro by SP6 RNA polymerase. Reassortment between genetically distinct isolates may produce pseudorecombinant virus (Miyanishi *et al* 2002).

Chen and Chen (2000) observed that repeated mechanical passaging of SBWMV or growth at 25-30°C results in rapid deletion of part of RNA2. The stable mutant was deleted between nucleotides 1420 and 2180, resulting in the loss of 759 nucleotides (253 amino acids) from the CP-RT gene. Spontaneous stable deletion mutant is associated with increased symptom severity.

39. Hordeivirus [Barley stripe mosaic virus (BSMV) (ss RNA+)]

The genome is a linear ss RNA with three or four segments and the length of the genome 9000-15 6000 nucleotides. The 5′ end has a methylated cap and the 3′ end has a Poly (A) tail (8-40 nucleotides in length). 3′ end may also have a tRNA-like structure (236-238 nucleotides that accepts tyrosine)

The virion is associated with a helper virus, but independent from its functions during replication (ICTVdB Management 2006).

Johnson *et al* (2003) observed that BSMV contains three positive sense single stranded genomic RNAs, designated alpha, beta and gamma, that encode seven major proteins and one minor protein. Three proteins are translated directly from the genomic RNA and the remaining proteins encode on RNA beta and RNA gamma are expressed via sub-genomic mRNAs (sgRNAs). SgRNA beta 1 directs synthesis of the triple gene block 1 (TGB 1) protein. The TGB 2 protein, the TGB 2′ minor translational read through protein, and the TGB 3 protein are expressed from sgRNA beta 2, which is present in considerably lower abundance than sgRNA beta 1 sgRNA gamma is required for expression of the gamma protein. Johnson *et al* (2003) observed the sequence elements controlling expression of BSMV sub-genomic RNAs *in vitro*.

Lawrence *et al* (2001) observed that the three proteins of the TGB have not only an essential role in infectivity but are also important in cell-to-cell movement. Kalinina *et al* (2001) made a similar observation with Poa semi-latent virus (PSLV). A 63 kDa protein encoded by the triple gene-block of the virus that comprises C-terminal NTPase/helicase domain and N-terminal extension domain containing two positively charged sequence motifs A and B. RNA-binding activity the N-terminal extension domain of 63 kDa protein is associated with long distance movement of the virus. Kalimina *et al* (2002) observed the necessity of coordinated action of three movement proteins encoded by theTGB. The largest of TGB protein TGB p1 is a member of the superfamily I of DNA/RNA helicase and possesses a set of conserved helicase sequence motifs necessary for virus movement.

Gorshkova *et al* (2003) found that TGB p3 protein of PSLV had a tendency to form large protein complexes of an unknown nature. This was found to represent an integral membrane protein and probably co-localized with an endoplasmic reticulum derived domain. Microscopy of epidermal cells in transgenic plants demonstrated that GFP (green fluorescence protein)-TGB p 3 localized to cell wall-associated punctate bodies which often formed pairs of opposing discrete structures, that co-localized with callose indicating their association with plasmodesmata-enriched cell wall fields.

Yelina *et al* (2002) observed complementary functions between hordei- and potyviruses-controlled by a single protein in long distance movement, virulence and RNA silencing suppression. A mutant hordeivirus BSMV, lacking the gamma b gene was confined to inoculated leaves in *Nicotiana benthamiana*, but systemic infection was observed in transgenic *N. benthamiana* expressing the potyviral silencing suppressor protein Hcpro, suggesting that the gamma b protein may be a long distance movement factor and have anti-silencing activity. Potyviral HCPro and hordeivirus gamma b proteins contribute to systemic viral infection, symptom severity and RNA silencing suppression.

40. Macluravirus [Maclura mosaic virus (MacMV) (ss RNA+)]

The genome is mono-partite, linear ss RNA with a probable length of 10000 nucleotides. The 5′ terminal end of the genome has a genome-linked protein (VPg) and the 3′ end has a Poly A tail. The genome translated into a single poly-protein that subsequently processed into functional products with the coat protein at the C-terminus (Anonymous 2002).

The size of the genome of MacMV ranges from 8.0-9.5 kb and the coat proteins have a molecular weight 38-40 kDa (Bowen *et al* 2003). Sabanadzovic *et al* (2001) studied the complete nucleotide sequence and genome organization of grapevine fleck virus (GFKV) an important member of the macluraviruses. The length of the genome of GFKV is 7564 nucleotides excluding 3′-terminal Poly (A) tail. Its cysteine content is extremely high (ca 50%). The genome contains four putative ORFs and an untranslated region of 291 and 35 nucleotides at the 5′ and 3′ ends respectively. ORF1 potentially encodes a 215.4 kDa polypeptide (P215) that has the conserved motifs of replication-associated proteins of ss RNA viruses. ORFs2 encodes a 24.3 kDa polypeptide (P24) identified as the coat protein. ORFs3 and 4 are located at the extreme 3′ end of the viral genome. These ORFs encode proline-rich proteins of 31.4 kDa (P31) and 15.9 kDa (P16) proteins respectively. The function of these proteins is unknown.

41. Ophiovirus [Citrus psorosis (CPsV) (ss RNA−)]

The genome is segmented into three segments of circular, negative sense ss RNA and ds RNA relative intermediates. The length of the genome is 1100-1200 nucleotides. 1600-1800 nucleotides of RNA1 have been sequenced. RNA2 is 150 nucleotides long and has been sequenced; RNA3 encodes coat protein and yet to be sequenced.

The 3′ ends of the nucleotides are complementary to those at the 5′ end and form a panhandle (ICTVdB Management 2006). CPV isolate 4 has a genome with three single-stranded RNAs. Sanchez *et al* (2002) completely sequenced RNA2 (1643 nucleotides). Most of encapsidated RNA2 is of negative polarity. The RNA2 complementary strand contained a single ORF encoding a protein of 476 aminoacids that includes a motif resembling a nuclear localization signal. In the 3′ terminal untranslated region there is a putative polyadenylation signal.

42. Pecluvirus [Peanut clump virus (PCV) (ss RNA+)]

The genome is bi-partite linear ss RNA and 10400 nucleotides long. The length of RNA1 is 5900 nucleotides and that of RNA2 is 4500 nucleotides. The 3′ terminus of the genus has a tRNA-like structure. There are several ORFs on each of the RNAs. The coat protein is at the 5′ end of the RNA2 and there is a triple gene block (TGB) at the 3′ end of the RNA2.

The products of RNA1 are: (i) 131 kDa replication protein; (ii) 191 kDa polymerase read through ORF1 including RNA polymerase motif; and (iii) 15 kDa protein of unknown function. The products of RNA2 are: (i) 23 kDa coat protein; (ii) 39 kDa of unknown function; (iii) 51 kDa protein of unknown function; (iv) 14 kDa. TGB probably involved in cell-to-cell movement and (v) 17 kDa protein of unknown function.

Dunoyer *et al* (2001) observed that RNA1 of CPV is able to replicate in absence of RNA2 by the helicase-like P131 and polymerase-like P191 domains. The P15 kDa product of this RNA is expressed via a sub-genomic RNA. This protein does not act primarily at sites of viral replication but intervenes indirectly to control viral accumulation. Dunoyer *et al* (2002) observed that the P15 of pecluvirus, a small cysteine-rich protein that has several novel properties particularly for suppressing anti-silencing activity. This protein possesses four C-terminal proximal heptad repeats that can potentially mediate a coiled-coil interaction and is targetted to peroxisomes via a C-terminal SKF motif. The coiled-coil sequence is necessary for its anti-PTGS (post-transcriptional gene silencing) activity.

Hemmer *et al* (2003) mapped the viral RNA sequences required for the assembly of PCV particles. Encapsidation of RNA1 was found to require a sequence domain in the 5′ proximal part of the P15 gene and the 3′ proximal proximal gene of RNA1. On the other hand, the sub-genomic RNA that encodes P15 was not encapsidated suggesting that other features of RNA1 are also important. Two sequences that could drive encapsidation of RNA2 deletion mutants were located. One was in the 5′ proximal coat protein gene and the other in the P14 gene of the RNA3′ terminus. No sequence homology was found between the different assembly initiation sequences.

RNA2 of pecluviruses shows diversity. Naidu *et al* (2003) conducted a comparative study on the complete nucleotide sequences of RNA2 genome of six isolates of PCV collected from geographically different locations and observed a high degree of variability in size and nucleotide sequence identities. Amino acid sequence alignments of the five ORFs showed that ORF4 which encodes second of the triple geneblock protein is highly conserved whereas the protein encoded by ORF2 whose function is unknown, is less conserved.

43. Pomovirus [Potato mop top virus (PMTV) (ss RNA+)]
The genome is a linear ssRNA in 3 segments. The 3′ terminus of all RNAs has a tRNA-like structure. The respective lengths of RNA1, RNA2 and RNA3 are ca 6050 nucleotides, ca 3100 nucleotides and ca 3000 nucleotides. The products of the RNAs are: RNA1—148 kDa replication protein, methyltransferase and helicase domain and RNA2—20 kDa coat protein, 91 kDa coat protein, involved in fungus transmission, subject to the deletion in some isolates and RNA3-51 kDa protein of unknown function, 13 kDa 3-5 TGB probably involved in cell-to-cell movement, 21 kDa protein of unknown function, and 8 kDa cysteine rich protein. PMTV is capable of establishing systemic infection without the CP-encoding RNA and also without the putative CRP (cysteine rich protein) (Savenkov *et al* 2003).

Natural deletions are found in RNA2 depending on the geographical locations. All deletions start within a short distance (18 nucleotides) and removed 558-940 nucleotides from the 3' end of the RT region (Sandgren *et al* 2001) and the genome shows a very high mutation rate. Cerovska *et al* (2003) observed that the nucleotide sequences of the coat protein (CP)-coding region is highly conserved suggesting that the differences in virulence and biological properties found in different isolates are not due to CP but to some other part of the genome.

Cowan *et al* (2002) established the sub-cellular localization, protein interactions and RNA-binding of the TGB proteins facilitate delivery and localization of the ribonucleoprotein complex to the plasmodesmata. The process was facilitated by RNA-protein rather than protein-protein interactions between the TGBP1 in complex with viral RNA and membrane-localized TGBP2. Lukhovitskaya *et al* (2005) observed that RNA3 contains a triple gene-block (TGB) encoding viral movement proteins and an ORF for a putative 8 kDa cysteine-rich protein (CRP). The PMTV CRP affects virulence of the virus or might suppress a host defence mechanism, but did not show RNA silencing suppression activity. Viral proteins and/or virus induced cell components are probably involved in the plant reaction to CRP. PMTV CRP was predicted to possess a transmembrane segment. CRP fused to the green fluorescent protein was associated with endoplasmic reticulum-derived membranes and induced dramatic rearrangements of the endoplasmic reticulum structure, which might account for protein functions as a virulence factor of the virus.

44. Tenuivirus [Maize stripe virus (MSpV) (ss RNA–)].

The genome is linear ss RNA– and has 4 or 5 segments. It may also contain ds RNA (in its replicative intermediate form). The length of the complete genome is 15460-18600 nucleotides (ICTVdB Management 2006).

Zhang *et al* (1999) sequenced 4, 6, 3 and 4 genomic segments of Rice stripe virus, Rice grassy stunt virus, Rice hoja blanca virus, and maize stripe virus respectively. All the RNA segments share conserved terminal sequences, each forming a stable base-paired panhandle structure by the complementation between the 5' and 3' termini. The genomic segments mainly use the ambisense coding strategy, but the functions of most putative proteins are unknown. A cap-snatching mechanism might be involved in mRNA transcription. While studying the coding strategy Saito (2002) found two proteins from RSV-infected rice that were encoded on two different segments in opposite polarity. Segment 3 encoded coat protein in the negative sense while segment 4 encoded the major nonstructural protein in the positive sense.

Chomchan (2002) observed the translation of genomic RNAs of Rice grassy stunt virus (TGSV). The genome of this virus consists of 6 ambisense RNA segments of which RNAS1, 2, 5 and 6 are equivalent to RNAS1, 2,

3 and 4 respectively of the type species Rice stripe virus. The RGSV 36 kDa nucleocapsid protein (N) is encoded on the complementary strand of RNA5 where accumulation of three non-structural proteins occurs: (i) a 23 kDa P2, 22 kDa P5 and (iii) 21 kDa P6 proteins. Virus-sense strand (P2 on RNA2, P5 on RNA5, P6 on RNA6) encode all these proteins.

The Rice stripe virus (RSV) genome contains 7 ORFs. Little is known of the products of 4 of these ORFs including the 23.9 kDa protein encoded by the virus-sense ORF on RNA3. This protein aggregates in vivo and forms inclusion bodies in infected plant tissue (Takahashi *et al* 2003).

Tenuivirus L protein shares with the Bunyaviridae three conserved regions corresponding to the so-called polymerase module. Tenuiviruses have an additional fourth conserved region that contains at least two active sites indicating that this region has an important role in the function of this protein. It is believed that the tenuiviruses could have cell-derived capped primer mechanism as found in influenza virus (Wei *et al* 2004).

45. Tobamovirus [Tobacco mosaic virus (TMV) (ss RNA+)]

The genome is mono-metric or poly-metric. The genome is linear and contains one molecule of ssRNA. The length of the genome is 6355-6500 nucleotides and the genome has been completely sequenced. The encapsidated nucleic acid is mainly genomic in origin but virions may also contain nucleic acid of host origin and sub-genomic RNA including rRNA found in the short particles of some species. There is a methylated cap at the 5′ end and tRNA-like structure at the 3′ end (ICTVdB Management 2006).

Genomes of several tobamoviruses have been completely sequenced. Genome expression takes place through five open reading frames (ORF1—126 kDa, ORF2—183 kDa, ORF3—54 kDa, ORF4—30 kDa and ORF5—17.6 kDa) following 2 different strategies, sub-genomic RNAs 1(2991 nucleotides), 2 (1500 nucleotides) and 3 (692 nucleotides) together with a read-through or frame shift protein. Sub-genomic RNA1 expresses through ORFs 3, 4 and 5; sub-genomic RNA2 expresses through ORFs4 and 5 and the sub-genomic RNA3 expresses through ORF5. These genes code for the production of 5 proteins. ORF1 (126 kDa) and ORF2 (183 kDa) translate the codes for polymerase; ORF4 (30 kDa) translates the code for the protein responsible for the cell-to-cell movement and the ORF5 (17.6 kDa) translates the code for coat protein synthesis (Solis and Garcia-Arenal 1990, Meulewaeter *et al* 1990).

Molnar *et al* (2005) showed that TMV carrying a short inverted repeat of the phytoene desaturase (PDS) gene triggered more accumulation of PDS siRNAs than the corresponding antisense PDS sequence indicating that virus-derived siRNAs originate predominantly by direct DICER cleavage of imperfect duplexes in the most folded regions of the positive strand of the viral RNA.

46. Tobravirus [Tobacco rattle virus (TRV) (ss RNA+)]

The genome is bi-partite, linear ss RNA with a length of 8600-11300 nucleotides. Both RNA1 (6791 nucleotides) and RNA2 (1905 nucleotides) genomes have a methylated nucleotide cap at the 5' end and OH at the 3' end. RNA-1 has 4 open reading frames (ORF1—134 kDa, ORF2—184 kDa, ORF3—29 kDa and ORF4—16 kDa). RNA2 has 1 reading frame (ORF5—25 kDa). Multi-partite genomes, sub-genomic RNAs (sub-genomic RNA1A, 1B and sub-genomic RNA2 (1413 nucleotides) and read through protein or frame shift strategies are followed to produce functional proteins. Genes in RNA1 are translated to produce polymerase (194 kDa), protein for cell-to-cell movement (29 kDa) and for other functions (134 kDa and 16 kDa). Genes of genomic RNA2 and sub-genomic RNA2 are translated to produce coat protein (25 kDa). Expression of sub-genomic RNA1A and 1B operates through ORF3 and ORF4 (16 kDa) to produce other functional proteins (Angenent *et al* 1990).

In general RNA1 encodes proteins involved in virus multiplication, cell-to-cell movement and pathogenicity. RNA2 encodes proteins that are involved in particle production and transmission by the nematode vector, but this RNA shows wide genetic variability depending upon the isolates (Crosslin *et al* 2003). A small cysteine-rich 16 kDa protein is a pathogenicity determinant that suppresses RNA silencing. The Pea early browning virus (PEBV) 12 kDa protein is also responsible for passage of the virus in seeds indicating that invasion of reproductive tissue by the virus may involve silencing suppression. RNA1 can systemically infect in the absence of RNA2 but the virus lacks the coat protein. In highly susceptible infections RNA1 proceeds rapidly via the vascular system rather than from cell-to-cell via plasmodesmata (ICTVdB Management 2006). The cysteine-rich 16 kDa protein can be replaced by a pathogenicity protein from unrelated viruses (Liu 2002).

Serial passage of TRV under different selection conditions results in deletion of structural and non-structural genes in RNA2 (Hermandez *et al* 1996). Crosslin *et al* (2003) showed genetic variability of RNA2 for rapid adaptations to non-field condition. A nematode non-transmissible isolate encodes a truncated version of ORF2c, the deletion that causes the non-transmissibility.

47. Varicosavirus [Lettuce big vein virus (LBVaV) (ss RNA−)]

The genome is bi-partite, linear, negative sense ss RNA−. Lengths of the two segments are 6350-7000 nucleotides and 5630-6500 nucleotides. The genome structure is closely related to that of rhabdoviruses particularly in the sequences of both the CP and L proteins and possession of conserved transcription termination/polyadenylation on signal-like Poly (U) tail and in transcribing mono-cistronic RNAs. The presence of Poly (U) tail in the NCRs of LBVaV RNA1 and RNA2 suggest that the transcription of LBVaV is regulated by a mechanism similar to that of rhabdoviruses. The major difference between the two genera is the presence of two negative RNAs in LBVaV and only one negative sense RNA in rhabdoviruses.

Wilk *et al* (2002) observed that the genome consisted of 4 RNA molecules of approximately 7.8, 1.7, 1.5 and 1.4 kb. The 5′ terminal and 3′ ends of the RNA molecules are largely, but not perfectly, complementary to each other. There are homologies with the RNA-dependent RNA polymerase of rhabdoviruses and Ranunculus white mottle virus, and the capsid protein of Citrus psorosis virus. The gene encoding the viral polymerase appears to be located on the RNA segment 1, while the nucleocapsid protein is encoded by RNA3.

48. Ourmiavirus [Ourmia melon virus (OuMV) (ss RNA+)]

The genome organization and expression is not adequately known. The virus is non-enveloped bacilliform and contains 3 ssRNAs of molecular weight of 0.9, 0.35 and 0.32 × 106 combined with a single CP of 21-25 kDa (ICTVdB 2006).

(e) Isometric Particles Containing ss RNA

(i) Family: Tombusviridae

49. Aureusvirus [Cucumber leaf spot virus (CLSV) (ss RNA+)]

The genome is mono-partite, linear ssRNA. The encapsidated nucleic acid is mainly genomic in origin but virions may contain sub-genomic RNA (mRNA of 1100 nucleotides). The length of the complete genome is 4400-4500 nucleotides. The 5′ end has a methylated cap (ICTVdB Management 2006).

Rubino *et al* (1995) sequenced the genome of the aureusvirus. ORF1 produces 2 proteins involved in the production of RNA dependent polymerase complex. ORF2 produces coat protein. The product of ORF3 is associated with the cell-to-cell movement of the virus and the product of ORF4 is associated with the development of symptom severity (Rubino and Russo 1997) and is regulated by the coat protein.

Miller *et al* (1997) found the length of the genome of CLSV was 4432 nucleotides and contains 5 long ORFs. The 5′ proximal ORF encoded a 25 kDa product that terminated in an amber codon which might be read through to produce a 84 kDa protein (ORF2). ORF3 coded for the 41 kDa coat protein (CP). ORFs4 and 5 were completely overlapping at the 3′ terminus and coded for 27 and 17 kDa products, respectively.

Martelli *et al* (1998) confirmed that aureusvirus is a new genus of the plant viruses and made similar observations. The 84 kDa product was with the conserved motifs of RNA dependent RNA polymerase. ORF4 coded for the movement protein and the product of ORF5 was responsible for symptoms. Replication probably followed the strategy based on direct expression of the 5′ proximal ORF and expression of downstream ORFs through subgenomic RNAs.

Reade *et al* (2003) constructed full-length clones of the genome and produced infectious T7 polymerase derived synthetic transcripts. Mutational analysis of the genome indicated a role for p84 in viral RNA replication, CP in systemic movement, p27 in viral cell-to-cell movement and p17 in symptom induction.

50. Avenavirus [Oat chlorotic stunt virus (OCSV) (ss RNA+)]

The genome is mono-partite, linear ssRNA. The encapsidated nucleic acid is mainly genomic in origin but may also contain sub-genomic RNA. The length of the complete genome is 4100 nucleotides. There is a methylated cap at the 5' end (ICTVdB Management 2006).

The size of the genome is 4.4 kb. Genomic RNA is translated into 3-4 putative products. The four products of the type member are: (i) 23 kDa polymerase, (ii) 84 kDa polymerase produced by the read through of (i), (iii) 48 kDa coat protein and (iv) 8 kDa protein of unknown function.

The genome is expressed following the sub-genomic mRNA strategy. Boonham *et al* (1995) located 3 ORFs. ORF1 produces two proteins that are involved in virus replication. ORF2 encodes the coat protein. The product of ORF3 is probably associated with the cell-to-cell movement of the virus.

51. Carmovirus [Carnation mottle virus (CarMV) (ss RNA+)]

The genome is mono-partite, linear ssRNA. The length of the complete genome is 3800-4300 nucleotides. Encapsidated nucleic acid is mainly genomic in origin but there may be certain sub-genomic RNAs and satellite RNA. Virions may provide helper functions to dependent virus replication. It acts as a helper for a satellite RNA (ICTVdB Management 2006).

5' terminus end of the genome has a methylated nucleotide cap. The genome has 5 ORFs on it. 3 ORFs are translated from 2 sub-genomic RNA. In the type member the ORFS are translated into 5 products: (i) ca 27 kDa polymerase (ii) ca 88 kDa polymerase read through (iii) 7 kDa movement protein (cell-to-cell transport translated from sub-genomic RNA1-1689 nucleotides long) (iv) 8 kDa movement protein 2 (v) ca 38 kDa coat protein translated from sub-genomic RNA2—1431 nucleotides). Vilar *et al* (2001) observed that CarMV P7 is an alpha/beta RNA-binding soluble protein with three putative domains. The central region from the residue 17-35 (represented by peptide P7—17-35) is responsible for the RNA binding properties of CarMV P7. The RNA recognition appears to occur via an "adaptive binding" mechanism. The amino acid sequence and structural properties of the RNA-binding domain of P7 seem to be conserved among carmoviruses and some other RNA-binding proteins and peptides. The low conserved N-terminus of P7 (peptide P7—1-16) is unstructured in solution. Highly conserved C-terminus motif (peptide 7—40-61) adopts a beta sheet conformation in aqueous solution. Positively charged amino acids are more important in the RNA binding process and hydrophobic amino acid side chains participate in the stabilization of the protein-RNA complex.

Studies have also been made on genome expression in other carmoviruses; cowpea mottle virus (CpMoV), turnip crinkle (TCV), and Hibiscus chlorotic ringspot virus (HCRSV). Guan *et al* (2000a) observed the requirement of a 5'-proximal linear sequence of minus strands for plus-strand synthesis of a satellite RNA-associated TCV. Guan *et al* (2000b) further analyzed the cis-acting sequences involved in plus strand synthesis of the satellite RNA (Sat C-a 356-base sub-viral RNA-associated with TCV) and identified a new replication element. Sat-C associated with TCV intensifies the symptoms on all symptomatic hosts but attenuates the symptom of a mutant that expresses low levels of a defective coat protein. TCV virions are substantially fewer in the presence of Sat-c or when two amino acids are inserted at the N-terminus of the coat protein resulting in similar enhanced symptoms. Since CP is a suppressor of RNA silencing, it results in enhanced colonization of the plant (Zhang and Simon 2003). Thus Sat-c enhances the ability of the virus to colonize plants by interfering with stable virion accumulation. Motif 1-hairpin (M1H), a replication enhancer on minus strands forms a plus strand hairpin flanked by a CA-rich sequence that may be involved in enhancing systemic infection. Sequence non-specific plus strand hairpin brings together flanking CA-rich sequences in the M1H region that confer fitness to Sat-c by reducing the accumulation of stable virions (Sun and Simon 2003).

Nagy *et al* (2001) characterized an RNA replication enhancer in the satellite RNA associated with TCV. The 3' end stem-loops of the sub-viral RNAs associated with TCV are involved in symptom modulation and coat protein binding. Qu and Morris (2000) studied the role of 5' and 3' untranslated regions (UTRs) of TCV genomic RNA (4 kb), 5' UTR of the coat protein sub-genomic RNA (1.45 kb) in translational regulation. All three UTRs enhanced translation of the luciferase reporter gene to different extents. Optimal translational efficiency was achieved when mRNAs contained both 5' and 3' UTRs. The translational enhancement of TCV mRNAs occurred in a cap-independent manner. Koh *et al* (2002) made a similar observation with HCRSV.3' UTR (3' untranslated region) region of this virus. It significantly enhanced the translation of several open reading frames on genomic RNA or sgRNA and a viral gene in a bi-cistronic construct with an inserted internal ribosome entry site. A six-nucleotide sequence was required for the enhancement of translation induced by the 3'UTR.

RNA recombination occurs frequently during replication. The most common recombinants generated are either defective interfering (DI) RNAs or chimeric satellite RNAs that are thought to be generated by template switching of the viral RNA-dependent RNA polymerase (RdRp) during the viral replication process (Cheng and Nagy 2003).

52.   Dianthovirus [Carnation ringspot virus (CRSV) (ss RNA+)]

The genome is bi-partite, linear ssRNA. The length of the complete genome is 5250-5600 nucleotides. The lengths of RNA1 and RNA2 are 3750-4000 and

1448-1600 nucleotides. 5′ end has a methylated cap. Sub-genomic mRNA is found in the infected cells. Virions are associated with helper virus but independent from its functions during replication.

Mizumoto *et al* (2003) found cap-independent translational enhancement by the 3′ untranslated region of red clover necrotic mosaic virus (RCNMV). In this virus both RNA1 and RNA2 lack a 5′ cap structure. 5′ cap and 3′ Poly (A) tail play important roles in the translation of mRNAs. As both these structures are absent in the genome of RCNMV, the translational activity is done by some other means. The 3′ untranslated region (UTR) of RNA1 alone significantly enhances translation of the luciferase reporter gene in the absence the 5′ cap structure; the middle region (between the nucleotides 3596 and 3732) in 3′ UTR, designates the 3′ translation element of dianthovirus. RNA1 (3′TE-DRL) plays an important role in cap-independent translation. This region contains a stem-loop structure conserved among members of the genera Dianthovirus and Luteovirus. A five-base substitution of the loop abolishes cap-independent translational activity indicating that this stem-loop is one of the functional structures in the 3′ TE-DRI involved in the cap independent translation.

53. Machlomovirus [Maize chlorotic mottle virus (MCMV) (ss RNA+)]

The genome is mono-partite, linear ssRNA. The length of the complete genome is 4400 nucleotides. There is a methylated cap at the 5′ end. The encapsidated nucleic acid is mainly of genomic origin but may also contain sub-genomic RNA (mRNA) (ICTVdB Management 2006).

54. Necrovirus [Tobacco necrosis virus A (TNV-A) (ss RNA+)]

The genome is mono-partite, linear ss RNA. The length of the complete genome is 3660-3760 nucleotides (ICTVdB Management 2006). A methylated cap is present at the 5′ end. There may be 5 or 6 ORFs on the genome. ORF1 and ORF2 (a read through of ORF1) are believed to be polymerase. The ORF3 and ORF4 are probably translated from the larger of the two sub-genomic RNAs and the coat protein (ORF5). The six products in the type member are: (i) 23 kDa polymerase (ii) 82 kDa polymerase (iii) 8 kDa movement protein (iv) 6 kDa putatitive movement protein (v) 30 kDa coat protein and (vi) 7 kDa of unknown function.

The TNV genome is an uncapped, nonpolyadelynated RNA whose translational mechanism was ill understood. Computational analysis by Shen and Miller (2004) predicted a cap-independent translation element (TE) within the 3′ untranslated region (3′ UTR) of TNV RNA that resembles the TE of BYDV, functioning both at *in vitro* and *in vivo*. Both have a similar secondary structure and the conserved sequence, inactivated by a four-base duplication in the conserved sequence and can function in the 5′UTR. When the TE is located in its natural 3′ location, may form long-distance base pairing with the viral with viral 5′ UTR that is conserved and probably required. The TNV TE differs from the BYDV TE by having only three helical domains instead of four. Similar structures were found in all members of the necroviruses except satellite tobacco necrosis virus that harbours a different 3' cap-independent translation domain.

55. Panicovirus [Panicum mosaic virus (PMV) (ss RNA+)]

The genome is mono-partite, linear ssRNA. The length of the complete genome is 4300 nucleotides. There is a methylated cap at the 5' end. The encapsidated nucleic acid is mainly genomic origin but virions may also contain sub-genomic RNA (mRNA—1100 nucleotides) (ICTVdB Management 2006).

The genome has 6 ORFs, the products of these ORFs are: (i) 48 kDa polymerase (ii) 112 kDa polymerase (iii) 8 kDa movement protein (iv) 6.6 kDa involved in virus movement: suggestion of possible frame-shift (extending the product of ORF3) (v) 26 kDa coat protein and (vi) 15 kDa protein also involved in virus movement.

56. Tombusvirus [Tomato bushy stunt virus (TBSV) (ss RNA+)]

The genome is mono-partite, linear ss RNA. The length of the complete genome is 4800 nucleotides. The encapsidated nucleic acid is mainly genomic in origin but the virions may also contain sub-genomic RNA or sub-genomic RNA and satellite RNA or nucleic acid of host origin and sub-genomic mRNA and defective RNA species arising from a deletion of full length genomic RNA (ICTVdB Management 2006).

The 5' end of the genome may have a methylated cap. 3-4 virus-specified ds RNA species are found in the infected cells. There are three virus-coded non-structural proteins. The virion is associated with a helper virus but independent from its functions during replication. It may act as a helper for a satellite RNA (Anonymous 2002).

Hearne *et al* (1990) sequenced the genome of tombusviruses and Rochon (1999) reviewed their genome organization, expression and function. There are 4 ORFS on the genome. ORF1 encodes 2 proteins associated with the synthesis of viral polymerase. ORF2 encodes coat protein. The product of ORF3 is associated with the cell-to-cell movement of the virus and the product of the ORF4 regulates symptom development and the host dependent induction of systemic infection (Chu *et al* 2000). The coat protein (CP), movement associated proteins P22 and P19 and RNA sequences influence the accumulation of RNA at very early stages of infection. Three regulatory sequences are involved in RNA accumulation. These sequences are located within the 3'proximal nested movement protein genes P22 and P19. The first is a 16 nucleotides element termed III-A is positioned at the very 3' end of P32 and is essential for RNA accumulation. Approximately 300 nucleotides-upstream of III-A residues, ~80 nucleotides inhibitory elements (IE) obstructive to replication only in the absence of the third regulatory element of ~30 nucleotides (SOR-III) positioned immediately upstream of III-A (Park *et al* 2002).

P19, TBSV movement protein is a pathogrnicity determinant with host dependent effects on virus spread and symptom inducton (Qiu *et al* 2002). Desvoyes *et al* (2002) observed that P22 provides valuable tools to study trafficking of macromolecules through plants. Wild P22 amino acid substitution mutants were found to be equivalent to biochemical features associated with MPs (i.e. RNA binding, phosphorylation and membrane partitioning). A new plant homeodomain leucine-zipper protein interacts specifically in a functionally relevant manner with the virus MP. Desvoyes and Scholthof (2002) further observed that truncated capsid subunits might benefit systemic spread of the virus. Mutation of the host dependent recombination of the coat protein (deletion of the substantial portion of the RNA-binding domain) can produce such truncated virus-like complex.

Two of the five viral-coded proteins of TBSV that are small, non-segmented plus-strains, are required for replication of the virus in the infected cells. These replicase proteins, namely P32 and P92 of cucumber necrosis tombusvirus are expressed directly from the genomic RNA via a read through mechanism. Their overlapping domains contain an argenine- and proline-rich RNA-binding motif (RPR). This motif is essential for the replication function of both the proteins (Panavas *et al* 2003).

The sequence of P33 replicase protein overlaps with N-terminal domain of P92 that contains the signature motifs of RNA-dependent RNA-polymerases (RdRps) in its non-overlapping C-terminal portion. The binding of P33 to ss RNA is stronger than binding to ds RNA. P33 could bind to RNA in a cooperative manner. The RPR motif is highly conserved and non-overlapping C-terminal domain of P92 and contains additional RNA-binding regions (Rajendran and Nagy 2003, Panaviene *et al* 2003) confirmed that the overlapping RNA-binding domains of P33 and P92 replicase proteins are essential for tombusvirus replication.

Zamore (2004) provided the first glimpse into how tombusviruses can defeat their host's anti-viral RNAi defenses by locating the three dimensional structure of a si RNA bound to the P19 protein, the suppressor of gene-silencing activity.

RNA silencing (known as RNA interference) is a conserved biological response to ds RNA that regulate gene expression and has evolved in plants as a defense against viruses. Small interfering RNAs (si RNAs) mediate the response and guide the sequence-specific degradation of cognate messenger RNAs. As a counter defense, viruses encode proteins that specifically inhibit the silencing machinery. P16 protein from the tombusvirus is such a viral suppressor of RNA silencing that specifically binds to si RNA. Ye *et al* (2003) found that the crystal structure of P19 bound to a 21 nucleotide si RNA where the 19-base-pair RNA duplex is cradled within the concave face of a continuous eight-stranded beta sheet, formed across the P19 homo-dimer interface.

In the process of replication of tombusviruses, enhancement of RNA synthesis may take place by promoter duplication. Panavas *et al* (2003) demonstrated that the minus- strand RNA of tombusviruses contains, in addition to the 3'-terminal minimal plus-strand initiation promoter, a second cis-acting element, termed the promoter proximal enhancer (PPE). The PPE element enhanced RNA synthesis by almost three-fold from the adjacent minimal promoter that may have been the mechanism leading to the generation of the PPE.

(ii) Family: Bromoviridae

57. Bromovirus [Brome mosaic virus (BMV) (ss RNA+)]

The genome is segmented, tri-partite, linear ssRNA. The length of the complete genome is 8620-8800 nucleotides. All the RNAs have methylated nucleotide cap at the 5' end and a highly conserved 3' terminal sequence of about 200 nucleotides. The terminal 135 nucleotides of this sequence may be folded into a three-dimensional tRNA-like structure that accepts tyrosine.

Genomic RNAs are designated as RNA-1 (3158 nucleotides), RNA-2 (2799-3400 nucleotides) and RNA-3 (2117-2600 nucleotides). There are 4 ORFs on the RNAs encoding 4 proteins. These are: 109-110 kDa replication protein 1a (from RNA1) containing methyltransferase and helicase domains; 92-94 kDa polymerase 2a (from RNA2) RNA-dependent RNA polymerase; 32-35 kDa movement protein 3a (from 5' half of RNA3) and 20-21 kDa coat protein 3b (from 3' half of RNA3) expressing from sub-genomic RNA (RNA4) (ICTVdB Management 2006).

Between the cistron of the 2 proeins of the RNA3, there is an inter-cistronic non-coding region (250 nucleotides long). An internal Poly (A) sequence of heterogeneous length occurs in this region (Ahlquist *et al* 1990, Ahlquist 1999). RNA1 and RNA2 encode 1a and 2a proteins for viral replication. RNA3 encodes 3a movement protein and the coat protein that are essential for systemic infection in plants but dispensable for RNA3 replication.

1a and 2a co-localize in an endoplasmic reticulum (ER)-associated replication complex that is the site of RNA dependent RNA synthesis. 1a also localizes to ER in the absence of 2a and a viral RNA template. 1a also interacts independently with the ER and viral RNA and is a key organizer of RNA replication complex assembly (Chen and Ahlquist 2000). In the absence of 2a and RNA replication, 1a acts through an inter-genic replication signal in RNA3 to stabilize RNA3 and induce RNA3 to associate with the cellular membrane. 1a-induced RNA3 stabilization reflects interactions involved in recruiting RNA3 templates into replication (Chen *et al* 2001). The 3' end of the subgenomic RNA that contains tRNA-like sequence directs the synthesis of RNA. There is a loop-like structure in this sequence called stem loop C (SLC) consisting of two discrete domains, a flexible stem with an internal loop and a rigid stem containing a 5'-AUA-3' triloop. Efficient RNA synthesis requires only one side of the flexible stem and a specific compact conformation of the tri-loop for replicase binding and initiation of the synthesis (Sivakumaran *et al* 2000). Viral RNA synthesis is initiated by the viral replicase in association with RNA core promoters.

A sub-set of 250-bases intergenic region (IGR), promoter or the replication enhancer (RE) contain all cis-acting signals necessary for a crucial, early template selection step, the 1a dependent recruitment of RNA3 into replication. One of these signals is a motif matching the conserved box B sequence of RNA polymerase III transcripts. The box B motif is presented as a hairpin loop of seven nucleotides closed by G:C base pairs in perfect analogy to the TC-stem loop in tRNA (Baumstark and Ahlquist 2001).

Haasnoot and Olsthoorn (2002) observed that the trinucleotide hairpin loops play an important role in RNA transcription. In the absence of the viral replicase the hexa-nucleotide loop 5 (prime or minute) of this RNA structure adopts s pseudo-nucleotide loop conformation by trans-loop base pairing between C1 and G5. Other base pairs contribute to sub-genomic (sg) transcription, probably by stabilizing the formation of this pseudo-tri-loop. This structure strongly resembles iron-responsive elements in cellular messenger RNAs and may represent a general protein-binding motif. The BMV sg hairpin is functionally equivalent to the minus-strand core promoter hairpin stem-loop C at the 3′ or minute end of BMV RNAs. Replacement of sg hairpin by stem-loop C yields increases sg promoter activity whereas replacement of stem-loop C by the sg hairpin results in reduced minus strand promoter activity.

Boon *et al* (2001) identified the sequence of BMV replicase protein 1a that mediate association with endoplasmic reticulum membranes. BMV RNA replication occurs with peri-nuclear region of the endoplasmic reticulum (ER). The only viral component in the BMV RNA replication complex that localizes independently to the ER is 1a. The other viral replication components: protein 2a (RNA polymerase) and the RNA template, depend on 1a for recruitment to the ER. Sequences in the N-terminal RNA capping module of 1a mediate membrane association. In particular, a region C-terminal to the core methyltranferase homology is sufficient for high affinity ER membrane association. 1a localized to ER membranes recruit 2a and RNA templates and form membrane-bound, capsid-like spherules in which RNA replication occurs. Lee and Ahlquist (2003) reported preferential interaction of 1a with one or more types of membrane lipids.

Chen *et al* (2003) determined an alternative pathway for recruiting template RNA to the BMV RNA. Normally box B motif is crucial to 1a responsiveness of the RNA2 and RNA3. Unexpectedly they found that some chimeric mRNAs expressing the 2a polymerase ORF from RNA2 were recruited by 1a to the replication complex and served as templates for negative strand RNA synthesis, despite lacking the normally essential box B-containing 5′ signal. This template recruitment required high efficiency translation of the RNA templates. Moreover multiple frame shifting insertion or deletion mutations throughout the N-terminal region of the ORF inhibited this template recruitment, while an in-frame insertion did not. Providing 2a in trans did not restore template recruitment of RNAs with frame-shift mutations. Only those deletions in the N-terminal region of 29 that abolished 2a interaction with 1a abolished template recruitment of the RNA. This alternate pathway for 1a dependent RNA recruitment involves 1a interaction with the translating mRNA via the 1a-interactive N-terminal region of the nascent 2a polypeptide. Interaction with nascent 2a also may be involved in 1a recruitment of 2a polymerase to membranes.

Ranjith *et al* (2003) demonstrated the enhancer like activity of a BMV promoter. RNA sequences named specificity and initiation of determinants allow recognition and coordinated interaction with the viral replication enzyme. Using enriched replicase from BMV-infected plants and variants of the promoter template for minus strand and sub-genomic RNA initiation, it was observed that specificity determinant for minus strand and sub-genomic RNA initiation could function at variable distances and positions from the 3' initiation site similar to enhancers of transcriptions from DNA templates. This determinants addition could convert a cellular tRNA into a template for RNA synthesis by the BMV replicase in vitro. Sivakumaran *et al* (2003) characterized the cis-acting elements for RNA synthesis and its mechanism extracting replicase from the BMV infected plants. Minus RNA synthesis in vitro requires stem-loop C (SLC) that contains a clamped adenine motif. In vitro, there are several specific requirements for SLC recognition. The BMV replicases extracted from plants at different times after infection have different levels of recognition of minimal promoters for plus and minus strand RNA synthesis.

Choi and Rao (2003) observed that packaging of BMV RNA3 is mediated through a bi-partite signal. The three genomic and a single sub-genomic RNA of BMV are packaged by a single coat protein (CP) into three morphologically indistinguishable icosahedral virions with T=3 quasi-symmetry. Genomic RNA1 and RNA2 are packed individually into separate particles whereas genomic RNA3 and sub-genomic RNA (coat protein mRNA) are co-packaged into a single particle. Packaging of di-cistronic RNA3 requires a bi-partite signal. A highly conserved 3 tRNA-like structure postulated to function as a nucleating element (NE) for CP subunits and a cis-acting, position dependant packaging element (PE) of 187 nucleotides present in the non-structural movement protein gene are the integral components of the packaging core. Efficient incorporation into BMV virions of non-viral RNA chimeras containing NE and the PE provides confirmatory evidence that these two elements are sufficient to direct packaging. Damayanti *et al* (2003) also observed separate packaging of the virions. There is an element in the 3a ORF of RNA3 that regulates the packaging of RNA3 and RNA4 (sub-genomic RNA).

Sasaki *et al* (2003) observed that the movement protein gene is involved in the virus specific requirement of the coat protein in cell-to-cell movement of the BMV. Fujisaki *et al* (2003) observed that RNA2 mediated interaction between viral components may be required for efficient cell-to-cell movement of BMV. Takeda *et al* (2004) observed that 3a movement protein (MP) plays a central role in the movement of BMV. The central region of the BMV MP is important for viral movement and both termini of the BMV MP have effects on the development of systemic symptoms. C-terminus of the MP is involved in the requirement for CP in cell-to-cell movement and plays a role in long distance movement.

Interactions with other viruses and host systems

Okinaka *et al* (2003) characterized a novel barley protein HCP1 that interacts with BMV coat protein (CP). The CP-HCP1 binding controls BMV infection of barley interacting directly with CP, probably in the cell cytoplasm. Viivanco and Turner (2003) demonstrated translation inhibition of capped and uncapped viral RNAs mediated by ribosome inactivating protein. Pokeweed antiviral protein (PAP) does not depurinate each capped-RNA and the PAP can inhibit translation of viral RNAs in vitro without causing detectable depurination at multiple sites. Thus the capped structure is not the only one determinant for inhibition of translation by PAP.

The 3a movement protein (MP) plays a central role in the movement of the virus. Takeda *et al* (2004) showed that the central region of the MP is important for viral movement and both the termini of this protein have effects on the development of systemic symptoms. The C-terminus of the MP is involved in the requirement for CP in cell-to-cell movement and plays a role in the long distance movement. Furthermore, the ability to spread locally and form virions is not sufficient for the long distance movement of BMV.

There may be variability in the RNA genome arising from the homologous recombination in BMV RNA1 and RNA2 components during infection. Basal RNA-RNA cross-over could occur within the coding region of both RNAs. Urbanowicz *et al* (2005) observed that although recombination frequencies are much lower than the rate observed for the inter-cistronic recombination spot in BMV RNA3 and the probability calculations accounted for at least one homologous cross-over per RNA molecule per replication cycle.

58. Cucumovirus [Cucumber mosaic virus (CMV) (ss RNA+)]

The genome is segmented, tri-partite, linear ss RNA. The length of the complete genome is 8500-8698 nucleotides. The 5′ end has a methylated cap or genome-linked protein (VPg). At the 3′ end there is a conserved region of 200 nucleotides and a tRNA-like structure. The encapsidated nucleic acid is mainly genomic in origin but may also contain sub-genomic RNA and satellite RNA (mRNA) derived from RNA3.

The lengths of the RNA1, RNA2, RNA3 and RNA4 are: 3355-3410, 2946-3074, 2186-2214 and 100 nucleotides respectively (ICTVdB Management 2006).

The genome organization of the members of the cucumovirus is more or less similar to that found in bromoviruses but the three genomic RNAs and the sub-genomic coat protein RNA are slightly longer and there is an additional ORF on RNA2. The genomes encode five proteins with similar functions as found with bromoviruses. RNA3 of all members of the genus can be exchanged yielding viable virus whereas RNAs 1 and 2 can only be exchanged within a species. The RNAs are highly labile and depend upon RNA-protein interactions for integrity. RNA1, RNA2 and RNA3 are packaged in individual particles. A sub-genomic RNA, RNA4 is packaged with RNA3 making all the particles roughly equivalent in composition. In some strains additional sub-genomic RNA, RNA4A is also encapsidated at a low level. RNAs with 5′ capped structures, and 3′ conserved regions that can be folded into tRNA-like structures.

CMV, the type species, can also harbour "molecular parasites" or satellite RNAs (sat-RNA) that can dramatically alter the symptom phenotype induced by the virus. These sat-RNAs do not encode any protein but rely on the RNA for their biological activity (Roossinck 1999a, b, 2001). The satellite consists of 368 nucleotides. The sequence similarity and grouping are not related to the sizes of the sat-RNAs. Chen *et al* (2001) observed the sites of nucleotide insertion or deletion on the 13 sat-RNAs between 90 and 255 nucleotides. The 80 nucleotide at 5′ end, the 45 nucleotide at the 3′ end and unpaired sections in the satellite RNA structure are highly conservative.

Minimum sequence of RNA4A required for specific initiation from cytididylate (T1) used in vivo consists of 31nucleotide, 3′ of T1 and a 13 nucleotide-template sequence. This 44 nucleotide RNA provides 3 elements that contribute to efficient initiation of RNA4A synthesis by the CMV replicase (Sivakumaran *et al* 2002). Some satellite RNAs may cause programmed cell death in some plants. Xu (2000) observed such programmed cell death caused by D satellite RNA in tomato plants.

Zhang *et al* (1999a) constructed a full-length cDNA clone and complete nucleotide sequences of RNA2 of the CMV (SD strain). It consists of 3048 nucleotides and there are two partial overlapping ORF1 that is located in the 5′ portion (79-2652 nucleotides), coded for 2a protein of 858 amino acids (aa). ORF2 located in the 3′ portion (2414-2746 nucleotides), codes for 2b protein of 111 aa. Kwon *et al* (2000) observed that 5′ un-translated region (UTR) of CMV RNA4 confers a highly competitive translational advantage on a heterologous luciferase ORF. Stabilization of the 5′ UTR RNA secondary structure inhibits competitive translational activity. The translation of chimeric CMV RNA4 is about 40 fold higher than that CMV chimeric RNA3 in a competitive environment. The competitive translational activities of the 5′UTRs are: RNA4 (coat protein) > RNA2 and RNA1 (2a and 1a protein or replicases)

> RNA3 (3a protein). In nature, there may be emergence of new satellite RNA from CMV. Perez Alvarez *et al* (2003) observed that CMV (P1 strain) after several passages over five years on tobacco and tomato caused unusually severe symptoms. Sequence analysis showed the presence of two sat-RNAs (sat-P1-1 and sat-P1-2). Sat P1-1 contained 335 nucleotides and sat-P1-2 contained 394 nucleotides. Based on the differences in the sequence and secondary structure between these two sat-RNAs, sat-P1-2 represents the emergence of a new satellite (necrotic satellite) from attenuated sat-RNA populations.

Canto *et al* (2001) observed that in some cases RNA1 with 3' termini regenerate during replication by a template switching mechanism between the inoculated transcript RNAs and the mRNA originating from RNA2 and RNA3. These chimeras occur naturally due to recombination between wild type viral RNAs.

CMV has a very wide host range. It infects more than 1000 species belonging to 85 families. This species has been subdivided into groups IA, 1B and II on the basis of its coat protein (CP) and 5' non-translated region (NTR) of RNA3. Roossinck (2002) observed the continuous evolution of this virus due to frequent reassortment of the genomes. Different RNAs have independent evolutionary history. 3' NTR is conserved among all RNAs indicating that the evolutionary constraints on this region are specific to the RNA component rather than virus isolates. The evolutionary tree of the 1a and 2a and CP ORFs are more compact and display more branching than those of the 2a and CP ORFs indicating host interactive constraints exerted on the 1a and 2a ORFs.

Cillo *et al* (2002) observed in situ localization and tissue distribution of the replication associated 1a and 2a proteins. They are predominantly localized to the vacuolar membranes, the tonoplast, while plus-strand RNAs are found distributed throughout the cytoplasm. 2a proteins are found at higher densities than 1a proteins. Both the proteins show quantitative differences with regard to tissue distribution depending upon the type of the hosts. When cucumber and tobacco are infected with CMV, about three times as much as 2a protein is detected in cucumber in comparison to that found in tobacco. Moreover 1a and 2a proteins in high densities are found in mesophyll cells of cucumber whereas in tocacco cells they are located in vascular tissues of minor veins. It therefore appears that relative rates of tissue infection and virus replication in vascular tissues depend on the hosts.

Nagano *et al* (2001) observed that CMV requires its own coat protein (CP) in addition to the movement protein (MP) for intercellular movement. CMV-MP truncated in its C-terminal 33 amino acid has the ability to mediate viral movement independently of CP. Sequential deletion analyses of the CMV MP reveals that the dispensability of CP occurs when C-terminal deletion ranges between 31 and 36 amino acids and that shorter deletion impairs the ability of the MP to promote viral movement. A region of MP determines the requirement of

CP in cell-to-cell movement. Shalitin *et al* (2002) observed that viral MPs cause a significant elevation in the proportion of sucrose in the phloem sap collected from petioles of the source leaves indicating alterations in the mechanism of phloem loading. Matsushita *et al* (2002) observed that the 3a protein is essential for the cell-to-cell movement of the viral RNA through the plasmodesmata and the phosphorylation of this protein occurs to mediate the movement. Huppert *et al* (2002) observed that CMV-MP is exchangeable with that of other viruses. A hybrid virus (CMV-Cym MP) constructed by replacing MP of CMV with that of Cymbidium ringspot tombusvirus (CymRSV) could efficiently spread both cell-to-cell and long distance in host plants. Kobori *et al* (2002) observed that the amino acid at position 129 of CMV CP is the determinant for systemic movement and symptom expression in different hosts.

Studies show that CMV-CP is required for cell-to-cell movement that is mediated by 3a MP. Deletion of C-terminal 33 amino acids, and the CMV 3a MP results in CP independent cell-to-cell movement, but not long distance movement. Kim *et al* (2004) demonstrated that the C-terminal 33 amino acids of CMV 3a protein affect virus movement, RNA binding and inhibition of infection and translation. Kim *et al* (2006) observed that CMV-encoded 1a protein plays a role in the replication of the viral genome along with 2a and one or more host factors. One of the cDNA clones encoded a protein homologous to the Arabidopsis putative protein kinase and was designated as Tcoi2 (Tobacco CMV 1a interacting protein2). This protein specifically interacted with methyltransferase (MT) domain of CMV 1a protein targetting endoplasmic reticulum. Tcoi2 protein appears to be a novel host factor capable of interacting and phosphorylating MT domain of CMV 1a protein.

Aphids transmit CMV in a non-persistent manner. A conserved capsid protein surface domain is essential for efficient aphid transmission (Liu *et al* 2002). There is a negatively charged loop structure (beta H-beta I loop) on the surfaces of the virions. Six of eight amino acids in this capsid protein loop are highly conserved and facilitate aphid vector transmission. Bowman *et al* (2002) also demonstrated that beta H-beta I loop of the capsid subunit (pentons not hexons) is potentially the key motif responsible for the interactions with the insect vector.

59. Ilarvirus [Tobacco streak virus (TSV) (ss RNA+)]

The genome is tri-partite, linear ss RNA. The complete length of the genome is 86,000 nucleotides. The encapsidated nucleic acid is mainly genomic in origin but may also contain sub-genomic RNA or mRNA derived from RNA3. Lengths of RNA1, RNA2, RNA3, RNA4, and mRNA are: 2901-4300, 2366-3700, 1605-2700, 845-1409 and 760-871 nucleotides respectively. 5′ end of the genome has a methylated cap and there is intergenic PolyA tail (ICTVdB Management 2006).

RNA1 codes for 1 protein whereas RNA2 and RNA3 codes for 2 proteins each. The protein from RNA1 and 1 protein from RNA2 are involved in the replication process. One protein of the RNA3 is involved in cell-to-cell movement and the other is involved in coat protein synthesis along with the protein from the RNA4. Coat protein is required for the activation of the replication process. The coat protein of Alfamovirus and the Ilarvirus is interchangeable. Reusken *et al* (1995) observed that CP of TSV substitutes for the early function of alfalfa mosaic alfamovirus (ALMV) CP in genome activation. Replacement of the CP gene in ALMV RNA3 with the TSV CP gene and analysis of the replication of the chimeric RNA indicated that the TSV CP could not substitute for the function of ALMV CP I asymmetric plus-strand RNA accumulation but could encapsidate the chimeric RNA and permitted a low level of cell-to-cell transport. Xin and Ding (2003) identified and characterized a spontaneous mutant of TSV that became defective in triggering recovery in tobacco plants. Genetic and molecular analyses identified an A>G mutation in the TSVnr genome that was sufficient to confer nonrecovery when introduced to TSV. The mutation was located in the intergenic region of RNA3 upstream of the mapped transcriptional start site of the coat protein mRNA.

60.   Oleavirus [Olive latent virus 2 (OLV-2) (ss RNA+)]

The genome is tri-partite, linear ss RNA. The length of the complete genome is 8301 nucleotides. The encapsidated nucleic acid is mainly genomic in origin but may also contain sub-genomic RNA, mRNA derived from genomic RNA3. Lengths of RNA1, RNA2, RNA3, RNA4 are: 3300, 2800, 2450 and 2100 nucleotides. RNA4 is not an mRNA (ICTVdB Management 2006). There are four ORFs in the type member, one each on RNA1 and RNA2 and two on RNA3. RNA1 encodes 103 kDa replication protein 1a containing methyltransferase and helicase domains; RNA2 encodes 90 kDa polymerase 2a, RNA-dependent RNA polymerase; 36 kDa movement protein 3a is encoded by 5' half of the RNA3 involved in cell-to-cell transport; 3' half of RNA3 encodes 20 kDa 3b protein expressed from the sub-genomic RNA4. .

61.   Alfamovirus [Alfalfa mosaic virus (AMV) (ss RNA+)]

The genome is tri-partite, linear, ssRNA. The length of the complete genome is 8274 or 9155 nucleotides. The encapsidated nucleic acid is mainly of genomic origin but may also contain sub-genomic RNA derived from genomic RNA3. Lengths of RNA1, RNA2, and RNA3 are: 3644, 2593 and 2037 nucleotides respectively. The length of RNA4, an encapsidated mRNA is 881 nucleotides. It is derived from the RNA3 negative strand template. The genome sequence at 3' terminus, with RNA3 has cross-linked hairpin ends.

The 5' end has a methylated cap and the 3' end has a conserved sequence (ICTVdB Management 2006).

Virus preparations contain spheroidal particles each containing two copies of RNA4. RNA of the particles is mainly stabilized by protein-RNA interactions. 3'-termini of RNA contain a homologous sequence of 145 nucleotides that can adopt two alternative conformations and represents a high-affinity binding site for CP, the other resembles a tRNA-like structure and is required for minus-strand promoter activity. RNAs require addition to the inoculum of a few molecules of coat protein (CP) per RNA molecule. RNAs 1 and 2 encode the replicate proteins P1 and P2. RNA3 encodes movement protein and CP.RNAs 1 and 2 are involved in cis- and trans-acting functions of the encoded helicase-like and polymerase-like domains (Vlot *et al* 2003).

RNA1 may encode one or two proteins. Both RNA2 and RNA3 encode 1 protein each. Proteins from RNA1 and RNA2 are involved in replication process. The protein of RNA3 is involved in cell-to-cell movement. RNA4 is involved in the process of coat protein synthesis. Coat protein is required for activation of the replication.

Vlot *et al* (2001) established the role of 3'-untranslated regions (UTR) of AMV RNAs in the formation of the transiently expressed replicase in plants and in the assembly of the virions. 3'-UTR is required for the formation of a complex with in vitro enzyme activity. RNAs1 and 2 with the 3'-UTRs deleted are encapsidated into virions by CP expressed from RNA3. Olsthoorn and Bol (2002) observed that the minus-strand promoter of the AMV is located within the 3' terminal 145 nucleotides that can adopt a tRNA-like structure (TLS). This contrasts with the sub-genomic promoter for RNA4 synthesis that requires ~ 40 nucleotides and forms a single tri-loop hairpin. A similar tri-loop hairpin, hairpin E (hpE) is crucial for minus-strand synthesis. The loop sequence of hpE is not essential for RNA synthesis whereas base-pairing capability of bases below the tri-loop is essential. Reducing the size of the bulge loop of hpE, triggers transcription from an internal site similar to the process of sub-genomic transcription. Minus-strand promoter hpE and the sub-genomic promoter hairpin are equivalent in binding the viral ploymerase. The major role of the TLS is to enforce the initiation of transcription by polymerase at the very 3' end of the genome.

Jaspars and Houwing (2002) demonstrated that a N-terminal genome-activating peptide of 25 amino acid residues of the coat protein is unable to activate the incomplete viral genome consisting of RNAs 1 and 2. RNA3 complements such an inoculum in order to produce RNA4 that triggers the process. Volt *et al* (2002) established the role of AMV methytransferase-like domain on negative-strand RNA synthesis. P1 contains a methyltransferase-like domain in its N-terminal half with a putative role in capping the viral RNAs. Six residues of this domain that are highly conserved in the methyltransferase domains. This domain has distinct roles in replication-associated functions required for negative-strand RNA synthesis. Vlot and Bol (2003) observed that the replacement of the 5'- UTR RNA1 by that of 2 or 3 yielded infectious replications. The sequence of a putative 5' stem-loop structures in RNA1 was found to be required for negative-strand RNA synthesis. Similar putative 5' stem-loop structure is present in RNA2 but not in RNA3.

Aparicio *et al* (2001) observed that the replicase of AMV recognizes the cis-acting sequences in RNA3 of the Prunus necrotic ringspot virus (PNRSV), a member of the Ilarvirus genus. Initiation of infection by AMV and PNRSV requires binding of a few molecules of coat protein (CP) to the 3' termini of the inoculum RNA, and the CPs of the two viruses are interchangeable in this early step of the replication cycle. Cis-acting sequences in PNRSV RNA3 are recognized by the AMV replicase. The AVM replicase recognized the promoter for minus-strand RNA synthesis in PNRSV movement protein and CP genes accumulated in tobacco that is a non-host for PNRSV. Neelemann *et al* (2004) observed that AMV CP stimulates translation of AMV RNA in vivo 50-100 fold. The 3'-UTR subgenomic CP messenger RNA4 contains at least two CP binding sites. A CP binding site in the 3'-terminal 112 nucleotides of RNA4 was found to be required for efficient translation of the RNA whereas an upstream binding site was not. Binding of CP to the AMV 3'-UTR induces a conformational change of the RNA but this change alone was not sufficient to stimulate translation.

Several workers studied the operations of the movement protein (Sanchez Navarro and Bol 2001, Huang *et al* 2001). It was found that C-terminal deletions up to 21 amino acids did not interfere with the function of the CP in cell-to-cell movement, although some of these mutations interfered with the virion assembly. Deletion of N-terminal 11 or C-terminal 45 amino acids did not interfere with the ability of MP to assemble into tubules consisting of a wild-type MP. All MP-deletion mutants that showed cell-to-cell and systemic movement in plants form tubular structures on the surface of protoplasts. When C-terminal 48 amino acids were replaced by the C-terminal 44 amino acids of the AMV MP, AMV/BMV chimaeric proteins permitted wild-type levels of AMV transport. The C-terminus of AMV MP, although dispensable for cell-to-cell movement, confers specificity to the transport.

The order of events in AMV transport is: the formation of tubular structures on the surface of the protoplast, association with the endoplasmic reticulum (ER), targeting to punctate structures in the cell wall of epidermis cells, movement from transfected cells to adjacent cells in the epidermis tissue, cell-to-cell movement, or long distance movement in plants.

(iii) Family: Comoviridae

62. Comovirus [Cowpea mosaic virus (CPMV) (ss RNA+)]

The genome is bi-partite, linear, ssRNA. The length of the complete genome is 9570-13500 nucleotides. The 5' end has a genome-linked protein (VPg) and the 3' end has a conserved nucleotide sequence (30-150 nucleotides in RNA1 and 50-300 nucleotides in RNA2). Poly (A) tail of variable length is present in the 3' end (ITCVdB Management 2006).

Genome expression takes place through two open reading frames (ORFs 1 and 2), one is located in RNA1 and the other is located in RNA2. Bi-partite genomes and poly-protein strategies are followed in the translation processes. ORF1 of the RNA1 (105/95 kDa) is translated to two proteins (105 kDa and 95 kDa). 95 kDa protein is further cleaved to produce two proteins (58 kDa and 60 kDa). Another protein (48 kDa) is also produced in this process. RNA-1 genes code for the proteins for cell-to- cell movement (58 kDa and 48 kDa) and coat proteins (37 kDa and 23 kDa). ORF 2 (202 kDa) of the RNA2 translates 1 protein. It is cleaved to produce 2 proteins one of which is 32k. The other part is further cleaved to produce 4 proteins (110 kDa, 84 kDa, 87 kDa and unassigned one). 110 kDa is cleaved again to produce 2 proteins (24 kDa and 87 kDa), 84 kDa is again cleaved to produce 3 proteins (58 kDa, 4 kDa and 24 kDa). The unassigned one also cleaved again to produce 2 proteins (58 kDa and 4 kDa). Genes of RNA2 codes for the proteins involved in the modulation of protease (32 kDa), VPg, protease (58 kDa), replicase (110 kDa) and RNA replication complex (58 kDa) (van Kammen and Eggen 1986). Lomonossoff and Shanks (1999) described the details of the organization, expression and function of the expressed proteins of this genome.

CPMV replication occurs in close association with small membranous vesicles in the host cell. Replication induces an extensive proliferation of endoplasmic reticulum (ER) membranes leading to the formation of these vesicles where replication takes place. Early in the infection plus-strand RNA accumulates at numerous distinct subcellular sites distributed randomly throughout the cytoplasm which rapidly coalesce into a large body located in the centre of the cell, often near the nucleus. During the course of an infection, CPMV RNA co-localizes with the 110 kDa viral polymerase and other replication proteins and is always found in close association with proliferated ER membranes. There is a role of actin (not tubulin) in establishing the large central structure. Unfolded protein response is not involved in this process (Carette *et al* 2002a). CPMV 32 kDa and 60 kDa replication proteins normally target and change the morphology of the ER membrane (Carette *et al* 2002b). CPMV RNA-1 encoded 60 kDa nucleotide-binding protein (60 kDa) and 5 host proteins also play different roles in the formation of these vesicles and replication cycle. Five cellular proteins interact with different domains of 60 kDa. Two of these host proteins interact specifically with C-terminal domain of 60 kDa and another is picked up using the central domain of this protein (Carette *et al* 2002 c).

Within the host plants viruses spread from the initially infected cell through plasmodesmata to neighbouring cells (cell-to-cell movement) until reaching the phloem.

MP protein forms tubules through plasmodesmata in infected plants and enable the virus particles to move from cell to cell. Several functional domains within the MP are involved in distinct steps during tubule formation. Dimeric or multimeric MP is first targetted to the plasma membrane where the MP quickly accumulates in peripheral spots from which the tubule formation is initiated. The breakdown or disassembly of tubules in neighbouring uninfected cells is required for cell-to-cell movement (Pouwels *et al* 2003). MP shows affinity to bind specifically to intact virions and to the large CP but not to the small CP. C-terminal 48 amino acids constitute the virion-binding domain of the MP (Carvalho *et al* 2003). N-terminal and central regions of the MP are involved in the tubule formation and the C-terminal domain has a role in the interactions with virus particles. C-terminal border of the tubule forms a domain to a small region between amino acids 292 and 298 involved in the incorporation of the virus particles in the tubule (Bertens *et al* 2003). But the process of the cell-to-cell movement is not involved in the long distance movement or phloem loading. For vascular movement CPMV is loaded into both major and minor veins but unloaded exclusively from the major veins. The virus establishes infection in all vascular cell types with the exception of companion cells (CC) and sieve elements (SE). Tubular structure indicative of virion movement are never found in plasmodesmata connecting phloem parenchyma cells and CC or CC and SE. SE is symplastically connected only to the CC-CPMV suggesting that a different mechanism is involved in phloem loading and unloading (Silva *et al* 2002).

Liu *et al* (2004) observed that CPMV RNA1 acts as amplicon whose effects can be counteracted by a RNA2 encoded suppressor of silencing. *Nicotiana benthamiana* transgenic for full-length copies of CPMV genomic RNAs either singly or together shows distinct differences. Plants transgenic for both RNAs develop symptoms characteristic of a CPMV infection. When plants transgenic for RNA1 are agroinoculated with RNA2, no infection develops and the plants are also resistant to challenge inoculation. By contrast, plants transgenic for RNA2 become infected when agroinoculated with RNA1 and are fully susceptible to CPMV infection. The resistance of RNA1 transgenic plants is related to the ability of RNA1 to self-replicate as an amplicon. The ability of the transgenetically expressed RNA2 to counteract effect suggests that it encodes a suppressor of post-transcriptional gene silencing (PTGS). Small coat protein as the CPNV RNA2 encodes suppressor of PTGS.

CPMV in the production of a new generation of vaccines
A new generation of human and animal vaccines is under development based on CPMV. The viral capsid is employed as a physical support for multiple antigen presentation.

CPMV may be considered a bi-molecular platform to display heterologous peptide sequences. Such peptide chimaeras can be easily and inexpensively produced in large quantities from experimentally infected plants. CPMV chimaeras may be a platform to create antiviral agents. The most prominent feature at the virus surface is a region of the smaller of the two coat protein(s) which has been extensively used for the insertion of foreign peptides. Given the availability of the three-dimensional strucure of the native virus and the amenability of foreign peptide-expressing CPMV chimaeras to crystallization, immunological data can be correlated with the conformational state of the foreign insert. The latter is influenced by proteolysis that occurs within the foreign inserts. A second region of the protein has tolerance with respect to smaller insertions. Techniques have been developed to make CPMV suitable for a wider spectrum of presentation, which allow surface coupling of a peptide that can serve as the anchoring point for a range of proteins. The new approach is also widely applicable for the direct chemical cross-linking of peptides and full-length protein domains to the viral capsid (Chatterjee *et al* 2002).

Khor *et al* (2002) created such an antiviral agent against the measles virus (MV) by displaying a peptide known to inhibit MV infection. This peptide sequence corresponds to a portion of the MV binding site on the human MV receptor CD46. The CPMV-CD46 chimaeras efficiently inhibit MV infection of Helea cells in vitro, while wild-type CPMV does not.

The inhibitory CD46 peptide expressed on the surface of CPMV retains virus-binding activity and is capable of inhibiting viral entry, both in vitro and in vivo. CD46 peptide present in the context of CPMV is also up to 100-fold more effective than the soluble CD46 peptide at inhibiting MV infection in vitro.

Nicholas *et al* (2002) vaccinated MIH mice sub-cutaneously or intra-nasally with chimaeric CPMV constructs expressing a 17-mer peptide sequence from canine parvovirus (CPV) as numerous dimmers on the small or large protein surface subunits. Presentation of peptides on viral particles is useful to bias the immune response in favour of TH1 response. A mucosal immune response to CPV can be generated by systemic immunization with CPV vaccines. Erum antibody from both sub-cutaneously vaccinated and intra-nasally vaccinated mice shows neutralizing activity against CPV in vitro.

CPMV has also the potentiality for its application as a building block for supra-molecules or nano-chemicals. CPMV can be isolated in gram quantities and possesses a structure known as "atomic resolution" that is quite stable. It is therefore of potential use as a molecular entity in synthesis, particularly as a building block on a nano-chemical scale. CPMV capsid possesses a lysine residue per asymmetric unit and thus 60 such lysines per virus particle. Under forcing conditions, up to four lysine residues per asymmetric unit can be addressed. In combination with engineered cysteine reactivity, this provides a powerful platform for the alteration of the chemical and physiological properties of the CPMV particle (Wang *et al* 2002).

63. **Fabavirus** [Broad bean wilt virus 1 (BBWV-1) (ss RNA+)]

The genome is bi-partite, linear ssRNA. The length of the complete genome is 10800 nucleotides. The lengths of the RNA1 and RNA2 are: c 6000 and c 3600 nucleotides respectively. The 3' terminus of the genome has a Poly A and the 5' may have a genome-linked protein (VPg). Each of the RNAs is translated into a poly-protein that is processed by a series of steps into functional proteins. RNA1 produces a 206 kDa poly-protein that is processed into the following functional proteins: (i) 35 kDa processing regulator (protease co-factor) (ii) 66 kDa nucleotide binding protein with helicase activity (iii) 3.1 kDa VPg protein (iv) 23 kDa protease and (v) 79 kDa polymerase. RNA2 produces a 114 kDa poly-protein that processed into the following functional proteins: (i) 47 kDa movement protein (ii) 44 kDa large coat protein and 23 kDa small coat protein (ICTVdB Management 2006).

Kuroda *et al* (2000) sequenced the nucleotide of the infectious RNA2 of BBWV-2. It is 3593 nucleotides in length containing a single ORF that has the potential to encode a poly-protein of 119 kDa. Direct amino acid sequencing of the virus coat proteins suggest that the poly-protein is processed to produce 3 proteins of 52 kDa, 44 kDa and 22 kDa. A full-length cDNA constructed under the control of the T7 promoter could be successfully transcribed to produce biologically active RNA. Qi *et al* (2000a) characterized RNA2 nucleotide sequence of BBWV-2 (isolate P158) and made more or less similar observations. They observed the length of RNA2 as 3.6 kb (the length of the other RNA, RNA1 is 6.0 kb). RNA2 was composed of 3597 nucleotides residues excluding Poly (A) tail and a single long ORF extending from 230 to 3424 nucleotides encoding a poly-protein of Mr 119002 (119 kDa) that cleaved at Q/G (465/466) and Q/A (867/868) to release 3 mature proteins.

Qi *et al* (2000b) made complete nucleotide sequence of RNA1 of BBWV-2 (Chinese isolate). It is 5956 nucleotides in length excluding the 3'-terminal Poly (A) tail and contains a single long ORF of 5613 nucleotides extending from nucleotide 234 to 5846. A repeated motif has been found in the 5'-non-coding region. The predicted poly-protein encoded by the long ORF is 1870 amino acid in length with a molecular weight of 210 K. Based on the determined sequence, full-length cDNA clone of RNA-1 designated as pU2FL, transcripts from pU1FL infected *Chenopodium quinoa* successfully.

Koh *et al* (2001) found that RNA1 of the BBWV (Singapore isolate) is 5951 nucleotides in length that encodes a putative protease co-factor, nucleotide triphosphate (NTP)-binding domain (helicase), viral genome linked protein VPg, protease and RNA-dependent RNA polymerase (RdRp). RNA2 (3607 nucleotides in length) encodes a putative movement protein (MP) and coat protein (CP).

Qi *et al* (2002) observed in vivo accumulation of BBWV-2 VP 37 protein and its ability to bind single-stranded nucleic acid. They found that VP 37 protein encoded by RNA2 was over-expressed in *Escherichia coli*. The protein was purified and polyclonal antibody specific for the protein was produced. Time course studies by western blot assays in BBWV-2 infected *Chenopodium quinoa* leaves showed that VP 37 protein was present in cells of the inoculated leaves by 12-h post-inoculation and in cells systemically infected leaves by 2-d post inoculation. The protein was able to be at high level in infected leaves at the late infection stage. Gel retardation and UV-cross-linking assays demonstrated that the VP 37 protein bound preferentially single stranded RNA and DNA in a non-sequence specific manner. Sequence analysis of VP 37 protein and its ability to bind ssRNA and ssDNA suggest that the protein plays a role similar to the movement proteins.

64. Nepovirus [Tobacco ring spot virus (TRSV) (ss RNA+)]

The genome is bi-partite, linear ssRNA. The length of the complete genome is 10100-16900 nucleotides. The encapsidated nucleic acid is mainly genomic in origin but may also contain satellite RNA. The 5′ end has a genome-linked protein VPg and the 3′ end has a conserved nucleotide sequence and Poly (A) tail. The lengths of RNA1 and RNA2 are 5300-9500 and 3774-7500 (ICTVdB Management 2006).

Wang *et al* (2000a) observed the diversity in the coding regions of the coat protein, VPg (genome-linked virion protein), Pro (protease) and Pol (putative RNA dependent RNA polymerase) among different isolates of tomato ring spot nepovirus ToRSV). The deduced amino acid sequence was more conserved than the nucleotide sequence. The entire VPg sequence and conserved amino acid motifs within the Pro, Pol and CP (coat protein) domains were identical in all isolates. Wang *et al* (2000 b) observed that ToRSV RNA1-encoded poly-protein (P1) contains the domains for the putative NTP-binding protein VPg, 3-C like protease and a putative RNA dependent RNA polymerase in its C-terminal region. Using partial cDNA clones, it was shown that the C-like protease processes the N-terminal region of P1 at a novel cleavage site in vitro. Chisholm *et al* (2001) observed that the 3C-like proteinase (Pro) is responsible for the processing of the RNA1-encoded (P1) and RNA2-encoded (P2) poly-proteins. Neither the Pro nor the VPg-Pro could cleave in trans P1-derived substrates containing the three cleavage sites delineating the X1, X2, putative NTP-binding protein and VPg domains and possibly controls the regulation of proteinase activity during virus replication.

Wang *et al* (2004) found that the putative NTP-binding protein of ToRSV contains 9 hydrophobic regions at a C-terminus consisting of 2 adjacent stretches of hydrophobic amino acids separated by a few amino acids. In infected plants the NTB-VPg poly-protein (containing the domain for the genome-linked protein) is associated with the endoplasmic reticulum-derived membranes that are active in ToRSV replication. The NTB-VPg poly-protein acts as a membrane anchor for the replication complex.

(iv)   Family: Luteoviridae

65.   Luteovirus [Barley yellow dwarf (BYDV) (ss RNA+)]

The genome contains one molecule, linear ss RNA. The length of the complete genome is 5641-6900 nucleotides. The 5'end of the genome has a genome-linked protein VPg. The genome encodes three or six non-structural proteins. Members of this genus help some other viruses for their vector transmission.

Mayo and Miller (1999), Smith and Barker (1999) and Miller (1999) reviewed the organization of the genomes of the members of Luteoviridae. The size of the genomes ranges from 5.6-5.7 kb. The length of the genomes also differs from member to member. Genomes of BYDV-PAV, PLRV and PEMV-1 consist of 5677, 5882 and 5705 nucleotides respectively. The number of ORFs on the genomes of different members is six and their expression depends upon the type of the member. The products of these ORFs are involved in different functions, viral replicase, proteinases, aphid transmission and expression in the vector cells, virus accumulation, etc. (Sadowy *et al* 2001) but true dissection of the functions are yet to be ascertained. Non-structural P17 and structural RT luteovirus proteins determine the biologically active regions of the capsid and RT proteins that regulate the uptake and transmission of the virus by the aphids.

The largest ORF encodes the coat, 17 kDa and readthrough proteins from 2 initiation codons. Moon *et al* (2001) investigated the role of intergenic and 3' proximal noncoding regions (NCRs) in coat protein (CP) expression and BYDV-PAV replication. A full-length infectious cDNA of the RNA genome was constructed downstream of the cauliflower mosaic virus-35S promoter. Linear DNA molecules of these cDNAs were infectious, expressed the 22 kDa CP, and produced both genomic RNA and sgRNAs in the ratios similar to those observed in protoplasts inoculated with viral RNA. The portion of 5'-NCR of sgRNA1 between ORFs 2 and 3 was not required for, but enhanced translation of CP from ORF3. Mutants containing deletions in the NCR downstream of ORF5 failed to replicate in oat protoplasts indicating that an intact 3'-NCR is required by BYDV-PAV replication.

66.   Enamovirus [Pea enation mosaic virus-1 (PEMV-1) (ss RNA+)]

The genome is tri-partite linear ssRNA. The length of the complete genome is 9958-10675 nucleotides. The 5' end has a genome-linked protein VPg. The lengths of the RNA1, RNA2 and RNA3 are 5705, 4253 and 717 nucleotides. The encapsidated nucleic acid is mainly genomic in origin but may also contain satellite RNA (RNA3) (ICTVdB Management 2006).

The genome of the type member has 5 ORFs. ORF1 produces 84 kDa replication protein; ORF2 produces 67 kDa polymerase; ORF3 produces 21 kDa coat protein and ORF5 produces 54 kDa coat protein. The VPg is involved in the replication and not the movement of the virus (Skaf *et al* 2000).

67. Polerovirus [Potato leaf roll virus (PLRV) (ss RNA+)]

The genome is bi-partite, linear ss RNA. The length of the complete genome is 5880-5990 nucleotides. The encapsidated genome is mainly genomic in origin but may also contain sub-genomic RNA, mRNA (ICTVdB Management 2006). Virus accumulation depends upon the ORF0 product in case of potato leaf roll virus (Sadoway *et al* 2001).

The gene of the viral capsid protein resides in the 3′ terminal half of the genomic RNA (Tacke *et al* 1989). The nucleotide sequence of the 3′ portion of the genome contains 2361 bases. The coat protein gene contains 624 nucleotides and encodes a protein of 208 amino acids with a calculated MW of 23.2 kDa. This gene is located internal to the 3′ end of the genome, contains the complete overlap of an ORF encoding 17.4 kDa protein, and is followed in-frame by an ORF encoding 56.4 kDa protein. There is 3′ positioned viral subgenomic RNA, not encapsidated and likely to be the mRNA for expression of the 3′ portion of the genome (Smith and Harris 1990).

(v) Family: Tymoviridae

68. Tymo virus [*Turnip yellow mosaic virus* (TYMV) (ss RNA+)]

The genome is of one molecule, linear ss RNA. The length of the complete gemone is 2000-6500 nucleotides. The 5′ end has a methylated cap; 3′ end may or may not have a Poly A tail and tRNA-like structure. The genome replicates in association with the chloroplast. The virus is associated with a helper virus but independent from its functions during replication (ICTVdB Management 2006).

Bink *et al* (2002) observed that the 5′ UTR of TYMV RNA contains two conserved hairpins with internal loops consisting of C.C and C.A mismatches. The importance of these mismatches lies in their pH-dependent protonation and stable base pair formation. Encapsidation efficiency is adversely affected in mutants lacking the protonable mismatches in the internal loop of the 5′ proximal hairpin. Stabilization of the 5′ proximal hairpin also leads to a delay in the development of symptoms on plants. This delay in symptom development for both locally and systemically infected leaves depends on a change in the free energy of the hairpin caused by introduced mutations. The accumulation of plus strand full-length RNA and sub-genomic RNA, as well as protein expression levels are affected by hairpin stability. Stabilization of the hairpin inhibits translation. The hairpin may exist in close proximity to the 5′ cap as long as its stability is low enough to enable translation. In acidic pH the hairpin structure become more stable and is functionally transformed into the initiation signal for viral packaging. Slight acidic condition can be found in the chloroplasts, whereas TYMV assembly is driven by a low pH generated by active photosynthesis (Bink *et al* 2003). Matsuda and Dreher (2004) studied the function of tRNA-like structure found in many members of TYMV. The presence of a translational enhancer is observed in the 3' UTRs of TYMV. 3' terminal 109 nucleotides comprising the tRNA-like structure and an upstream pseudo-knot act in synergy with a 5' cap to enhance translation. The process has a minor contribution in stabilizing the RNA. It seems to be a new role of a viral tRNA-like structure in production of a novel type of translational enhancer.

Barends *et al* (2003) reported a surprising interaction between TYMV RNA programmed ribosome and 3'-valydated tRNA-like structure (TLS) that yields poly-protein with the valine N terminally incorporated by a translational mechanism resistant to regular initiation inhibitors. Disruption of tRNA-like structure exclusively abolishs poly-protein synthesis that can be restored by adding excess TLS *trans*. This observation seems to be a novel eukaryotic mechanism for internal initiation of mRNA translation.

The 82 nucleotide long TLS (previously 109 nucleotides) form a stable complex with the translation elongation factor eEFIA.GTP. Transcription of the RNA by RNA-dependent polymerase to yield minus strands is initiated within 3'-CCA sequence. But the minus strand synthesis is strongly repressed upon the binding of eFFIA to the valylated viral RNA. eEFIA. GTP has no effect on RNA synthesis templated by non-aminoacylated RNA. Higher eEFIA. GTP levels are needed to repress minus strand synthesis templated by valyl-EMV TLS RNA that binds eEFIA. GTP with lower affinity than does the valyl-TYMV RNA (Matsuda *et al* 2004).

Genome expression takes place through 3 open reading frames (ORF1—206 kDa, ORF2—69 kDa, ORF3—20 kDa) following sub-genomic RNA strategy. ORF1 translates 1 protein that further cleaves to produce 2 proteins (150 kDa, 70 kDa). ORF2 and ORF3 produce 1 protein each (69 kDa and 20 kDa) respectively. The sub-genomic RNA is translated through ORF 3 and produces 20 kDa protein. The genes encode the production of 4 proteins. 150 kDa and 70 kDa proteins are involved in RNA replication and 20 kDa is the coat protein. The remaining one (69 kDa) may be involved in cell-to-cell movement (Ding *et al* 1990). This protein is recognized as a substrate for the attachment of polyubiqutin chains and for subsequent rapid and selective proteolysis by the proteasome, the ATP-dependent proteolytic system present in the reticulocytate lysate. There are 2 degradation signals within the sequence of the 69 kDa protein. This process contributes to the transient nature of their accumulation during infection (Drugeon and Jupin 2002). TYNV RNA produces a sgRNA for the expression of its coat protein. Sub-genomic promoter sequence is located on a 494-nucleotide fragment containing highly conserved sequence elements that are essential for the promoter function.

The proteins responsible for replication are 140 kDa (earlier 170) and 66 (earlier 70) proteins. The 140 kDa protein contains domains indicative of methyltransferase and NTPase/helicase. The 66 kDa protein localizes to RNA-dependent RNA polymerase domain. During viral infection, the 66 kDa protein localizes to virus-induced chloroplastic membrane vesicles closely associated with TYMV RNA replication. 140 kDa protein is a key organizer of the assembly of TYMV replication complexes and a major determinant of their chloroplastic localization and retention (Prod'homme *et al* 2003).

There is also a novel mechanism for expression of TYMV virulence. The virulence factor P69 that suppresses the si RNAs (small interferin RNAs) pathway but promotes the mi RNA (micri-RNA). P69 suppression of the si RNA pathway is upstream of ds RNA production. P69-expressing plants contain elevated levels of a dicer mRNA and miRNAs including a correspondingly mi-RNA guided cleavage of two host RNAs. P69 expressing plants exhibit disease-like symptoms in the absence of viral infection (Chen *et al* 2004).

69. Marafivirus [*Maize raydo fino virus* (MRFV) (ss RNA+)]
The genome consists of one molecule of linear ssRNA. The length of the complete genome is 6500-7500 nucleotides. There is a sub-genomic promoter, a conserved region called "Tymobox" (ICTVdB Management 2006).

(vi) Family: Sequiviridae

70. Sequivirus [*Parsnip yellow fleck* (PYFV) (ss RNA+)]
The genome consists of one molecule, linear ssRNA. The length of the total genome is 9871 nucleotides. 5' end of the genome has a genome-linked protein and the 3' end has a Poly (A) tract. The virion is associated with a helper virus but independent of its function during replication (transmission is dependent on the presence of a helper protein encoded by *anthriscus yellows* waikavirus) (ICTVdB Management 2006).

71. Waikavirus [*Rice tungro spherical virus* (RTSV) (ss RNA+)]
[*Maize chlorotic dwarf virus* (MCDV) (ss RNA+)]
The genome is mono-partite, linear ssRNA. The length of the complete genome is 10600-16670 nucleotides. There is a genome-linked proteib VPg at the 5' end and a poly A tract at the 3' end. Virions are associated with a helper virus but independent from its functions during replication. Virions act as helpers for another virus ICTVdB 2006). Shen *et al* (1993) first described nucleotide sequence and genomic organization of RTSV.

The RNA genome of RTSV is predicted to be expressed as a large poly-protein precursor. The poly-protein is processed by at least one virus-encoded proteinase located adjacent to the C-terminal putative RNA polymerase that shows sequence similarity to viral serin-like proteinases. Besides acting in *cis*, the proteinase had activity in trans on precursors containing regions of the 3' half of the poly-protein but did not process a substrate consisting of a precursor of the protein (Thole and Hull 2002).

(vii) Family: Bunyaviridae

72. **Tospovirus** [*Tomato spotted wilt virus* (ToSPWV) (ss RNA–)]

The length of the complete genome is 17200-17600 nucleotides. Lengths of the RNA of different segments are: 8800-8897 (ambisense RNA L), 2916-3500 (ambisenseRNA S) and 5400 (RNA-M) nucleotides respectively; each RNA may form a pseu-circular structure by non-covalent bonds. The genome has terminally redundant sequences. Nucleotide sequences at the 3′ terminus are complementary to the 5′ end and form a panhandle. 5′ and 3′ terminal sequence has conserved region (Büchen-Osmand 2003). de Haan *et al* (1990) recorded the ambisense character of sRNA segment of TSWV.

Non-structural NSm protein of TSWV represents nine putative viral movement protein involved in cell-to-cell movement of nonenveloped ribonucleocapsid structures through modified plasmodesmata (Soelkek *et al* 2000).

TSWV is transmitted by thrips. There may be a host factor necessary for the entry of the virus from the vector. A50 kDa protein, a viral receptor for the vector is considered as the candidate host factor for this entry (Medeiros *et al* 2000).

N-protein of the virus is highly conserved but the intergenic region of the sRNA is highly variable. Twenty- six independent sequences reveal seven variable amino acid positions. The type of amino acids in these positions seems to be independent of the geographical origin. In contrast to the structural N protein, comparison of the related IGR-sequences led to the cluster correlated to the origin of the isolates. There are three conserved parts within this region. The outstanding part is a central area of 31 nucleotides with a significantly increased G-C content. This viral RNA stabilized structure which might serve as a transcription terminator during the synthesis of the two mRNA from the ambisense segment (Heinze *et al* 2001). Takeda *et al* 2002, Sonoda 2003 demonstrated the presence of RNA silencing suppressor in TSWV. The NSs protein was found to suppress sense transgene induced PTGS (post-transcriptional gene silencing) when tested in *Nicotiana benthamiana* but did not suppress inverted repeat transgene-induced PTGS. TSWV NSs protein is the first RNA silencing suppressor identified in negative strand RNA viruses. TSWV supports both both genome replication and translation (Knippenberg *et al* 2002) but the transcription is independent of translation.

Naidu *et al* (2004) established the differences in lactin-binding properties and sensivities to glycosidases of the two envelope membrane glycoproteins of the TSWV. The N-terminal (GN) and C-terminal (GC) contain several tripeptide sequences suggesting that the proteins are N-glycosylated. GC showed strong binding with five mannose-binding lectins, four N-acetyl lactosamine-binding lectins and one frucose-binding lectin. GN was resolved into two molecular masses with three of the five mannose-binding lectins. GC is more heavily N-glycosylated than GN.

Currently 18 *tospovirus* species have been identified on the basis of serological properties and amino acid sequence identity of the nucleocapsid protein (NP) gene (Whitefield *et al* 2005; http://www.oznet.ksu.edu/tospovirus/tospo_list. htm). These viruses have a wide host range mostly species of Cucurbitaceae, fabaceae and solanaceae (Fauquet *et al* 2005; http://www.oznet.ksu.edu/ tospovirus/hostlist.htm). It appears from the comparative NP gene sequence analysis that the species varied with different hosts (Jain *et al* 2007).

(viii) Family: Pseudoviridae

73. Sirevirus [*Lycopersicon esculentum To RTLt virus* (LesTORV) (ss RNA-RT)]

> [*Glycine max SIRE virus* (GmaSIRV) (ss RNA-RT)]
>
> [*Zea mays Opie-2 virus* (ZmaOp2V) (ss RNA-RT)]
>
> [*Zea mays Prem-2 virus* (ZmaPr2V) (ss RNA-RT)]

The genome consists of a single molecule, linear ssRNA+ RT. The length of the complete genome is 4200-9700 nucleotides. The genome produces RNA-dependent DNA polymerase and repelcase non-structural proteins. Translation of the genes takes place through RT and the transcription requires host cell enzymes (ICTVdB Management 2006).

74. Pseudovirus [*Arabidopsis thaliana Evelknievel virus* (AthEveV) (ss RNA-RT)]

> [*Arabidiopsis thaliana Ta 1 virus* (ss RNA-RT)]
>
> [*Arabidiopsis thaliana AtRE1* (ss RNA-RT)]
>
> [*Arabdiopsis thaliana Art1* (ss RNA-RT)]
>
> [*Brassica oleracea Melmoth virus* (ss RNA-RT)]
>
> [*Cajanus cajan Panzee virus* (ss RNA-RT)]
>
> [*Hordeum vulgare BARE-1 virus* (ss RNA-RT)]
>
> [*Nicotiana tabacum TnT1 virus* (ss RNA-RT)]
>
> [*Nicotiana tabacum Tto1 virus* (ss RNA-RT)]
>
> [*Oryza australiensis RIRE1 virus* (ss RNA-RT)]
>
> [*Oryza longistaminata Retrofit virus* (ss RNA-RT)]
>
> Solanum tuberosum Tst1 virus (ss RNA-RT)]
>
> [*Zea mays Sto-4 virus* (ZmStonorV) (ss RNA-RT)]

The genome consists of one molecule, linear ssRNA-RT. The length of the complete genome is 4300-8900 nucleotides. The genome produces RNA dependent DNA polymerase and replicase enzymes. The infective genome integrates into the host genome (ICTVdB Management 2006). The integration is mediated by the integrase and long terrminal repeat sequences.

(ix) Family: Metaviridae

75. **Metavirus** [*Arabidopsis thaliana Athila virus* (AthAthV) (ss RNA-RT)]

[*Arabidiopsis thaliana Tat4 virus* (AthTat4V) (ss RNA-RT)]

[*Lilium henryi Del 1virus* (LheDel 1V)(ss RNA-RT)]

The genome consists of a single molecule, linear ssRNA-RT. The genome codes for RNA dependent DNA polymerase (ICTVdB Management 2006)

(x) Family: Unknown

76. **Cheravirus** [*Cherry rasp leaf virus* (CRLV) (ss RNA+)]

The genome is bi-partite, linear, ssRNA. The length of the genome is 10,000 nucleotides. The 5′ terminus has a genome-linked protein VPg and the 3′ end has a Poly A tract. Both RNA1 and RNA2 produce poly-proteinss that are cleavaged to functional proteins. RNA1 (6.5 kb) produces 235 kDa poly-protein that is processed into five polypeptides: (i) processing regulator (ii) nucleotide banding protein (iii) VPg genome-linked protein (iv) protease and (v) polymerase. RNA2 (3.4 kb) produces 108 kDa poly-protein that is processed into four functional proteins: (i) movement protein (ii) coat protein (iii) coat protein-2 and (iv) coat protein-3 (ICTVdB Management 2006).

77. **Idaeovirus** [*Raspberry bushy dwarf virus* (RBDV) (ss RNA+)]

The genome is bi-partite, linear ss RNA. The length of the complete genome is 86 nucleotides. The encapsidated nucleic acid is mainly genomic in origin but may also contain sub-genomic RNA mRNA. The lengths of RNA1, RNA2 and RNA3 are: 5400, 2200 and 1000 respectively.

Virions are associated with helper viruses but independent from its function during replication (ICTVdB Management 2006).

RNA1 encodes proteins that are involved in virus replication. Near the 3′ end of the major ORF of RNA1, there is a second overlapping but out-of-frame ORF that encodes a 12 kDa protein comparable to the 2 b protein of CMV that is not essential for infection but plays a role in long distance movement of CMV in infected plants and in the suppression of post-transcriptional gene silencing (Wood *et al* 2001).

78. **Sadwavirus** [*Satsuma dwarf virus* (SDV) (ss RNA+)]

The genome is bi-partite, linear ssRNA. The length of the complete genome is 11000-12000 nucleotides. 3′ terminus of each RNA has a Poly (A) tract and 5′ end has a genome-linked protein VPg. RNAs are translated into poly-proteins that are processed by a series of steps into functional proteins.

RNA1 (6.8 kb) produces 230 kDa poly-protein that is cleaved into 5 peptides with the following functions: (i) processing regulator (ii) nucleotide binding protein (iii) VPg genome-linked protein (iv) protease (v) polymerase. RNA2 (5.3 kb) produces 174 kDa poly-protein that is cleaved into 3 functional peptides: (i) movement protein (ii) large coat protein (iii) small coat protein (ICTVdB Management 2006).

79. Sobemovirus [*Subterranean clover mottle virus* (SCMoV) (ss RNA+)]

The genome consists of one molecule of linear ssRNA. The length of the complete genome is 4100-5700 nucleotides. The 5' end of the genome has a genome-linked protein. Encapsidated nucleic acid is mainly genomic in origin but may also contain sub-genomic satellite RNA.

The virus is associated with a helper virus but independent from its functions during replication. It acts as a helper for a satellite virus or satellite RNA (ICTVdB Management 2006).

SCMoV RNA encodes four overlapping ORFs. ORF1 and ORF4 are predicted to encode a single protein the function of which remains unclear. ORF2 is predicted to encode 2 proteins that are derived from a –1 translational frame shift between two overlapping reading frames (ORF2a, ORF2b). ORF2a contains a motif typical of chynotrypsin-like serine proteases and ORF2b has motifs characteristically present in ss RNA-dependent RNA polymerases. ORF4 is likely to be expressed from a sub-genomic RNA and encodes the viral coat protein (Dwyer *et al* 2003).

Translation of the poly-protein from two overlapping ORFs in *cocksfoot mottle virus*, by a –1 ribosomal frameshift mechanism was observed by Makinen *et al* (2000). The VPg domain was located between the serine proteinase and replicase motifs. The N-terminus of the VPg is cleaved from the poly-protein between the glutamic acid and asparagine residues. The poly-protein is processed at several additional sites. Fully processed replicase is entirely encoded by ORF2b. The potential ribosome frameshift signal and the down stream stem-loop structure are conserved in RNA sequence indicating that ORF3 might be expressed via –1 frameshifting mechanism (Lokesh *et al* 2001).

80. Umbravirus [*Pea enation mosaic virus* (PEMV) (ss RNA+)]

The genome consists of one molecule of ssRNA. The length of the complete genome is 4203-6900 nucleotides. Virions are associated with a helper virus dependent or independent of its function during replication. Transmission of the virus depends specifically up on a helper virus (commonly a *luteovirus*).

Genome of the type species has four ORFs producing the following functional proteins: (i) 37 kDa replication protein (ii) 64 kDa polymerase (probably expressed as 90-100 kDa protein) (iii) 27 kDa protein of unknown function but apparently not a coat protein probably expressed from a 29 kDa sub-genomic RNA (iv) 28 kDa movement protein probably expressed from a sub-genomic RNA.

The movement protein of the virus induces tubule formation on the surface of protoplasts and binds RNA incompletely and non-cooperatively (Nurkyanova *et al* 2001). This virus is unusual in that it lacks genetic information for a capsid protein. In mixed infection with *luteovirus*, RNA is encapsidated by the capsid protein provided from the *luteovirus* partner and the virion so composed is readily transmissible by the vector (Syller 2003). Taliansky and Robinson (2003) observed that *umbravirus* genomes encode two other proteins from almost completely overlapping ORFs. One of these two is a cell-to-cell movement protein that can mediate the transport of homologous and heterologous viral RNAs through plasmodesmata without the participation of a coat protein. The other RNA3 protein binds to viral RNA to form filamentous ribonucleoprotein (RNP) particles that have elements of helical structure. It serves to stabilize the RNA that facilitates its transport through the vascular system of the plant. It may also be involved in protection of the viral RNA from the plant's defensive RNA-silencing. The ORF3 protein also enters the cell nucleus specifically targeting the nucleolus. RNP particles represent a novel structure that may be used by umbraviruses as an alternative to classical virions (Taliansky *et al* 2003).

81. Cilevirus [(*Citrus leprosis virus* (CiLV) (ss RNA-)]
There are two types of this virus Cytoplasmic (CiLV-C) and Nuclear (CiLV-N) type.

The genome is bi-partite, linear negative sense ssRNA. The lengths of CiLV-C RNA1 and RNA2 are 3745 and 4986 nucleotides respectively. 5′ end of the genome has a methylated cap and the 3′ end has a poly A tract. There are 2 ORFs on RNA1 encoding 286 kDa poly-protein and 29 kDa protein. The poly-protein is putatively involved in replication having 4 conserved domains: methyltranferase, protease, helicase and polymerae. There are ORFs on RNA2 encoding 15, 61, 32 and 24 kDa proteins respectively. 32 kDa protein is involved in cell-to-cell movement (Locali-Fabris *et al* 2006).

References

Adams, M.J., Antoniw, J.F., Bar-Joseph, M., Brunt, A.A., Candresse, T., Foster, G.D., Martelli, G.P., Milne, P.G. and Fauquet, C.M. 2004. The new plant virus family Flexiviridae and assessment of molecular criteria for species demarcation. Archiv. Virol. 149: 1045-1060 (Published on line 2004 c Springer-Verlag 2004)

Aguilar, J.M., Franco, M., Marco, C.F., Berdiales, B., Rodriquez Cerezo, E., Truniger, V. and Aranda, M.A. 2003. Further variability within the genus *Crinivirus* as revealed by determination of the complete RNA genome sequence of cucurbit yellow stunting disorder virus. J. Gen. Virol. 84: 2555-2564

Ahlquist, P. 1999. Bromoviruses (Bromoviridae). In: Encyclopedia of Virology. A. Granoff and R.G. Webster (eds), 2nd edit. Acad. Press, San Diego, USA. pp. 198-204

Ahlquist, P., Allison, R., and Dejong, W. 1990. Molecular biology of bromovirus replication and host specificity. In: Viral and Plant Pathogens. T. P. Pirone and J.G. Shaw (eds.). Springer-Verlag, New York, USA. pp. 145-153

Angenent, G.C., van den Ouweland, J.M.V. and Bol, J.F. 1990. Susceptibility to vitro infection of transgenic tobacco plants expressing structural and non-structural genes of tobacco rattle virus. Virology 175: 191-195

Anindya, R. and Savithri, H.S. 2003. Surface exposed amino- and carboxy-terminal residues are crucial for the initiation of assembly in pepper vein-banding virus: a flexuous rod-shaped virus. Virology 316: 325-336

Anindya, R., Joseph, J., Gowri, T.D.S. and Savitri, H.S. 2004. Complete genomic sequence of pepper vein-banding virus (PVBV): a distinct member of the genus Potyvirus. Archiv. Virol. 149: 625-632

Aparicio, F., Sanchez Navarro, J.A., Olsthoorn, R.C.L., Pallas, V. and Bol, J.F. 2001. Recognition of cis-acting sequences in RNA3 of Prunus necrotic ringspot virus by the replicase of alfalfa mosaic virus. J. Gen. Virol. 82: 947-951

Ariye, O.A., Koerbler, M., Dixon, A.G.O., Atiri, G.I. and Winter, S. 2005. Molecular variability and distribution of Cassava mosaic begomoviruses in Nigeria. J. Phytopath. 153: 226-231

Aronson, M.N., Meyer, A.D., Gyorgyey, J., Katal, L., Vetten, H.J., Gronenborn, B. and Timchenko, T. 2000. Clink, a nanovirus encoded protein, binds both pRB and skP1. J. Virol. 74: 2967-2972

Aronson, M.N., Complainville, A., Clerot, D., A.l. Calde, H., Katul, L., Vetten, H.J., Gronenborn, B. and Timchanko, T. 2002. In planta protein-protein interactions assessed using a nanovirus-based replication and expression system. Plant J. 31: 767-775

Astier, S., Albouy, J., Maury, Y., Robaglia, C. and Lieoq, H. 2007. Principles of Plant Virology: Genome, Pathogenicity, Virus Ecology. Science Publishers Inc. Enfield, USA

Bai, F.W., Zhang, H.W., Yan, J., Qu, J.C., Xu, J., Wen, J.G., Ye, M.M. and Shen, D.L. 2002. Selection of phage- display peptides that bind specifically to the outer coat protein of the rice black streaked dwarf virus. Acta Virologica 46: 85-90

Baliji, S., Black, M.C., French, R., Stenger, D.C. and Sunter, G. 2004. Spinach curly top virus: a newly described Curtovirus species from Southwest Texas with congruent gene phylogenies. Phytopath. 94: 772-779

Barends, S., Bink, H.H.J., van den Worm, S.H.E., van den Pleij, C.W.A., Kraal, B. 2003. Entrapping ribosomes for viral translation: tRNA mimicry as a molecular Trojan horse. Cell Cambridge 112: 123-129

Bar-Joseph, M., Yanmg, G., Gafney, R. and Mawassi, M. 1997. Sub-genomic RNAs: The possible binding blocks for modular recombination of Closteroviridae genomes. Sem. Virol. 8: 113-119

Baumstark, T. and Ahlquist, P. 2001. The brome mosaic virus RNA3 intergenic replication enhancer folds to mimic a tRNA T (Psi) C-loop and is modified in vivo. RNA 7: 1652-1670

Bertens, P., Heijme, W., van der Wel, N. van der, Wellink, J. and van Kammen, A. 2003. Studies on the C-terminus of the cowpea mosaic virus movement protein. Arch. Virol. 148: 265-279

Bink, H.H.J., Hellendoorn, K., van der Meulen, J. and ven der Pleij, C.W.A. 2002. Protonation of non-Watson-Crick base pairs and encapsidation of turnip yellow mosaic virus RNA. Proc. Nat. Acad. Sci. USA 99: 13465-13470

Bink, H.H.J., Schirawski, J., Haenni, A.L. and Pieij, C.W.A. 2003. The 5-proximal hairpin of turnip yellow mosaic virus RNA: its role in translation and encapsidation. J. Virol. 77: 7452-7458

Boccardo, G. and Candresse, T. 2005a. Complete sequence of the RNA2 of an isolate of *White clover cryptic virus1*, type species of the genus *Alphacryptovirus*. Archiv. Virol. 150: 403-405

Boccardo, G. and Candresse, T. 2005b. Complete sequence of the RNA1 of an isolate of White clover cryptic virus1, type species of the genus Alphacryptovirus. Archiv. Virol. 150: 339-402

Bol, J.F. 2003. Alfalfa mosaic virus: coat protein-dependent initiation of infection. Mol. Plant Path. 4: 1-8

Boon, J.A. den, Chen, J.B. and Ahlquist, P. 2001. Identification of sequences in brome mosaic virus replicase protein 1a that mediate association with with endoplasmic reticulum membranes. J. Virol. 75: 12370-12381

Boonham, N., Henry, C.R. and Wood, K.R. 1995. The nucleotide sequence and proposed genome organization of oat chlorotic stunt, a new soil-borne virus of cereals. J. Gen. Virol. 76: 2025-2034

Bowen, A.J., Clarke, I.T. and Foster, G.D. 2003. Review of the genus Macluravirus of Potyviridae. Bulgarian J. Agric. Sci. 9: 277-286

Bowman, V.D., Chase, E.S., Franz, A.W.E., Chipman, P.R., Zhang, X., Perry, K.L., Baker, T.S. and Smith, T.J. 2002. An antibody to the putative aphid recognition site on cucumber mosaic virus recognizes pentons but not hexons. J. Virol. 76: 12250-12258

Braithwaite, K.S., Geijskes, R.J. and Smith, G.R. 2004. A variable region of the Sugarcane Bacilliform Virus (SCBV) genome can be used to generate promoters for transgene expression in sugarcane. Plant Cell Reports 23: 319-326

Briddon, R.W. and Markham, P.G. 2001. Complementation of bipartite begomovirus movement functions by topocuvirus and curtoviruses. Arch. Virol. 146: 1811-1819

Briddon, R.W., Watts, J., Markham, P.G. and Stanley, J. 1989. The coat protein of beet curly top virus is essential for infectivity. Virology 172: 628-633

Briddon, R.W., Bedford, J.D., Tsai, J.H. and Markham, P.G. 1996. Analysis of the nucleotide sequence of the tree hopper transmitted geminivirus, tomato pseudocurly top virus suggests a recombinant origin. Virology 219: 387-394

Briddon, R.W., Mansoor, S., Bedford, I.D., Pinner, M.S., Saunders, K., Stanley, J., Jafar, Y., Malik, K. and Markham, P.G. 2001. Identification of DNA components required for induction of cotton leaf curl disease. Virology 285: 234-243

Briddon, R.W., Bull, S.E., Amin, I., Idris, A.M., Mansoor, S., Bedford, I.D., Dhawan, P., Rishi, N., Siwatch, S.S., Abdel-Salam, A.M., Brown, J.K. and Jafar, Y. 2003. Diversity and DNA-β, a satellite molecule associated with some monopartite begomoviruses. Virology 312: 106-121

Briddon, R.W., Bull, S.E., Amin, I., Mansoor, S., Bedford, I.D., Rishi, N., Siwatch, S.S., Zafar, Y., Abdel-Salam, A.M. and Markham, P.G. 2004. Diversity of

DNA 1: A satellite-molecule associated with monopartite-DNA-β complexes. Virology 324: 462-474

Büchen-Osmond, C. (ed.) 2003. ICTVdB Management. The Earth Institute and Department of Epidemiology. Mailman School of Public Health, Columbia University, New York, NY, USA

Buck, K.W. 1999. Geminiviruses (Geminviridae). In: Encyclopedia of Virology. A. Granoff and R.G. Webster (eds), 2nd edit, Academic Press, San Diego, USA. pp. 597-606

Bull, S.E., Tsai, W.S., Briddon, R.W., Markham, P.G., Stanley, J. and Green, S.K. 2004. Diversity of begomovirus DNA-β satellites of non-malvaceous plants in East and South East Asia. Arch. Virol. 149: 1193-1200

Bureau, M., Leh, V., Haas, M., Geldreich, A., Ryabova, L., Yot, P. and Keller, M. 2004. P6 protein of cauliflower mosaic virus, a translation re-initiator, interacts with ribosomal protein L13 from *Arabidopsis thaliana*. J. Gen. Virol 85: 3765-3775

Cabauatan, P.Q., Melcher, U., Ishikawa, K., Omura, T., Hibino, H., Koganezawa, H. and Azzam, O. 1999. Sequence changes in six variants of rice tungro bacilliform virus and their phylogenetic relationships. J. Gen. Virol. 80: 2229-2237

Calvert, L.A., Ospina, M.D. and Shepherd, R.J. 1995. Characterization of cassava vein mosaic virus: a distinct plant pararetrovirus. J. Gen. Virol. 76: 1271-1278

Canto, T., Choi, S., Palukaitis, P. and Choi, S.K. 2001. A subpopulation of RNA1 of cucumber mosaic virus contains 3' termini originating from RNA2 or 3. J. Gen. Virol. 82 : 941-945

Cao, X.S., Zhou, P., Zhang, X.M., Zhu, S.F., Zhong, X.H., Xiao, Q., Ding, B. and Li, Y. 2005. Identification of an RNA silencing suppressor from a plant double-stranded RNA virus. J. Virol. 79: 13018-13027

Caratte, J.E., van Lent, J., McFarlane, S.A., Wellink, J., van Kammen, A. 2002a. Cowpea mosaic virus 32- and 60-kilodalton replication proteins target and change the morphology of endoplasmic reticulum membrane. J. Virol. 76: 6293-6301

Cardoso, J.M.S., Felix, M.R., Clara, M.I.E. and Oliveira, S. 2005. The complete genome sequence of a new necrovirus isolated from *Olea europea* L. Archiv. Virol. 150: 815-823

Carette, J.E., Guhl, K., Wellink, J. and van Kammen A. 2002b. Coalescence of the sites of cowpea mosaic virus RNA replication into a cytopathic structure. J. Virol. 76 : 6235-6243

Carette, J.E., Verver, J., Martens, J., van Kampen, T., Wellink, J. and van Kammen, A. 2002c. Characterization of plant proteins that interact with cowpea mosaic virus '60 k' protein in the yeast two-hybrid system. J. Gen. Virol. 83: 885-893

Carvalho, C.M., Wellink, J., Rebeiro, S.G., Goldbach, R.W. and van Lent, J.W.M. 2003. The C-terminal region of the movement protein of cowpea mosaic virus is involved in binding to the large but not to the small coat protein. Virology 84: 2271-2277

Casado, C.G., Ortiz, G.J., Padron, E., Bean, S.J., McKenna, R., Agbandje, McKenna, M. and Boulton, M.I. 2004. Characterization of subgenomic DNAs encapsidated in "single" T=1 isometric particles of Maize streak virus. Virology 323: 164-174

Cercovska, N., Moravec, T., Rosecka, P., Filigarova, M. and Pecenkova, T. 2003. Nucleotide sequence of coat protein coding regions of six potato mop-top virus isolates. Acta Virologica 47: 37-40

Champagne, J., Benhamou, N. and Leclerc, D. 2004. Localization of the N-terminal domain of cauliflower mosaic virus coat protein precursor. Virology 324: 257-262

Chatterjee, A., Burns, L.L., Taylor, S.S., Lomonossof, G.P., Johnson, J.E., Lin, T. and Porta, C. 2002. Cowpea mosaic virus: from the presentation of antigenic peptides to the display of active biomolecules. Intervirology 45: 362-370

Chen, J.P. and Chen, J.P. 2000. Genome organization of soil-borne wheat mosaic virus and analysis of spontaneous deletion mutants of its RNA2. Acta Agriculturae Zhejiangensis 12: 241-250

Chen, J., Du, J. and Adams, M.J. 2000a. Sequence diversity in the coat protein coding region of wheat yellow mosaic bymovirus isolates from China. J. Phytopath. 148: 515-521

Chen, J., Chen, J.P., Yang, J.P., Cheng, Y., Diao, A. Adams, M.J. and Du, J. 2000b. Differences in cultivar responses and complete sequence analysis of two isolates of wheat yellow mosaic bymovirus in China. Plant Pathol. 49: 370-374

Chen, J.B. and Ahlquist, P. 2000. Brome mosaic virus polymerase-like protein 2a is directed to the endoplasmic reticulum by helicase-like viral protein 1a. J. Virol. 74: 4310-4318

Chen, J., Noueity, A. and Ahlquist, P. 2001. Brome mosaic virus protein 1a recruits viral RNA2 to RNA replication through a 5' proximal RNA2 signal. J. Virol. 75: 3207-3219

Chen, J.B., Noueiry, A. and Ahlquist, P. 2003. An alternative pathway for recruiting template RNA to the brome mosaic virus RNA. J. Gen. Virol. 77: 2568- 2577

Chen, J., Feng, M,G., Zhang, Y.Z., Chen, J.S., Feng, M.G. and Zhang, Y.Z. 2001. Cloning and sequencing of a cucumber mosaic satellite RNA isolated from *Raphanus sativus* and comparison of its sequence with 12 satellite RNAs. J. Zhejiang Univ. Agric. Life Sci. 27: 249-254

Chen, J., Li, X.W., Xie, D.X., Peng, J.R. and Ding, S.W. 2004. Viral virulence protein suppresses RNA-silencing mediated defence but upregulates the role of micro-RNA in host gene expression. Plant Cell 16:1302-1313

Chen, J., Zheng, H.Y., Antoniw, J.F., Adams, M.J., Chen, J.P. and Lin, L. 2004. Detection and classification of allexiviruses from garlic in China. Archiv. Virol. 149: 435-445

Cheng, C.P. and Nagy, P.D. 2003. Mechanism of RNA recombination in carom- and tombusviruses: evidence for template switching by the RNA dependent RNA polymerase in vitro. J. Virol. 77: 12033-12047

Cheng, J., Shi, N., Cheng, Y., Diao, A., Chen, J., Wilson, T.M.A., Antoniw, J.F. and Adams, M.J. 1999. Molecular analysis of barley yellow mosaic virus isolates from China. Virus Research 64: 13-21

Chisholm, J., Wieczorek, A. and Sanfacon, H. 2001. Expression and partial purification of recombinant tomato ring spot nepovirus 3C-like proteinase: comparison of the activity of the mature proteinase and the VPg-proteinase precursor. Virus Research 79: 153-164

Choi, I.R., Horken, K.M., Stenger, D.C., and French, R. 2002. Mapping of the P1 proteinase cleavage site in the polyprotein of wheat streak mosaic virus (genus Tritimovirus). J. Gen. Virol. 83: 443-450

Choi, Y.G. and Rao, A.L.N. 2003. Packaging of brome mosaic virus RNA3 is mediated through a bipartite signal. J. Virol. 77: 8750-9757

Chu, M., Doivoyes, B., Turina, M., Noad, R. and Scholthof, K.B.G. 2000. Genetic dissection of tomato bushy stunt p19-protein mediated host-dependent symptom induction and systemic invasion. Virology 266: 79-87

Cillo, F., Roberts, I.M. and Palukaitis, P. 2002. In situ localization and tissue distribution of the replication-associated proteins of cucumber mosaic virus in tobacco and cucmber. J. Virol. 76: 10654-10664

Comchan, P., Miranda, G.J. and Shirako, Y. 2002. Detection of rice grassy stunt virus tenuivirus non-structural proteins P2, P5, and P6 from infected rice plants and from viruliferous brown plant hoppers. Arch. Virol. 147: 2291-2300

Conci, L.R., Nishizawa, Y., Saito, M., Date, T., Hasegawa, A., Miki, K. and Hibi, T. 1993. A strong promoter fragment from the large noncoding region of soybean chlorotic mottle virus DNA. Annals. Phytopath. Soc. Japan 59: 432-437

Coutts, R.H.A. and Livieratos, I.C. 2003. Nucleotide sequence and genome organization of cucurbit yellow stunting disorder virus. Archiv. Virol. 148: 2055-2062

Cowan, G.H., Lioliopouiou, F., Ziegler, A. and Torrance, L. 2002. Potato mop-top virus triple geneblock proteins. Virology 298: 106-115

Crosslin, J.M., Thomas, P.E. and Hammond, R.W. 2003. Genetic variability of genomic RNA2 of four tobacco rattle tobravirus isolates from potato fields in the Northwestern United States. Virus Research 96: 99-105

Dai, S.H., Zhang, Z.H., Chen, S.Y. and Beachy, R.N. 2004. RF2b, a rice bZip transcription activator interacts with RF2a and is involved in symptom development of rice tungro disease. Proc. Natl. Acad. Sci. USA 101: 687-692

Dai, S., Zhang, Z., Bick, J. and Beachy, R.N. 2006. Essential role of the Box II cis element and cognate host factors in regulating the promoter of Rice tungro bacilliform virus. J. Gen. Virol. 87: 715-722

Damayanti, T.A., Taukaguchi, S. and Mise, T. 2003. Cis-Acting elements required for efficient packaging of brome mosaic virus RNA3 in barley protoplasts. J. Virol. 77: 9979-9986

Dasgupta, I., Hull, R., Eastop, S., Poggi-Pollini, C., Blakebrough, M., Boulton, M. and Davies, J.W. 1991. Rice tungro bacilliform virus DNA independently infects rice after *Agrobacterium*-mediated transfer. J. Gen. Virol. 72: 1215-1221

Davies, P.H., Townsend, R. and Stanley, J. 1987. The structure, expression, functions and possible exploitation of geminivirus genomes. Adv. Plant Sci. 4: 31-52

de Haan, P., Wagemakers, L., Peters, D. and Goldback, R. 1990. The sRNA segment of tomato spotted wilt virus has an ambience character. J. Gen. Virol. 71: 1001-1007

De Kochko, A., Verelagner, B, Taylor, N., Carcamo, R., Beachy, R. and Fauquet, C. 1998. Cassava vein mosaic virus (CsVMV) type species for a new genus of plant double stranded DNA viruses? Archiv. Virol. 143: 945-962

Dell'Orco, M., Salderelli, P., Minafra, A., Boscia, D. and Gallitelli, D. 2002. Epitope mapping of grapevine A capsid protein. Archiv. Virol. 147: 627-634

Desvoyes, B. and Scholthof, H.B. 2002. Host-dependent Recombination of a *Tomato bushy stunt virus* Coat Protein Mutant Yields Truncated Capsid Subunits that Form Virus-like Complexes which Benefit Systemic Spread. Virology 304: 434-442

Desvoyes, B., Faure Rabasse, S., Chen, M.H., Park, J.W. and Scholthof, H.B. 2002. A novel plant homeo domain protein interacts in a functionally relevant manner with a virus movement protein. Plant Physiol. 129: 1521-1532

Dinant, S., Ripoll, C., Pieper, M. and David, C. 2004. Phloem specific expression driven by *wheat dwarf geminivirus* V-sense promoter in transgenic dicotyledonous species. Physiol. Plantarum 121: 108-116

Ding, S.W., Howe, J., Keese, P., Mackenzie, A., Meek, D., Osorio-Keese, M., Skotnicki, M., Srijah, P., Torronen, M. and Gibbs, A. 1990. The tymobox, a sequence shared by most tymoviruses: its use in molecular studies of tymoviruses. Nucl. Acids Res. 18: 1181-187

Distefanco, A.J., Conci, L.R., Munoz, Hidalgo, M., Guzman, F.A., Hopp, H.E., del Vas M. 2003. Sequence and phylogenetic analysis of genome segments S1, S2, S3, S6 of Mal de Rio Cuarto virus, a new accepted Fiji species. Virus Res. 92: 113-121

Dolja, V.V., Karasev, A.B. and Koonin, E.V. 1994. Molecular biology and evolution of Closteroviruses: Sophisticated build-up of large RNA genomes. Annu. Rev. Phytopathol. 32: 261-285

Dry, I.B., Krake, L., Rigden, J.E. and Rezaian, M.A. 1997. A novel subviral agent associated with a geminivirus: The first report of a DNA satellite. Proc. Natl. Acad. Sci. USA 94: 7088-7093

Drucker, M., Froissart, R., Herbrad, E., Uzest, M., Ravallec, M., Esperandieu, P., Mani, J.C,, Pugniere, M., Roquet, F., Fereres, A. and Blanc, S. 2002. Intracellular distribution of viral gene products regulates a complex mechanism of cauliflower mosaic virus acquisition by its aphid vector. Proc. Natl. Acad. Sci. USA 99: 2422-2427

Drugen, G. and Jupin, I. 2002. Stability in vitro of the 69 K movement protein of Turnip yellow mosaic virus is required by the ubiquitin-mediated proteasome pathway. J. Gen. Virol. 83: 3187-3197

Dugdale, B., Becker, D.K., Harding, R.M. and Dale, J.L. 2001. Intron-mediated enhancement of the banana bunchy top virus DNA-6 promoter. Plant-cell Reports 20: 220-226

Dumoyer, P., Herzog, E., Hemmer, O., Ritzenthaler, C. and Fritsch, C. 2001. Peanut virus 1-encoded P15 regulates viral RNA accumulation but is not abundant at viral RNA replication sites. J. Virol. 75: 1941-1948

Dumoyer, P., Pfeffer, S., Fritsch, C., Hammer, O., Voinet, O. and Richards, K.E. 2002. Identification, subcellular localization and some properties of a cysteine-rich suppressor of gene silencing encoded by peanut clump virus. Plant J. 29: 555-567

Dwyer, G.I., Njeru, R., Williamson, S., Fosu Nyarko, J., Hopkins, R., Jones, R.A.C., Waterhouse, P.M. and Jones, M.G.K. 2003. The complete nucleotide sequence of subterranean clover mottle virus. Archiv. Virol. 148: 2237-2247

Erhardt, M., Dunoyer, P., Guilley, H., Richards, K., Jonards, G. and Bouzaubaa, S. 2001. Beet necrotic yellow vein virus particles localize to mitochondria during infection. Virology 286: 256-262

Fan, Z., Dahal, G., Dasgupta, I., Hay, J. and Hull, R. 1996. Variation in the genome of rice tungro bacilliform virus: molecular characterization of six isolates. J. Gen. Virol. 77: 847-854

Fauquet, C.M., Mayo, M.A., Maniloff, J., Dusselberger, U. and Ball, L.A. (eds). 2005. Virus Taxonomy. Classification and Nomenclature of Viruses. Eighth Report of the International Committee on the Taxonomy of Viruses. Elsevier/Academic Press, Amsterdam, The Netherlands.

Froissart, R., Uzest, M., Ruiz-Ferrer, V., Drucker, M., Hebrard, E., Hohn, T. and Blanc, S. 2004. Splicing of cauliflower mosaic virus 35 S RNA serves to down-regulate a toxic gene product. J. Gen. Virol. 85: 2719-2726

French, R. and Stenger, D.C. 2005. Genome sequences of Agropyron mosaic virus and Hordeum mosaic virus support reciprocal monophyly of the genera Potyvirus and Rymovirus in the family Potyviridae. Arch. Virol. 150: 299-312

Fujisaki, K., Kaido, M., Mise, K. and Okuno, T. 2003. Use of spring beauty latent virus (SBLV) to identify compatible interaction between bromovirus components required for virus infection. J. Gen. Virol. 84: 1367-1375

Fütterer, J., Rothnie, H.M., Hohn, T. and Potrykus, I. 1997. Rice tungro bacilliform virus open reading frames I and III are translated from polycistronic pregenomic RNA by leaky scanning. J. Virol. 71: 7984-7989

Galiakparov, N., Tanne, E., Mawassi, M., Gafny, R. and Sela, I. 2003a. ORF5 of grapevine virus A encodes a nucleic acid binding protein and affects pathogenesis. Virus Genes 77: 257-262

Galiakparov N., Goszczynski, D.E., Che, Xi, B., Batuman, O., Bar-Joseph, M., Mawassi, M. and Che, X.B. 2003b. Two classes of sub-genomic RNA of grapevine A virus produced by internal controller elements. Virology 312: 434-448

Galiakparov N., Tanne, E., Sela, I. and Gafny, R. 2003c. Functional analysis of the grapevine virus A genome. Virology 306: 42-50

Geri, C., Love, A.J., Cechini, E., Barrett, S.J., Laird, J., Covey, S.N. and Milner, J.J. 2004. Arabidopsis mutants that suppress the phenotype induced by transgene-mediated expression of cauliflower mosaic virus (CaMV) gene VI are less susceptible to CaMV infection and show reduced ethylene sensitivity. Pl. Mol. Biol. 56: 111-124

German-Retana, S., Candresse, T. and Martelli, G. 1999. Closteroviruses (Closteroviridae). In: Encyclopedia of Virology. A. Granoff and R.G. Webster (eds). 2nd edit. Acad. Press, San Diego, USA. pp. 266-273

Ghorbel, R., Lopez, C., Fagoaga, C., Moreno, P., Navarro, L., Flores, R. and Pena, L. 2001. Transgenic citrus plants expressing the citrus tristeza virus p23 protein exhibit virus-like symptoms. Molecular Plant Path. 2: 27-36

Gorshkova, E.N., Erokhina, T.N., Stroganova, T.A., yelina, N.E., Zamyatnin, A.A. Jr., Kalinina, N.O., Schiemann, J., Solovyev, A.G. and Morozov, S.Y. 2003. Immunodetection and fluorescent microscopy of transgenetically expressed hordeivirus TGBp3 movement protein reveals its association with endoplasmic reticulum elements in close proximity to plasmodesmata. J. Gen. Virol. 84: 985-995

Groneborn, B. 2004. Nanoviruses: genome organization and protein function. Vet. Microbiol. 98: 103-109

Guan, H.C., Carpenter, C.D. and Simons, A.E. 2000a. requirement of a 5'-proximal linear sequence of minus strands for plus-strand synthesis of a satellite RNA associated with turnip crinkle virus. Virology New York 268: 353-363

Guan, H.C., Capenter, C.D. and Simons, A.E. 2000b. Analysis of cis-acting sequences involved in plus-strand synthesis of a turnip crinkle virus-associated satellite RNA identifies a new carmovirus replicon element. Virology New York 268: 345-353

Guerra Peraza, O., de Tapia, M., Hohn, T. and Hemmings-Mieszczak, M. 2000. Interaction of the cauliflower mosaic virus coat protein with pregenomic RNA leader. J. Virol. 74: 2067-2072

Guetierrez, C. 2002. Strategies for Gemini virus DNA replication and cell cycle interference. Physiol. Mol. Pl. Pathol. 60: 219-230

Haas, M., Bureau, M., Geldreich, A., Yot, P. and Keller, M. 2002. Cauliflower mosaic virus: still in the news. Mol. Pl. Pathol. 3: 419-429

Haasnoot, P.C.J. and Olsthoorn, R.L.C. 2002. The Brome mosaic virus subgenomic promoter hairpin is structurally similar to iron-responsive elements and functionally equivalent to the minus-strand core promoter stem-loop C. RNA 8: 110-122

Harper, G., Hull, R., Lockhart, B. and Olszewski, N. 2002. Viral sequences integrated into plant genomes. Annu. Rev. Phytopath. 40: 119-136

Harper, G., Richert-Poggeler, K.R., Hohn, T. and Hull, R. 2003. Detection of petunia vein-clearing virus: model for the detection of DNA viruses in plants with homologous endogenous pararetrovirus sequences. J. Virol. Methods 107: 177-184

Hasegawa, A., Verver, J., Shimada, A., Saito, M., Goldbach, R., van Kamen, A., Miki, K., Kameya-Iwaki, M. and Hibi, T. 1989. The complete sequence of soybean chlorotic mosaic virus DNA and the identification of a novel promoter. Nucleic Acids Res. 17: 9993-10013

Hay, J..M., Jones, M.C., Blakebrough, M.L., Dasgupta, I., Davies, J.W. and Hull, R. 1991. An analysis of the sequence of an infectious clone of rice tungro bacilliform virus, a plant retrovirus. Nucleic Acids Res. 19: 2615-2621

Hayano Saito, Y. 2002. Coding strategy of rice stripe virus and genetic analysis of rice stripe resistance. Res. Bull. Hokkaido Natl. Agric. Wept. Sta. 175: 1-45

He, X., Fütterer, J., Hohn, T. and He, X. 2002. Contribution of downstream promoter elements to transcriptional regulation of rice tungro bacilliform virus promoter. Nucleic Acids Res. 30: 497-506

Hearne, P.Q., Knorr, D.A., Hillman, B. and Norris, T.J. 1990. The complete genome structure and synthesis of infectious RNA from clones of tomato bushy stunt virus. Virology 177: 141-151

Hebrard, E., Drucker, M., Leclerc, D., Hohn, T., Uzest, M., Froissart, R., Strub, J.M., Sanglier, S., van Dorssalaer, A., Padilla, A., Labesse, G. and Blanc, S. 2001. Biochemical composition of the helper component of *Cauliflower mosaic virus*. J. Virol. 75: 8538-8546

Hefferon, K.L. and Dugdale, B. 2003. Independent expression of Rep and Rep A and their roles in regulating bean yellow dwarf virus replication. J. Gen. Virol. 84: 3465-3472

Heinze, C., LetSchert, B., Hristova, D., Yankulova, M., Kauadjouor, O., Willingmann, P., Atanassov, A. and Adam, J. 2001. Variability of the N-protein and the inter-genic region of SRNA of tomato spotted wilt virus TSWV. Microbiologica 24: 175-187

Heinze, C., Willingmann, P., Schwach, F. and Adam, G. 2003. An unusual large inter-genic region in S-RNA of a Bulgarian tomato spotted wilt virus isolate. Arch. Virol. 148: 199-205

Hemmer, O., Dunoyer, P., Richards, K. and Fritsch, C. 2003. Mapping of viral RNA sequences required for assembly of peanut clump virus particles. J. Gen. Virol. 84: 2585-2594

Hermandez, C., Carette, J.E., Brown, D.J.F. and Bol, J.F. 1996. Serial passage of tobacco rattle virus under different selection conditions results in deletion of structural and nonstructural genes in RNA2. J. Virol. 70: 4933-4940

Herzog, E., Guerra Peraza, O. and Hohn, T. 2000. The rice tungro bacilliform virus gene II product interacts with the coat protein domain of the viral gene III poly-protein. J. Virol. 74: 2073-2083

Hii, G., Pennington, R., Hartson, S., Taylor, C.D., Lartey, R., Williams, A., Lewis, D. and Melcher, U. 2002. Isolate-specific synergy in disease symptom between cauliflower mosaic and turnip vein clearing viruses. Arch. Virol. 147: 1371-1364

Hillman, B.I. and Nuss, D.L. 1999. Phytoreoviruses (Phytoreoviridae). In: Encyclopedia of Virology, 2nd edn. A. Granoff and R.D. Webster (eds.). Academic Press, San Diego, USA. pp. 1262-1267

Hirata, H., Lu, X.Y., Yamaji, Y., Kagiwada, S., Ugaki, M., Namba, S. and Lu, X.Y. 2003. A single silent substitution in the genome of Apple stem grooving virus causes symptom attenuation. J. Gen. Virol. 84: 2579-2583

Hohn, T. 1999. Plant pararetroviruses—Caulimoviruses: molecular biology. In: Encyclopedia of Virology A. Granoff and R.G. Webster (eds). 2nd edit. Acad. Press, San Diego, USA. pp. 1281-1285

Horser, C.L., Harding, R.M. and Dale, J.L. 2001. Banana bunchy top nanovirus DNA1 encodes the 'master' replication initiation protein. J. Gen. Virol. 82: 459-464

Huang, Z., Han, Y., Howell, S.H., Huang, Z. and Han, Y. 2000. Formation of surface tubules and fluorescent foci in *Arabidopsis thaliana* protoplasts expressing a fusion between the green fluorescent protein and the cauliflower mosaic virus movement protein. Virology 271: 58-64

Huang, M., Jongejan, L., Bol, J.F., Huan, M., Zheng, H. and Zhang, L. 2001. Intracellular localization and movement of phenotypes of Alfalfa mosaic virus movement protein mutants. Mol. Plant Microbe Interactions 14: 1063-1074

Hughes, A.L. 2004a. Birth and death evolution of protein-coding regions in the multi-component genomes of nanoviruses. Molecular Phytogenetics and Evolution 30: 287-291

Hughes, A.L. 2004b. Birth and death evolution of protein-coding regions and concerted evolution of non-coding regions in the multi-component genomes of nanoviruses. Molecular Phylogenetics and Evolution 30: 287-294

Hull, R. 1996. Molecular biology of rice tungro viruses. Annu. Rev. Phytopathol. 34: 275-297

Hull, R. 2002. Matthew's Plant Virology. Acad. Press, San Diego, USA

Hull, R., Covey, S.N. and Dale, P. 2000. Genetically modified plants and the 35 S promoter: assessing risks and enhancing the debate. Microbial Ecology in Health and Disease 12: 1-5

Huppert, E., Szilassy, D., Salanki, K., Diveki, E. and Balazs, E. 2002. Heterologous movement protein strongly modifies the infection phenotypes of cucumber mosaic virus. J. Virol. 76: 3554-3557

ICTVdB Management 2006. In: ICTVdB—The Universal Database version 4. B. Osmond (ed). Columbia University, New York, USA. virus open reading frame 2 product. Virology 239: 352-359

Ikeda, K., Nagaoka, S., Winkler, S., Kotani, K., Yagi, H., Nakanishi, K., Miyajima, S., Kobayashi, J. and mori, H. 2001. Molecular characterization of *Bombyx mori* cytoplasmic polyhydrosisi virus genome segment 4. J. Virol. 75: 988-995

Jacquot, E., Keller, M. and Yot, P. 1997. A short basic domain supports a nucleic acid bonding activity in the rice tungro bacilliform virus open reading frame 2 product. Virology 239: 352-359

Jain, R.K., Bag, S., Umamaheswaran, K. and Mandal, B. 2007. Natural Infection by Tospoviruses of Cucurbitaceous and Fabaceous vegetable Crops in India. J. Phytopathol. 155: 22-25

Jaspars, E.M.J. and Houwing, C.J. 2002. A genome-activating N-terminal coat protein peptide of alfalfa mosaic virus is able to activate infection by RNAs 1, 2 and 3 but not by RNAs 1 and 2. Further support for the messenger release hypothesis. Arch. Virol. 147: 857-863

Jelkmann, W. 1994. Nucleotide sequences of apple stem pitting virus and of the coat protein gene of a similar virus of pear associated with vein yellows disease and their relationship wit Potex- and Carlaviruses. J. Gen. Virol. 75: 1535-1542

Johnson, J.A., Bragg, J.N., Lawrence, D.M. and Jackson, A.O. 2003. Sequence elements controlling expression of Barley stripe mosaic virus subgenomic RNAs in vivo. Virology 313: 66-80

Kagowada, S., Yamaji, Y., Komatsu, K., Takahashi, S., Mori. T., Hirata, H., Suzuki, M., Ugaki, M. and Namba, S. 2005. A single amino acid residue of RNA-dependent RNA polymerase in the potao virus X genome determines the symptoms in *Nicotiana* plants. Virus Research 110: 177-182

Kakul, L., Timchenko, T., Gronenborn, B. and Velten, H.J. 2000. Genome organization, Variability and Control of Faba bean necrotic yellows virus, a nanovirus. Beitrage zur Zuchtungsforschung Bundesanstalt fur Zuchtangsforscung an kulturpflanzen 6: 87-91 2002. RNA helicase activity of the plant virus movement

Kalinina, N.O., Rakitina, D.V., Solovyev, A.G., Schiamann, J. and Morozov, S.F. 2002. RNA helicase activity of the plant virus movement proteins encoded by the first gene of the tripple geneblock. Virology 296: 321-328

Kalinina, N.O., Rakitina, D.A., Velina, N.E., Zamyatnin, A.A. Jr., Stroganovo, T.A., Klinov, D.V., Prokhorov, V.V., Ustinov, S.V., Chernov, B.K., Schiemann, J., Solovyev, A.G., and Morozov, S.Y. 2001. RNA-binding properties of the 63 kDa proteins encoded by the tripple geneblock of poa semilatent hordeivirus. J. Gen. Virol. 82: 2569-2578

Kanyuka, K.V., Vishnichenko, V.K., Levay, K.E., Kondrikov, D.Y., Ryabov, E.V. and Zavriev, S.K. 1992. Nucleotide sequence of shallot virus X RNA reveals a 5′ proximal cistron closely related to those of potexviruses and a unique arrangement of the 3′- proximal cistrons J. Gen. Virol. 73: 2553-2560

Karsies, A., Merkle, T., Szurek, B., Bonas, U., Hohn, T. and Leclerc, D. 2002. Regulated nuclear targeting of cauliflower mosaic virus. J. Gen. Virol. 83: 1783-1790

Kashiwazaki, S. and hibino, H. 1996. Genomic reassortment of barley mild mosaic virus: evidence for the involvement of RNA1 in pathogenesis. J. Gen. Virol. 77: 581-585

Khor, I.W., Lin, T.W., Langedijk, J.P.M., Johnson, J.E. and Manchester, M. 2002. Novel strategy for inhibiting viral entry by use of a cellular receptor-plant virus chimaera. J. Virol. 76: 4412-4415

Kim, K.H. and Kim, K.H. 2001. Regulatory viral and cellular elements required for Potato virus X replication. Plant Path. J. 17: 115-122

Kim, C., Kao, C.C., Tinoco, Jr., I. and Kim, C.H. 2000. RNA motifs that determine specificity between a viral replicase and its promoter. Nature Structural Biology 7: 415-423

Kim, S.H., Kalinina, N.O., Andreev, I., Ryabov, E.V., Fitzgerald, A.G., Taliansky, M.E., Palukaitis, P. and Kim, M.J. 2004. The C-terminal 33 amino acids of cucumber mosaic virus 3a protein affect virus movement, RNA binding and inhibition of infection and translation. J. Gen. Virol. 85: 221-230

Kim, M.J., Ham, B.K. and Paek, K.H. 2006. Novel protein kinase interacts with the Cucumber mosaic virus 1a methyltransferase domain. Biochem. Biophys. Res. Comm. 340: 228-235

Klaassan, V.A., Boeshore, M.L., Koonin, E.V., Tian, T. and Falk, B.W. 1995. Genome structure and phylogenetic analysis of lettuce infectious yellows virus, a whitefly-transmitted bipartite crinivirus. Virology 208: 99-110

Klerks, M.M., Lindner, J.L., Vaskova, D., Spak, J., Thompson, J.R., Jelkmann, W. and Schoen, C.D. 2004. Detection and tentative grouping of strawberry crinkle virus isolates. Euro. J. Pl. Path. 110: 45-52

Knippenberg, I., van, Goldbach, R. and Kormelink, R. 2002. Purified Tomato spotted wilt virus particles support both genome replication and transcription in vitro. Virology 303: 278-286

Knippenberg, I. van, Goldbach, R. and Kormelnk, R. 2004. *In vitro* transcription of Tomato spotted wilt virus is independent of translation. J. Gen. Virol. 85: 1335-1338

Kobayashi, K. and Hohn, T. 2004. The avirulence domain of cauliflower mosaic virus transactivator/viroplasmin is a determinant of viral virulence in susceptible hosts. Mol-Plant-Microbe-Interactions 17: 475-483

Kobori, T., Miyagawa, M., Nishioka, K., Ohki, S.T. and Osaki, T. 2002. Amino acid 129 of cucumber mosaic virus coat protein determines local symptom expression and systemic movement in *Tetragonia expansa*, *Momordica charantia* and *Physalis floridana*. J. Gen. Plant Path. 68: 81-88

Koh, L.H., Cooper, J.I. and Wong, S.M. 2002. Complete sequences and phylogenetic analyses of a Singapore isolate of broad bean wilt fabavirus. Arch. Virol. 146: 135-147

Koh, C.Y.K., Liu, D.X. and Wong, S.M. 2002. A six nucleotide segment within the 3′ untranslated region of hibiscus chlorotic ring spot virus plays an essential role in translational enhancement. J. Virol. 76: 1144-1173

Kouchkovasky, F. 2000. The master rep concept in nanovirus replication: identification of missing genome components and potential for natural genetic re-assortment. Virology 274: 189-195

Kreuze, J.F., Savenkov, E.I. and Valkonen, J.P.T. 2002. Complete genome sequence and analyses of the sub-genomic RNAs of sweet potato chlorotic stunt virus reveal several new features for the genus Crinivirus. J. Gen. Virol. 76: 9260-9270

Krishnamurthy, K., Mitra, R., Payton, M.E. and Verchot Lubicz, J. 2002. Cell-to-cell movement of the PVX 12 K, 8 K, or coat protein may depend on the host, leaf developmental stage, and PVX 25 K protein. Virology 300: 269-281

Kuhne, T., Shi, N.N., Proeseler, G., Adams, M.J., Kanyuka, K. and Shi, N. 2003. The ability of a bymovirus to overcome rym-4-mediated resistance in barley correlates with a codon change in the VPg coding region on RNA1. J. Gen. Virol.84: 2853-2859

Kuroda, T., Okumura, A., Takeda, I., Miura, Y. and Suzuki, K. 2000. Nucleotide sequence and synthesis of infectious RNA from cloned cDNA of broad bean wilt virus 2 RNA2. Arch. Virol. 145: 787-793

Kwon, C.S. and Chung, W.I. 2000. Differential roles of the 5′ untranslated regions of cucumber mosaic virus RNAs 1, 2, 3 and 4 in translation competition. Virus Research 60: 175-188

Kwon, S.J., Park, M.R., Kim, K., Plante, C.A., Hemenway, C.L. and Kim, K.H. 2005. cis-acting sequences required for coat protein binding and in vitro assembly of Potato virus X. Virology 334: 83-97

Laco, G.S., Kent, S.B.H. and Beachy, R.N. 1995. Analysis of proteolytic processing and activation of rice tungro bacilliform virus reverse transcriptase. Virology 208: 207-214

Lawrence, D.M. and Jackson, A.O.2001. Requirements for cell-to-cell movement of Barley stripe mosaic virus in monocot and dicot hosts. Molecular Plant Path. 2: 65-75

Lazarowitz, S.G. 1987. The molecular characterization of geminiviruses. Plant Mol. Biol. Rep. 4: 177-192

Leclerc, D., Stavolone, L., Mier, E., Guerra Peraza, O., Herzog, E. and Honh, T. 2001. The product of ORF III in cauliflower mosaic virus interacts with the viral coat protein through its C-terminal proline rich domain. Virus Genes 22: 159-165

Lee, W.M. and Ahlquist, P. 2003. Membrane synthesis, specific lipid requirements, and localized lipid composition changes associated with a positive-strand RNA virus RNA replication protein. J. Virol. 77: 12819-12828

Leh, V., Jacquot, E., Geldreich, A., Hass, M., Blanc, S., Keller, M. and Yot, P. 2001. Interaction between the open reading frame III product and the coat protein is required for transmission of cauliflower mosaic virus by aphids. J. Virol. 75: 100-106

Li, Y.Z. and Leisner, S.M. 2002. Multiple domains within the cauliflower mosaic virus gene VI product interact with the full-length protein. Mol. Pl. Microb. Interactions 15: 1050-1057

Li, Y., Bao, Y.M., Wei, C.H., Kang, Z.S., Zhong, W.Y., Mao, P., Wu, G., Chen, Z.L., Schiemann, J. and Nelson, R.S. 2004. Rice dwarf phytoreovirus segment 6-encoded nonstructural protein has a cell-to-cell movement function. J. Gen. Virol. 78: 5382-5389

Liu, H., Boulton, M.I., Thomas, C.L., Prior, D.A.M., Operka, K.J. and Davies, J.W. 1999. Maize streak virus coat protein is karyophyllic and facilitates nuclear transport of viral DNA. Mol. Plant Microbe Interactions 12: 894-900

Liu, H., Licy, A.P., Davies, J.W., Boulton, M.I. and Liu, H. 2001. A single amino acid change in the coat protein of Maize streak virus abolishes systemic infection, but not interaction with viral DNA or movement protein. Mol. Plant Path. 21: 223-228

Liu, H., Reavy, B., Swanson M. and Macfarlane, S.A. 2002. Functional replacement of tobacco rattle virus cysteine rich-protein by pathogenicity protein from unrelated viruses. Virology 298: 232-239

Liu, D., Li, D.W., Yu, J.L. and Han, C.G. 2002. Sequence analysis of BNYVV RNA2 "triple block" and RNA3 sequence variation of different isolates. J. China Agri. Univ. 7: 1-4

Liu, S.J., He, X.H., Park, G., Josefsson, C. and Perry, K.L. 2002. A conserved capsid protein surface domain of cucumber mosaic virus is essential for efficient aphid transmission. J. Virol. 76: 9756-9762

Liu, L., Grainger, J., Canizares, M.C., Angell, S.M., Lomonossoff, G.P. and Liu, L. 2004. Cowpea mosaic virus RNA1 acts as amplicon whose effects can be counteracted by a RNA2 encoded suppressor of silencing. Virology 323: 37-48

Livicratos, L.C. and Coutts, R.H.A. 2002. Nucleotide sequence and phylogenetic analysis of cucurbit yellow stunting disorder virus RNA2. Virus Genes 24: 225-230

Livicratos, L.C. and Coults, R.H.A. 2002. Nucleotide sequence and phylogenetic analysis of cucurbit yellow stunting disorder virus RNA2. Virus Genes 24: 225-230

Locali-Fabris, E.C., Freitas-Astúa, J., Souza, A.A., Takita, M.A., Astúa-Monge, G., Antonioli-Luizon, R., Rodrigues, V., Targon, M.L.P.N. and Machado, M.A. 2006. Complete nucleotide sequence, genomic organization and phylogenetic analysis of citrus leprosis virus cytoplasmic type. J. Gen. Virol. 87: 2721-2729

Lockhart, B.E. and Olszewski, N.E. 1999. Plant pararetroviruses: Badnavirus. In: Encyclopedia of Virology. A. Granoff and R.G. Webster (eds). 2nd edit. Academic Press, San Diego, USA. pp. 1296-1300

Lockhart, B.E., Menke, J., Dahal, G. and Olszewski, N.E. 2000. Characterization and genomic analysis of tobacco vein clearing virus, a plant pararetrovirus that is transmitted vertically and related to sequences integrated in the host genome. J. Gen. Virol. 81: 1579-1585

Lokesh, G.L., Gopinath, K., Satheskumar, P.S. and Savithri, H.S. 2001. Complete nucleotide sequence of sesbania mosaic virus: a new virus species of the sobemovirus. Archiv. Virol. 146: 209-223

Lomonossof, G.P. and Shanks, M. 1999. Comoviruses (Comoviridae). In: Encyclopedia of Virology. A. Granoff and R.G. Webster (eds). 2nd edit. Academic Press, San Diego, USA. pp. 285-291

López-Moya, J.J. and Gracia, J.A. 1999. Potyviruses (Potyviridae). In: Encyclopedia of Virology. A. Granoff and R.G. Webster (eds). 2nd edit. Academic Press, San Diego, USA. pp. 1369-1375

Lukhovitskaya, N.I., Solovyev, A.G., Koshkina, T.E., Zavriev, S.K. and Morozov, S.Y. 2005a. Interaction of the Carlavirus cysteine-rich protein with the plant defense system. Molecular Biol. New York 39: 785-791

Lukhovitskaya, N.I., Yelina, N.E., Zamyatnin, Jr., A.A., Schepetilnikov, M.V., Solovyev, A.G., Sandgren, M., Morozov, S., Valkonen, J.P.T. and Savenkov, E.I. 2005. Expression, localization and effects on virulence of the cysteine-rich 8 kDa protein of Potato mop-top virus. J. Gen. Virol. 86: 2879-2889

Luque, A., Sanz Burgos, A.P., Ramirez Parra, E., Castellano, M.M. and Gutierrez, C. 2002. Interaction of geminivirus Rep protein with replication factor C and its potential role during geminivirus DNA replication. Virology 302: 83-84

Magome, H., Yoshikawa, N. and Takahashi, T. 1999. Single strand conformation polymorphism analysis of apple stem grooving capillovirus sequence variants. Phytopathology 89: 136-140

Makinen, K., Makelainen, K., Arshava, N., Tamun, T., Merits, A., Truve, E., Zavriev, S. and Saarma, M. 2000. Characterization of VPg and the poly-protein processing of cocksfoot mottle virus (genus Sobemovirus). J. Gen. Virol. 81: 2783-2788

Mansoor, S., Khan, S.H., Bashir, A., Saeed, M., Zafar, Y., Malik, K.A., Briddon, R., Stanley, J. and Markham, P.G. 1999. Identification of a novel circular single stranded DNA associate with cotton leaf curl disease in Pakistan. Virology 259: 190-199

Mansoor, S., Briddon, R.W., Bull, S.E., Bedford, I.D., Bashir, A., Hussain, M., Saeed, M., Zafar, Y., Malik, K.A., Fauquet, C. and Markham, P.G. 2003. Cotton leaf curl disease is associated with multiple monopartite begomoviruses supported by single DNA-β. Arch. Virol. 148: 1969-1986

Marmey, P., Bothner, B., Jaquot, E., De Kochko, A., Ong, C.A., Yot, P., Siuzdak, G., Beachy, R.N. and Fauquet, C.M. 1999. Rice tungro bacilliform virus open reading frame 3 encodes a single 37 kDa coat protein. Virology 253: 319-326

Martelli, G.P., Candresse, T., and Namba S. 1994. Trichovirus, a new genus of plant viruses. Archiv. Virol. 134: 451-455

Martelli, G.P. and Jelkmann, W. 1998. Foveavirus: a new plant virus genus. Archiv. Virol. 143: 1245-1249

Martelli, G.P., Minafra, A. and Sal darelli, P. 1997. Vitivirus: a new genus of plant viruses. Archiv. Virol. 142: 1929-1932

Martelli, G.P., Russo, M., Rubino, L. and Sabanadzovic, S. 1998. Aureusvirus, a novel genus in the family Tombusviridae. Archiv. Virol. 143: 1847-1851

Martelli, G.P., Agranovsky, A.A., Bar-Joseph, M., Boscia, D., Candresse, T., Coutts, R.N.A., Dolja, V.V., Falk, B.W., Gonsalves, D., Jelkmann, W., Karasev, A.V., Minafra, A., Namba, S., Vetten, H.J., Wisler, G.C. and Yoshikawa, N. 2002. The family Closteroviridae (revised). Archiv. Virol. 147: 2039-2044

Matthews, R.E.F. 1992. Fundamentals of Plant Virology. Academic Press, San Diego, USA. pp. 93-122

Matsuda, D. and Dreher, T.W. 2004. The tRNA-like structure of Turnip yellow mosaic virus RNA is a 3′ translational enhancer. Virology 321: 36-46

Matsuda, D., Yoshinari, S. and Dreher, T.W. 2004. eEFIA binding to aminoacylated viral RNA represses minus strand synthesis by TYMV RNA-dependent RNA polymerase. Virology 321: 47-56

Matsushita, Y., Yoshioka, K., Shigyo, T., Takahashi, H. and Nyunoya, H. 2002. Phosphorylation of the movement protein of the Cucumber mosaic virus in transgenic tobacco plants. Virus Genes 24: 231-234

Mayo, M.A. and Miller, W.A. 1999. The structure and expression of luteovirus genomes. In: The Luteoviridae. H.G. Smith and H. Barker (eds). CAB International, Wallingford, UK. pp. 23-67

McGivern, D.R., Findlay, K.C., Montague, N.P. and Boulton, M.I. 2005. An intact RBR-binding motif is not required for invasion of mesophyll cells. J. Gen. Virol. 86: 797-801

McQualter, R.B., Smith, G.R., Dale, J.L. and Harding, R.M. 2003. Molecular analysis of Fiji disease Fiji virus genome segments 1 and 3. Virus Genes 26: 283-289

McQualter, R.B., Burns, P., Smith, G.R., Dale, J.L. and Harding, R.M. 2004. Molecular analysis of Fiji disease virus genome segments 5, 6, 8 and 10. Arch. Virol. 149: 713-721

Medeiros, R.B., Ullman, D.E., Sherwood, J.I. and German, T.L. 2000. Immunoprecipitin of a 50 kDa protein: a candidate receptor component for tomato spotted wilt tospovirus (Bunyaviridae) in its main vector, *Frankliniella occidentalis*. Virus Research 67: 109-118

Meulewaeter, F., Seurinck, J. and van Emmelo, J. 1990. Genome structure of tobacco mosaic virus A. Virology 117: 699-709

Miller, E.D., Plante, C.A., Kim, K.H., Brown, J.W., Hemenway, C. and Kim, K.H. 1998. Stem-loop structure in the 5' region of potato virus X genome required for plus-strand RNA accumulation. J. Mol. Biol. 284: 591-608

Miller, J.S., Damude, H., Robbins, M.A., Reade, R.D. and Rochon, D.M. 1997. Genome structure of cucumber leaf spot virus: sequence analysis suggests it belongs to a distinct species within the Tombusviridae. Virus Research 52: 51-60

Miller, W.A. 1999. Luteovirus (Luteoviridae). In: Encyclopedia of Virology. A. Granoff and R.G. Webster (eds). 2nd edit. Academic Press, San Diego, USA. pp. 901-908

Milne, R.G. and Marzachi, C. 1999. Criptovirus (Partitiviridae). In: Encyclopedia of Virology. A. Granoff, and R.G. Webster (eds.). Academic Press, San Diego, USA. pp. 312-315

Mitra, R., Krishnamurthy, K., Blancaflor, E., Payton, M., Nelson, R.S. and Verchot Lubicz, J. 2003. The Potato virus X TGBp2 protein association with the endoplasmic reticulum plays a role in but not sufficient for viral cell-to-cell movement. Virology 312: 35-48

Miyanishi, M., Roh, S.H., Yamamiya, A., Ohsato, S. and Shirako, Y. 2002. Reassortment between genetically distinct Japanese and US strains of soil-borne wheat mosaic virus RNA1 from a Japanese strain and RNA2 from a US strain make a pseudorecombinant virus. Archiv. Virol. 147: 1141-1153

Mizumoto, H., Tatsuta, M., Kaido, M., Mise, K. and Okuno, T. 2003. Cap-independent translational enhancement by the 3' untranslated region of red clover necrotic mosaic virus RNA1. J. Virol. 77: 12113-12121

Molnar, A., Csorba, T., Lakatos, I., Varallyay, E., Laccome, C. and Burgyan, J. 2005. Plant virus-derived small interfering RNAs originate predominantly from highly structured single-stranded viral RMAs. J. Virol. 79: 7812-7818

Moon, J.S., McCoppin, N.K. and Domier, L.L. 2001. Role of intergenic and 3' proximal noncoding regions in coat protein expression and replication of Barley yellow dwarf virus—PAV. Plant Path. J. 17: 22-28

Moyer, J.W. 1999. Tospoviruses (Bunyaviridae). In: Encyclopedia of Virology. A. Granoff and R.G. Webster (eds). 2nd edit. Academic Press, San Diego, USA. pp. 1083-1087

Mukasa, S.B., Rubaihayo, P.R. and Valkonen, J.P.T. 2003. Sequence variability within 3'-proximal part of the sweet potato mild mottle virus genome. Arch. Virol. 148: 487-496

Munoz Martin, A., Collins, S., Herreros, E., Mullineaux, P.M., Fernandez Lobato, M. and Fenoll, C. 2003. Regulation of MSV and WDV virion-sense promoters by WDV non-structural proteins: a role for their retinoblastoma protein-binding motif. Virology 306: 313-323

Nagano, H., Furusawa, I. and Okuno, T. 2001. Conversion in the requirement of coat protein in cell-to-cell movement mediated by the cucumber mosaic movement protein. J. Virol. 75: 8045-8053

Nagy, P.D., Pogamy, J. and Simons, A.E. 2001. In vivo and in vitro characterization of an RNA replication enhancer in a satellite RNA associated with turnip crinkle virus. Virology New York 288: 315-324

Naidu, R.A., Sawyer, S. and Deom, C.M. 2003. Molecular diversity of RNA2 genome segments in pecluviruses causing peanut clump diseases in West Africa and India. Archiv. Virol. 148: 83-98

Naidu, R.A., Ingle, C.J., Deom, C.M. and Sherwood, J.I. 2004. The two envelope membrane glycoproteins of Tomato spotted wilt virus show differences in lectin-binding properties and sensitivities to glycosidases. Virology 319: 107-117

Nath, N., Mathur, S. and Dasgupta, I. 2002. Molecular analysis of two complete rice tungro bacilliform virus genomic sequences from India. Arch. Virol. 147: 1173-1187

Ndowora, T., Dahal, G., LaFleur, D., Harper, G., Hull, R., Olszewski, N.E. and Lockhart, B. 1999. Evidence that badnavirus infection in Musa can originate from integrated pararetroviral sequences. Virology 255: 214-220

Neeleman, L., Linthorst, H.J.M. and Bol, J.F. 2004. Efficient translation of alfamovirus RNAs requires the binding of coat protein dimers to 3' termini of the viral RNAs. J. Gen. Virol. 85: 231-240

Nicholas, B.L., Brennan, F.R., Martinez-Torrecuadrada, J.L., Casal, J.I., Hamilton, W.D. and Wakelin, D. 2002. Characterization of the immune response to canine parvovirus by vaccination with chaemeric plant viruses. Vaccine 20: 2727-2734

Nikovics, K., Simidjieva, J., Peres, A., Ayadin, F., Pasterna, K.T., Davies, J.W., Boulton, M.I., Dudits, D. and Horvath, G.W. 2001. Cell-cycle, phase-specific activation of Maize streak virus promoters. Mol. Plant Microbe Interactions 14: 609-617

Nukyanova, K.M., Ryabov, E.V., Kalinina, N.O., Fan, Y.C., Andreev, I., Fitzgerald, A.G., Palukaitia, P., Taliansky, M. and Fan, Y.C. 2001. Umbravirus encoded movement protein induces tubule formation on the surface of protoplasts and binds RNA incompletely and non-cooperatively. J. Gen. Virol. 82: 2579-2588

Oiu, W., Park, J.W. and Scholthof, H.B. 2002. Tombusvirus P19 mediated suppression of virus induced gene silencing is controlled by genetic and dosage features that influence pathogenicity. Mol. Plant Microbe Interactions 15: 269-280

Okinaka, Y., Mise, K., Okuno, T. and Fuvusawa, I. 2003. Characterization of a novel barley protein HCP1, that interacts with the Brome mosaic virus coat protein. Mol. Plant Microbe Interactions 16: 352-359

Olsthoorn, R.C.L. and Bol, J.F. 2002. Role of an essential triloop hairpin and flanking structures in the 3' untranslated region of alfalfa mosaic virus RNA in vitro transcription. J. Virol. 76: 8747-8756

Osaki, H., Wei, C.Z., Arakawa, M., Iwanami, T., Nomura, K., Matsumoto, N., Ohtsu, N. and Ohtsu, Y. 2002. Nucleotide sequences of double-stranded RNA segments from hypovirulent strain of the white ring rot fungus *Rosellinia necatrix*: possibility of the first member of the reoviridae from fungus. Virus Genes 25: 27-36

Palanichelvam, K. and Schoelz, J.E. 2002. A comparative analysis of the avirulence and translational transactivator functions of gene VI of cauliflower mosaic virus. Virology 293: 225-233

Palmer, K.E. and Rybicki, E.P. 1998. The molecular biology of mastreviruses. Adv. Virus Res. 50: 183-234

Panavas, T., Panaviene, Z., Pogamy, J. and Nagy, P.D. 2003. Enhancement of RNA synthesis by promoter duplication in tombusviruses. Virology 310: 118-129

Panaviene, Z., Baker, J.M. and Nagy, P.D. 2003. The overlapping RNA binding domains of P33 and P92 replicase proteins are essential for tombus virus replication. Virology 308: 191-205

Park, H.S., Himmelbach, A., Browning, K.S., Hohn, T., Ryabava, L.A. and Park, H. 2001. A plant viral "re-initiation" factor interacts with the host translation machinery. Cell-Chembridge 106: 723-733

Park, H. S., Browing, K.S., Hohn, T. and Ryabova, L.A. 2004. Eucaryotic initiation factor 4B controls eiF-3-mediated ribosomal entry of viral re-initiation factor. EMBO J. 23: 1381-1391

Park, J.W., Desvoyes, B. and Scholthof, H.B. 2002. Tomato bushy stunt virus genomic RNA accumulation is regulated by interdependent cis-acting elements within the movement protein open reading frame. J. Virol. 76: 12747-12757

Patil, B.L. and Dasgupta, I. 2006. Defective interfering DNAs of plant viruses. Crit. Rev. Plant Sci. 25: 47-64

Peerenboom, E., Jacobi, V., Cartwright, E.J., Adams, M.J., Steinbiss, H.H. and Antoniw, J.F. 1997. A large duplication in the 3'-untranslated region of a subpopulation of RNA2 of the UK-M isolate of barley mild mosaic bymovirus. Virus Research 47: 1-6

PerezAlvarez, S., Xue, C.Y., Zhou, X.P., Xue, C.Y. and Zhou, X.P. 2003. Emergence of a new satellite RNA from cucumber mosaic virus isolate. J. Zhejiang Univ. Sci. 4: 336-339

Petruccelli, S., Dai, S.H., Carcamo, R., Yin, Y.H., Chen, S.Y., Beachy, R.N. and Chen, S.Y. 2002. Transcription factor RF2a alters expression of the rice tungro bacilliform virus promoter in transgenic tobacco plants. Proc. Natl. Acad. Sci. USA 98: 7635-7640

Pfeiffer, P., Gordon, K., Filtterer, J. and Hohn, T. 1987. The life cycle of caulimomosaic virus. In: Plant Molecular Biology. D. von Wettstein and N.H. Chua (eds). Plenum, New York, USA. pp. 443-458

Pooggin, M.M., Hohn, T. and Futterer, J. 2000. Role of a short open reading frame in ribosome shunt on the cauliflower mosaic virus RNA leader. J. Biol. Chem. 275: 17288-17296

Pouwels, J., Kornet, N., van Bers, N. van Guigheiaar, T., Lent, J., van Bisseling, T., Wellink, J. and van Lent, J. 2003. Identification of distinct steps during tubule formation by the movement protein of cowpea mosaic virus. J. Gen. Virol. 84: 3485-3494

Preiss, W. and Jeske, H. 2003. Multitasking in replication is common among geminiviruses. J. Virol. 77: 2972-2980

Prod'homme, D., Jakubiec, A., Tournier, V., Drugen, G. and Jupin, I. 2003. Targeting of the Turnip yellow mosaic virus 66 K replication protein to the chloroplast envelope is mediated by the 140 K protein. J. Gen. Virol. 77: 9124-9135

Qi, Y.J., Zhou, X.P., Yu, Y.J., Xue, C.Y. and Li, D.B. 1999. Processing sites of polyprotein encoded by RNA-2 of broad bean wilt virus. Acta Phytopathologica Sinica 29: 280-282

Qi, Y.Z., Zhou, X.P., Tao, X.R. and Li, D.B. 2000a. Characterization and RNA 2 nucleotide sequence of broad bean wilt virus 2 isolate P158. J. Zhejiang Univ. Sci. 1: 437-441

Qu, F. and Morris, T.J. 2000. Cap-independent translation enhancement of turnip crinkle virus genomic and subgenomic RNs. J. Virol. 74: 1085-1093

Qu, Z.C., Lu, Y.Z., Chen, D., Su, D.M., Hull, R., Qu, Z., Lu, Y.Z., Shen, D.L. and Su, D.M. 2001. Cloning and sequence analysis of rice stripe virus genome segment 3 of the Chinese isolate Y. Chinese J. Virol. 16: 44-48

Qui, Y.J., Zhou, X.P. and Li, D.B. 2000b. Complete nucleotide sequence and infectious cDNA clone of the RNA 1 of a Chinese isolate of broad bean wilt virus 2. Virus Genes 20: 201-207

Qui, Y.J., Zhou, X.P., Muang, X.Z. and Li, G.X. 2002. In vivo accumulation of Broad bean wilt virus 2 VP 37 protein and its ability to bind single-stranded nucleic acid. Arch. Virol. 147: 917-928

Rabenstein, F., Seifers, D.L., Schubert, J., French, R. and Stenger, D.C. 2002. Phylogenetic relationships, strain diversity and bio-geography of tritimovirus. J. Gen. Virol. 83: 895-906

Rajendran, K.S. and Nagy, P.D. 2003. Characterization of the RNA-binding domains in the replicase proteins of tomato bushy stunt virus. J. Virol. 77: 9244-9258

Ranjith Kumar, C.T., Zhang, X. and Kao, C.C. 2003. Enhancer like activity of a brome mosaic virus RNA promoter. J. Virol. 77: 1830-1839

Rao, S., Carner, G.R., Scott, S.W., Omura, T. and Magiwara, K. 2003. Comparison of the amino acid sequences of RNA-dependent RNA-polymerases of cypoviruses in the family Reoviridae. Archiv. Virol. 148: 209-219

Reade, R., Miller, J., Robbins, M., Xiang, Y. and Rochon, D. 2003. Molecular analysis of the cucumber leaf spot virus genome. Virus Research 91: 171-179

Reusken, C.B.E.M., Neeleman, L. and Bol, J.F. 1995. Ability of tobacco streak virus coat protein to substitute for late functions of alfalfa mosaic virus coat protein. J. Virol. 69: 4552-4555

Remans, T., Grof, C.P.L., Ebert, P.R. and Schenk, P.M. 2005. Identification of functional sequence in the pregenomic RNA promoter of the Banana streak virus Cavendish strain (BSV-Cav). Virus Research 108: 177-186

Richert-Poggeler, K.R., Noreen, F., Schwarzacher, T., Harper, G. and Hohn, T. 2003. Induction of infectious petunia vein clearing (pararetro) virus from endogenous provirus in petunia. EMBO J. 22: 4836-845

Rochon, D'A.M. 1999. Tombusviruses. In: Encyclopedia of Virology. A. Granoff and R.G. Webster (eds). 2nd edit. Academic Press, San Diego, USA. pp. 1789-1798

Roossinck, M.J. 1999a. Cucumoviruses (Bromoviridae)—general features. In: Encyclopedia of Virology. A. Granoff and R.G. Webster (eds). 2nd edit. Academic Press, San Diego, USA. pp. 315-320

Roossinck, M.J. 1999b. Cucumoviruses (Bromoviridae)—molecular biology. In: Encyclopedia of Virology. A. Granoff and R.G. Webster (eds). 2nd edit. Academic Press, San Diego, USA. pp. 320-324

Roossinck, M.J. 2001. Cucumber mosaic virus, a model for RNA virus evolution. Mol. Plant Path. 2: 59-63

Roossinck, M.J. 2002. Evolutionary history of Cucumber mosaic virus deduced by phylogenetic analysis. J. Virol. 76: 3382-3387

Rothnie, H.M., Chen, G., Fütterer, J. and Hohn, T. 2001. Polyadenylation in rice tungro bacilliform virus: *cis*-acting signals and regulation. J. Virol. 75: 4184-4194

Rubino, L. and Russo, M. 1997. Molecular analysis of Pothos latent virus genome. J. Gen. Virol. 78: 1219-1226

Rubino, L., Russo, M. and Martelli, G.P. 1995. Sequence analysis of Pothos latent virus genomic RNA. J. Gen. Virol. 76: 2835-2839

Rush, C.M. 2003. Ecology and epidemiology of Benyviruses and plasmodiphoroid vectors. Annu. Rev. Phytopath. 41: 567-592

Rustici, G., Milne, R.G. and Accotto, G.P. 2002. Nucleotide sequence, genome organization and phylogenetic analysis of Indian citrus ringspot virus. Archiv. Virol. 147: 2215-2224

Ryabova, L.A., Pooggin, M.M., Dominghez, D.I. and Hohn, T. 2000. Continuous and discontinuous ribosome scanning on the cauliflower mosaic virus 35 S RNA leader is controlled by short open reading frames. J. Biol. Chem. 275: 37278-37284

Ryabova, L., Park, H.S. and Hohn, T. 2004. Control of translation re-initiation on the cauliflower mosaic virus (CaMV) polycistronic RNA. Biochem. Soc. Transaction 32: 592-596

Sabanadzovic, S., Sabanadzovic, N.A., Saldararelli, P. and Martelli, G.P. 2001. Complete nucleotide sequence and genome organization of grapevine fleck virus. J. Gen. Virol. 82: 2009-2015

Sadowy, E., Maasen, A., Juszczuk, M., David, C. Zagorski-Ostoja, W., Gronenborn, B. and Hulanicka, M.D. 2001. The ORFO product of *potato leafroll virus* is indispensable for virus accumulation. J. Gen. Virol. 82: 1529-1532

Saenz, P., Salvador, B., Simon Mateo, C., Kasschau, K.D., Carrington, J.C. and Garcia, J.A. 2002. Host-specific involvement of the HC protein in the long-distance movement of potyviruses. J. Virol. 76: 1922-1831

Sanchez Navarro, J.A. and Bol, J.F. 2001. Role of the Alfalfa movement protein and coat protein in virus transport. Mol. Plant Microbe Interactions 14: 1051-1062

Sanchez de la Torre, M.E., Lopez, C., Grau, O. and Garcia, M.L.2002. RNA2 of citrus psorosis virus is of negative polarity and has a single open reading frame in its complementary strand. J. Gen. Virol. 83: 1777-1781

Sandgren, M., Savenkov, E.I., and Velkonen, J.P.T. 2001. The readthrough region of potato mop-top virus (PMTV) coat protein encoding RNA, the second largest RNA of PMTV genome, undergoes structural changes in naturally infected and experimentally inoculated plants. Archiv. Virol. 146: 467-477

Sasaki, N., Arimoto, M., Nagano, H., Mori, M., Kaido, M., Mise, K. and Okuno, T. 2003. The movement protein gene is involved in the virus specific requirement of the coat protein in cell-to-cell movement of bromoviruses. Arch. Virol. 148 : 803-812

Saunders, K. and Stanley, J. 1999. A nanovirus-like DNA component associated with yellow vein disease of *Ageratum conyzoides*: evidence for interfamilial recombination between plant DNA viruses. Virology 264: 142-152

Saunders, K., Bedford, I.D., Briddon, R.W., Markham, P.G., Wong, S.M. and Stanley, J. 2000. A unique virus complex causes *Ageratum* yellow vein disease. Proc. Natl. Acad. Sci. USA 97: 6890-6895

Saunders, K., Belford, I.D. and Stanley, J. 2001. Pathogenicity of a natural recombinant associated with ageratum yellow vein disease: Implications for begomovirus evolution and disease. Virology 282: 38-47

Saunders, K., Belford, I.D. and Stanley, J. 2002. Adaptation from whitefly to leafhopper transmission of an autonomously replicating nanovirus-like DNA component associated with ageratum yellow vein disease. J. Gen. Virol. 83: 907-913

Savenkov, E.I., Germundsson, A., Zamyatruin, A.A. Jr., Sandren, M. and Valkonen, J.P.T. 2003. Potato mop-top virus: the coat protein-encoding RNA and the gene for cysteine-rich protein are dispensable for systemic virus movement in *Nicotiana benthomiana*. J. Gen. Virol. 84: 1001-1005

Seal, S.E., Jeger, M.J. and van den Bosch. 2006. Begomovirus evolution and disease management. Adv. Virus Res. 67: 297-315

Shalitin, D., Wang, Y., Omid, A., Gal On, A. and Wolf, S. 2002. Cucumber mosaic virus movement protein affects sugar metabolism and transport in tobacco and melon. Plant Cell and Environment 25: 989-997

Shelly, P., Barthe, C.A., Derrick, K.S. and Ahlawat, Y.S. 2003. Comparison of the isolates of Indian citrus ringspot virus Citris psorosis virus. Indian Phytopath. 56: 230-232

Schoen, C.D., Limpens, W., Moller, I., Groeneveld, L., Klerks, M.M. and Linder, J.L. 2004. The complete genomic sequence of strawberry crinkle virus: a member of the Rhabdoviridae. Acta Horticulturae 656: 45-50

Schunmann, P.H.D., Surin, B. and Waterhouse, P.M. 2000b. A suite of novel promoters and terminators for plant biotechnology II. The pPPLEX series for use in monocots. Functional Pl. Biol. 30: 453-460

Schunmann, P.H.D., Llewellyn, D.J., Surin, B., Boevink, P., Feyter, R.C. de and Waterhouse, P.M. 2003a. A suite of novel promoters and terminator for plant biotechnology. Functional Pl. Biol. 30: 443-458

Serazev, T.V., Nadezhdina, E.S., Shanina, N.A., Leshchiner, A.D., Kalinina, N.O. and Morozov, S.Y. 2003. Virions and the the the coat protein of the potato virus X interact with microtubules and induce tubulin polymerization in vitro. Molecular Biol. New York 37: 919-925

Shao, C.G., Wu, J.H., Zhou, G.Y., Sun, G., Peng, B.Z. *et al.* 2003. Ectopic expression of the spike protein of Rice Ragged Stunt Oryzavirus in transgenic rice plants inhibits transmission of the virus to insects. Mol. Breeding 11: 295-301

Shen, P., Kaniewski, M.B., Smith, C. and Beachy, R.N. 1993. Nucleotide sequence and genomic organization of rice tungro spherical virus. Virology 193: 621-630

Shen, R.Z. and Miller, W.A. 2004. The 3′ untranslated region of tobacco necrosis vrus RNA contains a barley yellow dwarf virus-like cap-independent translation element. J. Virol. 78: 4655-4664

Shepherd, D.N., Martin, D.P., McGivern, D.R., Boulton, M.I., Thomson, J.A. and Rybicki, P. 2005. A three nucleotide mutation altering the Maize streak virus Rep pRBR-interaction motif reduces symptom severity in maize and partially reverts at high frequency without restoring pRBR-Rep binding. J. Gen. Virol. 86: 797-801

Shim, H.K., Min, Y.J., Hong, S.Y., Kwon, M.S., Kim, D.H., Kim, H.R., Choi, Y.M., Lee, S.C. and Yang, J.M. 2004. Nucleotide sequences of a Korean isolate of apple stem grooving virus associated with black necrotic leaf spot disease on pear (*Pyrus pyrifolia*). Molecules and Cells 18: 192-199

Shirako, Y. and Wilson, T.M.A. 1993. Complete nucleotide sequence and organization of the bipartite RNA genome of soil-borne wheat mosaic virus. Virology 195: 16-32

Silva, M.S., Wellink, J., Goldbach, R.W. and van Lent, J.W.M. 2002. Phloem loading and unloading of cowpea mosaic virus in *Vigna unguiculata*. J. Gen. Virol. 83: 1493-1504

Sit, T.L., Laclere, D. and Abou Haidar, M.C. 1994. The minimal 5' sequence for *in vitro* initiation of papaya mosaic potex virus assembly. Virology 199: 238-242

Sivakumaran, K., Bao, Y., Rossink, M.J. and Kao, C.C. 2000. Recognition of the core RNA promoter for minus-strand RNA synthesis by the replicases of brome mosaic virus and cucumber mosaic virus. J. Virol. 74: 10323-10331

Sivakumaran, K., Chen, M.H., Rossinck, M.J. and Kao, C.C. 2002. Core promoter for initiation of cucumber mosaic virus subgenomic RNA4A. Mol. Plant Path. 3: 43-52

Sivakumaran, K., Hema, M. and Kao, C.C. 2003. Brome mosaic virus RNA synthesis *in vitro* and in barley protoplast. J. Virol. 77: 5703-5711

Skaf, J.S., Schultz, M.H., Hirata, H. and de Zoeten, G.A. 2000. Mutational evidence that the VPg is involved in the replication and not the movement of pea enation mosaic virus. J. Gen. Virol. 81: 1103-1109

Smith, H.G. and Barker, H. (eds). 1999. The Luteoviridae. CAB International, Wallingford, UK

Smith, O.P. and Harris, K.F. 1990. Potato leafroll virus 3' genome organization: sequence of the coat protein gene and identification of a viral subgenomic RNA. Phytopathology 80: 609-614

Stenger, D.C. and French, R. 2004a. Functional replacement of wheat streak mosaic virus H-C Pro with the corresponding citron from a diverse array of viruses in the family Potyviridae. Virology 323: 257-267

Stenger, D.C. and French, R. 2004b. Compete nucleotide sequence of oat necrotic mottle virus: a distinct Trimovirus species (family Potyviridae) most closely related to wheat mosaic virus. Archiv. Virol. 149: 633-640

Soellick, T.R., Uhrig, J.F., Bucher, G.L., Kellmann, J.W. and Schreier, P.H. 2000. The movement protein NSm of tomato spotted wilt tospovirus (TSWV): RNA binding interaction with the TSWV N protein, and identification of interacting plant proteins. Proc. Natl. Acad. Sci. USA 97: 2373-2378

Solis, I. and Garcia-Arenal, F. 1990. The complete nucleotide sequence of the genomic RNA of the tobamovirus tobacco mild green mosaic virus. Virology 177: 553-558

Sonoda, S. 2003. Analysis of the nucleocapsid protein gene from Tomato spotted wilt virus as target and inducer for post-transcriptional gene silencing. Pl. Sci. 164: 717-725

Stanley, J. 2004. Subviral DNAs associated with geminivirus disease complexes. Vet. Microbiol. 98: 121-129

Stanley, J. and Gay, M.R. 1983. Nucleotide sequence of cassava latent virus DNA. Nature (London) 301: 260-262

Stanley, J., Markham, P.G., Callis, R.J. and Pinner, M.S. 1986. The nucleotide sequence of an infectious clone of the geminivirus beet curly top virus. EMBO J. 5: 1761-1767

Stanley, J., Frischmuth, T. and Ellwood, S. 1990. Defective viral DNA ameliorates symptoms of geminivirus infection in transgenic plants. Proc. Natl. Acad. Sci. USA 87: 6291-6295

Stavolone, L., Herzog, E., Leclerc, D. and Hohn, T. 2001. Tetramerization is a conserved feature of the virion-associated protein in plant pararetroviruses. J. Virol. 75: 7739-7743

Stenger, D.C. 1998. Replication specificity elements of the worland strain of beet curly top virus are compatible with those of the CFH strain but not those of the Cal/Logan strain. Phytopathology 88: 1174-1178

Stenger, D.C. and French, R. 2004. Complete nucleotide sequence of oat necrotic mottle vrus: a distinct Tritimovirus species (family Potyviridae) most closely related to wheat mosaic virus. Arch. Virol. 149: 633-640

Subr, Z., Fomitcheva, V.W. and Kuhne, T. 2000. The complete nucleotide sequencee of RNA1 of barley mild mosaic virus (ASL1) and attempts at bacterial expression of the P3 protein. J. Phytopath. 148: 461-467

Sun, L.Y., Xu, J.L., Fang, S.G., Wang, Z.H., Han, C.G., Li, D. and Yu, J.L. 2004. Sequence analysis and prokaryotic expression of genomic segments 8 and 9 of rice black-streaked dwarf virus isolated from maize. J. Agric. Biotech. 12: 306-311

Sun, X.P. and Simon, A.E. 2003. Fitness of a turnip crinkle virus satellite RNA correlates with a sequence-nonspecific hairpin and flanking sequences that enhance replication and repress the accumulation of virions. J. Virol. 77: 7880-7889

Syller, J. 2003. Molecular and biological features of umbraviruses, the unusual plant viruses lacking genetic information for a capsid protein. Physiol. Mol. Pl. Path. 63: 35-46

Tacke, E., Sarkar, S. and Rohde, W. 1989. Cloning of the gene for the capsid protein of potato leaf roll virus. Archiv. Virol. 105: 153-163

Tadu, G., Winter, S., Gadelseed, A.M.A. and Dafalla, G.A. 2006. Association of the East African cassava mosaic virus-Uganda (EACMV-UG) with cassava mosaic disease in Sudan. Plant Pathology 55: 287

Takahashi, M., Goto, C., Ishikawa, K., Matsuda, I., Toryama, S. and Tsuchiya, K. 2003. Rice stripe virus 23.9k protein aggregates and forms inclusion bodies in cultured insect cells and virus infected plant cells. Archiv. Virol. 148: 2167-2169

Takeda, A., Sugiyama, K., Negano, H., Mori, M., Kaido, M., Mise, K., Tsuda, S. and Okuno, T. 2002. Identification of a novel RNA silencing suppressor, NSs protein of tomato spotted wilt virus. FEBS-Letters 532: 75-79

Takeda, A., Kaido, M., Okuno, T. and Mise, K. 2004. The C terminus of the movement protein of Brome mosaic virus controls the recruitment for coat protein in cell-to-cell movement and plays a role in long distance movement. J. Gen. Virol. 85: 1751-1761

Takemoti, Y. and Hibi, T. 2001. Genes Ia, II, III, IV and V of Soybean chlorotic mottle virus are essential but the gene Ib product non-essential for systemic infection. J. Gen. Virol. 82: 1481-1489

Takemoto, Y. and Hibi, T. 2005. Self interaction of ORF II protein through the leucine zipper essential for Soybean chlorotic mosaic virus infectivity. Virology 332: 199-225

Taliansky, M.E. and Robinson, D.J. 2003. Molecular biology of umbraviruses: phantom warriors. J. Gen. Virol. 84: 1951-1960

Taliansky, M.E., Roberts, I.M., Kalinina, N., Ryabov, E.V., Raj, S.K., Robinson, D.J. and Operka, K.J. 2003. An umbraviral protein in long distance RNA movement, binds viral RNA and forms unique protective ribonucleoprotein complexes. J. Virol. 77: 3031-3040

Thole, V. and Hull, R. 1996. Rice tungro spherical virus: nucleotide sequence of 3′ genomic half and studies on the two small 3′ open reading frames. Virus Genes 13: 239-246

Thole, V. and Hull, R. 2002. Characterization of a protein from Rice tungro spherical virus with serine proteinase-like activity. J. Gen. Virol. 83: 3179-3186

Thomas, C.L. and Maule, A.J. 1999. Identification of inhibitory mutants of cauliflower mosaic virus movement protein function after expression in insect cells. J. Virol. 73: 7886-7890

Timchenko, T., Katul, L., Sano, Y., Kouchkovsky, F., de Vetten, H.J., Gronenborn, B., Tsuge, S., Okuno, T., Furuswa, I., Kubo, Y. and Horino, O. 2001. Stabilization of cauliflower mosaic virus P3 tetramer by covalent linkage. Microbiol. Immunology 45: 367-371

Tsuge, S., Okuno, T., Furusawa, I., Kubo, Y. and Horino, O. 2001. Stabilization of cauliflower mosaic virus P3 tetramer by covalent linkage. Microbiol, Immunology 45: 365-371

Urbanowicz, A., Alejska, M., Formanowicz, P., Błażewicz, J., Figlerowicz, M. and Bujaraski, J.J. 2005. Homologous crossovers among Molecules of Brome mosaic Bromovirus RNA1 or RNA2 segments in vivo. J. Virol. 79: 5732-5742

Van Kammen, A. and Eggen, H.I.L. 1986. The replication of cowpea mosaic virus. BioEssays 5: 261-266

Van Regenmortel, M.H.V., Fauquet, C.M. and Bishop, D.H.L. (eds.). 2000. Virus Taxonomy, Classification and Nomenclature of Viruses. Seventh Report of the International Committee on Taxonomy of Viruses. Academic Press, San Diego, USA

Varma, A. and Malathi, V.G. 2003. Emerging geminivirus problems: a serious threat to crop production. Annals Appl. Biol. 142: 145-164

Verdaguer, B., Konan, K., Kochko, A.-de., Beachy, R.N. and Fauquet, C. 1997. The cassava mosaic virus promoter: a new promoter for cassava genetic engineering. African J. Root and Tuber Crops 2: 204-208

Vilar, M., Esteve, V., Pallas, V., Marcos, J.F. and perezPaya, E. 2001. Structural properties of carnation mottle virus P7 movement protein and its RNA-binding domain. J. Biol. Chem. 276: 18122-18129

Vivanco, J.M. and Tumer, N.E. 2003. Translation inhibition of capped and uncapped viral RNAs mediated by ribosome inactivating proteins. Phytopath. 93: 588-595

Vlasak, J., Smahel, M., Pavlik, A., Pavingerova, D. and Briza, J. 2003. Comparison of hCMV immediate early and CaMV 35 S promoters in both plant and human cells. J. Biotech. 103: 197-202

Vlot, A.C. and Bol, J.F. 2003. The 5' untranslated region of alfalfa mosaic virus RNA3 is involved in negative-strand RNA-synthesis. J. Virol. 77: 11284-11289

Vlot, A.C., Neeleman, L., Linthorst, H.J.M. and Bol, J.F. 2001. Role of the 3'-untranslated regions of alfalfa mosaic virus RNAs in the formation of a transiently expressed replicase in plants and in the assembly of virions. J. Virol. 75: 6440-6449

Vlot, A.C., Menard, A. and Bol, J.F. 2002. Role of the alfalfa mosaic virus methyltransferase-like domain in negative strand RNA synthesis. J. Virol. 76: 11321-11328

Vlot, A.C., Laro, S. and Bol, J.F. 2003. Coordinate replication of alfalfa mosaic vrus RNAs 1 and 2 involves cis- and trans-acting functions of the encoded helicase-like and polymerase-like domains. J. Virol. 77: 10790-10798

von Bargen, S., Salchert, K., Paape, M., Piechulla, B., Kellmann, J.U. and von Bargen, S. 2001. Interaction between the tomato spotted wilt virus movement protein and plant proteins showing homologies to myosin, kinesin and DnaJ-like chaperones. Pl. Physiol. Biochem. 29: 1083-1093

Wang, A.M. and Sanfacon, H. 2000a. Diversity in the coding regions for the coat protein, VPg, protease, and putative RNA-dependent RNA polymerase among tomato ring spot nepovirus isolates. Can. J. Plant Path. 22: 145-149

Wang, A.M. and Sanfacon, H. 2000b. Proteolytic processing at a novel cleavage site in the N-terminal region of the tomato ring spot nepovirus RNA1-encoded polyprotein in vitro. J. Gen. Virol. 81: 2771-2781

Wang, A.M., Han, S.M. and Sanfacon, H. 2004. Topogenesis in membranes of the NTB-VPg protein of tomato ring spot nepovirus: definition of the C-terminal transmembrane domain. J. Gen. Virol. 85: 535-545

Wang, H., Hao, L.H., Shung, C.Y., Sunter, G. and Bisaro, D.M. 2003. Adenosine kinase is inactivated by geminiviruses AL2 and L2 proteins. Plant Cell 15: 3020-3032

Wang, Q., Kaltgrad, E., Lin, T.W., Johnson, J.E. and Finn, M.G. 2002. Natural supramolecular binding blocks: Wild-type cowpea mosaic virus. Chem. and Biol. 9: 805-811

Wanit Chakorn, R., Hafner, G.J., Harding, R.M. and Dale, J. 2000. Functional analysis of proteins encoded by banana bunchy top virus DNA4 to -6. J. Gen. Virol. 81: 297-306

Wei, C.Z., Osaki, H., Iwanami, T., Matsumoto, N. and Ohtsu, Y. 2004. Complete nucleotide sequences of genome segments 1 and 3 of Rosellinia anti-rot virus in the family Reoviridae. Arch. Virol. 149: 773-777

Wei, T.Y., Lin, H.X., Wu, Z.J., Lin, Q.Y. and Xie, L.H. 2001. Molecular variability of intergenic regions of rice stripe virus RNA4. Chinese J. Virol. 17: 144-149

Wilk, F. van der, Dullemans, A.M., Verbeek, M. and Heuvel, J.F.J.M. van den. 2002. Nucleotide sequence and genomic organization of an ophivirus associated with lettuce big-vein disease. J. Gen. Virol. 83: 2869-2877

Wood, N.T., McGavin, W.J., Mayo, M.A., Jones, A.T. and Martin, P.R. 2001. Studies on a putative second gene in RNA-1 of raspberry bushy dwarf virus. Acta Horticulturae No. 551: 19-22

Wu, P. J. and Zhou, X.-P. 2005. Interaction between a nanovirus-like component and tobacco curly shoot virus/satellite complex. Acta Biochim. Biophys. Sin. 37: 25-31

Xin, H.W. and Ding, S.W. 2003. Identification and molecular characterization of a naturally occurring RNA virus mutant defective in the initiation of host recovery. Virology 317: 253-262

Yamamiva, A., Shirako, Y.K. and Shirako, Y.K. 2000. Construction of full-length cDNA clones to soil-borne wheat mosaic virus RNA1 and RNA2, from which infectious RNAs are transcribed in vitro: virion formation and systemic infection without expression of the N-terminal extensions to the capsid protein. Virology 277: 66-75

Yelina, N.E., Sevenkov, E.I., Solovyev, A.G., Morozov, S.Y. and Valkonen, J.P.T. 2002. Long distance movement, virulence, and RNA silencing suppression controlled by a single protein in hordei- and potyviruses: complementary functions between virus families. J. Virol. 76: 12981-12991

Ye, K.Q., Malinina, L. and Patel, D.J. 2003. Recognition of small interfering RNA by a viral suppressor of RNA silencing. Nature, London 426: 6968: 874-878

Yoshikawa, N., Sasaki, E., Kato, M. and Takahashi, T. 1992. The nucleotide sequence of apple stem grooving capillovirus genome. Virology 191: 98-105

Yoshikawa, N., Oogake, S., Ter a da, N., Miyabayashi, S., Ikeda, Y., Takahashi, T. and Ogawa, K. 1999. Apple chlorotic leaf spot virus 50 kDa protein is targeted to plasmodesmata and accumulates in sieve elements in transgenic plant leaves. Archiv. Virol. 144: 2475-2483

Yoshikawa, N., Gotoh, S., Umezawa, M., Satoh, N., Satoh, H., Takahashi, T., Ito, T. and Yoshida, K. 2000. Transgenic *Nicotiana occidentalis* plants expressing the 50 kDa protein of apple chlorotic leaf spot virus display increased susceptibility to homologous virus, but strong resistance to grapevine berry inner necrosis virus. Phytopathology 90: 311-316

Yoshikawa, N., Matsuda, H., Oda, Y., Isogai, M., Takahashi, T., Ito, T., Yoshida, K. and Clark, M.F. 2001. Genome heterogeneity of apple stem pitting virus in apple trees. Acta Horticulturae No. 550 (Vol. 1): 285-290

Xu, P., Roossinck, M.J. and Xu, P. 2000. Cucumber mosaic virus D satellite RNA-induced programmed cell death in tomato. Plant Cell 12: 1079-1092

Zamore, P.D. 2004. Plant RNAi: How a viral silencing suppressor inactivates siRNA. Current Biol. 14: R198-R200

Zhang, C.M., Wu, Z. J., Lin, Q. Y., Xie, L. H., Zhang, C.M., Wu, Z.J., Lin, Q.Y., Zhang, G.H., Xu, Y., Cai, W. and Fang, R.X. 1999a. Construction of full-length cDNA clone and complete nucleotide sequence of RNA2 of cucumber mosaic virus SD strain. Acta Microbiologica Sinica 39: 8-14

Zhang, C.M., Wu, Z.J., Lin, Q.Y., Xie, L.H., Zhang, C.M., Wu, Z.J., Lin, Q.Y. and Xie, L.H. 1999b. Advances in research on molecular biology of tenuiviruses. J. Fujian Agric. Univ. 28: 445-451

Zhang, F.L. and Simon, A.E. 2003. Enhanced virus pathogenesis associated with a virulent mutant virus or a virulent satellite RNA correlates with reduced virion accumulation and abundance of free coat protein. Virology 312: 8-13

Zhang, H.M., Chen, J.P. and Adams, M.J. 2001. Molecular characterization of segments 1 to 6 of Rice black-streaked dwarf virus from China provides the complete genome. Arch. Virol. 146: 2331-2339

Zhang, H.M., Chen, J.P., Xue, Q.Z. and Lei, J.L. 2002. CDNA cloning and sequence analysis of genome segment S10 from rice black-streaked dwarf virus. Chinese J. Rice Sci. 16: 24-28

Zhang, S., Jones, M.J., Barker, P., Davis, J.W. and Hull, R. 1993. Molecular cloning and sequencing of coat protein-encoding c DNA of rice tungro spherical virus: a plant picorna virus. Virus Genes 7: 121-132

Zhang, Z.C., Li, D.W., Zhang, L., Yu, J. and Liu, Y. 1999a. Virus movement protein gene mediated resistance against cucumber mosaic virus infection. Acta Botanica Sinica 4: 585-590

Zhang, G.H., Xu, Y., Cai, W.Q. and Fang, R.X. 1999b. Construction of a full length cDNA clone and complete nucleotide sequence of RNA2 of cucumber mosaic virus SD strain. Acta Microbiologica Sinica 39: 8-14

Zhao, S.L., Liang, C.Y., Hong, J.J. and Peng, H.Y. 2003. Genomic sequence analyses of segments 1 to 6 of *Dendrolimus punctatus* cytoplasmic polyhydrosis virus. Archiv. Virol. 148:1357-1368

Zhou, X., Xie, Y., Tao, X., Zhang, Z., Li, Z. and Fauquet, C.M. 2003. Characterization of DNA-β associated with begomoviruses in China and evidence for co-evolution with other cognate viral DNA-A. J. Gen. Virol. 84: 237-247

Appendix IV-VIII

APPENDIX IV.1
Principles of electron microscopy

Electrons are the source of illumination in electron microscopy. An electron is a unit of electricity associated with atoms. Normally it is a particle with negative electric charge of magnitude 4.801×10^{-10} statcoulomb or 1.601×10^{-19} coulomb. Its rest mass is 9.107×10^{-28} g. Its radius is about 1×10^{-12} cm. The potential difference of 10000 volt when applied between the electrodes within a glass tube under a vacuum of 0.01 mm of Hg, cathode electrons are generated. The properties of these electrons are analogous to light quanta. The wavelength, which determines the resolution of the microscope, of an electron is usually 0.05 Å, compared with the wavelength of visible light of 3400-7800 Å, of UV light of around 2400 Å and of x-rays 1.0 Å

In an electron microscope the source of electrons is an electron gun consisting of a tungsten wire bent into a V or hairpin, a shield with a circular aperture of 1-3 mm in diameter centred at the filament apex and the anode. The bias voltage controls the angular spread of the issuing beam or brightness. The electron beam passing through the aperture of the anode falls on the condenser lens that controls the intensity and convergence of the illuminating beam by transmitting a part of the beam through a limiting aperture (diameter ranges from 0.25 to 1.0 mm) at its centre to the specimen plane. The condenser lens is adjusted by varying the strength of the field of the lens. The lenses consist of electrostatic or magnetic fields. Electrostatic lenses are parallel metallic plates with co-axial circular holes through which electrons pass. The voltage is applied to the plates so that electrons are accelerated or decelerated and as a result their paths are bent as they pass through the system close to the axis. Magnetic lenses consist of a cylindrically symmetrical gap in a magnetized iron circuit. Electrons passing through the gap close to the axis are deflected by lines of force experiencing a net deflection towards the axis. In the majority of electron microscopes magnetic lenses are used.

The lenses are continuously variable in strength as a variable resister can smoothly control the flowing current. Focusing involves no mechanical motion but only the variation of a current that changes the focal length of the lenses. Both types of lenses suffer from a variety of aberrations similar to those found in optical lenses. The most serious types of aberrations in practice are spherical and chromatic aberrations and astigmatism. Aberrations are more prevalent in electrostatic lenses.

When the microscope is switched on, the electron beam passes through the aperture of the anode then falls on the condenser lens and to the specimen plane. The beam after passing through the object enters the field of objective lens that produces a magnified image of the object. The aperture of the objective lens differs according to the thickness of the specimen. In the case of thick specimens, it is around 25 to 100 μ in diameter to intercept scattered electrons. The objective lens usually magnifies an object to a range of 100 to 200 × but with a lens to image distance one or more projector lenses are used to magnify the first image. An alternative viewing screen is usually incorporated for alignment purposes. The final image can be viewed on a fluorescent screen that is moved aside to allow the electrons to strike a photographic plate (Figure 4.1).

The production of the detail in the electron image depends on differential scattering of the illuminating beam in different parts of the specimen. The probability of scattering depends on the amount of matter in the path of the electrons so that the contrast in the image displays directly the distribution of matter in the cross section of the object presented to the beam. Since the formation of an electron image is related to the mass density in the specimen, production of visible picture depends on the thickness or density of the specimen and enhancement by staining with heavy metals such as tungsten. For example an accelerating voltage of 100 kV does not produce any visible picture if the specimen thickness is 0.25 μ or more. Appropriate thickness of the specimen is also required as the electrons have poor penetrability even at higher accelerating voltages. The ready scattering of electrons also necessitates the evacuation of air from their path as completely as possible. Too much background "fog" in the image is liable to be developed if the residual vacuum is poorer than $5 \cdot 10^{-4}$ mm Hg. In practice a biological material can be examined up to a dry thickness of about 0.1 μ at a voltage of 50 kV so long a resolution more than 100 Å is not required. At 100 kV voltage the tolerable thickness of the specimen is about 0.2 μ. To examine thicker materials, ultra thin sections of them have to be prepared.

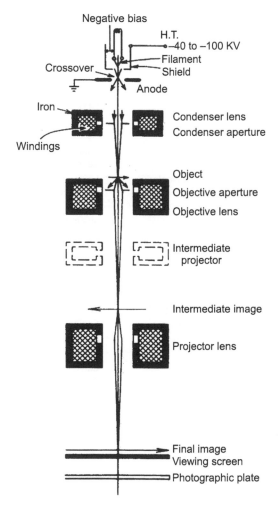

Figure 4.1 Schematic diagram of the optical system of an electron microscope (Hall 1964); reprinted with permission of the University Press of Florida, USA

APPENDIX IV.3

Local lesion hosts of some viruses and the nature of the lesions formed on them

Virus	Local lesion host	Nature of the lesion
1. *Tomato mosaic*	*Nicotiana glutinosa*	Brown with a dark brown edge, 2.3 mm diameter

Contd..

Contd..

2. Tobacco rattle	Chenopodium amaranticolor	Lesion spreads but infection does not become systemic
3. Tobacco necrosis	Phaseolus vulgaris	Lesions red, gradually spreading
4. Tomato bushy stunt	Datura stramonium	Lesion bright or greenish-yellow
	C. amaranticolor	Lesions brown and round
5. Tomato spotted wilt	N. glutinosa	Lesion gradually spreads and infection becomes systemic
	N. tabacum	As above
6. Arabisc mosaic	C. quinoa	Lesion round, dry or rot, gradually becomes systemic
7. Raspberry ringspot	C. amaranticolor	Lesion dry or rot (see above) does not become systemic
8. Tobacco blackring	C. amaranticolor	Lesion dry or rot, gradually becomes systemic
9. Tobacco ringspot	C. quinoa	Lesion dry or rot, does not spread
	Cucumber	Lesion rot, gradually becomes systemic, shows mottling symptom
10. Potato Virus X	D, stramonium	Lesion rot, gradually produces mosaic or mottling symptom
	Gomphrena globosa	Lesions dry or rot, edges red
11. Potato Virus Y	Chenopodium	Lesions dry, edges red
12. Tobacco mosaic	N. glutinosa	Lesions dry brown, edges dark brown
13. Cucumber mosaic	Cucumber	Lesion dry, gradually becomes systemic, shows mosaic symptoms
	Chenopodium	Lesions, innumerable minute grey dots, dry, gradually rots
	N. glutinosa	Lesion dry, gradually rots, becomes systemic

APPENDIX IV.2

Some detection methods used to detect viruses in some economic plants

Virus	Plant	Vector	Method	Reference
Wheat dwarf master virus	Wheat, barley	Psmmotettix alienus	**AMP-ELISA**	Vacke and Cibulka (2000)
Tomato spotted wilt tospovirus	–	Frankliniella occidentalis	**RT-PCR**	Mason et al (2003)
Soybean mosaic potyvirus	–	Soybean aphid	**RT-PCR**	Wang and Ghabrial (2002)
Rice black streak dwarf fiji virus	–	Loadelphax striatellus	**RT-PCR**	Wang et al (2002)
Rice ragged stunt oryzavirus	Rice	Insect vector	**RT-PCR**	Wang et al (2002)

Contd..

Contd..

Banana streak mosaic potyvirus; banana bunchy top babuvirus; cucumber mosaic cucumovirus (simultaneous detection)	Banana, plantain	–	**Multiplex immunocapture PCR (M-IC-PCR)**	Sharmon et al (2000)
Strawberry vein banding caulimovirus	Strawberry	–	**Nucleic acid sequence based amplification (NASBA) and Real Time NASBA**	Vaskova et al (2004)
Maize streak mastrevirirus	Maize	–	**Computer-based image analysis**	Martin and Rybicki (1998)
Faba bean necrotic yellows nanovirus; bean ellow mosaic potyvirus; broad bean mottle bromovirus; broad bean stain comovirus	Vicia faba	–	**Penicillinase-based ELISA**	Salama et al (2003)
Grapevine virus A vitivirus; Grapevine virus B vitivirus; Grapevine leafroll associated virus-2 closterovirus	Grapevine	Psedococcus longispinus	**RT-PCR; immunocapture RT-PCR**	Minafra and Hadidi (1994)
Cucumber mosaic cucumovirus; Tomato aspermy cucumovirus; tobacco etch potyvirus; potato Y potyvirus; alfalfa mosaic alfamovirus; potato leafroll polerovirus	Potato, Capsicum, Tomato	Aphis gossypii; A. craccivora; Myzus persicae	**DAS-ELISA**	Raboudi et al (2002)
Rice stripe tenuivirus	–	Laodelphus striatellus	**RT-PCR**	Cai et al (2003)
Tomato spotted wilt tospovirus	Ranunculus asiatica	–	**TIBA**	Whitefield et al (2003)
Tomato spotted wilt tospovirus & vector	Tomato	–	**RT-PCR**	Roberts et al (2000); Mason et al (2003)
Strawberru crinkle cytorhabdovirus	Frageria spp.	–	**Real Time RT-PCR**	Mumford et al (2004a)
Strawberry crinke cytorhabdovirus; strawberry mild yellow edgepotexvirus; straeberry mottle sadwavirus; strawberry	Frageria spp.	–	**Pentaplex Rt-PCR**	Thomson et al (2004)

Contd..

Contd..

veinbanding caulimo-virus (simultaneous detection)				
Citrus psorsis ophiovirus	Citrus	–	**Hemi-nested RT-PCR**	Legarreta *et al* (2000)
Zinnia leaf curl, Solanum yellow leaf curl, Ageratum yellow vein begomovirus	Cotton	–	**Dot Blot Hybridization, PCR**	Haider *et al* (2003)
Banana streak badnavirus	Banana, plantain	–	**Real Time RT-PCR**	Delanoy *et al* (2003)
Tobamoviruses (CGMMV, KGMMV, ORSV, PMMoV, BMV, SHMV, TMV, ToMV)	Host plants	–	RT-PCR	Maroon *et al* (2002)
Carlaviruses (BBScV, CLV, KLV, LSV, PotLV, PMV, SLV)	Leaves, stems, fruits of infected plants	–	RT-PCR	Maroon *et al* (2002)
Allexiviruses	Garlic	–	RT-PCR	Chen *et al* (2004)
Apple stem grooving cappilovirus	Apple	–	RT-PCR	Marinho *et al* (2003)
Pearvein yellows (Apple stem pitting foveavirus)	Pear	–	RT-PCR	Niu *et al* (2003)
Apple stem pitting foveavirus	Apple	–	RT-PCR	Salmon *et al* (2002)
Potato viruses (PVS, PLRV, PVX, PVA, PVY)	Potato	–	**Multiplex RT-PCR**	Nie and Singh (2001)
Potato viruses	Wide range of cultivars	–	**Multiplex Real Time PCR**	Boonham *et al* (2000)
Potato moptop pomovirus	Potato	–	**Multiplex Real Time fluorescent RT-PCR**	Mumford *et al* (2000)
Beet necrotic yellow vein benyvirus	Sugar beet	–	**DAS-ELISA**	Uhde *et al* (2000)
Potato moptop pomovirus	Potato	–	RT-PCR	Xu *et al* (2004)
Potato viruses (PVA, PVX, PVS, PVY, PNTV, PLRV)	Potato	–	**DNA Microarray**	Bystricka *et al* (2003)
Pelargonium flower-break comovirus	Pelargonium	–	**Non radioactive DOTBLOT Hybridization**	Ivars *et al* (2004)
Carnation mottle carmovirus	Carnation	–	RT-PCR	Kong *et al* (2002)

Contd..

Contd..

Six RNA viruses of olive (Cucumber mosaic cucumovirus, Cherry leaf roll nepovirus, strawberry latent ringspot sadwavirus; Arabic mosaic nepovirus; olive latent virus-1 necrovirus, olive latent virus-2 oleavirus)	Olive	–	**Multiplex RT-PCR**	Bertolini *et al* (2001)
Strawberry pallidosis disease	Strawberry	–	**RT-PCR**	Tzanetakis *et al* (2004)
Cucurbit yellow stunting crinivirus	Cucurbits	–	**TIBIA**	Hourani and Aboujawdah (2003)
Tomato yellowing (Tomato chlorosis) crinivirus	Tomato	–	**Multiplex RT-PCR**	Dovas *et al* (2002)
Carrot motley dwarf disease; carrot mottle umbravirus	Carrot	–	**RT-PCR**	Vercruysse *et al* (2000)
Cymbidium mosaic potex virus; Odonto-glossum ringspot tobamovirus	Orchid	–	**Real Time RT-PCR**	Eun *et al* (2000)
Sweet potato chlorotic stunt crinivirus	Sweet potato	–	**Non-radioactive Nucleic acid Spot Hybridization (NASH)**	Muller *et al* (2002)
Apple stem gooving cappilovirus	Citrus	–	**IC-IT-PCR**	Kusano and Ibi (2003)
Apple viruses (Apple stem pitting foveavirus; apple mosaic ilarvirus; apple chlrotic leafspot trichovirus; apple stem grooving cappilovirus)	Apple	–	**RT-PCR, ELISA**	Menzel *et al* (2003)
Apple chlorotic leaf spot trichovirus; apple stem grooving cappilo-virus	Apple	–	**RT-PCR**	Jan *et al* (2003)
Citrus viroids; apple stem grooving cappilovirus	Apple	–	**Multiplex RT-PCR**	Ito *et al* (2002)
Apple chlorotic leaf spot trichovirus; apple stem grooving cappilovirus; apple stem pitting foveavirus; cherry cappilovirusA	Fruit trees	–	**PDORT-PCR**	Foissac *et al* (2001)

Contd..

Contd..

Apple stem grooving cappilovirus	Apple, Pear	–	**RT-PCR**	James (1999)
Pear vein yellowing virus	Pear	–	**RT-PCR**	Niu et al (2003)
Grapevine leafroll; rugose wood of grapevine	Grapevine	–	**Spot-Multiplex nested RT-PCR**	Dovas and Kotis (2003a)
Vitivirus; foveavirus (simultaneous detection)	Grapevine	–	**Spot nested RT_PCR**	Dovas and Katis (2003b)
Barley yellow and mild mosaic viruses	Barley	–	**Real Time PCR**	Mumford et al (2004b)
Oat mosaic bymovirus	Oat	–	**RT-PCR**	Clover et al (2002)
Soilborne wheat mosaicfurovirus	Wheat	–	**ICRT-PCR**	Clover et al (2001)
Cucumber vein yellowing ipomovirus	Cucurbits	–	**Tissue Hybridization with dogoxigenin-labelled cDNA**	Rubio et al (2003)
Cassava brown streak ipomovirus	Cassava	–	**RT-PCR (Diagnostic kit)**	Monger et al (2001)
PLRV, PVX, PVY (simultaneous detection)	Potato	–	**Multiplex RT-PCR**	Verma et al (2003)
Broadbean wilt-2 fabavirus; cucumber mosaic cucumovirus; clover yellow vein potyvirus (simultaneous detection)	Gentian	–	**Multiplex RT-PCR**	Kuroda et al (2002)
Squash mosaic commovirus; zucchini yellow mosaic potyvirus; cucumber mosaic cucumovirus	Squash seeds	–	**Antigen coated plate (ACP)-seeds**	Dikova and Hristova (2002)

APPENDIX VII.1

Emerging approaches to climate change and global warming

(a) Climate Change

Chakraborty (2005) reviewed current research and development in climate change. Solar radiation mediates between atmosphere, hydrosphere and cryosphere to drive global climate. Infrared radiation heating up the earth's surface is partly radiated back to space. Relatively active water vapour, CO_2,

CH_4, N_2O and O_3 in the atmosphere trap reflected radiation to warm the earth that supports life. Covering over 70% of the earth's surface, the oceans absorb solar energy and transports heat from the equator towards the poles, regulating global temperature. Snow and ice reflect 60-90% of the solar energy they receive to cool the earth. The climate is influenced by the biosphere through CO_2 sequestration and shielding the earth surface from radiation. Global climate is modelled using rules that describe interactions between the atmosphere, land, water, ice and vegetation.

Palaeoclimatic records show an ever-changing global climate (Sabin and Pisias 1996) and continental glaciers have melted rapidly between 8,000 and 15,000 years ago to warm the earth (Cheddadi *et al* 1996). Human activities such as fossil fuel burning and deforestation have accelerated some changes. 'Vostok' ice core data show atmospheric CO_2 fluctuated between 180 and 280 ppm for the past 420,000 years (Petit *et al* 1999, Souchez *et al* 2003), but increased to 367 ppm in just over 200 years. Since the late 19th century, the average earth surface temperature has risen by 0.6 +/- 0.2°C with minimum temperatures increasing as twice the rate of maximum temperature (0.2 v. 0.1°C/decade) (IPCC 2001). Consequently, glaciers that cover 10% of the land area and contain some 70% of the global freshwater have shrunk and the average sea level has risen. Rainfall has increased in the middle and high latitudes of the Northern Hemisphere and decreased over the sub-tropics (Chakraborty 2005).

Questions naturally arise: "Is the atmospheric/oceanic circulation changing? Is the climate becoming more variable or extreme?" A rather abrupt change in the *El Niño*-Southern oscillation behaviour was observed around 1976/77 and this new regime has persisted. *El-Niño* episodes are more frequent and persistent rather than the cool *La Niñas*. Changes in precipitation over the tropical Pacific are related to the change in the *El Niño*-Southern oscillation, which has also affected the pattern of magnitude of surface temperature.

In areas where a drought or excessive wetness usually accompanies an *El Niño*, these dry or wet spells have been more intense in recent years. In some areas where overall precipitation has increased (the mid-high northern latitudes), heavy and extreme precipitation events also increase here. Even in areas such as eastern Asia, increased precipitation events increased despite total precipitation remaining constant or even decreasing. There has been a clear trend to fewer extremely low minimum temperatures in several widely separated areas in the recent decades. There are some indications of a decrease in day-to-day temperature variability in recent decades. The projected change of temperature over the next century is 3-7°F (1.5-4°C) in northern high latitudes (mostly in the cold season) and of the precipitation in northern mid-high latitudes, though the trends may be more variable in the tropics (Anonymous 2006).

Early warning signs of global warming are found in Africa, Asia, Central America, Europe, Russia, North America, Oceania and South America. Heat waves and unusually warm weather, ocean warming, sea-level rise and coastal flooding, glaciers melting, Artic and Antartic warming, spreading of diseases, early spring arrival, plant animal range shifts, population changes, coral reef bleaching, etc., are some of the early warning signs (http: //www. climate hotmap.org/).

Global warming also changes hydrological cycle (evaporation and precipitation). Land precipitation for the globe has increased ~2% since 1900. Precipitation changes have been spatially variable over the last century. Instrumental records show that there has been a general increase in precipitation of about 0.5-1.0%/decade over land in northern mid-high latitudes except in parts of eastern erstwhile USSR. However, a decrease of about ~0.3%/decade in precipitation has occurred during the 20th century over land in sub-tropical latitudes, though the trend has weakened in recent decades.

The Northern Hemisphere annual snow covers extent remained below average since 1987 and has decreased by about 10% since 1966. This is mostly due to decrease in spring and summer snowfall over both the Eurasian and North American continents since the 1980s. However, winter and autumn snow cover extent has shown no significant trend for the Northern Hemisphere over the same periods. Improved satellite data shows that the general trend of increasing cloud amount over both land and ocean (http://www.ncdc.roaa.gov/oa/climate/global warming.html).

The IPCC (Intergovernmental Panel on Climate Change) jointly established by the WMO (World Meteorological Organization) and UNEP (United Nation's Environment Program) in 1988, has the responsibility for assessing information relevant to climate change and summarizing the information for policy makers and the public. It has published major assessment reports most recently in 1995 and 2001 (IPCC 2001). Since the 1995 report, there have been a number of advances, including improvements in the Atmosphere-Ocean General Circulation Models (AOGCM) used to predict climate change. These predictions are based on scenarios that describe greenhouse gas emissions from potential resource use patterns, technological innovations and demographics. The results from modelling experiments based on the emission scenarios give a range of predictions, depending on the assumptions quantified by each scenario.

There are a total of 21 global climate modelling groups participating in the IPCC 4th Assessment. Different groups using various models have carried out the same set of experiments specifying different scenarios of atmospheric CO_2 concentration changes as well as control simulations for both pre-industrial and present day climates. Out of these 21 models only 15 include soil moisture data in their output. The official names of these models and their originating groups are listed in the Table.

The official names of the models that include soil moisture and their originating groups (Wang 2005)

Official Name	Originating Groups
CCSM3	US, National Center for Atmospheric Research
ECHAM5/MPI-OM	Germany, Max Planck Institute for Meteorology
FGOALS-gl.0	China, LASG/Institute of Atmospheric Physics
GFDL-CM2.0	US Department of Commerce/NOAA/Geophysical Fluid Dynamics Laboratory
GFDL-CM2.1	As above
GISS-AOM	US NASA/Goddard Institute for Space Studies
GISS-EH	Same as above
GISS-ER	Same as above
INM-CM3.0	Russia, Institute for Numerical Mathematics
IPSL-CM4	France, Institut Pierre Simon Laplace
MIROC3.0 (medres)	Japan, Center for Climate System Research in the University of Tokyo, National Institute for Environmental Studies, and Frontier Research Center for Global Change
MIROC3.0 (hires)	Same as above
MRI-CGCM2.3.2	Japan, Meteorological Research Institute
PCM	US National Center for Atmospheric Research
UKMO-HadCM3	UK Hadley Center for Climate Prediction and Research/Met Office

Predictions derived so far from these models have many uncertainties. Sources of uncertainties in predictions include inability to fully predict human resource use and incomplete understanding of climate processes. In addition to the predicted increases in temperature for much of the world, changes in extremes are also predicted. For temperature, more frequent high temperatures and less frequent extreme low temperatures are predicted. Likewise, increased intensity of precipitation events is predicted in some region (Garrett *et al* 2006). Recent analyses have concluded that there have been changes in storm patterns (Emanuel 2005, Webster *et al* 2005) that could influence the global movement of pathogens (Brown and Hovmoller 2002).

Additional predictive variability comes into play for modelling of regional climates. All the forms of uncertainty about global processes are still a factor, with added uncertainty due to lack of data from some regions. Remote regions with complex topography, could produce rapid climatic variation over small areas. While water vapour, evaporation, and precipitation are predicted to increase on an average, prediction of increased or decreased precipitation is region specific. In general, precipitation is predicted to increase in both summer and winter in high latitude regions. In northern latitudes, the Antarctica, and tropical Africa, precipitation is predicted to increase in winter. In southern and eastern Asia, precipitation is predicted to increase in summer. Decreases in winter rainfall are predicted for southern Africa, Central America and Australia. Supplementary material on IPCC websites supplies finer scale predictions. Decreased snow cover and land-ice extent are expected to follow from the trend in increasing temperature (Garrett *et al* 2006).

International approaches

Global change encompasses changes in atmospheric composition, climate and climate variability, and land cover and land use. The occurrence of these changes and their interactive effects on biological systems are worldwide; thus effective global change research and impact assessment programmes are to be based on international and interdisciplinary research and communication. Accordingly several collaborative research networks with a focus on global change have been established in the biological sciences. These include the Global Change and Terrestrial Ecosystem (GCTE). Core project of the International Geosphere-Biosphere Program (IGBP); it aims to predict the effects of global change on terrestrial ecosystems, including agriculture and production forestry. GCTE initiated a network activity on 'Global Change Impacts on Pests, Diseases and Weeds' because of their importance as yield-reducing factors in agriculture and as early indicators global change (Scherm *et al* 2000).

Several collaborative research and impact assessment programmes with a focus on global change have been established in the physical, biological and social sciences. These programmes were initiated through partnerships of non-governmental or intergovernmental organizations. But they differ in their specific objectives and organizational structure. The World Climate Research Program (WCRP) focusses on research to enhance the basic knowledge of the climate system; the Global Change and Terrestrial Ecosystems (GCTE) and International Human Dimensions Programs on Global Environmental Change (IHDP) feature, respectively biological and socioeconomic research and impact assessment. The third type of international collaboration is the Intergovernmental Panel on Climate Change (IPCC); which focusses chiefly on the synthesis of published information for scientists and policy makers. Some international global change programmes are more formal than others. For example, participation in IPCC is by nomination only, while other programmes [DIVERSITAS, GCTE, and Land-Use and Cover Change (LUCC) are open to all scientists in relevant areas who are willing to exchange ideas and conduct joint research.

APPENDIX VIII.1

Plant Quarantines in the State of Washington, USA (with special reference to a few viruses)

Nursery Inspection

All plant materials shipped into or within the state of Washington must be accompanied by an Inspection Certificate and/or any Certification tag/label of the state of origin and be free from injurious pests, diseases and obnoxious weeds. As for the viruses concerned, quarantine is strongly established with different viruses of fruit crops.

Blueberry scorch

A virulent strain of blueberry scorch virus exists, if introduced into the state, would have a severe economic impact on the blueberry industry

Area under quarantine: All states and districts of the USA.

Provisions: All regulated plants sold, offered for sale, transported within the state, or planted must be accompanied by a Phytosanitary Certificate issued from the place of origin, that verifies freedom from viruses through one or more of the following methods: (1) They must originate from a pest-free area, (2) must be certified by an official certification programme that includes inspection and testing, (3) must test free of viruses by methods approved by Washington, and (4) plants are micro-propagated and/or grown in an insect-proof greenhouse, and originate from plants that have been tested and are free from viruses. Special permits or exemptions issued by the director allowing entry must be obtained prior to transport, sale, and planting.

Grape virus

Risk or quarantined viruses: Fan leaf, Leaf roll, Stem pitting, and Corky bark

Area under quarantine: All states and districts of the USA and all areas outside the territorial border of the state of Washington.

Provisions: Grape plants from areas under quarantine may enter provided that: (1) They have been grown under an official grapevine certification programme that includes inspection and testing by indexing or suitable indicator hosts for freedom from the quarantined virus diseases and (2) All shipments are clearly marked on the package and container.

Similar specifications are there for all the quarantined viruses in the state of Washington.

Index

8-Azaguanine 263

A

A-9 CMV 268

ACLSV 255, 333, 398

Accumulated degree-days 74, 291

Aceria tulipae 119

Aceris tosichella 119, 269

Acquired resistance 266, 296

Acquisition threshold 118

Acrogenaceae 3

Actiguard 296

Active take-off 161

Acyrthosiphon pisum 94, 95, 102, 193

Advection 167

Aedeagus 76

Aerial netting 148, 157, 168

African cassva mosaic virus 188, 296, 332, 338, 347, 387

Agallinae 76

Agglutination 38, 40, 43, 45, 55

Agglutinin 38

Agglutinogen 38

AGRAPHID 154

Agronomic/cultural strategies and tactics 288

Aircraft-based air borne radar 157

Aleyrodidae 69, 181, 182

Alfalfa 95, 179, 189, 209, 268, 296, 327, 328, 330, 332, 426, 473

Alfalfa mosaic virus 189, 268, 327, 328, 330, 332, 426

Alfamovirus 7, 9, 18, 20, 26, 31, 32, 35, 37, 60, 95, 296, 330-332, 357, 371, 374, 426, 473

Alginate 259

Allergenicity 283, 284

Allexivirus 20, 26, 36, 119, 330, 331, 339, 347, 360, 367, 395, 396, 474

Almond 260

Alphacryptovirus 9, 27, 332, 334, 335, 340, 346-348, 351, 394

Alternenthera sessilis 191

Am 268

Amaranthus spp. 189

American finger millet 213

Ampelovirus 26, 36, 117, 331, 339, 341, 345, 367, 400, 401

Amphorophora idaei 280

Amplicons 52, 54

AMV 179, 189, 262, 274, 275, 328, 332, 426-428

Aucuba strain of tomato mosaic virus 256

Andean potato latent tymovirus 60, 332

Andean potato mottle comovirus 60, 332

Anholocyclic life cycle 71

Annual sowthistle 191

Annulaceae 3

Anti-sence RNA technology 276-278, 281

Anti-viral 263, 280, 283, 418

Antibody 38-41, 43-50, 189, 190, 431, 433

Anticyclone 162

Antigen 3, 5, 30, 38-49, 379, 430, 476

Antiserum 40-43, 45, 46, 49, 99

Aphididae 68, 69, 181, 182

Aphis craccivora 95, 193, 221

Aphis gossypii 95, 96, 166, 197, 221, 222, 268

Aphis pisum 95, 101

Aphis spiraecola 192, 221

Apple chlorotic leaf spot virus 255, 256, 263, 398

Apple stem grooving virus 256, 263, 328, 330, 333, 395

Apricot 260, 333

Ar MV 333, 121, 122, 255

Arabidopsis sp. 178, 333, 377, 425, 439, 440

Arachis hypogea 276

Architecture of viruses 25

ASGV 263, 333, 395, 396

Asparagus adscendens 262

Aspermy virus 57, 349

Aster yellows virus 1, 56, 255

Avena sativa 209, 213

Avenavirus 7, 27, 330, 331, 343, 359, 369, 414

AYV 332

B

Babuvirus 7, 28, 95, 331, 333, 354, 360, 390, 473

Bacilliform particles 19, 35

Bacteriocytes 101

Badnavirus 7, 26, 35, 117, 178, 181, 194, 329, 331, 332, 333, 335-339, 341, 345-349, 353, 365, 381, 474

BaMMV 269, 333, 403

Banana bunchytop virus 1, 333, 390

Banana streak virus 196, 381

BARI-T1 (Manik) 228

BARI-T11 228

BARI-T12 228

BARI-T2 (Ratan) 228

BARI-T4 228

BARI-T5 228

BARI-T6 (Apurba) 228

BARI-T7 (Chaity) 228

Barley mild mosaic virus 269, 333, 403

Barley yellow dwarf 57, 97, 154, 191, 207, 208, 269, 327, 330, 333, 434

Barley yellow dwarf disease 97

Barley yellow mosaic virus 269, 333, 404

Barley yellow striate mosaic virus 120, 334

BaYMV-2 280, 404

BaYmv-1 269, 404

BaYmv-II 269, 280, 404

BaYmv-III 269

BCMV 271, 278, 334

BCTV 105, 112, 224, 225, 278, 334, 386

Bean common mosaic necrosis virus 226

Bean common mosaic virus 226, 334

Bean leaf roll virus 94, 334

Bean yellow mosaic virus 191, 193, 263, 334

Bean yellow vein banding virus 94

Beet chlorosis 226, 334

Beet curly top virus 180, 224, 226, 297, 329, 334, 386, 387

Beet leaf curl virus 118, 334

Beet leafhopper 224, 225

Beet mosaic virus 186, 334

Beet necrotic yellow vein virus 124, 187, 264, 334, 405

Beet necrotic yellow vein virus 124, 187, 264, 334, 405

Beet necrotic yellows virus 264

Beet pseudo yellows virus 110

Beet yellows 289, 328, 330, 334, 399

BG2814-6 268

Begomovirus 7, 9, 20, 28, 35, 109, 111, 112, 181, 194, 196, 199, 226, 228, 296, 329, 331, 332, 334-344, 346-351, 354, 387, 388, 474

Bemisia argentifolia 110

Bemisia tabaci 73-75, 109, 110, 187, 192, 195, 227

Bemisia tabaci biotype A 75, 110, 280

Bemisia tabaci biotype B 75, 110, 195, 227, 228

Bemisia tabaci spenish Q 75, 110, 227, 228

Benyvirus 9, 20, 26, 32, 35, 123, 126, 331, 332, 334, 362, 367, 405, 474

Benzo-(1,2,3)-Thiodiazole-7-Carbothionic acid 296

Bessia scoparia 191

Beta vulgaris 180

Betacryptovirus 9, 20, 28, 178, 329, 331, 335, 340, 346, 351, 394, 395

Bhendi yellow vein mosaic virus 334, 366

Biden spp. 189

Bimodal transmission 96

Biogel P 200 259

Biological fingerprint of climate change 205

Biosafety issues 281

Bipartite viruses 9, 328

Bitter gourd 189

Black nightshade 191

Blister beetle 69

Blueberry scorch 303, 334, 481

Brachiaria chloris 214

Brachiaria deflexa 214

Brachiaria lata 214

Brachycaudus helichrysi 192

Brachypterous 76, 78, 79, 81, 108, 190, 192

Brassica campestris 189

Brassica kaber 227

Brassica spp. 225, 271

Brevipalus phoenicis obovatus 120

Brinjal 1, 189, 292

Brinjal streak virus 1

Broad bean wilt virus 263, 335, 432

Bromovirus 20, 28, 32, 115, 194, 327, 328, 330, 331, 335-337, 342, 348, 357, 419, 423, 473

Bromuis arvensis 190

Bromuis commutatus 190

Bromuis hordaceus 190

Bromuis japonicas 190

Bromuis sterilis 190

BSMV 179, 254, 267, 333, 406, 407

BSV 262, 333, 381, 382

BSV 262, 333, 381, 382

BtMV 186

BNYVV 124-126, 187, 334, 405, 406

BBTV 7, 333, 390, 391

Buchnera 101

Bud mite 69

Bugs 53, 68, 69, 82-84, 110, 117, 118, 147, 149, 401

Burbank plum 179

BYDV-1 208

BYDV-2 208

C

Cab-1 268

Cab-2T 268

Cacao swollen shoot virus 117, 335

Cantaloupe 274, 286, 296

Cappilovirus 333, 336, 474-476

Capsella acutum 227

Capsella bursa-pastoris 180

Capsicum annuum 227, 268, 270, 287

Capsicum baccatum 227

Capsicum chacoens 270

Capsicum chinensis 227, 271

Capsicum frutescens 227, 268, 270

Capsicum pubescens 227

Capsid 15, 18, 19, 96, 99, 108, 109, 110, 125, 224, 272, 273, 279, 353-364, 376, 380, 381, 383, 394, 397, 413, 418, 420, 425, 430, 431, 434, 435, 442

Cardemine hirsuta 190

Carlavirus 7, 9, 20, 26, 36, 57, 95, 109, 182, 194, 326, 328, 330-332, 334-348, 351, 360, 367, 395, 396, 474

Carmovirus 9, 28, 115, 116, 123, 289, 327, 330-332, 334, 335, 337, 339-342, 344, 347, 351, 359, 414, 415, 474

Cartagena protocol 282

Carrot mottle virus 94, 331, 335

Carrot red leaf virus 94

Cassava brown streak virus 109, 336

Cassava mosaic virus 187, 188, 332, 338, 340, 347, 348, 387

Cauliflower mosaic virus 375

Caulimovirus 9, 18, 20, 28, 30, 94, 181, 194, 296, 327, 329, 331, 335, 336, 338-340, 342, 348, 349, 353, 366, 375, 377, 381, 473, 474

Cavariella aegopodii 94

Cavemo virus 20, 28, 30, 178, 194, 336, 349, 353, 380, 381

CCMV 255, 278, 337

cDNA probe 51, 110

Cerastium vulgatum 190

Ceratovacuna spp. 166

Cereal yellow dwarf virus 97, 336

Cf-2/Cf-5' 268

Chemical control of vectors 294

CHM-47 267

Chemo-therapeutant 263

Chenopodium 59, 60, 121, 126, 180, 189, 286, 336, 398, 432, 433, 472

Chenopodium amaranticolor 59, 60, 286

Chilli 189, 336, 366

Chitinase 296

Chlorogenaceae 3

Chrysanthemum 115, 259, 262, 335, 396

Chrysanthemum stem necrosis virus 115

Chrysomelidae 69, 80, 115

Cibenzolar 296

Cicadellidae 69, 76, 181, 182

Cicadulina arachidis 105, 214

Cicadulina bipunctata 106, 197, 214

Cicadulina bipunetella bimaculata 106, 197

Cicadulina ghaurii 214

Cicadulina lateens 214

Cicadulina mbila 105

Cicadulina niger 214

Cicadulina parazeae ghauri 214

Cicadulina similis 214

Cicadulina storeyl 214, 215

Circular genome 213

Circulative virus 95, 102, 103, 271

Circulifer tenellus 102, 180, 224, 390

Cire virus 7

Citrus leprosis 7, 120, 336, 363, 442

Citrus tristeza 57, 197, 220, 223, 255, 285, 337, 399

CLCuV 193

Clerodendron inerme 275

Closterovirus 9, 13, 20, 26, 36, 75, 95, 110, 182, 194, 221, 328, 330, 331, 334-337, 339, 351, 359, 367, 399, 400, 473

Clover wound tumour virus 31, 57, 103, 106

Clover yellow vein virus 263, 337

CMV 58, 96, 156, 189, 195, 199, 262, 266-268, 271-276, 278-282, 287, 293, 296-298, 337, 382, 422-425, 440

CMV-C 273

CMV-Y 296

Co-virus 200

Coccinellidae 69, 115

Coccinia grandis 189

Coccoidea 69, 117

Cold front 162

Colletotrichum lagenarium 296

Common chickweed 191

Community epidemiology 201

Comovirus 7, 9, 12, 17, 27, 28, 32, 57, 60, 115, 182, 327, 328, 330, 332, 334, 335, 337, 339, 344, 346, 348, 351, 357, 428, 473, 474

Complex viruses 18, 25, 30

Compound interest type progression 183

Convective updraughts 161

Conicum maculatum 189

Convection 161

Convergence 162, 167, 470

Corchorus olitorius 191

Costelao 268

Couch 209

Cowpea aphid-borne mosaic virus 268, 337

Cowpea mild mottle virus 109, 337

Cowpea mottle virus 270, 337, 415

CPM 110, 301

Crinivirus 20, 26, 36, 109, 330-332, 334, 338, 341, 349, 350, 360, 367, 400, 442, 475

Crinkle 57, 116, 333, 338, 348, 351, 405, 415, 473

Crip-31 275, 296

Cryptic virus 178, 332, 334, 335, 340, 347, 349, 351, 394, 395

Cryptogamic system 325

Cucumber 1, 58, 95, 110, 116, 123, 189, 195, 199, 263, 266, 268, 271, 273, 279, 280, 285-287, 292, 293, 296, 297, 326, 328, 330, 337, 338, 351, 413, 418, 422, 424, 472, 473, 475, 476

Cucumber mosaic virus 1, 58, 96, 189, 195, 199, 263, 268, 285, 286, 326, 328, 330, 337, 351, 422

Cucumis africanns 268

Cucumis melo 268, 270

Cucumis sativus 189

Cucumovirus 9, 18-20, 28, 32, 58, 95, 96, 98, 178, 287, 296, 326, 328, 330, 331, 337, 344, 349, 357, 374, 422, 423, 473, 475, 476

Cucurbit aphid-borne yellow virus 268, 338

Cucurbit yellow stunting disorder virus 110, 400

Cucurbita moschata 189, 267

Curtovirus 20, 28, 35, 102, 104, 105, 181, 224, 329, 331, 334, 340, 354, 366, 386, 387, 390

CyMV-Ca 273

Cyclone 162

Cynodon 214

Cyrtopeltis nicotianae 118

Cyrtorhinus lividipennis 165

Cytorhabdovirus 21, 25, 26, 30, 35, 95, 102, 106, 107, 119, 182, 329, 331, 334, 335, 339, 341, 343, 347, 348, 351, 371, 405, 473

CZW-3 274, 282

D

Dalbulus maidis 102, 107

Datura metel 189

Datura stramonium 114, 189, 227, 472

Deltocephalinae 76

Deoxyvira 4

Deutonymph 84, 86

DI 191, 200, 287, 415

Diabrotica undecimpunctata Howardi 116

Dianthovirus 9, 28, 328, 330, 331, 335, 346, 348, 363, 369, 415, 416

Dianthus caryophyllus 275

Digitaria horizontalis 214

Disease Index 191, 186

Dorylaimida 68, 69, 86, 120

Dot immuno-binding assay 48

ds DNA 5, 198, 327, 353, 365, 366, 373-375, 381, 382, 384, 386

ds RNA 277, 281, 354-356, 363, 365, 366, 374, 386, 391, 393-395, 398, 408, 410, 417, 418, 437

Dynamic website 75, 176, 182, 199, 212, 222, 251, 298

Dysmococcus brevipes 118

Dysmococcus neobrevipes 118

E

Echinochloa crus-galli 210

Ecological epidemiology 188

Ecological genomics 201

Eggplant mosaic tymovirus 60, 292, 338

ELISA 45-49, 53, 55, 107, 189, 190, 222, 261, 263, 472-475

Electrical Penetration Graph Technique 110

Electronic monitoring of the probing and feeding behaviour 191

Eleucine coracana 213

Eleusine indica 214

Elymus hispidus 269

Emergence of new viruses 194, 196

Endocytosis 100, 105

Endophyte 198

English grain aphid 72

Enveloped viruses 20-23, 29, 35, 95, 102, 107, 119, 327, 356

Environmental factors in epidemiology 191

Enzyme linked immuno sorbent assay 45

EPGT 75

EPPO 301, 302

Epilachna varivestis 116

Eryophydidae 69

Eupatorium capillifolium 190

Euphorbia sp. 227

Evolutionary epidemiology 196, 199, 200, 220

Extron 6

F

FAO 300, 301

Fabavirus 20, 27, 28, 95, 328, 330, 332, 335, 341, 357, 432, 476

False spider mite 120

Fiji disease of sugarcane 3, 57

Filamentous viruses 20, 26

Finger millet 213, 339

Flight trajectories 160

Fny-CMV 267, 268

Foregut-borne virus 92

Foveavirus 20, 26, 36, 330, 331, 333, 347, 367, 396, 4740-476

Frankliniella fusca 81, 114, 115, 185, 190, 193

Frankliniella schultzei 115

French bean 191, 292

Fulgoidea 69

Full moon 152, 153

Functional genomic analysis 201

Fundatrix 71

Furovirus 9, 20, 26, 32, 35, 123-126, 178, 328, 331, 332, 336, 343, 347, 361, 367, 406

G

Galea officinalis 189

Gallaceae 3

Gametangia 86

GATT 298, 301

Gel diffusion 42

Gelrite 259

Gene Yd 208

Genietic control of transmission 97

Genitalia 73, 86

Genomic technology 201

Geranium carolinianum 190

GFLV 122, 255, 262, 339

Gladiolus 258, 262, 263, 298

Global scale of wind 162

Global warming 202-207, 251, 476-478

Globulin 38, 39, 41, 43, 45, 47

Gnaphelium purpureum 190

Gompertz 223

Graminella nigrifrons 103

Grapevine fanleaf virus 339

Grapevine leafroll associated virus-3 117, 339, 473

Grapevine necrosis virus 398

Grapevine virus A 117, 262, 330, 339, 398, 473

Grapevine virus B 117, 340

Green tile trap 192

Greenhouse whitefly 75

Ground-based radar 148, 157

Groundnut ring spot virus 115, 340

Group concept of nomenclature 2

Grow-On-Test 59

GVA 117, 262, 339, 398, 399

H

Hairpin 32, 296, 367, 378, 387, 400, 415, 420, 426, 427, 435, 469

Hairy nightshade 190

Harmonic radar 160

Helical viruses 14

Helper component 96, 103, 104, 122, 379, 402, 404

Hemolymph 93, 97, 101, 102, 105, 108, 111, 112, 118

Heritable resistance 266

Heteroencapsidation 283, 284

High plains virus 270

Hin19 281

Hindgut 91-93, 100, 112, 115, 116

Holocarpic 86

Holocyclic life cycle 71

Hordeivirus 20, 26, 32, 35, 178, 194, 328, 331-333, 342, 345, 362, 368, 406, 407

Hordeum vulgare 209, 267, 269, 340, 439

Host in epidemiology 177, 182, 188

Host range 1, 5, 56, 58, 73, 81, 94, 123, 175, 180, 188, 196, 201, 211, 217, 225, 288, 289, 325, 388, 402, 424, 439

Hybridoma 39

Hydaphisi coriandri 192

Hyperomyzus lactuceae 192

I

Icosahedron 14, 15, 16, 357

Ig A 38

Ig G 38, 39, 41, 45-47, 49

Ig M 38

Ilarvirus 7, 10, 19, 28, 113, 194, 328, 330-334, 336-341, 344, 346, 348-350, 357, 369, 425, 426, 428, 475

Immunosorbent electronmicroscopy 37

Immuno-diffusion 40

Immuno-electrophoresis 40, 42

Immuno-fluorescense assay 44

Immuno-osmophoresis 40, 43

Immuno-tissue printing assay 44, 48

Indicator plant 58-60, 261, 296

Inducer protein 275, 296

Inoculation threshold 92, 93

Insect viruses 197

Inter-African Phytosantiary Council 302

Internet-based decision support system 212

Intron 6, 390, 391

Inversion layer 149

IPCC 202, 207, 477, 478, 480

Ipomovirus 20, 26, 35, 109, 182, 330, 331, 336, 338, 349, 358, 368, 402, 476

ISEM 37, 49, 55, 261

Isometric viruses 12, 15, 27, 37, 40, 369, 372

Isopathic contour map 189

ISPM 301

IT 86D-716 268

IT-85F-2687 268

J

Japanese plum 260

K

Ka-1 269

Kai 271

Kikuyu 209

Kochina scoparia 191

Kytoon 157, 158

L

Lactuca scariola 190

Lactuca serriola 191

Lady bird beetle 69

Laminar Boundary Layer 161

Laminum amplexicaule 191

Laodelphax striatellus 102, 186, 205, 207, 472, 473

Latency 105

Latex test 43, 44

LBVD 193

Leaf Dip method 36

Leaf beetle 69, 80, 116, 297

Leafhopper 69, 76, 77, 102, 103, 105, 106, 153, 169, 215, 218, 220

Leafhopper virus A 106

Leersia oryzoides 210

Lettuce bigvein disease 193

Lettuce chlorosis virus 117, 341

Lettuce infectious yellows virus 110, 341

Lettuce mosaic virus 192, 254, 289, 341

Light trap 149-153, 166, 220

Linaria canadense 190

Linear genomic 372

LIDAR 15

Local lesion host 57, 471

Logistic model 223

London rocket 225

Long distance migration 157, 163-166, 180, 220, 226, 295

Longidoridae 69, 86, 120

Longidorus 87, 88, 120-122

Luffa acutangula 189

Lunation induced, variation 152

Lycopersicon chmielewsnii 267

Lycopersicon esculentum 267, 296, 342, 439

Lycopersicon hirsutum 267

Lycopersicon hirsutum f sp. Glabrum 268

Lycopersicon pimpinellifolium 267

M

Machlomovirus 7, 28, 116, 178, 330, 331, 342, 359, 370, 416

Macropterous 78, 79, 81, 108, 190, 192

Macrosiphum euphorbiae 100, 193

Maculavirus 7, 339, 364

Maize chlorotic dwarf virus 103, 270, 327, 342, 437

Maize dwarf mosaic virus 270, 342

Maize fine streak virus 270, 342

Maize necrotic streak virus 270

Maize rayado fino virus 102, 107, 197, 270, 327, 331, 342, 437

Maize streak virus 180, 195, 196, 213, 270, 329, 342, 384, 386

Maize stripe virus 108, 342, 410

Malus domestica 263

Malus pumila 263

Malva parviflora 75, 227

Malva spp. 75, 189

Mandarivirus 7, 20, 26, 36, 331, 340, 364, 396

Mandible 70, 81

Marmaraceae 3

Marmor tabaci 2

Mastrevirus 20, 28, 35, 102, 104, 105, 182, 194, 213, 329, 331, 334, 336, 338, 342, 344, 348, 349, 351, 353, 384, 385, 388

Mathematical model 200, 206, 218, 223

MAV 99, 101, 191, 208-210, 212, 333

MDMV 166, 270, 342

Mealybug 110, 117

Measures against spread 290

Melilotus indica 179

Meloidae 69, 115

Melon necrotic resistance-2 270

Melon necrotic resistance-1 270

Melon necrotic spot virus 123, 270, 279, 342

Mercuralis ambigua 227

Meristem culture 255-259, 261-263

Meristem immunity 256, 257

Meso scale of wind 162

Messenger 31, 32, 55, 296, 297, 375, 405, 418, 420, 428

Metavirus 22, 28, 30, 332, 333, 341, 363, 440

Metopolophilum dirhodum 210

Micro RNA pathway 277

Micro-grafting 260

Micro-precipitin 40

Microarray technology 55

Micrutalis melleifera 105

Migratory flights of insects 159, 161, 163, 164

Milli-metric radar 159

Mirabilis jalapa 275

Miridae 69, 84, 118

MnSOD gene 276

Mnr-1 270

Mnr-2 270

MNSV 124, 178, 264, 271, 342

MSV-Kom 195

MSV-sat 195

MSV105, 195, 196, 207, 213-215, 342, 384-386

Moericke yellow water trap 192

Mokusekko 3 269

Molecular beacons 53, 54

Molecular ecology and epidemiology 193, 194

Mollungo verticillata 190

Momordica cheraritia 189

Monoclonal antibody 38, 39, 46, 47, 222

Monocyclic progression 183

Monopartite viruses 327, 398

Moonlight interference 152

Mosaics 56, 376

Movement protein mediated resistance 275

Multicyclic progression 183

Multiplex PCR 53

Mung bean yellow mosaic virus 191

Mycetocytes 101

Mycorrhyzae 198

Myzus ascalonicus 72

Myzus dirhodum 101

Myzus persicae 70, 71, 95, 96, 166, 184-186, 192, 221, 268

N

Na 2 269

New generation vertical looking compact entomological Radar 160

New moon 152, 153

N-gene 279, 396

Nicotiana benthamiana 60, 390, 399, 407, 430, 438

Nicotiana glutinosa 256, 267, 471

Nicotiana plumbaginifolia 276

Nicotiana rustica 57, 255, 262

Night vision equipment 148

Nilaparvata lugens 79, 108, 163, 164, 180, 343, 391-394

NSV 270, 279

Num Mex Twilight 270

Num Mex Bailley 270

O

Oat 56, 197, 330, 343, 391, 403, 414, 434, 476

Oat blue dwarf 197, 343

Oleavirus 7, 19, 20, 29, 35, 330, 331, 343, 357, 426, 475

Olpidiaceae 69, 86

Olpidium bornovanus 178

Olpidium brassicae 123

Olpidium radicale 123

Operational tactics 147, 252, 290

Ophiovirus 20, 26, 30, 35, 330, 332, 337, 341, 342, 346, 351, 356, 368, 408, 474

Ornithopus cornessus 179

Oryza glaberima cv. Tog 5681 270

Oryza sativa 209, 213, 343, 395

Oryza sativa Japonica cv. Azucena 280

Oryza sativa var. indica cv. Oidenta 270

Oryzavirus 20, 22, 27, 29, 106, 181, 194, 327, 329, 331, 338, 346, 354, 365, 393, 472

Ourmiavirus 9, 25, 26, 35, 178, 331, 332, 336, 338, 344, 362, 371, 413

Oviparous 72, 74

Ovipositor 73, 76, 84

P

P19 281, 409, 417, 418

P2 protein mediated resistance 272, 275

Panicovirus 7, 29, 330, 331, 344, 359, 417

Paracalyx scubiosus 191

Paratrichodorus 86, 87, 120-122, 191, 298

Paratrichodorus pachydermis 122

Parthenogentic progeny 72

Paspalum 209, 214

Passive take-off 161

Patchouli mild mosaic virus 273

PAV 91, 95, 101, 191, 208-212, 233, 234

PCR 51-55, 219, 220, 472-476

Pea early browning 56, 122, 344, 412

Pea enation mosaic virus 264, 327, 328, 330, 344, 434, 441

Pea seed-borne mosaic virus 96, 264, 344

Pea tip yellows 57

PEBV 122, 179, 344, 412

Peach 119, 260, 303, 344

Peach mosaic 119, 303, 344

Peach mosaic virus 119, 344

Peach rosette 303, 344

Peach yellows 303

Pearl millet 213, 215

Pergola stunt virus 108, 123, 344

Pennisetum americanum 213

Peregrinus maidis 108

Periodicity of flight activity 152

Perkinsiella saccharicida 102, 180

Persistent virus 89, 92, 95, 99, 116, 119, 120, 156, 166, 177, 180, 184, 193, 268, 289

Petuvirus 7, 20, 29, 30, 331, 345, 363, 366, 381

Pf-CMV 268

PI 241675 270

PI 273419 270

Phaginae 3

Phalaris aurandinacea 190

Phalaris austrialis 210

Phalaris communis 210

phalaris canariensis 210

Phase group analysis 152

Phasiolus vulgaris 191, 227, 267, 345, 472

Phytolacca americana 275, 296

Phytophaginae 3

Phytoreovirus 9, 12, 18, 20, 22, 29, 103, 106, 181, 194, 273, 327, 329, 331, 346, 351, 393

Phytosanitation 288, 298, 300

Piesma quadratum 84, 118

Pineapple mealy bug wilt associated virus 117, 345

Piquin 270

Plan position indicator 159

Plant hopper 69, 78, 79, 102, 104, 108, 154, 157, 163-165, 205, 207, 291, 394

Plantago 189, 190, 345

Plantago rugelii 190

Plasmodiophorales 69, 86

Plasmodium 86

PLRV 100, 101, 166, 190, 271-273, 294, 345, 434, 435, 474, 476

Plum pox virus 263, 345

Poaceae 78, 208, 210, 211, 213

Pointed gourd 189

Pokeweed antiviral protein 296, 345, 422

Polerovirus 29, 58, 95, 97, 182, 330, 332, 334, 336, 338, 345, 348, 351, 360, 370, 435, 473

Pollizo plum 260

Poly A 32, 365-371, 408, 432, 435, 437,440, 442

Polyacrylamide gel electrophoresis 49, 55

Polyclonal antibody 433

Polyhedral symmetry 14

Polymerase chain reaction 51

Polymyxa betae 123, 187

Polymyxa graminis 87, 123, 180

Polyphagus 81, 200, 221

Pomovirus 9, 20, 27, 35, 123, 126, 331, 332, 334, 335, 345, 362, 368, 409, 474

Popia 7

Population biology of gene expression 202

Positive electron stain 37

Positive sense RNA 19

Post-transcriptional gene silencing 277, 279, 396, 409, 430, 438, 440

Potato 3, 57-60, 89, 95, 102, 109, 123, 154, 156, 170, 180, 189, 190, 254, 255, 257, 258, 262, 265, 268, 271-273, 276, 279, 289, 293, 294, 296, 303, 326-332, 345, 346, 349, 351, 379

Potato T capillovirus 60

Potato black ring spot nepovirus 60, 345

Potato leaf roll 57, 58, 190, 255, 271, 294, 435

Potato mop top 123, 180, 331, 345, 409

Potato virus Y 57, 189, 190, 296, 326, 328, 330, 345, 346, 349, 397, 401, 472, 474

Potato yellowing alfamovirus 60

Potexvirus 7, 9, 20, 27, 32, 36, 178, 289, 296, 326, 328, 330-333, 335-341, 343-349, 351, 360, 368, 395, 397, 473

Potyvirus 7, 9, 12, 20, 25, 27, 32, 35, 57, 95-97, 186, 278, 296, 326, 328, 330-352, 358, 368, 374, 401-404, 407, 472, 473, 476

PP & Q 300

Predictive displacement model 162

Prickly lettuce 191

Prociphilus spp 166

Pronotum 80

Propagative virus 90-93, 104, 106

Protein A latex linked antiserum test 43

Protista 68, 69, 86, 123, 126

Protonymph 84, 86

PRSV 271, 273, 274, 282, 285, 344-346

PRSV resistant papaya 282

PRSV-HA 5-1 274

Prunus amygdatus 260

Prunus americanum 260

Prunus domestica 260

Prunus insititia 260

Prunus necrotic spot virus 255

Prunus padus 210, 211

Prunus persica 260

Prunus salicina 260

Pseudo-recombination 195, 265, 279, 388

Pseudococcus njalensis 118

Pseudovirus 7, 22, 29, 30, 332, 333, 335, 339, 340, 343, 347, 350, 352

PTGS 277, 280, 281, 284, 287, 409, 430, 438

PSTV 262

Pulse Radar system 159

Pumpkin 189

Puparia 74

PVS 262, 272, 345, 474

PVS-2 262

PVX 262, 267, 271-273, 279, 346, 397, 398, 474, 476

PVY 156, 190, 262, 265, 271-273, 275, 281, 293, 294, 296, 333, 338, 346, 401, 474, 476

Pygofer 76

Pyrrhopappus carolianum 190

Q

Quadri-partite viruses 9

Quantitative epidemiology 182-184

Quantitative trait loci 270

Quincy 226

R

R-gene 264, 406

Radar 148, 149, 157-159, 160, 161, 164, 165

Radio-immuno assay 45

Ranunculus sarodus 190

Raphanistrum sp 190

Raphanus sativus 189

Raspberry aphid 280

Raspberry bushy dwarf virus 263, 346, 440

Raspberry ringspot virus 122, 346

RDV 106, 108, 273, 342, 346, 391, 393

Real-time PCR 53, 54

Reassortment in multipartite RNA or DNA 199

Recilia dorsalis 103, 181, 218

Rep protein 274, 386, 389

Replicase mediated resistance 274

Reproductive number 187

Resting sporangia 86

RgMV 273, 347, 404

Rhabdovirus 9, 18, 21, 22, 58, 106, 107, 412, 413

Rhizomania 264, 406

Rhopalosiphum maidis 72, 205, 209

Rhopalosiphum padi 102, 209, 210

Rhopalosiphum rufiabdomonalis 166

Ribavirin 263

Ribo-probe 51

Ribonuclease protection assay of c-RNA probe 195

Ribosome inactivating protein 275, 422

Ribozyme mediated resistance 276, 278

Rice 1, 7, 35, 77, 89, 103, 106-108, 153, 163, 169, 180, 181, 194, 196, 198, 204, 205, 207, 209, 210, 213, 216- 218, 220, 265, 270, 274, 290, 327, 329-331, 346, 382, 383, 390, 391, 393, 394, 410, 411, 437, 472, 473

Rice Tungro virus 181, 198, 265, 290, 331, 382

Rice dwarf virus 106, 108, 346, 393, 394

Rice grassy stunt virus 108, 163, 346, 410

Rice green leafhopper 77, 153, 169

Rice hoja blanca 180, 346, 410

Rice ragged stunt virus 108, 163, 180, 327, 346, 393

Rice stripe virus 204, 207, 329, 346, 410, 411

Rice transitory yellowing virus 107

Rice tungro bacilliform virus 35, 103, 216, 220, 346, 382

Rice tungro spherical virus 103, 194, 216, 330, 346, 437

Rice yellow mottle virus 196, 270, 346

Rice yellow stunt virus 107, 346

Ridge gourd 189

RMV 99, 208, 209, 211, 333, 346

RNA mediated resistance 275-278

RNA silencing 198, 277, 280, 283, 394, 407, 410, 412, 415, 418, 438

RNAi 281, 418

Rod-shaped viruses 20, 42, 126, 372

Rostrum 68, 73, 76, 84

Rothamsted Insect Survey 150, 154, 165, 205

RPV 97, 100, 101, 208, 209, 212, 336

RSV 207, 273, 346, 410, 411

RT-PCR 52-54, 472-476

RT-protein 99-101, 126

RTV 7, 198

Rugaceae 4

Rugose mosaic in potato 57

Rumex crispus 190

Russian thistle 225

Ryd-1 269

Ryd-2 269

Rye 209

Rym-10 269

Rym-11 269

Rym-13 269

Rym-4 269, 280, 404

Rym-5 269

Rym-6 269

Rym-8 269

Rym-9 269

RYMV 196, 274, 280, 346

Rymovirus 20, 27, 36, 119, 330-332, 340, 347, 358, 368, 402-404

S

Safety assessment 282

Salivary gland 91-93, 97, 101, 105, 106, 108, 111-114, 118, 119, 225

Salsola tiberica 225

Sardinia virus 199, 226, 350

Satellite Tobacco necrosis virus 31, 417

Satellite virus 16, 31, 194, 358, 441

Schizaphis graminum 97, 209, 211

Scleranthus annus 81

SCMV 270, 348

Secale cereal 209

Semi-persistent virus 116, 119, 166

Senecia vulgaris 180

Sequivirus 12, 27, 29, 95, 182, 330, 332, 338, 344, 359, 473

Setaria barbata 214

Setaria pumila 210

SGV 97, 208, 209, 333

Shadow casting 37, 49

Signature analysis 157

Simple interest type progression 183

Sirevirus 29, 30, 332, 333, 339, 342, 352, 363, 439

Sisymbrium irio 225

Sitobion avenae 72, 191, 209, 210, 212

SMYEV 262, 348

S-methyl 296

S-Methyl ester 296

Sogatella furcifera 108

Sogatodes cabanus 180

Sogatodes oryzicola 180

Soil-borne barley mosaic virus 269

Soil-borne cereal mosaic virus 270, 347

Soil-borne wheat mosaic virus 264, 347, 406

Sonchus asper 81, 189-191

Solanum lureum 227

Solanum luteum 227

Solanum nigrum 227

Solanum sp. 189

Solanum triflorum 191

Solid State Marine Radar 158, 159

Sonchus asper 81, 189-191

Sonchus oleraceous 191

Sorachi Dai Sako 264

Sorghum bicolour 213

Sorghum halepense 189

Sorghum vulgare 209

Sorghum189, 209, 213, 215, 347

Southern bean mosaic virus 118, 328, 331, 347

Southern blotting 50

Soybean 20, 150, 191, 192, 204, 209, 297, 329, 341, 348, 380, 472

Soymovirus 20, 29, 30, 35, 331, 334, 344, 348, 353, 380

Specificity 52, 55, 108, 385, 421

Spider mite 68, 69, 120

Spongospora subterranea 123, 125, 180

Sporangia 86, 124, 125

Sporobolus 214

Squall line 162

SPS 254, 299-303

ss DNA 354, 366, 384, 386-390

ss RNA 26, 29, 102, 198, 355-364, 383, 395-406, 408-419, 422, 425, 426, 428, 429, 432-435, 437-442

Stachys arvensis 179

Stanley plum 260

Statistical epidemiology 175

Stellaria media 81, 180, 190, 191

Sticky trap 149, 150, 192, 293, 297

Strategies and Tactics 251-254, 288

Strawberry edge yellowing 57

Strawberry latent A virus 262

Strawberry mottle virus 262, 348

Strawberry pallidosis associated virus 110

Stylet 26, 70, 71, 74-76, 81, 84, 87, 89, 92-94, 97, 98, 103, 110, 111, 118, 268, 271

Suction trap 149, 150, 153-156, 186, 205, 219, 220

Sugareane fiji virus 327, 329, 391

Sultan plum 179

Sun Up

SBCMV 270

SW-5 270, 271, 280

Sweet potato mild mottle virus 109, 349, 402

Symbionin 91, 101

Symbiont 111, 112, 198

Synchitriaceae 69, 86

Synergism 196, 283, 284, 286, 297

Synoptic scale of wind 162

T

Taraxacum officinalis 190

Tarrif and barrier 298

Tarsi 68, 73, 84

TBRV 122, 273, 349

TBSV 281, 349, 417, 418

TCV 281, 351, 415

TE-87-98-13G 268

Temperature inversion 161

Temporal spread 185

Teramnus labilis 189

Tetranychid mite 84

Tetranychidae 69, 84, 119, 120

TG-87-98-99-2 268

Thallus 86

Therioaphis trifolii 192

Thermotherapy 255, 262, 263

Thinopyrum intermedium 269, 270

Thrips 53, 68, 69, 81, 82, 113-115, 149, 178, 185, 190, 192, 193, 273, 297, 325, 438

Thrips tabaci 82, 113

Thysanoptera 69, 81, 182

Tissue culture 255-257, 259-264

Tm22 286

TMV 1, 14, 178, 186, 261, 266, 267, 271-273, 275, 280, 285, 286, 332, 335, 349, 380, 411, 474

TMV-1 1

TNV 124, 273, 349, 358, 359, 416

ToMV 190, 256, 266, 271, 273, 275, 286, 296, 350, 474

Tobacco leaf curl virus 1, 349

Tobacco mosaic virus 1, 2, 14, 23, 57, 326, 328, 330, 349, 411

Tobacco mosaic virus-1 1

Tobacco rattle virus 57, 58, 122, 191, 263, 298, 326, 328, 330, 349, 412

Tobacco ring spot nepo virus 60

Tobacco ring spot virus 327, 328, 330, 433

Tobacco thrips 115, 190, 192

Tobamovirus 7, 10, 20, 27, 31, 32, 35, 57, 58, 178, 275, 286, 289, 296, 326, 328, 330, 332, 337, 339-341, 343-352, 360, 368, 374, 411, 474, 475

Tobravirus 9, 11, 20, 27, 32, 35, 58, 120-122, 178, 279, 326, 328, 330, 332, 344, 345, 349, 360, 368, 374, 412

Tomato black ring virus 122, 349

Tomato bushy stunt virus 256, 327, 349, 417

Tomato chlorosis virus 110, 344, 475

Tomato chlorotic spot virus 115, 349

Tomato infectious chlorosis virus 110, 350

Tomato leaf curl virus 350, 366, 388

Tomato mosaic virus 190, 256, 285, 345, 350

Tomato spotted wilt virus 3, 113, 256, 257, 265, 270, 327, 329, 350, 438

Tomato yellow leaf curl virus 110, 111, 199, 226, 292

Tombusvirus 7, 17, 27, 29, 178, 194, 281, 327, 328, 330, 331, 333, 335, 337-339, 341-345, 347, 349, 358, 417-419, 425

Topocuvirus 7, 20, 29, 329, 331, 354, 387, 388

Toremicity 265

Tospovirus 9, 18, 19, 23, 27, 29, 35, 113, 115, 182, 327, 329, 332, 340, 349, 350-352, 356, 438, 439, 472, 473

Totipotency 256

Toxoptera aurantii 221

Toxoptera citricidus 221, 223

Transcapsidation 99

Transcellular movement 100

Transgenic CP-mediated resistance crops 282

Transgenic papaya 274, 282

Transmission barrier 97

Transmission electron microscopy 36

Transovarial transmission 90, 106-108, 228

Transposon 5, 7

Trialeurodes abutilonia 109

Trialeurodes vaporariorum 75, 109

Triangulation number 16

Trichodoridae 69, 86, 88, 120

Trichodorus pachydermus 180

Triodania perfoliata 190

Tripartite viruses 328

Triple gene block 276, 395, 397, 405, 407, 408

Triticum aestivum 209, 213, 270, 350

Triticum durum 209

Triticum intermedium 270

Triticum monococcum 270

TSV 178, 349, 425, 426

TSWV 113-115, 185, 190-193, 267, 270, 271, 273, 276, 279, 280, 296, 297, 350, 438

TSWV-A (NA) 280

TSWV-A 280

TSWV-D 280

TSWV-N 280

TuMV 271, 278, 351

Tube precipitin 40

Ty-1 268

TYCLV-Ch 226

TYCLV-Is 226, 227

TYCLV-Ng 226

TYCLV-SS 226

TYCLV-Sar 226, 227

TYCLV-Th 226

TYCLV-Tz 226

TYCLV-Ye 226

TYLC CNV-Y10 278

TYLCSV 111, 199, 350

TYMV 58, 178, 351, 435-437

Tymovirus 7, 16, 27, 29, 32, 58, 60, 115, 116, 178, 182, 194, 326, 331, 332, 334-338, 341-345, 347, 351, 362

U

Ultrathin section 37

Urtica urens 189

USDA-APHIS 274, 282

USDA-Grif-9303 270

USDA-Grif-9322 270

V

Varicosavirus 7, 9, 20, 27, 30, 35, 123, 125, 194, 329, 332, 341, 355, 363, 368, 412

Vat 268

Vector activity in epidemiology 191

Vein yellowing of lucerne 57

Velvet tobacco mottle virus 351

Verges 210

Vigna mungo 191

Vigna vexillata 270

Vinca rosea 255

VPg 32, 367-370, 401, 402, 404, 408, 422, 428, 429, 432-434, 437, 440, 441

Virazole 263

Virulence gene 201, 279

Virus dynamics at the tissue level 200

Virus-vector interaction 67, 92, 93, 108

Virus-vector-host recognition 99

Vitivirus 20, 27, 36, 117, 330, 331, 339, 340, 361, 368, 398, 473, 476

Viviparous progeny 71, 72

W

Waika virus 103

Water trap 150, 186, 192

Watermelon mosaic virus-1 58

Watermelon mosaic virus-2 58, 190

WCM 269, 270, 351

Western blot assay 44, 433

Wheat American striate mosaic virus 351

Wheat curl mite 269, 291

Wheat dwarf virus 185, 190, 351, 384

Wheat rosette stunt virus 186, 351

Wheat spindle streak mosaic virus 264, 351, 404

Wheat streak mosaic virus 119, 180, 269, 270, 351

White sticky trap 150, 297

Whitefly 74, 325

World Trade Organization 254, 298

WSMV 119, 269, 270, 273, 351, 402, 403

WTO 254, 298-301

X

Xiphinema americana 121

Xiphinema diversicaudatum 169

Y

Yellow dwarf of barley 434

Yellow dwarf of oat 56

Yellow mosaic disease of soybean 191

Yellow trap 150

Yellows1, 3, 56, 57, 95, 110, 185, 186, 221-224, 255, 264, 268, 289, 291, 303, 328, 330, 332, 334, 335, 338, 339, 341, 343, 351, 389-400, 405, 437, 473, 474

Z

Zea mays 209, 270, 352, 439

Zhong-1 269, 270

Zhong-2 269, 270

Zhong-3 269, 270

Zhong-4 269, 270

Zhong-5 270

Zoophaginae 3

Zoospore 86, 124-126, 169

Zucchini yellow mosaic virus 189, 285, 352

ZW-20 282

ZYMV 267, 271, 273, 274, 279, 282, 286, 352

Color Plate Section

Chapter 5

Figure 5.1   Wingless aphid (*Myzus persicae*)

Figure 5.2 Winged aphid (*Myzus persicae*)

Figure 5.3 Adult whitefly (*Bemisia tabaci*)

Figure 5.4 Female rice green leafhopper (*Nephotettix virescens*)

Figure 5.5 Macropterous brown plant hopper (*Nilaparvata lugens*). With permission from CABI, Wallingford, UK

Figure 5.6 Chrysomelid beetle